D1566084

Microcomputers and Electronic Instrumentation

MAKING THE RIGHT CONNECTIONS

Howard V. Malmstadt
University of the Nations

Christie G. Enke
University of New Mexico

Stanley R. Crouch
Michigan State University

American Chemical Society
Washington, DC, 1994

Library of Congress Cataloging-in-Publication Data
Malmstadt, Howard V., 1922–
Microcomputers and electronic instrumentation: making the right connections / Howard V. Malmstadt, Christie G. Enke, Stanley R. Crouch.
p. cm.
Includes index.
ISBN 0–8412–2861–2: $75.00
1. Electronic instruments. 2. Computer interfaces. 3. Microcomputers. 4. Automatic data collection systems. I. Enke, Christie G., 1933–. II. Crouch, Stanley, R. III. Title.
TK7878.4.M294 1994 94–1407
681'.2—dc20 CIP

Manager, Media Courses Cyrelle K. Gerson

Design and project management A. Maureen Rouhi
Copy editing Marie D. Smith
Production Felicia F. Dixon
Graphics James E. Ramirez, Jr.
Indexing Angelina E. Ramirez

Printed and bound by United Book Press, Baltimore, MD

Published by the American Chemical Society
PRINTED IN THE UNITED STATES OF AMERICA

Preface

Microcomputers seem to be everywhere: in our offices, our homes, our laboratories, and throughout industry. They have become essential parts of modern measurement and control instrumentation and are dramatically affecting the way measurements are made and systems are controlled. But do we have a "feel" for how they are related to our overall instrument systems? Do we really understand the multiple roles that microcomputers perform in our instruments? Could we make connections of new modules? Do we know how laboratory experiments and devices can be connected to a general-purpose microcomputer?

This book not only answers these questions but also presents the electronic concepts, principles, and technology that are impacting our lives, in both our professional and daily activities. Written expressly for chemists, physicists, engineers, biologists, medical researchers, students, and other technical personnel who can benefit from "making the right connections" to modern instrumentation, this book will enable you to gain better control and make better use of your microcomputers and laboratory instruments.

Beginning with the first chapter, which focuses on connecting to the new wave of microcomputer instrumentation, we use what we call the "top–down" approach. That is, we start at the *top* by considering overall objectives and complete instrument systems: we provide a view of the big picture and a broad perspective of electronic instrumentation. Then we work *down* to functional modules, devices, and detailed operations: in chapter after chapter, we discuss in detail functional units, principles, and techniques for making the right connections.

The discussions focus primarily on the encoding, decoding, and flow of data in the instruments—the aspects relevant to your applications. Hence, a particular circuit or device is studied as one of many techniques to obtain electrically encoded data or to decode data. The focus on transformation and flow of data rather than devices is the ideal approach to understanding the operating principles of measurement and control instruments. Descriptions of modern devices and circuits are used to illustrate the discussions of data transformation concepts. Following the data flow through functional modules is also an effective way of finding a malfunction (troubleshooting) in equipment. Therefore after learning about data operations and major functional units in earlier chapters, the reader is shown in the final chapter how to apply this information in troubleshooting instruments.

The top–down approach enables all of the pieces to fit together so that a working knowledge is developed as one proceeds through the chapters.

Visualizing the Connections

Each of the 10 chapters in the book is loaded with diagrams to help the reader visualize clearly the material described. Although by itself this book provides a wealth of information that will benefit many readers, it is also part of a comprehensive learning package comprising this book, a video program, and a laboratory electronics kit with accompanying manual, all available from the American Chemical Society (ACS).

The video program consists of 10 segments, with each video segment related to a chapter in this book. By capturing real-life measurement and control situations, these video segments provide special insight into and enhance the learner's understanding of major topics discussed in this book.

Experimentation provides another way of visualizing functional modules and concepts. The ACS learning package includes an inexpensive laboratory electronics kit and manual, which we also developed. The kit and manual provide the learner with systematic hands-on experience and allow experimental verification of concepts presented in this book.

The combination of book, video, and experiment kit is a powerful educational tool for university courses or short courses and for self-study by already-trained scientists and engineers.

Applying This Book

Learning by reading remains important even in this multimedia age. We believe that this book has the ingredients for successful self-study and use as a reference or textbook for many types of courses. It can be used as the reference for short courses ranging in duration from a few days to a few weeks or as the textbook for full-semester or quarter-term college or university courses.

The chapters are written for those with little or no background in electronics. However, because of the emphasis on electronic instrumentation, which integrates so many electronic concepts and devices, this book is equally useful to those who have basic electronics experience. For those who do not remember or have had little exposure to basic physics, we added three supplements that review the basic laws and principles of electricity and the characteristics of electronic components. The supplements are referred to throughout the chapters, and for some readers, they might be a good starting point for review.

We believe strongly that individual hands-on experiments are necessary for effective learning. Therefore we believe that it is best to use this book in conjunction with an adequate set of experiments. For short courses or university instrumentation courses, the laboratory electronics kit and manual referred to earlier have worked exceptionally well.

Acknowledging Those Who Helped

This book is the successor of a series of widely used books we conceived and wrote during many years of collaboration. For several decades now, we have been teaching courses and seminars, designing instruments and experimental systems, and writing books on electronic instrumentation for scientists. Over these many years, there have been dozens, even hundreds, of persons who have influenced our efforts and provided significant feedback. Acknowledging all those who, through our past works, have influenced this new book would require several pages. Therefore, we have chosen to give a heartfelt thank you to all of the unnamed former students and colleagues, editors, logistics persons, and families who helped us in so many ways as we prepared our earlier works. You were a blessing to us and, we believe, to the tens of thousands who benefited from the books, the equipment, and the experiments.

We pay special tribute to the late Professor Herbert Laitinen, renowned scientist and educator. Over 40 years ago, when he was chairman of the analytical chemistry division of the University of Illinois, he encouraged two of us (Malmstadt and Enke) to continue the development of the Electronics for Scientists materials and programs. He saw what most others failed to realize at the time: that this type of training would have major impact on the practical effectiveness of future generations of scientists. During the same period 40 years ago, Mr. Verle Walters, who operated an electronic repair shop at the University of Illinois, provided great help as we designed and built the first rapid-connect experimental systems. These systems enabled students and already-trained scientists to gain a working knowledge of a wide range of electronic instrumentation in a short period. Over the years, we have continued to strive for the same goals.

We were encouraged by Harry Walsh, head of the ACS Department of Continuing Education, to develop a short course on laboratory electronics that would encompass the major concepts in our books and university courses. We did, and these 2- to 3-day short courses have now run for over 15 years. This book was greatly influenced by what we learned in teaching these short courses and by the suggestions of Professor F. J. Holler, who worked with us on several of these presentations. The organization of the book is based on what we found to be a logical and very effective way of presenting so much material in such a short time.

We are grateful to the reviewers and the staff of the ACS Department of Continuing Education, especially Cyrelle K. Gerson, K. Michael Shea, and A. Maureen Rouhi for their efforts in producing this book and the accompanying video. We acknowledge also the use of some materials conceived by Professor Gary Horlick, coauthor of an earlier work, *Optimization of Electronic Measurements*, the fourth module in the series *Electronic Measurements for Scientists.*

We thank those named and the hundreds unnamed who influenced this work, which was built upon four decades of experiences and helpful and enjoyable relationships.

Howard V. Malmstadt
Christie G. Enke
Stanley R. Crouch

About the Authors

Howard V. Malmstadt has been a pioneer in several areas of science and technology, including applied spectroscopy, automated chemical measurement systems, clinical methods, kinetic methods of analysis, and instrumentation for scientists. He has received national and international awards for these developments, which are described in over 150 journal publications and patents. He is coauthor of a dozen books, including *Electronics for Scientists* (published in 1961), *Digital Electronics for Scientists* (published in 1969), and *Computer Logic* (published in 1970), which were the basis of many new courses and have been used in hundreds of universities and laboratories worldwide.

Dr. Malmstadt was born in Marinette, Wisconsin, in 1922. After receiving a B.S. degree, he served as radar electronics officer in the United States Navy from 1943 to 1946. He received an M.S. degree in 1948 and a Ph.D. in chemistry in 1950 from the University of Wisconsin. He joined the faculty of the University of Illinois in 1951, was promoted to full professor in 1961, and became professor emeritus in 1978.

He has lectured widely throughout the United States, Europe, Australia, England, Brazil, Singapore, India, Japan, and China. Recognitions for his research and teaching include a Guggenheim Fellowship (1960), the American Chemical Society (ACS) Chemical Instrumentation Award (1963), the ISA Award in Education (1970), the ACS Fisher Award in Analytical Chemistry (1976), the Pittsburgh Conference Outstanding Chemist Award (1978), the Fullbright-Hays Distinguished Professor Award (1978), the ACS Analytical Division Excellence in Teaching Award (1984), and the ANACHEM Award (1987).

Since 1978, Dr. Malmstadt has been involved primarily in the overall development of the University of the Nations, which he serves in several capacities, including International Provost and Vice President of Academic Affairs and Dean of the College of Science and Technology.

Christie G. Enke is a professor of chemistry at the University of New Mexico. He was born in Minneapolis, Minnesota, in 1933. He obtained a B.S. degree from Principia College in 1955. In 1959, he obtained a Ph.D. in chemistry from the University of Illinois, where he worked with Professor H. A. Laitinen. He was a faculty member at Princeton University from 1959 to 1966, first as an instructor and then as an assistant professor of chemistry. In 1966, he joined the faculty of Michigan State University (MSU) as an associate professor. He left MSU as full professor in 1994, when he moved to the University of New Mexico.

Dr. Enke's current research interests include mass spectrometric instrumentation, especially tandem mass spectrometry and time-of-flight mass spectrometry, development of expert systems for the interpretation of mass spectral data, clinical applications of mass spectrometry, microbiological characterization by mass spectrometry, and reactions of ions with molecules and photons. He has published research papers in *Analytical Chemistry, Science, Journal of the Electrochemical Society, Review of Scientific Instruments, Journal of Electroanalytical Chemistry, Journal of the American Chemical Society, Journal of Physical Chemistry, Journal of Chemical Physics, International Journal of Mass of Spectrometry and Ion Processes, Journal of Chemical Education, Analytica Chimica Acta, Journal of the American Society for Mass Spectrometry,* and others. Since 1957 he has been involved in the development of support materials for courses in the electronics of instrumentation. With H. V. Malmstadt and S. R. Crouch, he is coauthor of five textbooks in the electronics of chemical instrumentation.

He was an Alfred P. Sloan Foundation Fellow from 1964 to 1969. His achievements have been recognized by numerous awards, including the American Chemical Society (ACS) Award in Scientific Instrumentation (1974), the Sigma Xi Senior Research Award at Michigan State University (1982), the ACS Award for Computers in Chemistry (1989), the Michigan State University Distinguished Faculty Award (1992), and the ASMS (American Society for Mass Spectrometry) Award for Distinguished Contribution to Mass Spectrometry (1993). He has served on several ACS committees, was a member of the editorial advisory board for *Analytical Chemistry,* was chairman of the ACS Division of Computers in Chemistry, and is currently Vice President for Programs for the American Society for Mass Spectrometry. He has served on the Metallobiochemistry Study Section of the National Institutes of Health. He also serves as consultant for the chemical and instrumentation industry.

Stanley R. Crouch is a professor of chemistry at Michigan State University. He attended Stanford University from 1958 to 1963 and received his M.S. degree in 1963. From 1963 to 1967, he was a graduate student at the University of Illinois, where he worked with Professor H. V. Malmstadt; he received his Ph.D. degree in 1967. He was a National Science Foundation Fellow in 1963 and a University of Illinois Fellow in 1965. From 1967 to 1968, he was an instructor and a visiting assistant professor in general chemistry at the University of Illinois. He became assistant professor at Michigan State University in 1968, associate professor in 1972, and professor in 1977. He was an Alfred P. Sloan Foundation Fellow from 1973 to 1975. He became Associate Chairman for Undergraduate Studies in 1993.

Dr. Crouch is a member of the American Chemical Society, the Optical Society of America, the Society for Applied Spectroscopy, and the American Association for the Advancement of Science. He has been a member of the Instrumentation Advisory Panel of *Analytical Chemistry* and is currently on the editorial advisory boards of *Analytical Instrumentation, Talanta, Laboratory Microcomputer, Analytica Chimica Acta,* and *Intelligent Instruments and Computers.*

His research interests are in the areas of kinetics and mechanisms of analytical reactions; fast kinetics measurements; continuous-flow methods; flow injection analysis; laser spectroscopy in flames and plasmas; laser-assisted ionization methods; atomic, emission, absorption, and fluorescence spectroscopy; liquid chromatography; clinical chemistry; and chemical instrumentation. He has written six books and has over 110 publications in various areas of his research interests.

Contents

Chapter 1

Connecting to the New Wave of Microcomputer Instrumentation

For many of us who practice science and engineering, electronics technology is a mixed blessing. It provides marvelous new tools for the acquisition of greater amounts of often heretofore unobtainable data about the systems we study and design, but sometimes it can also result in increasingly mysterious and quixotic black boxes that stand between us and the phenomena we are trying to understand. Becoming comfortable with our instruments and confident of their processes would be problematic if it involved acquiring a working knowledge of semiconductor physics, microcircuit fabrication, and electronic circuit design. These are each complex, specialized, rapidly evolving fields. Fortunately, *there is an approach to understanding modern instrumentation that is readily attainable* and will greatly improve the quality of our "connection" with our instruments and computers. This approach is to *concentrate on the data and not on the devices*: to focus on the form the data take in various parts of the instrument and how the data can be processed to achieve the final result. From this perspective, a particular circuit or device is seen and studied as one of a variety of techniques available to perform one of these data operations, and subsequent technological advances simply give other, possibly more efficient and precise devices for performing the same operation. We have adopted this approach for this book. As you proceed through this book, you will create an enduring framework for your understanding of electronic instrumentation, a framework centered on the aspect of your instruments that is most relevant to your work—the data. You will then better understand and be able to take greater advantage of the capabilities available through modern electronic control and data processing. The limitations of the instruments you use will be clearer, and you will appreciate what is involved in extending the capabilities of these instruments. In current terminology, you will become a "power user."

1–1 Joining the Microcomputer Age

The first phase of the microcomputer revolution is over; virtually all new scientific instruments include a microprocessor. What was at first an attractive way of combining instrument control and data processing in instrument design is now revolutionizing the way classical techniques are implemented and is also providing the basis for previously unimaginable measurement systems. More and more instrument measurement and control functions are being computerized. The basic ways in which measurement and control processes are carried out, the ways in which we interact with instruments, and the ways in which the data are presented are all undergoing dramatic changes.

No longer an optional attachment for automated data logging, the microcomputer is now an integral part of the basic instrument. It manages the central data communication pathway, controls the sequence of operations, processes the data in a variety of ways, and responds and reports to the operator (or to other computers that provide further data processing or oversee functions). This is illustrated in

2861-2/94/0001$08.40/1 Published 1994 American Chemical Society

Figure 1–1, the block diagram of a modern spectrophotometer. All operator interactions with the instrument are through the computer. All functions are performed in response to computer commands, that is, steps in the computer program (software) running the system. One part of the system allows the operator to select experimental protocols and operating parameters; other parts set these parameters, direct the experiment, control various quantities, collect and process the data, and provide messages and graphical output. The extent and precision of the instrument control potentially available and the mechanics of operator/instrument interaction are principally a function of the instrument sensors and manipulators and the computer hardware. It is the software, however, that largely determines what operations are available to the operator, how these operations will be executed, and how the data will be processed and presented.

Computer-controlled instruments vary widely in the modes of operator interaction provided. In some cases, the presence of the computer in the instrument is not at all apparent. The operator interacts with the instrument through traditional knobs, switches, and indicator lights or through the more modern alphanumeric displays, keypads, and mice. Within the instrument, a microprocessor is connected to all these devices as well as to all the instrument's sensors and controls. In these systems, called **stand-alone** or **turnkey** systems, all programs required by the computer for the operation of the instrument are permanently installed in a portion of the computer memory. When the instrument is turned on, the computer automatically begins the program that responds to the operator commands. No special training in computer operation is required to operate such an instrument.

With larger and more complex instruments, such as a mass spectrometer, signs of the presence of the computer are often unavoidable. The computer keyboard, monitor, and disk drives are obvious. Upon start-up, the operator is often required to call from a disk the program(s) required to operate the instrument. Greater flexibility is afforded by such a system; the programs can be upgraded by the manufacturer or the user, and new instrument accessories can be brought under computer control. The operator of such an instrument, however, needs to know how to operate the computer as well as the instrument.

The extreme example of the "obvious computer" instrument is when a general-purpose laboratory computer is used in custom instrumentation or when it is used as a versatile laboratory tool for measurement and control. In these cases, the computer is the central unit and the system is built outward through interfaces to the device(s) to be computerized. In such systems, the software, too, is of the general-purpose type and must be customized to fit individual applications.

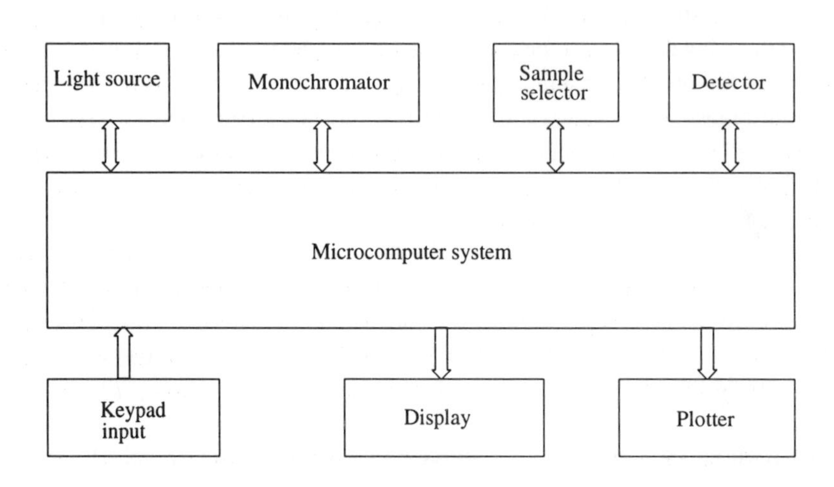

Figure 1–1. Block diagram of a computer-controlled spectrophotometer. All instrument elements are controlled or measured by computer operation. All settable instrument parameters and choices of operational mode are input to the computer by the operator. The computer collects and processes the data before directing the results to the output display or plotter.

In the current phase of the microcomputer revolution, more and more of the functions previously performed by the operator or an expert in data interpretation or experiment design are being performed by the instrument's computer. The enhanced instrument capabilities often provide much more information to interpret, and much more processing power is applied at the site of data generation. The remainder of this section introduces the functions of the microcomputer and the mechanisms of data transfer in modern instrumentation.

Processing Commands

The digital computer derives its great power from its ability to perform a sequence of very simple operations very quickly. Commands such as fetch a number from a particular place, fetch another number from another place, add the two numbers, and store the sum in a specified place are typical of the commands in the computer's elementary instruction set. These functions are carried out in the central processing unit (CPU). To carry out these operations, the CPU must have access to all the data sources and destinations as well as to the places where its instruction lists (the program) are stored. This access is provided by a central communication pathway, a set of parallel wires and sockets called the **CPU bus.** All devices that provide data, store data, receive data, or are controlled by the CPU are directly or indirectly connected to this bus. As shown in Figure 1–2, these devices can be divided into four categories: those required for the computer operation itself, those for interaction with the operator, those for interaction with the sensors and controlled parameters in the system, and those for interaction with other systems in a broader context. The microprocessor is the principal traffic controller for the communication of data along the CPU bus; it specifies which devices are to provide data on or receive data from the bus and when.

The data transferred on the bus include not only sensor output, control parameters, and processed numbers but also information on specific operations the various devices are to perform next. The CPU itself obtains its instructions one at a time through the bus from the programs stored in memory. The successful development of the common communications bus system between the CPU and its various peripherals, a system that is open to the connection of a variety of

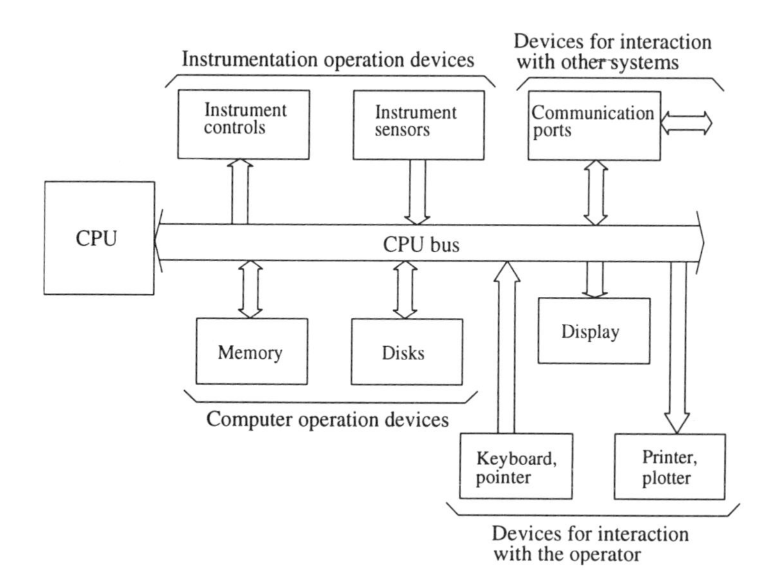

Figure 1–2. CPU–bus connections in a computer-controlled instrument. All data and control commands are transferred along the CPU bus. The CPU requires permanent and temporary memory for its operations. Other devices provide measurement and control functions within the instrument, devices for operator input and data display, and a means of exchanging data with other computer-based systems.

measurement and control devices, was an essential ingredient in the development of today's computer-intensive instrumentation.

Communicating with the Processor

The signals on the CPU bus wires, like the signals in the CPU itself, are digital. This means that the signal on each wire can represent one digit in a number. If a digital signal represented a digit in the normal decimal number system, it would need 10 distinguishable states, one for each of the numerals 0 through 9. The use of a number system with fewer numerals than the common base-10 system reduces the number of signal states that need to be distinguished and thereby decreases the likelihood of error. The number system with the smallest possible base is the binary system, which uses only 0 and 1. A signal representing one binary digit needs only two distinguishable states: higher than or lower than a prescribed threshold level. For maximum simplicity and reliability, the binary system has been chosen for digital signals, and the signal levels that represent the value of each binary digit, or **bit** of information, are called HI and LO.

The CPU bus wires and contacts are subdivided into three sections: the **data bus**, which specifies the content of the data communication; the **address bus**, which specifies the source or destination of the data; and the **control bus**, which contains signals relating to bus management timing and direction of flow on the data bus. The number of conducting paths in each of these sections of the bus determines the amount of information that can be on the bus at one time. For example, if the data bus is made up of eight conducting paths, it can transmit 8 bits at a time. Eight bits can represent all the binary numbers from 00000000 to 11111111, which in decimal form is 0 to 255. Many such data units can be sent along the bus each microsecond. Providing more bits in the data bus can shorten the communication time for larger numbers. Microprocessors with 8-, 16-, and 32-bit data buses are currently available.

The number of contacts in the address bus poses a different type of limitation. A 16-bit address bus provides 2^{16}, or 65 536, different address codes or addressable locations on the bus. Each location can contain one digital memory **register** that can store as many bits as there are contacts in the data bus. These registers can be used for program memory, data memory, data input and output, and device control. Thus, the number of contacts in the address bus limits the length of the program and the size of the data space available to the computer at any one time. Microprocessors with address buses of up to 32 bits (4 294 967 296 addresses), previously the domain of only the largest mainframe computers, are now available.

The control bus completes the CPU bus by providing a number of signals for synchronizing the operation of the data source and destination registers, indicating bus status, allowing for slowly responding registers, passing bus control to other devices, indicating a device's need for attention, and so forth. The more versatile and fail-safe a CPU bus is, the more signals are required in the control bus and the more complex is the task of connecting external device registers to the CPU bus.

Connecting to the CPU Bus

All devices that are to use data from or provide data to the central computer must be connected in some way to the computer bus. The data must be in the form of digital signals, with HI and LO signal levels appropriate to the computer circuits,

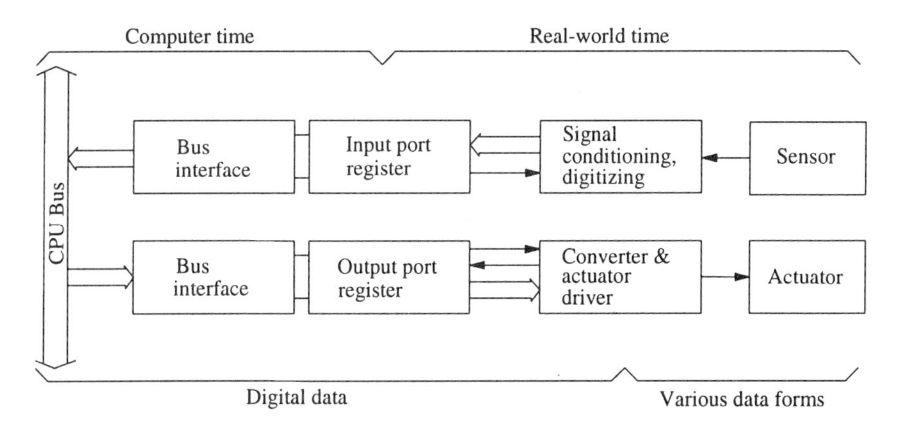

Figure 1–3. Interfacing real-world sensors and actuators to the CPU bus. At a time specified by a clock timer or a computer command, the digitized sensor output is stored in the input port register. A later CPU command transfers these data to an appropriate memory or CPU register. For control of an external device, the CPU transfers the control value to the output port register. When needed, this new control value is converted to the data form required by the actuator.

and divided into groups of bits equal to the number of bits in the data bus. The complete connection of a sensor or manipulator to the computer bus includes both data transformation and bus-tending devices, as shown in Figure 1–3. A sensor encodes the measured quantity (such as pressure) as some quality of an electrical signal, the digitizer converts the sensor output to a digital signal, and the bus interface manages the appropriate application of the digital signal to the bus. For the computer control of a heater or motor, the process is reversed. The bus interface circuitry is specific to each computer type.

When a computer is built into an instrument and dedicated to it, the specific data conversion and bus interface circuits needed are wired on custom circuit boards and connected directly to the computer bus. For more general-purpose laboratory applications, such custom interface designs are less practical. It may be desirable to change computer types or to use a particular computer in a variety of applications. Three interfacing options that are used in such situations are shown in Figure 1–4. The laboratory interface board provides the bus interface along with analog-to-digital and digital-to-analog data conversion circuits. The user needs only to modify the sensor and actuator levels to those appropriate for the laboratory interface board used. Laboratory interface boards that plug directly into the computer's CPU bus sockets are available for all widely used scientific computer

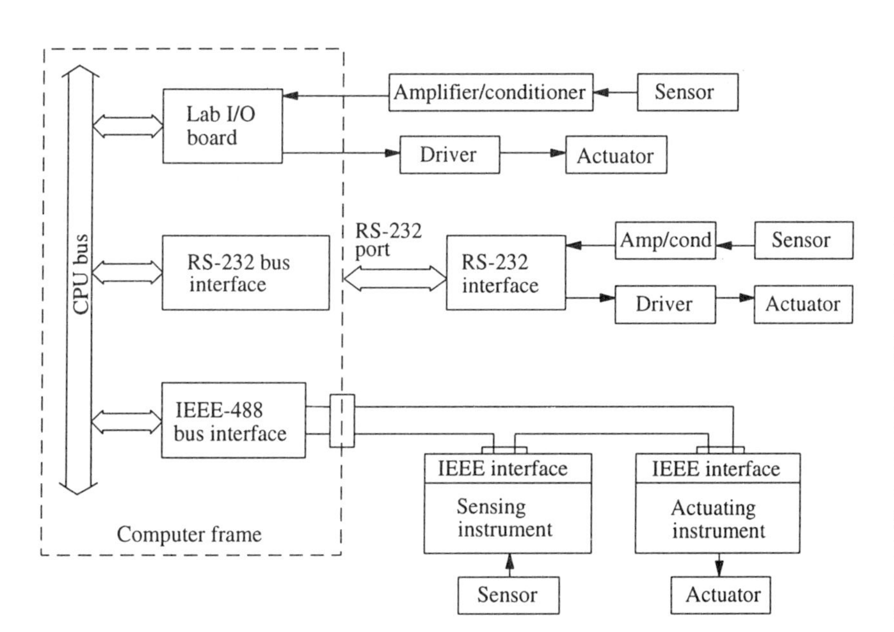

Figure 1–4. General-purpose interface systems. The laboratory I/O board includes data conversion and timing circuitry as well as the CPU–bus interface circuitry. Many sensors and actuators can be connected directly to this board. The laboratory I/O board is specific for each computer type. RS-232 port and IEEE-488 bus cards provide industry-standard communication links. Although these CPU–bus interface cards are computer-specific, the external interfaces connected to them can be moved to any computer system that provides these ports. Amp, amplifier; cond, conditioner.

systems. The other two options essentially substitute user interfacing to the computer bus for interfacing to a less computer-specific digital communication system. RS-232 and IEEE-488 interfaces (described in Chapter 2) are available for all scientific computers. Therefore, an interface to either of these communication standards will be generally applicable. Furthermore, many instruments and peripheral devices with direct connections to one or the other of these external bus standards are available.

Acquiring Data

Several distinct operations are involved in the computer acquisition of even a single measurement value. These processes are illustrated in Figure 1–5 for the measurement of temperature. The temperature is sensed by a thermistor (a device whose resistance is related to its temperature), and the resistance is converted to a related voltage level by a conversion circuit. The time that this voltage level is sampled defines the instant of data acquisition. The sampled voltage level is then converted to a group of 1-bit HI/LO signal levels that are stored in an input/output (I/O) register that has an address on the CPU bus. A program to transfer the data from this register to the microprocessor or some other memory location must then be run to complete the acquisition. Two operations in this process must be initiated in some way: the voltage sampling, which the conversion follows automatically, and the transfer of data from the port register to memory. In each case, the initiation can come from a variety of sources. The voltage sampling can be triggered by a command in a temperature acquisition program, by an operator or system request, or by a timer that triggers this action at regular intervals. The data transfer program can be started by the appearance of new data in the input port register or can be made to occur at the appropriate time in the overall temperature acquisition program.

If, in the above example, one wished to record the variation of temperature with time, successive temperature values would be acquired and stored in separate (generally consecutive) memory locations. This process is illustrated in Figure 1–6. When the process is complete, only the temperature values related to the acquisition times remain. Note that what was a continuous data stream in the voltage output from the thermistor circuit has become a discontinuous set of numbers through the process of digital sampling. The samples must be taken sufficiently often to show the variations in temperature with time to the desired detail. The criteria for the adequate sampling of varying quantities are discussed in Chapter 2.

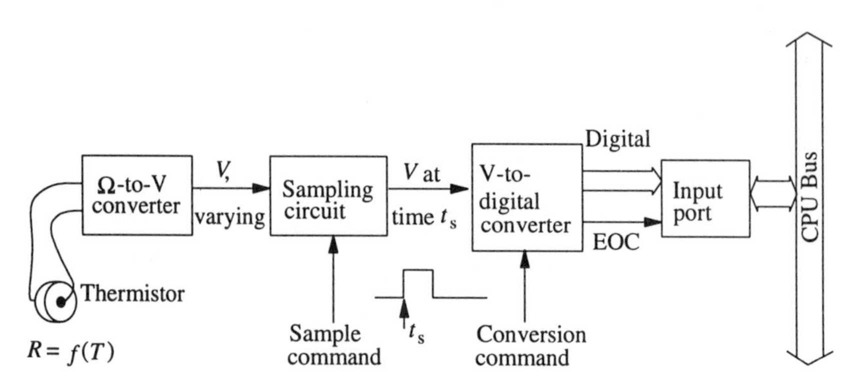

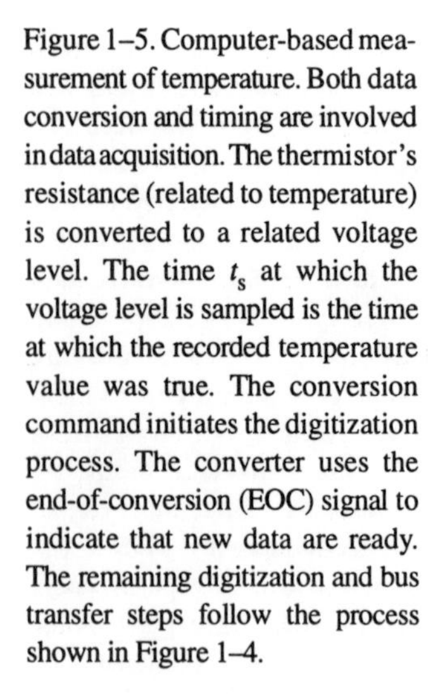
Figure 1–5. Computer-based measurement of temperature. Both data conversion and timing are involved in data acquisition. The thermistor's resistance (related to temperature) is converted to a related voltage level. The time t_s at which the voltage level is sampled is the time at which the recorded temperature value was true. The conversion command initiates the digitization process. The converter uses the end-of-conversion (EOC) signal to indicate that new data are ready. The remaining digitization and bus transfer steps follow the process shown in Figure 1–4.

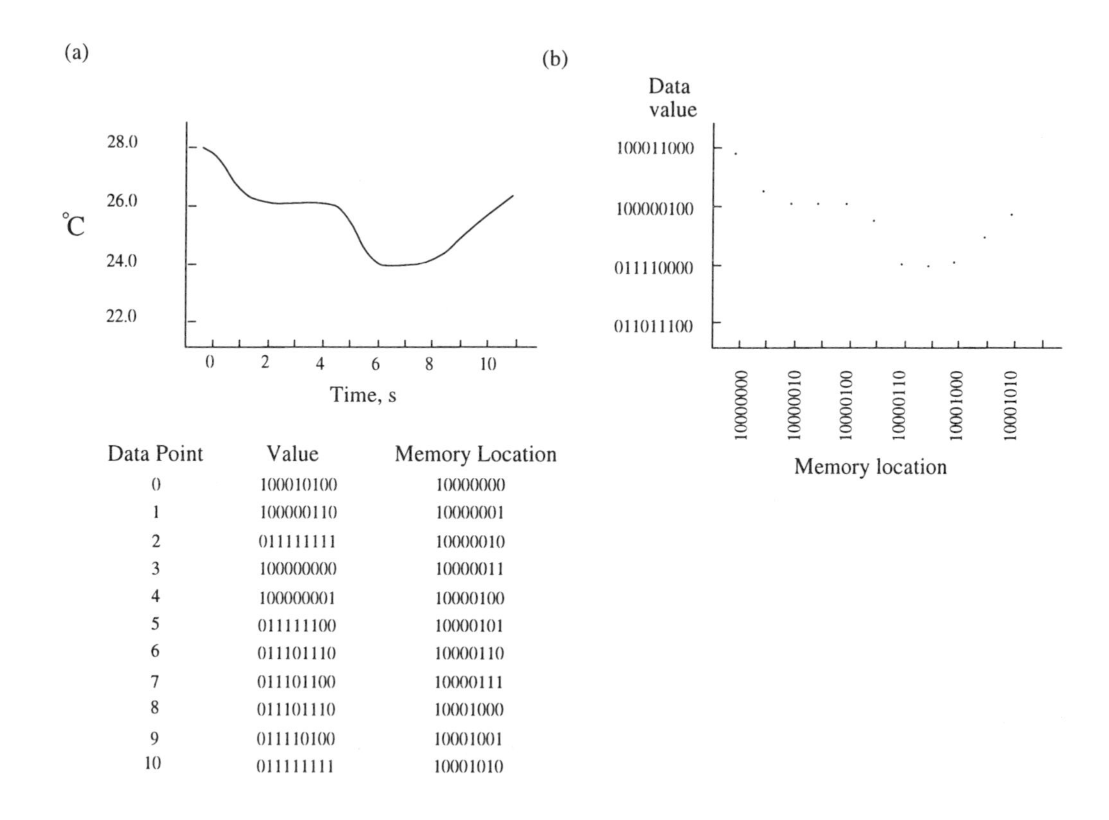

Data Point	Value	Memory Location
0	100010100	10000000
1	100000110	10000001
2	011111111	10000010
3	100000000	10000011
4	100000001	10000100
5	011111100	10000101
6	011101110	10000110
7	011101100	10000111
8	011101110	10001000
9	011110100	10001001
10	011111111	10001010

Figure 1–6. Acquisition of time-varying temperature. Ticks on the time axis of part a indicate sampling times. The plot in part b is the value at each memory location where the data were stored. Note that the data stored in memory are more like entries in a table than points on a plot and that they are not continuous. Information on the temperature between sampling times has not been acquired.

1–2 Working with Digital Data

Digitally encoded data are composed of bits. Since a single bit of information is simply a 1 or a 0, an on or an off, a true or a false, etc., a bit can be encoded by any signal, device, or object that has two clearly distinguishable states. With signals, the two states are simply HI and LO; the voltage (or current) of a HI signal is more positive than the threshold level that divides the HI and LO signal levels. Physical devices are also used to encode bits of information. A tiny circular magnet can be magnetized with its field clockwise or counterclockwise. A particular space in a paper tape or a thin metallic film can have a hole in it or not. These physical digital data storage devices can then be "read" by electrical detectors to convert their magnetism or transparency into an electrical digital signal. The important thing is that the encoding property must distinguish between the two states with very high reliability. The quality of digital encoding is determined by the number of correct bit readings per error bit; for modern encoding techniques, this number is often greater than 10^{15}. This incredibly high degree of reliability is the principal driving force behind the digital revolution. Digitally encoded data can be stored, processed, and transmitted essentially error free.

The high reliability associated with digital data affects the precision of digital data encoding. Data precision is a function of the reproducibility of the data value and the resolution with which the value is represented. For example, if multiple measurements of a light intensity show a relative variation of 1%, we say that the precision of the data is 1 part in 100. Similarly, if only three decimal digits can be used to represent a data value, the value is not known to more than 1 part in 999 (actually 1000, since zero is included as one possible data value). The overall precision of a data value is determined by the less precise of the two effects, reproducibility and resolution.

For digital data encoding, the resolution is determined by the number of distinguishable states afforded by the digital code used. Each bit has only two distinguishable states, giving a numerical precision of 1 part in 2, or 50%. For higher resolution, multiple bits must be used in combination. If two bits (two digital signals) are used to represent a value, there are four distinguishable states: LO–LO, LO–HI, HI–LO, and HI–HI. Table 1–1 lists the number of distinguishable states provided when various numbers of bits are used to represent a data value. As the table shows, for n bits, the number of states is equal to 2^n. These 2^n states can be used to represent all the numbers from 0 to 2^n–1, 2^n different operation codes, or 2^n different printer characters. If a resolution of 1 part in 10 000 (four significant figures) is required for a data value, a minimum of 14 bits is needed to represent it.

Since the reliability of digital encoding is so great, the reproducibility of transmitting and retrieving digital data is essentially perfect. Therefore, the overall precision of most operations performed on digitally encoded data is limited only by the number of bits used to represent the data value. A collection of bits used to represent one data value or function is called a **word**. The length of a word is given by the number of bits in it. Obviously, the longer a word, the greater the number of different states it can represent. A word that is 8 bits long is called a **byte**. A 32-bit word is 4 bytes long.

Transmitting Digital Data

A separate digital signal is required for each bit of data to be transmitted. To convey a 16-bit word of data, 16 HI or LO signals are required. If all the bits in the word are to be conveyed simultaneously, 16 data channels or wires will be needed. This is called **parallel transmission**. The number of simultaneously available channels in the connection determines the number of bits that can be transmitted simultaneously. A data bus with 32 channels can transmit 4 bytes simultaneously. There is also a minimum time the signal level corresponding to the transmitted bit must be present on the channel. This time is determined by the time required to establish a new level reliably, the time required for the data transmitter to activate the data ready signal in the control bus, and the time required for the receiver to determine the signal level on the channel. If a new signal level can be established and read in 1 µs, new words can be applied to the data bus at the rate of 1 million per s. Such a data bus has a rate of 1 MHz. By current standards, this is relatively slow; the data channels of current computers can reliably transmit data at 5 to 20 data words per µs. Even these high data transfer speeds can limit the performance of modern microcomputer systems. Data-transfer-intensive devices, such as the color monitor and the hard-disk drive controller, are now sometimes connected to the CPU through a special local bus that serves a limited number of devices but at speeds up to 50 MHz.

Words with more bits than the number of simultaneous data channels can be transmitted by sending portions of the word sequentially or serially. An 8-bit data channel can transmit a 32-bit word in four successive bytes. This form of data

Table 1–1. Number of States for Various Numbers of Bits

No. of Bits (n)	No. of States (2^n)
0	1
1	2
2	4
3	8
4	16
5	32
6	64
7	128
8	256
9	512
10	1 024
11	2 048
12	4 096
13	8 192
14	16 384
15	32 768
16	65 536
17	131 072
18	262 144
19	524 288
20	1 048 576
22	4 194 304
24	16 777 216
26	67 108 864
28	268 435 456
30	1 073 741 824
32	4 294 967 296
36	68 719 476 736
40	1 099 511 627 776

transmission is a combination of parallel and serial. It is frequently used in computers in which the data bus has fewer bits than the desired precision of the data value being conveyed. Thus, an 8-bit computer (one with an 8-bit-wide data bus) is not limited to working with values having a precision of 1 part in 256 or less; any desired precision is possible, but the data transmission on the bus will require multiple cycles.

The lower limit in the width of a data bus is one channel. Such a single connection can obviously transmit only 1 bit at a time. This kind of purely **serial transmission** is used over longer distances where the cost of providing multiple simultaneous channels is not warranted. Serial transmission is used for digital data transmission over telephone line connections. Serial transmission lines are rated for the maximum bit rate for reliable operation. The transmitter applies a stream of HI and LO signal levels, or other encoding, corresponding to the 1's and 0's in the data word to be transmitted. The receiver must know the rate at which the bits are being transmitted so that the bit level detection can be synchronized to the bit arrival rate. Serial transmission rates are expressed as so many **baud** (bits per second). Standard baud rates have evolved within the data communication industry. An early standard was 110 baud, limited not so much by the reliability of the lines as by the printing speed of the Teletype machines that were the most common receivers. More recently, the standard has evolved from 300 baud through 4800 baud to 19.2 kbaud.

Serial data transmission has become the standard method for intercomputer communication whether the connection is a single, temporary link (as when a modem and phone line are used to access a data service) or is part of a local network of computers. **Local-area networks** (LANs) are becoming increasingly popular because they allow all the data and processing capability on the network to be shared by every user. Intelligent instruments connected to a network can provide data to a central data coordination facility that can keep track of test results, sample status, billing information, experimental trends, instrument throughput, and many other items of interest to both managers and users. Such a system is called a **laboratory information management system** (LIMS). All new instruments should be able to be integrated into a LIMS. This requires not only the necessary hardware connections between the serial communication line and the instrument computer but also, and equally important, the software necessary to format the information correctly and handle the communication protocols.

Storing Digital Data

Once data are in digital form, it is desirable to retain them digitally for later access. The storage of each bit requires a two-state element that can be put in either state and whose state can be determined quickly and reliably. One state encodes a 1, and the other encodes a 0. An electrical circuit called a **flip-flop**, such as that shown

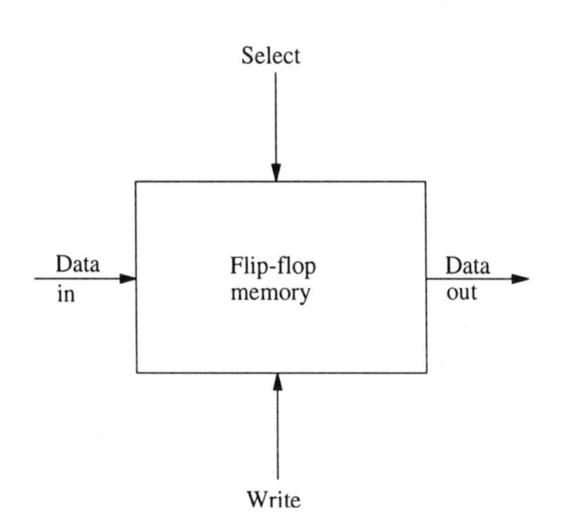

Figure 1–7. A one-bit memory element. When the Select and Write signals are HI, the level of the Data in signal is stored in the flip-flop. The state of the flip-flop is applied to the Data out signal when Select is HI and Write is LO.

in Figure 1–7, has only two stable states and is used to store 1 bit of data. The output signal from the flip-flop is always either HI or LO. When the flip-flop is in the write mode, the signal level at its input determines the state of the circuit and is thus stored. The value (1 or 0) of the stored bit is indicated by the level (HI or LO) of the output signal. A type of flip-flop digital storage is used for part of the memory in computers. Such memory can respond very quickly to the input signal level when in read mode, and its stored data are readily available in the form of a binary signal level whenever needed.

The retention of a complete digital word from a data bus requires the simultaneous storage of several bits. This is accomplished by the use of an assembly or combination of storage elements called a **register**. An 8-bit register, which can store 1 byte of data, will have 8 data inputs and 8 data outputs, but its flip-flops will share the enable and write connections. The inputs of an 8-bit flip-flop register can be connected to the lines of an 8-bit data bus. The enable input is activated by the appearance of the code for this register's address on the address pulse, and the write input is connected to a control line that is activated when the data bus contains the data to be received. In this way, the CPU can write data into any register so connected to its bus. The outputs of the register are connected to some device to which the CPU can send new data as needed. Similarly, an input register can supply data to the CPU bus. In this case, the register inputs are connected to the data source. When activated by the address code and control signals, the register outputs are connected to the bus to supply the data that another register on the bus may receive.

To retain or store a large number of digital words, an assemblage of multiple registers called a **memory** is required. Generally, each of the registers in a memory can store data from or provide data to a common data bus. Since both the source of and destination for the data stored in a memory are the CPU data bus, only one data connection per register bit is required between the register and bus. The direction of data transfer is determined by read and write signals from the control bus. When the multiple registers are contained in a single integrated circuit, the circuitry for decoding the address line to identify and activate the particular register addressed must also be in that same integrated circuit. To connect a given memory register to the CPU data bus for data storage or retrieval, the digital code for its address is applied to the address connections of the memory circuit. Such a memory is called **random access memory** (RAM), because the registers can be accessed in any order by simply applying the appropriate digital signals to the address connections. Memory circuits designed around the flip-flop circuit are called **static RAM**, because the information stored is stable as long as the circuit remains powered. The term RAM, as currently used, implies that each storage cell can be either read or changed (written) through the data bus. A more accurate descriptive term is read–write memory, which is also sometimes used.

To reduce the size of storage cells and thereby increase the memory capacity per integrated circuit, a non-flip-flop type of memory cell was developed. In this more compact device, information is retained as the presence or absence of a tiny charge rather than as the state of a more complex bistable circuit. Because the cell charge can leak off rather quickly, data bits stored in these cells are not stable indefinitely. The temporary nature of the data stored can be overcome by reading each cell periodically and restoring its appropriate charge state. A type of memory that operates in this manner is called **dynamic RAM**, and the circuit that performs the continual read-and-replace function is called the **refresh circuit**. At the time of this writing, dynamic RAM integrated circuits with capacities of up to 2^{22} bits are available. Memory size is generally spoken of in terms of groups of 2^{10} (1024), which, because it is so close to 1000, is called 1K. (Careful writers use K when they mean 2^{10} and k when they mean 10^3.) The number 2^{20}, loosely referred to as 1 megabit, is actually 1 048 576.

A 1-megabit dynamic RAM chip contains 1 048 576 separate 1-bit memory cells. These could be arranged into 131 072 8-bit registers or 1 048 576 1-bit registers. In the former case, 8 data and 17 address connections are required; in the latter, 1 data and 20 address connections are needed. For microcomputer memory, the same address lines would be connected to eight of these chips at once to provide eight simultaneous data bits, as shown in Figure 1–8. In practice, a ninth **parity** bit is often used for each byte of data. When the data are entered, the parity bit is set to a value that makes the number of HI bits in each set of 9 bits an even number. An error in the storage or reading of any single bit will result in an odd number of HI bits in the set. An error-detecting circuit that supplies the parity bit when writing to memory and checks the parity of the data when the memory is read is a common part of dynamic RAM memory systems. Note that the refresh and parity circuitry can be shared by whole sections of memory, which makes dynamic RAM an economical choice for relatively large amounts of memory. Because each cell must be refreshed within a particular time interval (e.g., 2 ms), the refresh circuit must intersperse refresh cycles with the read and write cycles for which the memory is being used. This essential housekeeping can sometimes delay access or limit the duration of continuous access to the memory.

Memory based on two-state semiconductor circuits, whether static or dynamic RAM, requires continuous electrical power to maintain the state in which it has been set. Such memory is called volatile, because when power is removed (the computer is turned off), the stored information is lost. Since most programs keep active data in RAM, where it can be accessed and changed most quickly, a power failure can cause a loss of all work performed after the last version of the information was saved in some nonvolatile medium.

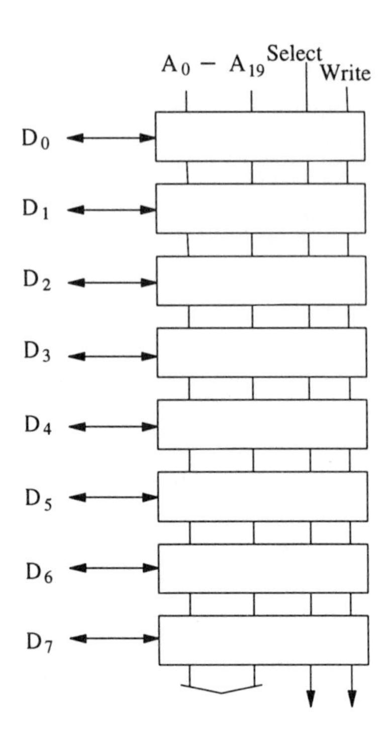

Figure 1–8. A 1-megabyte memory arranged as eight 1-megabit chips. The code on the address bus (A_0–A_{19}) determines which of the 1 048 576 cells in each chip is active. The Select and Write signals determine the timing and direction of data transfer along connections D_0 through D_7 between the cell and the bus. Collectively, the eight chips constitute an 8-megabit memory, accessible 1 byte at a time.

To achieve nonvolatile digital storage, the storage element must be able to remain in either of its two states when the circuit power is removed. This requires that the standby power source be switched on or the property providing the storage be physical rather than electrical. Resistance, magnetic field direction, and optical transparency or reflectance are physical properties often used for nonvolatile storage. The resistance of an element can be made permanently (or semipermanently) low or high, so that when that resistance is connected to a reading circuit, a LO or HI output signal is produced. The state of the storage elements in such a memory cannot be changed quickly, if at all, so that this type of memory is called **read-only memory** (ROM). Parts of computer programs and archival data are stored in ROM.

As the power requirements of electronic memory have decreased with more efficient designs, the standby battery solution to the volatility problem has become more practical. The battery is switched on to provide power to the memory circuits (or a portion of them) when the main power is off. The battery power required is greatly reduced by the use of memory chips that have a special low-power standby mode. This is the technique used in calculators with the "constant memory" feature.

Storing Large Amounts of Data

Most modern microcomputers use some form of mass storage, that is, a method for retaining quantities of digital information many times larger than the memory capacity of the computer. The digital data are stored on magnetic disks or tape or on optical disks. Mass storage devices serve several important functions: they provide a large database capacity that can be read into and out of CPU memory relatively quickly under computer control; they provide nonvolatile storage of programs and data; they can provide, with removable storage media, archival storage for permanent records or occasional applications; and they serve as a medium of data exchange with other computers.

A widely used storage device in microcomputer systems is the floppy disk. The 3.5-in.-square (ca. 8.9-cm-square) version of the floppy disk is shown in Figure 1–9. In the normal (720K) format, data are recorded as serial bits on 40 concentric circular tracks on each side of the disk. Each track is divided into nine sectors. A substantial fraction of the total bit capacity of each track is used for track and sector identification, gaps between the sectors, extra bits for error detection, etc. These data are written on the disk at the time of formatting. Different computers or

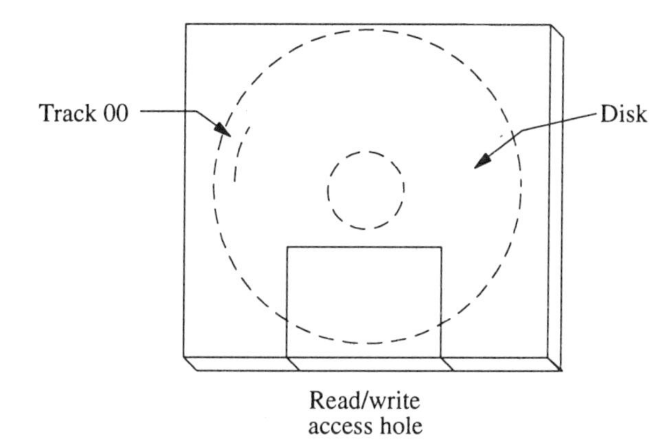

Figure 1–9. Floppy disk, 3.5 in. square. Data bits are arranged serially in concentric tracks on both sides of the magnetic disk. Floppy disks are convenient devices for off-system storage and data transfer between systems.

programs often use different disk formats, which makes information stored on one type inaccessible to another system. In the IBM format, data are arranged in nine sectors per track with 512 bytes per sector. Two of the 80 tracks are used for an index of the material stored on the disk. In all, 737 280 bytes of formatted data can be stored per disk (double sided). High-density disk drives using high-density disks provide double this amount by doubling the number of sectors per track to 18.

The advent of the Winchester-type drive has made fixed disks common in microcomputer systems. In a fixed-disk drive, the disk is not removable. In such systems the disk rotation is faster (3600 rpm) and the recording density is greater than those of the floppy disk. This results in a relatively high data transfer rate of 0.6 to 1 million bytes per s and a capacity of over 1 gigabyte for a 5-in. (ca. 13-cm) drive that fits in a standard drive bay. The fast access time and transfer rates allow large sections of memory to be loaded or dumped in fractions of a second. The read and write head of a fixed disk floats a few micrometers above the disk surface, so there is no wear of the recording surface. In contrast, floppy disks and tape have limited lifetimes.

The information on a fixed disk is not volatile, but neither is it completely safe. A speck of dust smaller than a smoke particle on a disk can cause a disk "crash," in which the read/write head grinds against the disk surface in catastrophic failure. Data may be lost if a record is unintentionally erased or written over. When this is an unacceptable risk, some off-line data storage is required as backup.

1–3 Making the Computer "Intelligent"

Digital computers are well known for their ability to solve problems whose length and complexity are well beyond the limits of manual computation. In this context, they seem to possess superhuman intelligence. On the other hand, computers have been responsible for some monumental blunders, and some computerized data systems will not allow certain desirable operations that would be very simply and quickly done if performed manually. The difference between these two perceptions of the computer lies not with the computer itself but with the type of application and the quality of the program the computer is running. Without software, the computer is only a piece of electronic hardware that can move data between locations in the system and perform simple arithmetic and logical operations on the data in response to the programmed instructions. Another part of the computer hardware is involved in managing and following the program, that is, the list of operations it has been given to do. While the hardware limits what the computer *can* do, it is the program that determines what the computer *will* do in any given application. The software package determines the range of options available to the operator and the nature of the system response to operator and instrument data input. Whether the computerized instrument treats the data in a sensible way, offers intelligent choices to the operator, and handles unusual situations appropriately depends very much on the acumen, thoroughness, and accuracy of the programmers.

The availability of faster computers and larger computer memories has made increasingly sophisticated programs practical. Although much of the increased complexity of modern software provides improved user friendliness, the longer the program and the greater the variety of interactions provided for, the greater the chance of some combinations of conditions that will produce undesirable results. In general, the fewer the tasks the program is expected to accomplish and the narrower the range of parameters and conditions it is given, the easier it is to make it reliable. The computer hardware capabilities and limitations determine what system goals are potentially realizable; the choices made by the programmer determine whether or not the system is convenient or seems intelligent.

In general-purpose microcomputer systems, the software exists on three definable levels: the disk-operating system, the application program, and the special functions written to enhance or customize the application program. The **disk-operating system** is a collection of programs that provides for efficient loading and execution of applications programs, management of program and data files in the mass storage devices, and interaction with the user input and output devices (keyboard, monitor, and printer). Since the operating system programs work quite directly with the system hardware, they are generally hardware specific; a particular operating system will work only with one type of microcomputer. However, the owner of a given microcomputer frequently has a choice of several operating systems.

The operating system along with the hardware imposes limits on the system programmer such as maximum file size, allowable memory allocations, maximum number of simultaneous users, and number of applications that can run concurrently. To perform an application such as writing within a word processor or operating a mass spectrometer, an operating system program is invoked to copy the **application program** from a disk to microcomputer memory and to direct the microprocessor to take its next instruction from the beginning of the application program's instruction list. From that point on, all operator interaction is with the *application* software.

If the application software is designed to be very versatile, such as sophisticated database or spreadsheet programs, setting up the application for specific tasks can be a major undertaking. To make this task easier, the application program may provide a set of commands that can be used in combination for customizing the application. In other words, the general-purpose capability is made more specific by the user (or professional customizer) writing special-purpose programs in an application-specific language. These three levels of software act as layers or shells,

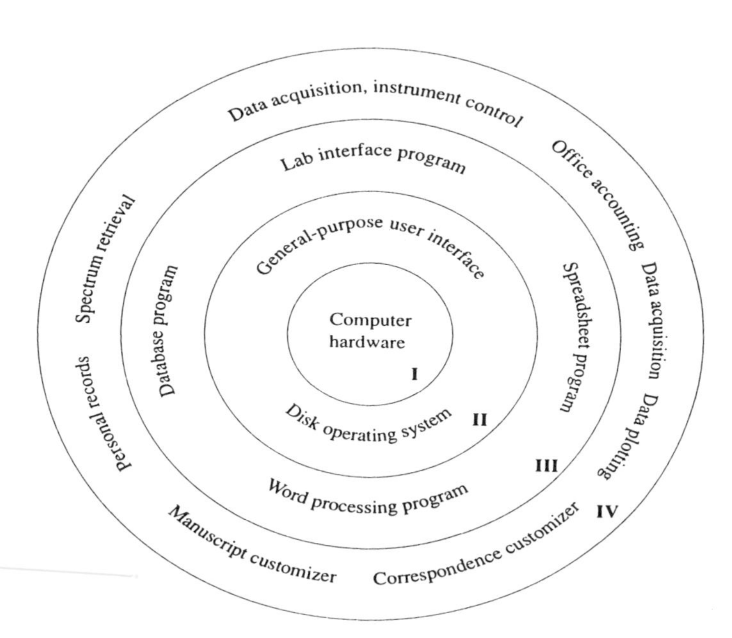

Figure 1–10. Levels of software. Basic to all operations is the disk-operating system (level II) that supervises the functions available in the computer hardware (level I). User interface programs (also level II) may add screen graphics, pointer operation, and printer control to the basic operating system. Application programs (level III) provide general tools for manipulating particular types of data, while subroutines within these applications (level IV) are customized to meet individual needs.

as illustrated in Figure 1–10. As one moves outward in this diagram, the operator is increasingly shielded from the details of hardware operations and program details. On the other hand, the environment becomes increasingly specific to a particular task. This diagram also illustrates the dependencies: the operating system is dependent on the specific hardware, the application software is written to operate with a particular operating system, and the customizing programs apply only to the specific application program for which they were written.

The programs that are used to control, operate, or interact with the peripheral devices such as special disk memory, printers, pointing devices, laboratory I/O cards, and specific instrument functions can be found in any of the software levels shown in Figure 1–10. Common peripheral devices may be supported by subroutines supplied with the operating system, user interface, or application program. These subroutines often include setup routines during which the program becomes configured to match the particular combination of hardware available to it. The closer to the core the program for interacting with a particular piece of hardware is, the more accessible it is to other programs. For example, if the printer device program is located in a user interface, such as Windows, all programs operating through this interface program can simply send their output to be printed by the user interface program. Conversely, if the mouse driver program is included in a specific application such as word processing, the spreadsheet program would not have access to it and would therefore need its own mouse driver program. Differences in the way in which each of these programs responds to manipulation of the mouse can cause operator confusion and frustration. This is the advantage of choosing a common user interface for both standard and custom application programs.

Some application programs require specific hardware peripherals that were designed to be present in the system. Certainly, instrument control and laboratory data acquisition programs are in this category. The program must contain the appropriate list of commands to achieve automatic sequencing of the experimental functions to be performed. The control of instrument function through the computer program can provide great sophistication and flexibility in control, acquisition, and data processing operations.

In order for the hardware part of the computer system to remain as widely useful as possible, the software that is permanently resident in the computer memory is generally kept to a minimum. However, without some resident program, the computer at start-up would have no instruction list and therefore nothing to do. Modern microcomputers have a program in ROM to which the microprocessor is directed at power-up and reset. This program generally performs some system diagnostics and then begins to search among the floppy and fixed disks for the operating system. When it is found, the resident part of the operating system is loaded into memory, and control of the microprocessor is transferred to the operating system executive program. Other programs in the computer ROM can include routines for exchanging data with the computer peripherals such as the disks, graphics display, keyboard, and external ports. There are many advantages in having such hardware-specific routines be part of the hardware.

As the purpose of the microcomputer system becomes increasingly specific, the shell system evolved for the general-purpose microcomputer can be greatly simplified. If only one application program will be run, it can be located in the system ROM and begin immediately on power-up. Such ROM-based microprocessor systems are often used in small instruments such as pH meters and simple spectrometers, particularly when disk data storage is not provided and the operator interaction is through special function keys and predefined displays. In these systems, the application program is the only program. Versatility and upgradability have been sacrificed for simplicity and economy.

Making Smart Instruments

A three-way interaction occurs, as shown in Figure 1–11, when a microcomputer-based instrument is being used. The operator interacts with the measurement system and the computer controls, and the computer interacts with the instrument and the operator. The choices made by the designer in the nature of these interactions greatly affects the capabilities of the system, the sophistication required of the operator, the apparent "intelligence" of the system, and ultimately, the real usefulness of the system. I once had a "smart" travel clock that was extremely versatile, but to change the indicated time or set the alarm, one needed to press, in various sequences, buttons on the bottom marked A, B, and C. I didn't travel often enough to remember the sequences for each operation and quickly tired of lengthy trial-and-error sessions when I was ready to go to sleep. The clock was soon replaced. Such experiences are more than irritations when a recalcitrant instrument costing several hundred thousand dollars is involved.

In modern computer-based instrumentation, virtually all aspects of the instrument's operation are monitored and controlled by the computer. All that remains of the operator's direct interaction with the instrument is the process of sample introduction. The more functions, parameters, and operations that are brought under the control of the computer, the more the programmability of the computer can be used to provide flexible automation, automatic recording of all selected parameters, and continual checks on critical instrument conditions and experimental results. The disadvantages are that the operator may feel that he or she has lost contact with the instrument, and the operator may not want to be so interactive with or dependent on the computer. These disadvantages can be largely overcome, or even reversed, by proper design.

In the early days of computer-based instrumentation, when computer resources were extremely limited, many operator inconveniences were accepted for the sake of applying the newly available computing power to the most difficult operations. Today, computing power is very inexpensive, so that more and more of it is used to increase the convenience to the operator and make the instrument more reliable. Through the use of trackballs, mice, and knobs, control of instrument parameters has regained much of its original feel (often with greater range and precision). This simulation of analog control provides more natural control action and better operator assurance. The development of high-resolution color displays allows operators to see the consequences of their adjustments better than was possible

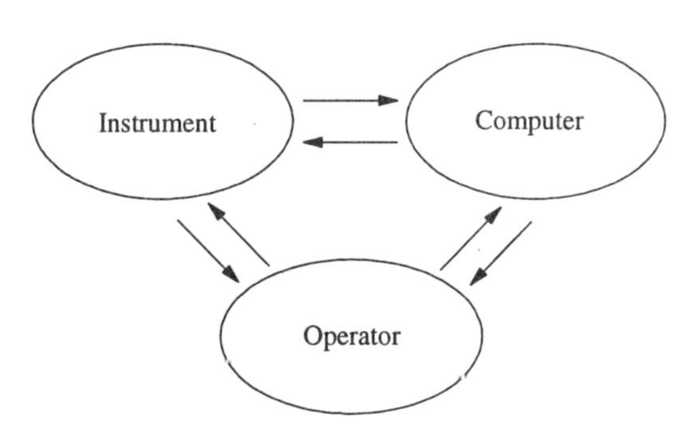

Figure 1–11. Interactions in a computer-based instrument. The interactions between the instrument and the computer are handled by the computer interface. The operator may interact with the instrument directly and/or indirectly through the computer for experimental setup. The mechanisms provided for operator interactions greatly affect the versatility and ease of use of the instrument.

before. Displays of system status can be much more readable than old instrument display panels, and undesirable combinations of operating conditions can be automatically detected and indicated or disallowed. Menus and dialog boxes guide the operator's choices for new experiments, whereas completely preprogrammed procedures are used for routine measurements. A welcome innovation in recent instruments is the inclusion of a "command" language in the application software that allows the user to customize and extend the routines provided by the manufacturer. The result of these developments is use of the computer to enhance the interaction between the operator and the instrument, providing versatility and simplicity concurrently. To accomplish this, designers need to discern the most natural forms for this interaction and allow for as many of them as possible. For our part as users, we must provide good feedback to the designers on what works and what feels awkward, and we must avoid products that require oversophistication in esoteric technology.

1–4 Using the Microcomputer in Science and Technology

The computer has become an indispensable tool in the scientific laboratory. Programs such as word processors and database managers are of very broad applicability, but some of the generally useful programs such as spreadsheets, mathematical equation solvers, statistical analyzers, and graphical presentation programs offer powerful new capabilities to those working with experimental data. Computer modeling and simulation of physical processes or instrument functions can greatly aid in instrument design and bring new insights into aspects of particle motion, chemical structure, and system dynamics that are difficult or impossible to observe directly. The combination of the computer with laboratory interface software can provide many of the functions of standard laboratory test instruments such as the oscilloscope, digital voltmeter, frequency meter, and signal generator. With some specialized programs, these test instruments can be specifically emulated, enabling the computer to become a set of convenient tools for diagnostics and troubleshooting. In this section, we will take a closer look at these areas of computer applications in the laboratory.

Processing and Presenting Data

Every year we see dramatic improvements in the power and convenience of general-purpose programs for the analysis and presentation of data. Even though many of these programs are designed principally for application in the business office, they are no less useful in processing experimental results. Scientists are discovering more and more applications for spreadsheet programs initially developed for accountants. Data in a spreadsheet are contained in cells that are arranged in rows and columns like a table. Some of the cells are used for the initial data, such as experimental results. Other cells can then be programmed to contain the results of mathematical operations on the original data. Thus, columns of cells containing the averages or differences of sets of input data values are readily generated. Additional columns can be programmed to automatically compare averaged experimental data with standard values, and so on. Experimental data, collected through a laboratory I/O board and data acquisition program, can be organized in a data file compatible with a spreadsheet program. When these data are input into a spreadsheet specifically preprogrammed to process these data in the desired way, a table of the final, calculated experimental results is produced automatically. In addition, all spreadsheet programs now provide extensive graphing and plotting capabilities so that a variety of graphs of the processed experimental results can easily be generated.

Mathematical problem solvers can perform powerful and complex operations on experimental data as well as solve mathematical equations and plot the results. Investigators formerly had to write custom programs to implement many of the data operations now available in these commercial packages. The time saved and the increased power available are tremendous. The disadvantage of using general-purpose programs for an oft-repeated mathematical operation is that more time and more of the computer's resources are required than if the specific function were programmed directly. With the decreasing cost and increasing speed of modern microprocessors, this disadvantage is becoming less significant. Sophisticated data treatment has also become much simpler through the use of packages of statistical programs. These sets of programs perform a variety of operations such as regression analysis, curve fitting, clustering, and confidence interval calculations. When these operations can easily be invoked to process experimental data already in computer data files, there is little reason for not taking advantage of the insights into the data that they can provide.

Once a difficult and time-consuming task, the presentation of data in graphical form is now readily accomplished through the use of graphical presentation programs. As indicated earlier, some graphical capability accompanies nearly all spreadsheet programs. In addition, designed principally for slick business graphs, presentation programs provide very convenient and sophisticated plotting functions. Axis labeling, data scaling, multiple curve plotting, nonlinear axis calculations, and error bar indicators are automatically accomplished with typeset quality. Output can be directed to ordinary printers and plotters but also to image-maker cameras for the generation of beautiful colored slides or overhead transparencies. The computer screen itself is increasingly becoming a useful option for the presentation of graphical results. Large computer screens and devices that adapt an overhead projector or a projection television for computer screen projection make printing to an intermediate medium unnecessary. For such applications, most presentation programs also include a slide show feature that will display a number of prepared screens in succession automatically.

Simulating Processes under Investigation

If a suitable mathematical model exists for a process under investigation, there is much to be gained by using a computer to simulate the process. The results of the simulated process can be compared with the actual experiment so that differences between them can be used either to improve the accuracy of the model or to remove artifacts in the experiment. Systematic variation of some of the parameters in the simulation can sometimes provide insights into the effects of these variables that might be difficult to observe in the process itself. Simulation programs can also be indispensable in the process of design. Programs for photon optics simulate the effects of various optical elements and even provide ray tracing. Similarly, programs for ion optics calculate the electric field created by various optical elements and then trace the paths of charged particles through the system. Such programs are far easier to experiment with than actual optical or ion transmission systems, and much more can be learned about the significant design elements than by "cut-and-try" experiments on the actual physical system.

Emulating Standard Test Instruments

Many functions of standard test instrumentation can be accomplished with the combination of a laboratory I/O board and suitable software. For example, the acquisition and display of an input waveform by a digital oscilloscope can also be accomplished by a suitably interfaced computer. Similarly, a voltage measurement can be performed by connecting the signal source to a digital voltmeter or to the analog input of the laboratory I/O board. The computer can even be used as a signal generator by loading the values of the desired successive signal voltages in memory and then transferring these values sequentially, at the desired rate, to the analog output of a laboratory I/O board. Of course, there are several significant differences between a computer with laboratory interface and a dedicated instrument, even though both may be microprocessor-based. In the dedicated instrument, microprocessor operations are controlled by front-panel controls, which are usually arranged in a way familiar to those who have used such instruments before they were computer-controlled. The interface hardware is specific for functions to be performed and may be more efficient in their accomplishment. Although the general-purpose computer with interface is more versatile than a single-purpose instrument, the cost of this flexibility is that the user has to set it up specifically to do each separate task.

Recently, computer programs that are intended to decrease the difference in ease of use between the dedicated instrument and a general-purpose computer used for a similar function have been developed. These programs simulate the controls and displays of various dedicated test instruments as nearly as possible. By a pointing device, the simulated controls on the computer screen are set at the same time that the appropriate parameters are set in the operating program. It certainly is overkill to use several thousand dollars worth of computing equipment to simulate a $100 digital voltmeter, but if one already has the requisite hardware, investment in a few programs may add the equivalent of several new useful instruments to the laboratory.

The increase in capability and flexibility that has resulted from the inclusion of more powerful microprocessors in dedicated instruments has been accompanied by changes in the operator interface that make these systems more like those of the interfaced computer. A computerlike screen, upon which is displayed not only the output data but menus of options available to the operator during the setup process, has become quite common. The once-familiar knobs that controlled each separate input parameter are being replaced with arrow keys that are used to increase or decrease the value of whatever parameter is currently being adjusted. Some very sophisticated dedicated instruments such as high-speed transient recorders even contain floppy- or hard-disk drives for the storage of acquired data and may contain the operating program in a replaceable ROM for easy field updating of instrument capability. Such instruments provide more of the flexibility afforded by the standard general-purpose microcomputer while retaining the predetermined functionality of the dedicated microcomputer system. Some may even provide a degree of user programmability. Similarly, a person using a general-purpose microcomputer in an instrument system can "harden" the instrument-operating programs in read-only memory to provide more of the operational simplicity of the dedicated microprocessor instrument. The use of an application-specific control panel can further "customize" a general-purpose system. Which approach is the better choice will depend on each environment and application. Being able to make the choice is always to the user's advantage.

Acquiring Data and Automating Experiments

The availability of so many tools for data manipulation and presentation within the microcomputer environment makes it highly desirable to bring the experimental data into an application program file. If the data come into the microcomputer environment through a directly coupled interface, the data acquisition is at least partly program-controlled. If data acquisition is under program control, it is often desirable to control other aspects of the experiment such as parameter setting, sequencing of operations, error checking, and overall timing by the computer's program. Once all the functions and parameters are interfaced to the microprocessor, all experimental processes can be initiated and followed by various steps in the computer program. Although this fact has been known and a large variety of interfaces have been available for some time, their adoption as a standard experimental approach has been limited by the time required to write the program(s) with which to automate experiments of even modest complexity. Inflexible programs are simpler to write than flexible ones, but they require much reprogramming when changes in the experiment are desired. Recently, however, some of the high-level (high-function) programming techniques developed to provide easy flexibility for database and other commercial programs have been adopted in programs designed to acquire data and control experiments. At setup, the program has to be instructed about the nature of the hardware interfaced to the computer and the operations it will be controlling (just as setting up a word processor program requires an indication of the type of printer available). Then, by means of a series of graphical or menu choices, the operator can design the complete, automated experiment. This capability has the potential to vastly increase the efficiency with which experiments can be devised and executed and the data from them recorded and evaluated. As a result, this trend is expected to continue. As with all application programs, there is a direct trade-off between flexibility, ease of implementation, and performance.

Generally, in any instrument or automated experiment, one microprocessor runs the program that coordinates the overall operation. However, the devices that carry out the individual functions (such as recorders, data acquisition boards, printers, spectrometers, and lasers) are likely to be themselves microprocessor-controlled. The interface link between a microprocessor-controlled function and the host computer must support the bilateral transfer of measurement and control data under the direction of the host computer. It is not unusual today for even a modestly complex experiment to involve a dozen or more microprocessors linked to each other and to their individual input and output devices in a variety of ways. This interconnection network provides the information fabric of the modern laboratory instrument and experiment. When using such instruments and developing such experiments, it is essential to understand data flow and know how to make the right connections.

Related Topics in Other Media

Video, Segment 1

- Digital data and digitization
- Use of computers with instruments (pH meter and FTIR spectrometer)
- Data flow in an instrument

- Computer-controlled instruments
- Hardware modules making up a computer system

Laboratory Electronics Kit and Manual

- Preface
  - ¤ Introduction to hands-on experiments
- Introduction
  - ¤ Experiments and microcomputer-controlled instrumentation
  - ¤ Data flow and top–down approaches

Chapter 2

Converting and Acquiring Data in Scientific Instrumentation

In modern electronic instruments, information exists in many forms. In computers, data are digitally encoded. The output signals from many of the sensors used are analog signals; that is, the encoded quantity is related to the amplitude of the electrical signal. Also, the timing of particular signal variations can be used to convey the desired information. From these examples, we see that an important part in the understanding of modern instrumentation is an awareness of the various ways in which electrical signals can convey information and an understanding of the techniques used to convert one form of electrical data encoding into another.

When measurement is involved, the desired result is a number: specifically, the number of standard units of the measured quantity (volts, milligrams, etc.) present in the sample. Because all measurement proceeds toward a digital result, conversion to digital data encoding is an essential part of modern electronic measurement systems.

Only three main categories of electrical data-encoding techniques exist: analog, time, and digital. In this chapter, the characteristics of each category of data-encoding techniques are explored, and the methods of interconversion among the categories are introduced. The emphasis is on digitization and computer acquisition of data from a variety of common sensor types.

2–1 Sensing Digitally

In electronically aided measurements, a sensor is used to convert the quantity to be measured into some property of an electrical signal, and this property is in turn converted into the corresponding electrically encoded number. The electrical signals produced by sensors are necessarily encoded by one or more of the three data-encoding techniques: analog, time, and digital. Sensors that produce an analog signal related to the measured quantity are by far the most common. The use of the time relationships of sensor output signals is also quite often encountered. However, there are relatively few sensors whose output is inherently digital.

Even though there are few examples, it is worthwhile to consider how measured quantities could be converted directly into an electrically encoded number. Direct digitization is the electronic equivalent of using a ruler (technically, a scale) to measure the length of something. One aligns the zero end of the scale with one end of the object and reads the length directly as the number on the scale aligned with the other end of the object. The number of length units that equals the dimension of the object is determined directly. This type of measurement is free of complicating conversions. Unfortunately, most measurements made today are quite indirect, involving a number of conversions from one representation of the measured quantity to another until a digitally encoded number is finally achieved.

A digitally encoded signal is composed of a number of two-level signals that can represent bits in a binary number. One way to achieve direct digital encoding

2861-2/94/0023$10.80/1 Published 1994 American Chemical Society

is to use a set of sensors, each one of which can contribute 1 bit to the resulting number. For example, an optical sensor can produce a signal that is either HI or LO depending on whether it is illuminated at a level greater or less than some preset value. In this way, the measured quantity is converted directly into a 1-bit digital signal. To obtain a direct digital conversion with more resolution (more bits), one must use more sensors, each responding at different levels (parallel), or the same sensor sequentially (serial). Both of these approaches are illustrated below.

Parallel Digital Sensing

A parallel direct digital measurement requires a data source that produces a parallel digital signal in which the pattern of HIs and LOs produced in the output connections indicates the state of the sensor or the value of the quantity being measured. An example of a direct digital measurement of position using parallel, or simultaneous, detection is shown in Figure 2–1a. Six sensors detect the transparency of a track made of alternating clear and opaque regions at a constant radius on the wheel. The pattern of the regions is such that no two positions on the wheel have the same pattern. Many such patterns are possible. One, called the gray code, avoids having more than one sensor output change at a time. The value of the 6-bit digital word produced by the sensor outputs is thus indicative of the position of the wheel. The resolution of this measurement is 1 part in 2^6, or 1 in 64. If more precision is needed, more tracks and sensors are required; as the pattern becomes finer, the mechanical tolerances become more stringent. Circular position detectors with a precision of 1 part in 2^{14} (16 384) per turn are available. This concept is also applied to detect linear position or movement, as shown in Figure 2–1b.

The direct conversion of a measured quantity to a parallel digital signal is rare, though possibly, as the digital revolution continues, more such conversions will be developed.

Counting

Perhaps the most time honored method of determining a quantity is to count the units of that quantity contained in the unknown. A single sensor can be used to produce a pulse for each unit in the unknown quantity, and a counting circuit can be used to count the pulses. For example, suppose we want to know the number of apples in a bushel. One could arrange to have the apples roll sequentially down a chute where they would interrupt the light to an optical sensor or temporarily load a pressure or strain sensor as they go past. Each apple would then produce a blip in the output signal from the sensor. A pulse counter connected to the sensor output would record the number of apples that passed the sensor. Suppose, instead, we want to know how many grams the bushel of apples weighs. Now we have to arrange to count grams. One way would be to put the apples on one pan of a balance and arrange to allow 1-g weights to slide past a sensor and then come to rest on the other pan. A counter would keep track of the number of weights added, and a mechanism would stop sending weights down the chute when the pans contained equal weights. The important point in this example is that when counting is used for measurement, one must arrange to have the units of the quantity being measured be the events that are counted. Events are sensed sequentially to produce a signal that is a series of pulses. The final count is equal to the number of pulses in the series. Thus, the count signal is a type of serial digital signal.

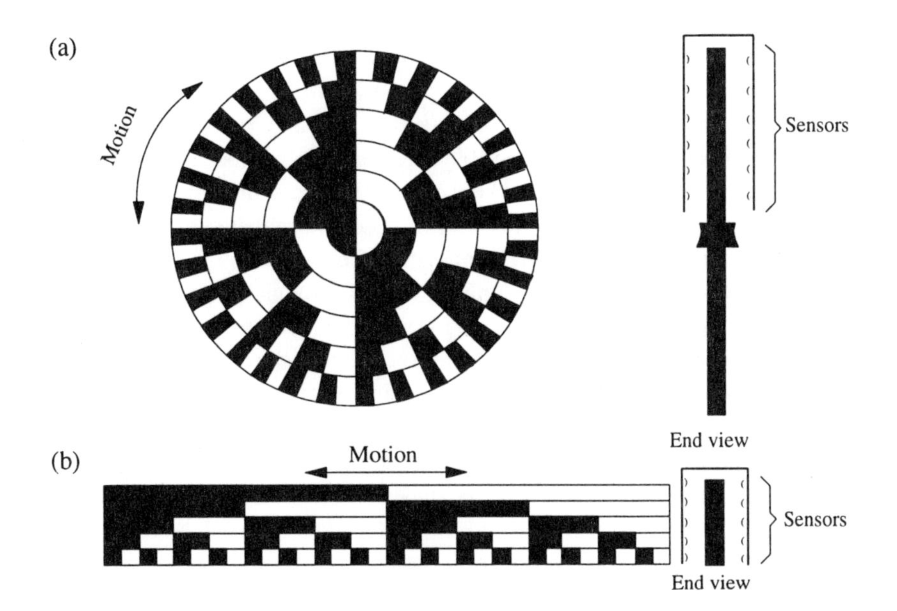

Figure 2–1. Optical position sensors: (a) circular; (b) linear. A separate light/detector combination is used to determine the transparency of each track of the movable part of the sensor. Each segment of the movable part produces a combination of sensor outputs that is distinctive for that segment.

The elements of a digital counting system are shown in Figure 2–2. The occurrence of an event or the presence of an object at the sensor causes a change in the level of the output signal. The signal shaper and discriminator converts the changes in the sensor output level into a digital signal (one with levels that are either HI or LO). Each event is thus converted into a pulse that can be counted electronically. The threshold control in the discriminator circuit determines the level of sensor output that will be interpreted as an actual event. For precise counting, it is important that the change in level due to true events be substantially greater than the background variations in the sensor output. This reduces the possibility of a nonevent adding to the count. The pulses proceed to the counting circuit through a counting gate that is used to control when the counting will begin and end. Control signals to the counting gate determine the boundary conditions of the count. For example, stopping the count exactly 1 h after it is started produces a count that is the number of events per hour. Starting the count when an empty

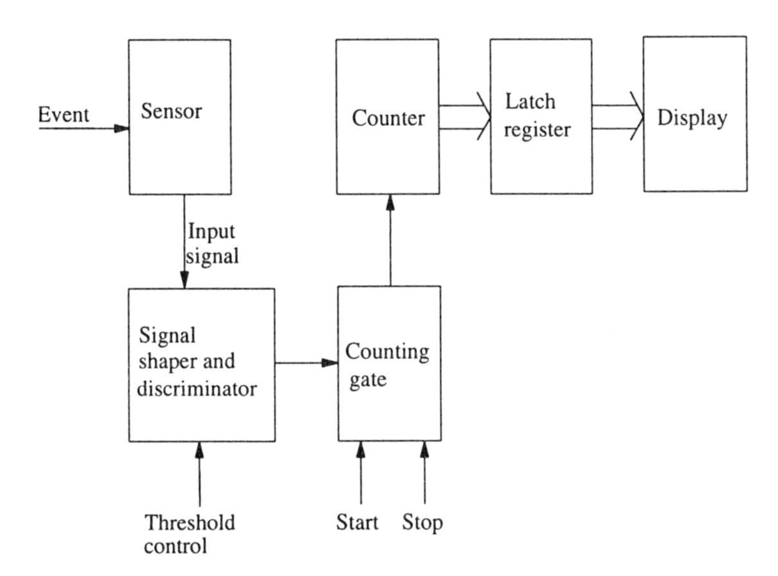

Figure 2–2. An electronic counting system. The sensor and the signal shaper and discriminator define the events to be counted. The Start and Stop signals define the counting interval. The counter is an incrementing register, the contents of which are transferred to the latch register for display at the end of the counting interval.

basket is placed under the apple chute and stopping it when the basket is removed gives a count that is the number of apples per basket.

The process or algorithm for counting is to add one to the previous value. Therefore, a counting circuit must be able to remember a number and increment it. The counting circuit is thus a special kind of register, that is, one that can increment the numerical value of the word it is storing. Each pulse that comes through the gate is used as a command to increment the number stored in the counter. Table 2–1 gives the sequence of binary numbers for 0 through 15 (decimal). In an electronic counter, incrementing flip-flop storage elements are used, one for each bit in the counter. A 4-bit flip-flop counter, such as that shown in Figure 2–3, can count from 0 to 15. The number of states of a counting circuit is called its **modulus**. The counter of Figure 2–3 is a modulo-16 counter. To increase the count capacity, more flip-flops are used to generate the additional bits. When the counting interval is over, the word stored in the counter is transferred to a memory register that can store the count value for output to a display or a computer data bus. The counter can then be cleared (set to all zeroes) by the application of a pulse at the clear input.

The output waveforms of the 4-bit counter are also shown in Figure 2–3. A comparison of the output states after each pulse with the binary count table demonstrates that the value of the word stored in the register increments after each count. When the count reaches the maximum (all the bits are 1), the next event returns all the bits to 0. The waveforms of Figure 2–3 reveal another interesting

Table 2–1. Binary Numbers (0000 through 1111)

	Binary			
Decimal	Bit 3	Bit 2	Bit 1	Bit 0 (LSB[a])
0	0	0	0	0
1	0	0	0	1
2	0	0	1	0
3	0	0	1	1
4	0	1	0	0
5	0	1	0	1
6	0	1	1	0
7	0	1	1	1
8	1	0	0	0
9	1	0	0	1
10	1	0	1	0
11	1	0	1	1
12	1	1	0	0
13	1	1	0	1
14	1	1	1	0
15	1	1	1	1
16	0	0	0	0

[a]Least significant bit.

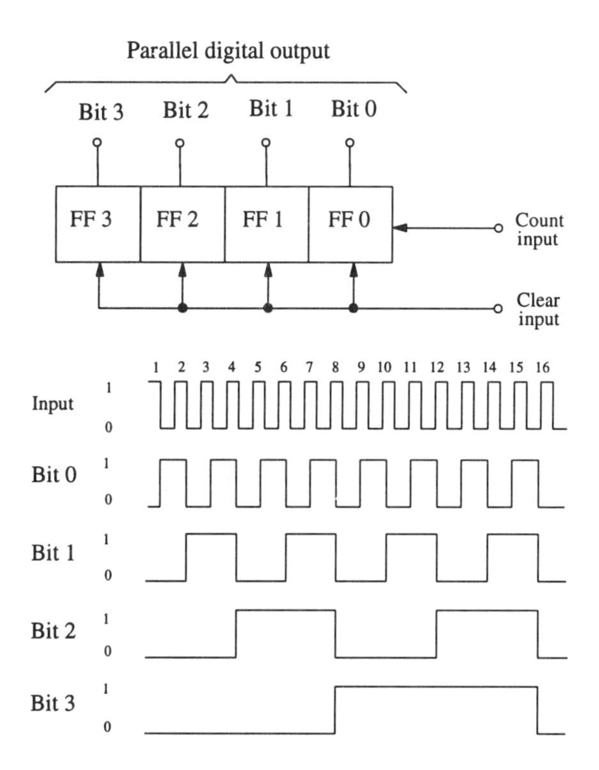

Figure 2–3. Four-bit binary counting register and waveforms. A counter is a memory register whose content can be incremented or reset on command. The number of flip-flops (FF) in the register determines the number of bits in the output and the capacity of the counter. The bit output levels follow the binary counting sequence.

and useful fact about binary counters. For the 16 input pulses, there are 8 pulses in the output of flip-flop 0 (FF 0), 4 from FF 1, 2 from FF 2, and 1 from FF 3. Thus, each flip-flop produces a number of output pulses that is exactly half the number that appeared at its input. Also, the output pulse rate is exactly half the pulse rate at the input. This characteristic can also be seen in the table of states (Table 2–1). Notice also that the output pulse of FF 3 (the most significant bit) uniquely terminates on the count that exceeds the counter capacity. Thus, the most significant bit output of a counter provides the overflow signal, that is, a falling edge that indicates that the counter has filled up and is starting over.

Such binary counters are naturally at home with computers, in which numbers are most often represented in binary-coded form. When a numerical value is to be presented to the screen or printer, a program is called to convert the binary-coded number to decimal form. In the case of small instruments, where the counter must provide its output directly to a human observer, it is often inconvenient to convert a binary counter output to decimal form. Special decade counters that are better suited for direct decimal display of the result have been invented. Four flip-flops are required to provide the 10 necessary states. The flip-flop output sequence follows that of the binary counter but terminates on count 10 instead of count 16. This is called **binary-coded decimal** or **BCD** code. The output of flip-flop 4 goes to 1 on count 8 and back to 0 on count 10. Thus it provides exactly one pulse for every 10 pulses at the input. The modulus of the decade counter is thus 10. A 2-decade BCD counter is shown in Figure 2–4. A separate 4-bit decade counter is needed for each decimal digit: the first one for the one's digit, the second for the ten's digit, and so on. Decimal counters with 8 decade counting circuits, with a count capacity of 99 999 999, are fairly common.

Still more sophisticated counting circuits that can preset the counter to any given starting value and count up or down are available. Such a counter is useful for signaling when a certain number of events have occurred by presetting the

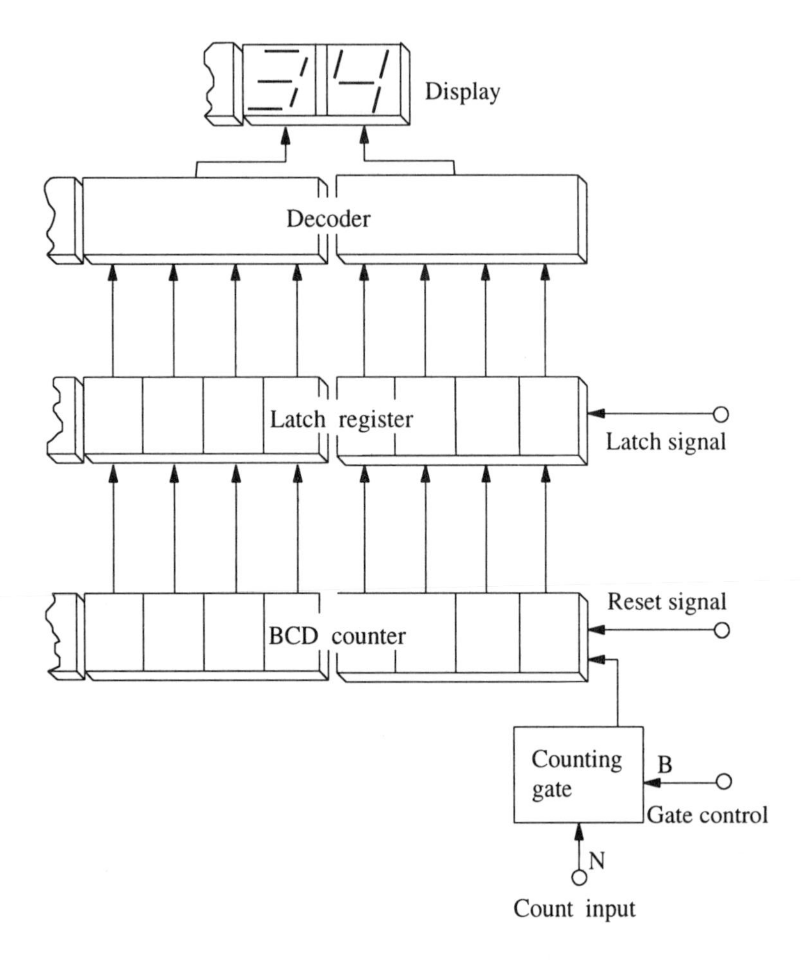

Figure 2–4. BCD counter with decimal display. The 4 bits in each BCD-encoded decade counter proceed from 0 through 9. On event 10, the counter state returns to 0 and passes an increment signal on to the decade counter to the left. When the counting interval is over (when the gate control closes the gate), a signal is applied to the latch register to store the count value. The latch register outputs are connected directly to the display decoders and the display.

counter to the number, counting down, and watching for the appearance of all zeroes in the output. Counting is the most basic of all the direct digital measurements. As we will see in later sections of this chapter, counting is often involved in converting data encoded in a variety of ways into digital signals.

2–2 A Closer Look at Analog Data Encoding

There is a popular misconception that the digital revolution has made analog electronics obsolete. This is far from the truth. Most sensors used to monitor physical and chemical quantities and properties produce analog-encoded electrical signals. Paradoxically, because of their ability to absorb and correlate data from multiple sources, computer-based instruments tend to monitor and control many more instrumental functions and parameters and therefore use many more sensors than the earlier equivalent analog instruments. Because computer-based instruments rely so heavily on data from analog transducers, it is essential to have a good understanding of analog-encoded electrical signals.

Analog-encoded data are represented by the magnitude of one of the four electrical quantities: charge, current, electrical potential, or power. The units for the magnitudes of these quantities are coulombs, amperes, volts, and watts, respectively. The magnitude of a physical quantity or property is converted into the magnitude of an electrical quantity by a sensor, or **input transducer.** The characteristic of an analog input transducer is that the magnitude of the measured quantity

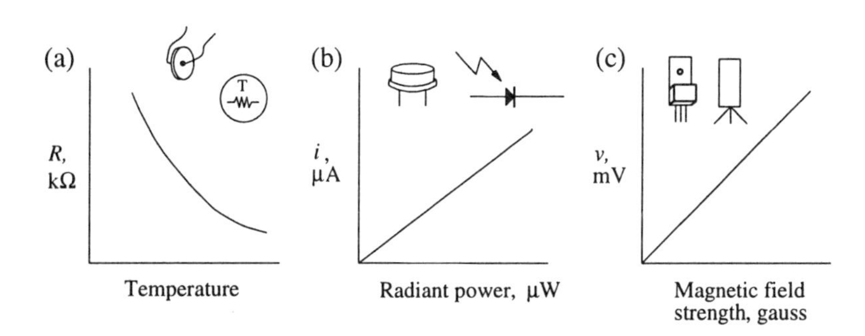

Figure 2–5. Analog input transducers, circuit symbols, and transfer functions: (a) thermistor (T) temperature sensor; (b) photodiode light sensor; (c) Hall effect magnetic field sensor. Each device converts an input quantity (x axis) into an output quantity of a different type (y axis). The relationship between input and output quantities is the transfer function.

can be determined from the magnitude of the electrical output signal. As the input quantity changes, the magnitude of the electrical signal changes correspondingly. The relationship between the input and output quantities is called the **transfer function**. Several examples of analog input transducers are shown in Figure 2–5 along with their input quantities, output quantities, and transfer functions. Note that the transducer converts a quantity expressed in one unit to a quantity expressed in another unit, for example, degrees (temperature) are converted to volts. Many other analog transducers are described in Chapter 7. All measurement systems require some form of input transducer unless the quantity to be measured is already some property of an electrical signal.

Devices that convert an analog signal into the magnitude of a physical quantity or property are called **analog output transducers**. Examples of several analog output transducers are shown in Figure 2–6. Many more output transducers are described in Chapter 8. Electronic systems that actively control some quantity or process in the physical world require both input and output transducers. A constant-temperature bath or oven requires a temperature sensor and a heater. The electronic part of the system compares the temperature-related voltage from the thermocouple with the thermocouple voltage expected when the system is at the set temperature and, on the basis of the difference between the sensed and desired temperatures, changes the power applied to the heater. In this case, the heater is the output transducer, converting the magnitude of the electrical power applied to heat.

Position controllers, motor speed controllers, and light intensity controllers are all examples of control systems in which both input and output transducers are used. Very sophisticated controllers, now called robots, are described in Chapter 8. The conversion of the quantity to be controlled into an electrical signal and back into the physical quantity allows all the sophistication of modern electronic signal-processing techniques to be applied to the control process. The rapid expansion of computer control in complex consumer goods such as the automobile is spurring the development of better and cheaper transducers, most of which still produce or require analog-encoded electrical signals.

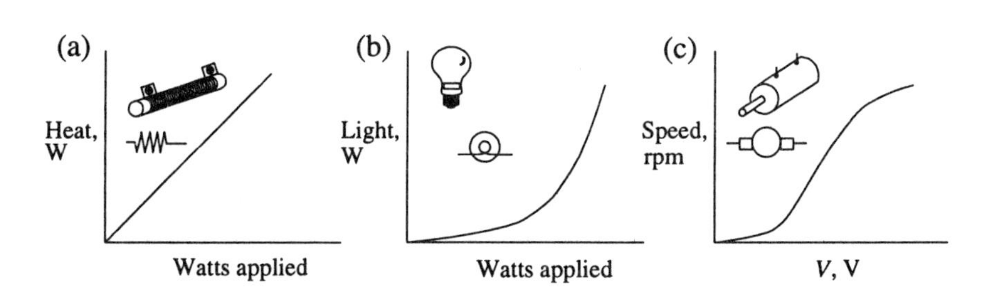

Figure 2–6. Analog output transducers: (a) heater; (b) incandescent lamp; (c) motor. For each device, the application of an electrical quantity at its input results in a related effect on the physical output quantity, that is, heat, light, or rotational velocity, respectively.

Characterizing Analog Signals

Each of the three types of data encoding (analog, digital, and time) has characteristics that are very different from those of the other two. These differences must be appreciated when choosing the encoding technique for a particular application and in understanding what advantages are gained and what information is lost in converting data from one encoding form to another. In particular, what limits the precision of data encoded electrically, and why is the world going digital for data recording and transmission in everything from instrumentation to high-fidelity audio and television?

With analog signals, it is generally the amplitude of the signal voltage or current that conveys the desired information, although the other two electrical quantities, charge and power, are sometimes used. Because the smallest unit of charge, the charge of an electron, is so small compared with the total charge stored or transferred by most signal sources, the amplitude of an analog signal is variable in essentially infinitesimal increments. In other words, the amplitude of an analog signal is continuously variable.

Electronic signals are potential differences or current amplitudes. Electrical potentials are generally measured with respect to a stable point of potential reference called circuit common (*see* discussion in Supplement 1). In a circuit diagram, circuit common is indicated by a downward-pointing triangular symbol, as used later in Figures 2–11, 2–13, and 2–14.

The amplitude of an analog signal can be recorded continuously over time or sampled at any instant. When continuous measurements are made, the variations in the signal amplitude can be plotted against time, wavelength, magnetic field strength, temperature, or other experimental parameters, as shown in Figure 2–7. Note that *analog data can be plotted continuously in both magnitude and time dimensions*. As we shall see, this continuity in both magnitude of encoding and time of access to the encoded value is peculiar to analog encoding. Additional information can often be obtained from such plots by correlating amplitudes at different times. Such information includes simple observations such as peak height, peak position, and number of peaks or more complex correlations such as peak area, peak separation, and comparison with data from other plots. The techniques of correlating data taken at different times, that is, data processing,

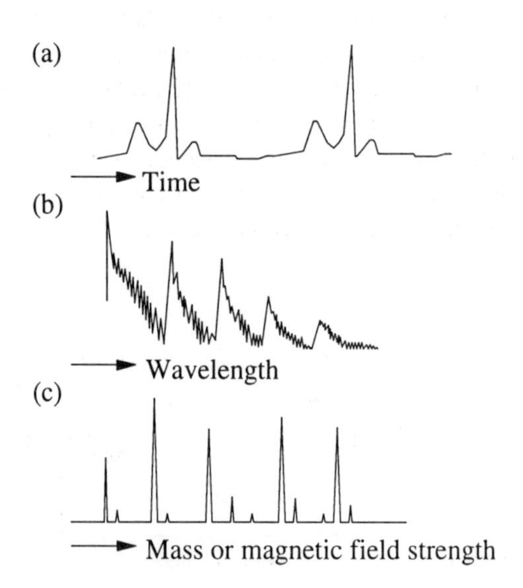

Figure 2–7. Analog data plots: electrocardiogram monitor output vs. time (a); emission spectrometer detector output vs. wavelength (b); ion sensor output versus magnetic field strength B in a mass spectrometer (c). The sensor output is a continuous function of the quantity measured and is infinitesimally variable.

should not be confused with the techniques of converting data from one manner of representation to another.

Analog signals are susceptible to electrical noise sources contained within or induced upon the circuits and connections of the instrument. The resulting signal amplitude at any instant is the sum of the data and noise components. The effect of the noise is to reduce the **precision** (reproducibility) of the data that are analog encoded. Even for a constant quantity encoded, the signal amplitude will vary because of the noise component, and this variation will be indistinguishable by subsequent devices and circuits from variations in the desired quantity. Ultimately, *electrical noise limits the degree of precision possible with analog encoding.* One millivolt of noise will introduce an uncertainty of 1 part per 1000 (three significant figures) in a 1-V signal. Increasing the precision of an analog-encoded signal often involves a painstaking attack on the various sources of noise (see Appendix A for a review of shielding and grounding techniques). Each additional significant figure of precision is increasingly difficult to obtain.

Analog signals can transfer data at a very high rate. Think of one piece of data as the result of obtaining the amplitude of the signal at a particular instant of time (a technique called **sampling**). In principle, one could sample the analog signal as often as one chose: a thousand, a million, or even hundreds of millions of times each second. If each of these samples really represented a new piece of data, the data rate would be equal to the sampling rate. However, if the time between samplings is less than the response time of the circuit being sampled, each new sample will not represent a new value for the encoded quantity and the real data rate will be less than the sampling rate. The maximum rate at which data can be transferred by an analog signal is determined by the signal **bandwidth** (the difference between the maximum and minimum frequencies of magnitude variations the signal contains).

Knowing the frequency composition of a signal is important if operations such as sampling or filtering are to be performed on the signal without loss of information. The analysis of the frequency composition of a waveform is called **Fourier analysis**. In Fourier analysis, an amplitude–time waveform is transformed into its **frequency spectrum**, which is the amplitude of each frequency component of the signal plotted against the frequency. Every amplitude–time waveform $v(t)$ has a related frequency spectrum $F(\omega)$. The two functions $v(t)$ and $F(\omega)$ are called Fourier transform pairs and are related by the Fourier integral

$$F(\omega) = \int_{-\infty}^{+\infty} v(t)\,[\cos \omega t - j \sin \omega t]\,dt$$

where j is the complex operator $\sqrt{-1}$. These two representations of a waveform are often called the **time domain** and **frequency domain** representations. Several Fourier transform pairs are illustrated in Figure 2–8. In many cases, the frequency spectrum of a waveform is plotted as the amplitude density (in volts per hertz) versus frequency (in hertz) to avoid mathematical problems in calculating the Fourier integral. Such a plot is called an **amplitude spectrum**. In some cases the power density (in watts per hertz) is plotted against frequency (in hertz). Such a plot is called a **power density spectrum** or simply a **power spectrum**. The spectrum of a band-limited signal is zero everywhere except in a limited frequency range. Band-limited signals fall into two classes: direct current (dc) signals and alternating current (ac) signals.

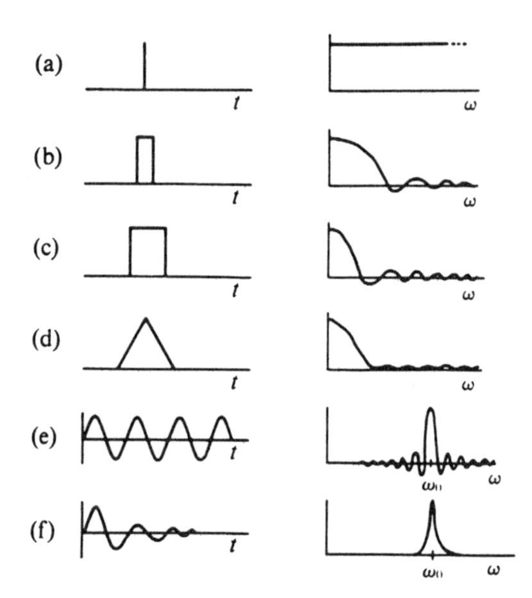

Figure 2–8. Pictorial Fourier transform pairs. An infinitely sharp amplitude–time signal (a) has equal amplitude at all frequencies (white spectrum). For the rectangular pulses (b and c), the frequency spectrum has the form of the function (sin x)/x. As the pulse widens, the frequency spectrum narrows. The triangular pulse (d) gives a frequency spectrum of functional form ($\sin^2 x$)/x^2. A finite-duration sine wave (e) and an exponentially decaying sine wave of frequency ω_0 (f) are also shown. If the sine wave of part e were continuous (rather than just four cycles), the transform pair would be a single line at the sine-wave frequency.

A dc signal is one in which the direction and magnitude of the current are constant over the period of interest. However, no signal can be constant indefinitely. Consider the output of a thermocouple used to monitor temperature. A typical plot of the transducer output voltage against time is shown in Figure 2–9 along with the signal power density that results from Fourier analysis of the waveform. The signal frequency composition at frequencies higher than 0 Hz (dc) may arise from actual temperature changes or from changes in the thermocouple transfer characteristics with time. If the variation of the temperature with time is of interest, the system used to amplify and transmit this signal should have an equal response to all frequencies throughout the bandwidth of the signal. Thus, the term bandwidth can also be used to indicate the range of frequencies over which the operation of a

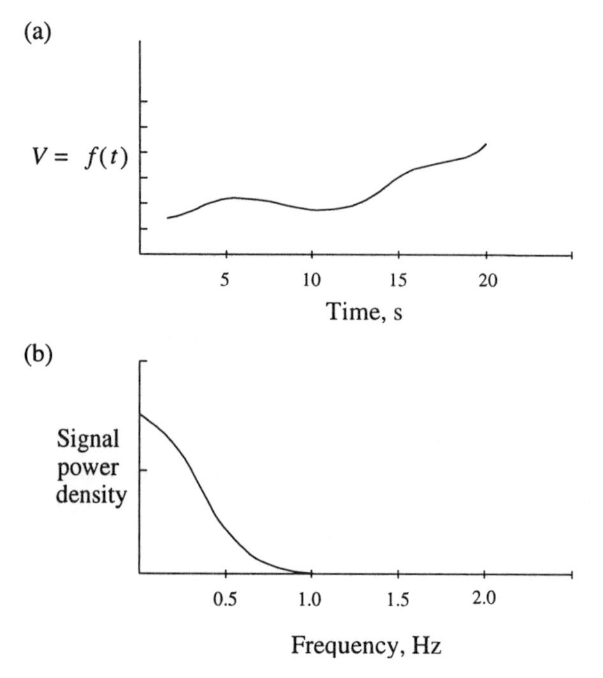

Figure 2–9. Output voltage versus time (a) and power spectrum for a thermocouple over a time of thermocouple temperature variation (b). Note that most of the signal power is at or near 0 Hz (dc) but that some information is present at higher frequencies.

device or system will be accurate. The bandwidth of a system is sometimes called the **bandpass**. Any electronic system for amplifying, modifying, and displaying a dc signal must have a low-frequency response that extends to 0 Hz. Spectra similar to that of the thermocouple output signal arise from other transducers that are usually considered to produce dc outputs. Because all signals have some bandwidth, a general definition of a **dc signal** is one whose power is concentrated in a band of frequencies near 0 Hz.

An ac signal can also be usefully characterized by its power spectrum. In contrast to a dc signal, the power density in an ac signal occurs at frequencies higher than 0 Hz. Often, dc signals are converted to ac signals by modulation in order to perform amplification and other operations at higher frequencies. Some typical ac signals and their power spectra are shown in Figure 2–10. Such ac signals may by their nature be band-limited, or they may be intentionally limited in bandwidth by filters as discussed in Chapter 7. Electrical noise induced on signals also has a power spectrum that is highly dependent on the nature of the noise source. The bandpass of a system is sometimes intentionally limited to favor the desired signal over the electrical noise. For example, most of the information from the chopped signal in Figure 2–10b is at the repetition frequency of the chopper f_0, and a bandpass filter or tuned amplifier could be used to increase the signal strength in this narrow frequency interval. Noise components outside this narrow frequency range would thus be rejected. As will be shown in the next section, it is important to know the bandwidth of a signal and to have a band-limited signal in order to carry out the sampling operation accurately.

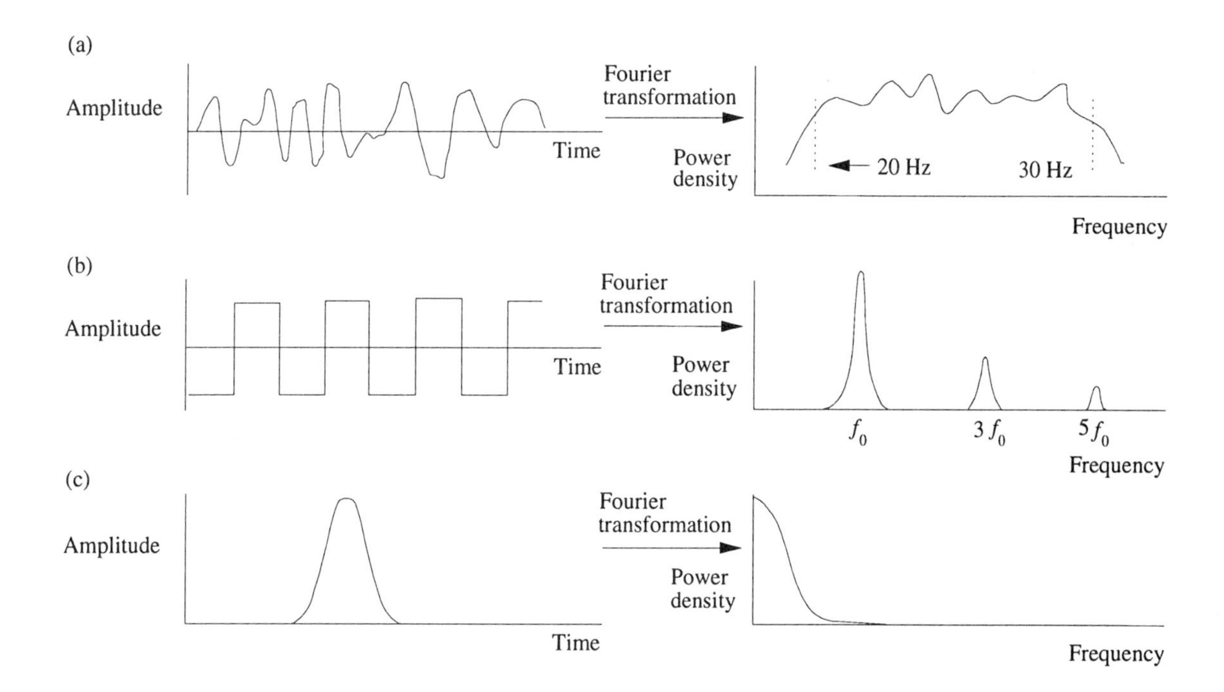

Figure 2–10. Power spectra of ac signals. (a) The audio range signal is a broad-based signal, as shown by its frequency spectrum. (b) The chopped signal contains odd harmonics of the fundamental frequency f_0, as does the square wave. (c) The peak signal is similar to a dc signal, but the signal power may extend to very high frequencies.

2–3 Converting between Analog and Digital Encoding

Data in computer-based systems are represented by digital signals. Because most transducers provide or require information encoded in analog form, a conversion in the manner of data encoding is often required between the transducer and the computer. The devices that perform the conversion between analog and digitally encoded signals are appropriately called **analog-to-digital** and **digital-to-analog converters** (ADCs and DACs). They are available in a variety of types and forms, some of which are introduced in the following sections and described in greater detail in Chapter 9. It is well to keep in mind that analog signals are continuous in time and magnitude, whereas digital signals are quantized in magnitude and represent the value of the encoded quantity at discrete points in time. If these limitations are not carefully considered, important information can be lost in the conversion process.

Converting from Digital to Analog Encoding

The digital-to-analog converter (DAC) is a circuit that converts an electrically encoded number into a directly related electrical analog quantity. The digital input signal is generally an 8- to 18-bit parallel signal. Binary coding is more common than BCD coding. The analog output signal is usually a voltage or current magnitude. The DAC simply converts the number at its input to a corresponding number of units of current or voltage at the output.

The most common approach for digital-to-analog conversion is to generate a current proportional in magnitude to the value of each bit in the digital word and then to sum the currents of all the bits that are 1 and ignore all those that are 0. This basic DAC takes the form shown in Figure 2–11. Since the relative amplitudes of the bit current generators must exactly match the bit weights of the digital signal, a DAC designed for signals of one code cannot be used with signals that are coded differently. A complete DAC is shown in Figure 2–12. The input latch samples the digital data source at appropriate times and holds the data in parallel digital

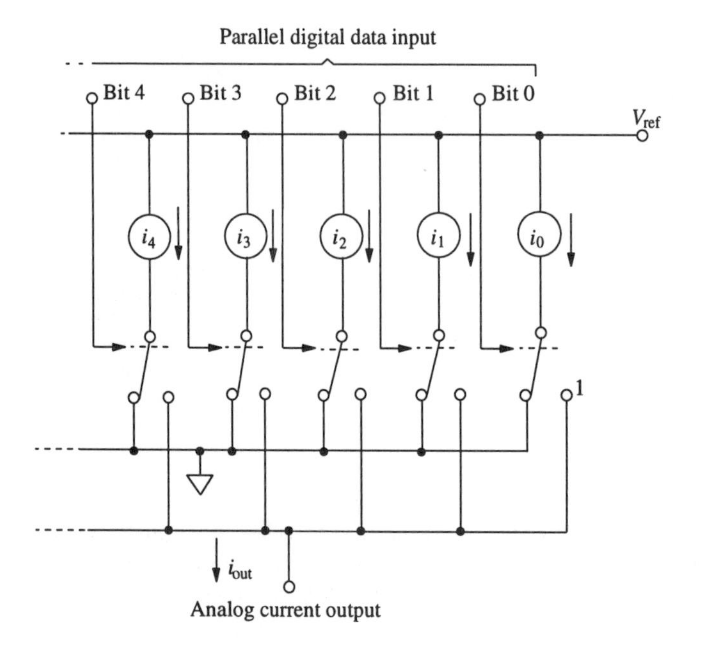

Figure 2–11. Functional basis of DACs. The current generators i_0 through i_n produce currents with magnitudes proportional to the weights of the digital input bits (1 for bit 0, 2 for bit 1, 4 for bit 2, etc.). The value of each bit (1 or 0) determines the destination of the current related to its weight. If a given input bit is 1, the current proportional to its weight is summed with the others at the analog current output. If the bit is 0, its related current is sent to common. The output current is thus proportional to the magnitude of the input number.

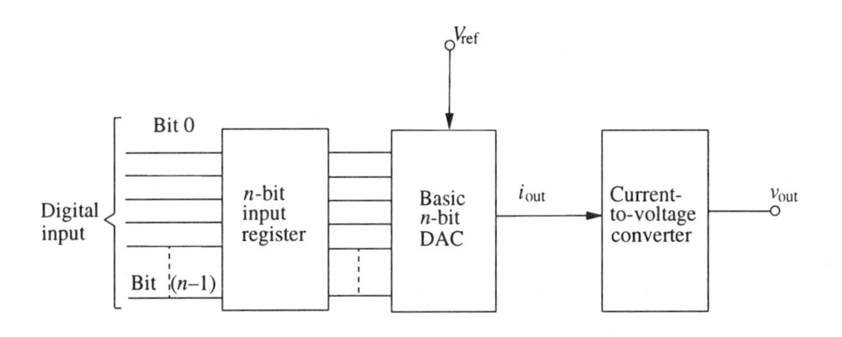

Figure 2–12. Complete DAC. The contents of the digital input register, which provides the input to the DAC, are updated as needed. The output of the current-summing unit of Figure 2–11 is generally converted to a proportional voltage at v_{out}. Additional signals allow fine control over gain (v_{out} for full-scale input) and offset (v_{out} for zero input).

form for a steady input to the DAC. In a microcomputer system, the latch may be part of the bus interface. The reference voltage V_{ref} is a voltage source used to supply power to the current generators. In many designs, the output current per bit weight is proportional to the reference voltage. The DAC output voltage is then the product of the analog reference voltage and the numerical input. Such a DAC is called a **multiplying DAC**.

The analog output of the DAC has 2^n states, where n is the number of bits in the digital input. For example, an 8-bit DAC will have 256 output voltage (or current) values. If its full-scale output voltage is 0.00 to 5.00 V, the output voltage will increase in 0.0195-V steps (5.00/256) from 0.00 V (for a binary 00000000 input) to 4.98 V (for a binary 11111111 input). The step size of a DAC is its **resolution**. To obtain finer steps in the output (higher resolution), one must use a DAC with more bits in the input so that more states are available. The resolution of a 12-bit DAC with a 5-V full-scale output is just a little over 1 mV. A 16-bit DAC could provide an output range from –10.00 to +10.00 V with a resolution of $20/2^{16}$, or 0.3 mV.

DACs have many uses. Interfaced to a computer bus, they convert digitally encoded numbers into proportional voltages or currents to control heaters, lamps, motors, and graphic displays. For example, in a cyclotron, mass spectrometer, or nuclear magnetic resonance instrument, a DAC is used to provide a control signal to the magnet power supply. If the magnetic field strength is to be controlled to have 10 000 discrete values, a 14-bit DAC with a resolution of 1 part in 16 384 (2^{14}) will be required.

A DAC is also used to control the intensity of the electron beam that illuminates the screens of many types of computer monitors. This application requires ultrafast DACs. For example, assume that the monitor has a resolution of 1024×768 pixels and is refreshed 60 times per s. This is a rate of almost 50 million pixels per s. A DAC for this application should have a response time of 20 ns or less.

Finally, as illustrated in the next section, the DAC is an essential part of many types of ADCs.

Converting from Analog to Digital Encoding

Saying that we want to convert an analog-encoded quantity to a digitally encoded quantity is the same as saying that we want to determine the number of units represented by the analog quantity or, simply, that we want to measure the voltage, current, or charge value of the analog signal. Circuits that perform this function do so by counting, in various ways, the number of volts, amperes, or coulombs it takes to equal the magnitude of the signal being measured. Many ingenious techniques have been and are still being developed to perform this process. Two very popular techniques, the dual-slope converter and the successive-approximation converter, are introduced in this section and described in greater detail in Chapter 9. These two converters differ widely in their characteristics and thus in their optimum applications. The dual-slope converter is used most often in digital meters intended for human observation; the successive-approximation converter is more often used in computer acquisition of data.

Dual-slope converter and digital meter. The operating principle of the dual-slope converter is illustrated in Figure 2–13. During the first phase of its operating cycle, in which the input switch is connected to v_{in}, the input voltage is converted to a related current (flow of charge). This charge is accumulated in the integrator over the time required for the counter (counting clock pulses) to go from 0 through full scale (N counts). The amount of the accumulated charge is now proportional to the input voltage. During the second phase, the input switch is connected to $-V_r$ and charge is withdrawn at a constant rate, with the counter starting at 0. The number of counts required to completely discharge the charge accumulated in phase one is proportional to that charge and therefore to the input voltage. The ratio of the counts required for discharge to the counter capacity N is equal to v_{in}/V_r. (If $v_{in} = 0$, no discharge counts are required; if $v_{in} = V_r$, N counts are needed.) The value of V_r is generally set so that the value of each count bit is a

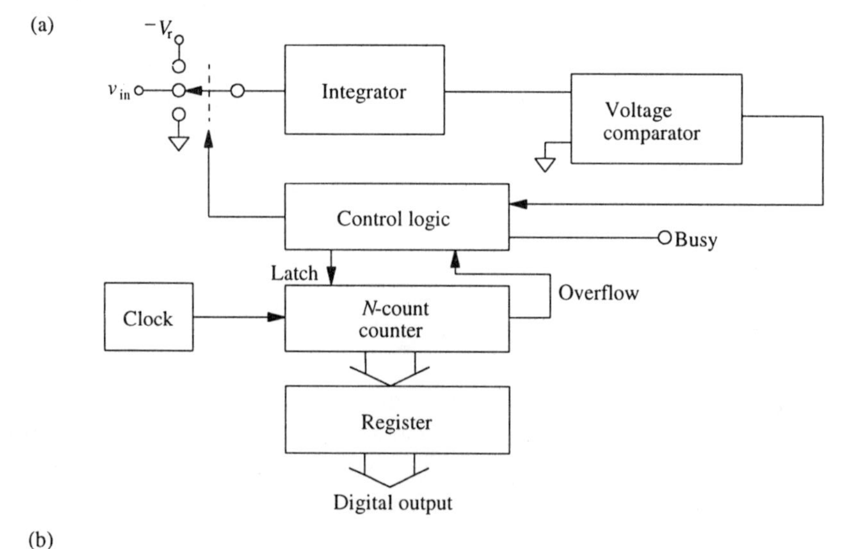

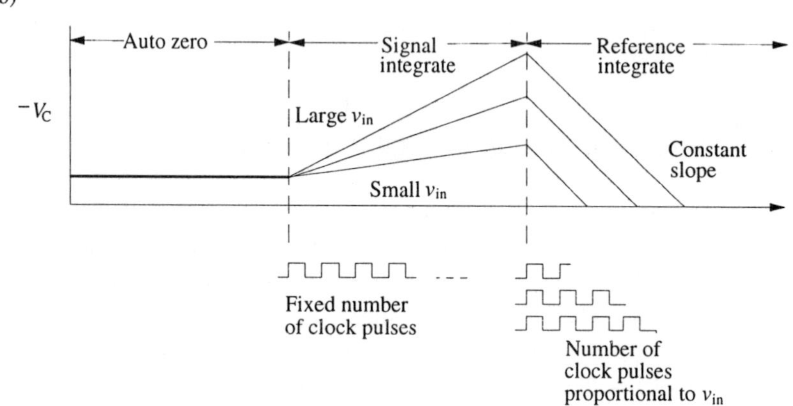

Figure 2–13. Dual slope ADC: block diagram (a); waveforms (b). The input signal is integrated for one full-scale count of the counter. Then a reference voltage of opposite polarity is applied to the integrator. The value of the count when the integrator output returns to zero is transferred to the register and thus the output. The readout count is the same fraction of the full-scale count as v_{in} is of V_r.

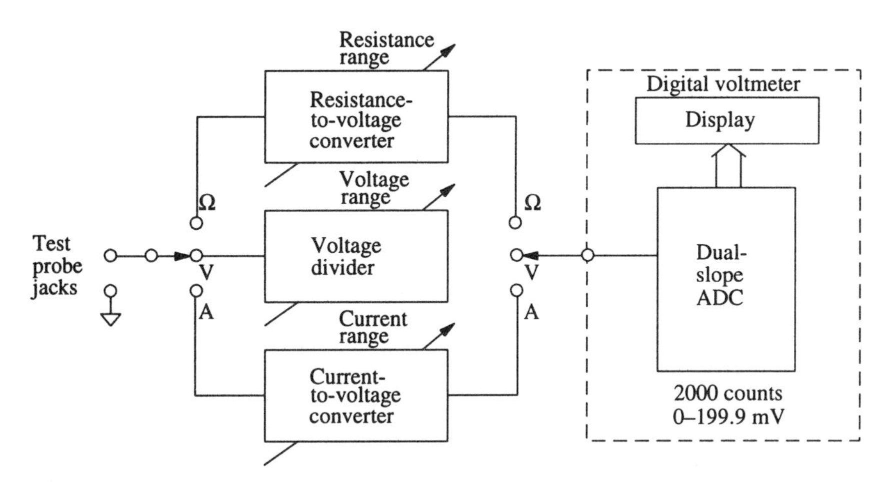

Figure 2–14. DMM. The digitizing device is a fixed-range digital voltmeter based on a dual-slope ADC. All measured quantities must be converted to a proportional voltage within the fixed range. The function selector switch engages the appropriate conversion circuit, each of which has a way of setting the measurement range.

round number such as 0.1 mV. The resolution of the conversion is equal to the voltage equivalent to 1 count and the maximum input voltage is V_r. A common type of dual-slope converter has a 2000-count counter (0–1999) and a V_r of 200 mV. This results in a resolution of 0.1 mV and a maximum input voltage of 199.9 mV.

The dual-slope ADC has the advantage of averaging the input voltage over the period of accumulation, which substantially reduces the effects of input noise. It is available in inexpensive integrated circuit form. The counters used are often BCD, so that the output can be connected directly to a decimal display. Its only disadvantage is that it is relatively slow, completing only a few conversions each second. Therefore, it is at its best with slowly changing signals. One of its most common applications is in direct-reading multimeters and digital panel meters.

The digital multimeter (DMM) is a versatile laboratory instrument for the measurement of voltage, current, and resistance. Unlike the analog multimeter, which is usually based on the current-actuated moving-coil pointer-type meter, the DMM is based on the digital voltmeter, which is actually a dual-slope ADC with associated numerical display. The measurement of current and resistance with the DMM then necessarily involves the conversion of these quantities to a related voltage. The overall scheme is shown in Figure 2–14.

The selector switch determines which of the conversion circuits will be connected between the test probes and the digital voltmeter. Each conversion circuit must convert its input quantity into an output voltage in the range of 0.00 to ±199.9 mV. When in the resistance measurement mode, the resistance-to-voltage converter may be set for 200-Ω full scale. The converter output would then be 1 mV for each ohm of unknown resistance. A reading of 138.3 would indicate a resistance of 138.3 Ω. If the probes are connected to a 1-kΩ resistance, the range of the converter is changed to 2-kΩ full scale. Now the converter output will be 0.1 mV/Ω of input resistance, and the meter reading for the 1-kΩ resistance will be 1.000 kΩ. The decimal point position and units indicator are changed automatically when the resistance range is changed. Note that as the scale increases from 200 to 2000 Ω, the resolution decreases from 0.1 to 1 Ω.

Successive-approximation converter. The successive-approximation converter operates by comparing a digitally generated voltage with the input voltage and adjusting the former until the two are equal. This is illustrated in Figure 2–15.

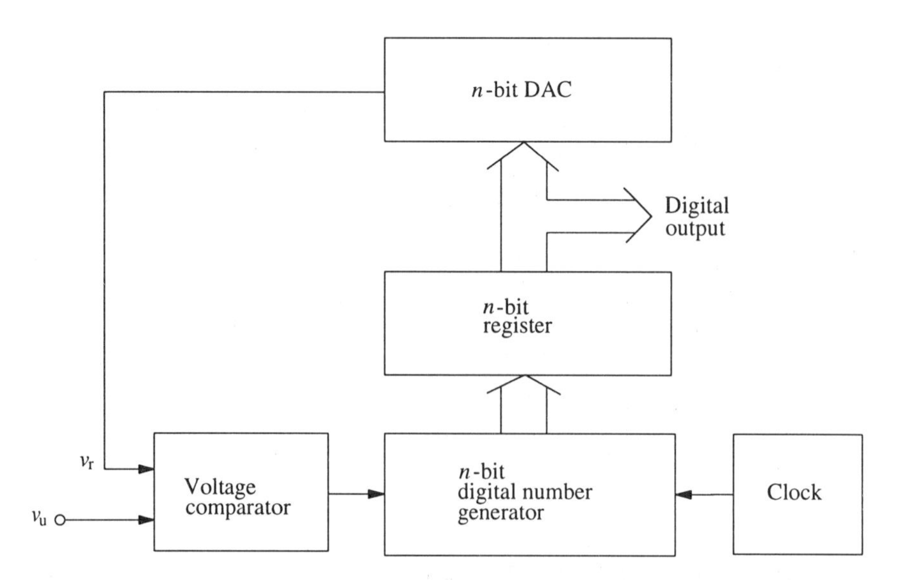

Figure 2–15. Successive-approximation ADC. The number generated by the digital-number generator and stored in the register generates a proportional voltage v_r at the DAC output. A voltage comparator compares the magnitudes of v_r and v_u; its output signal, HI or LO, indicates whether $v_r > v_u$ or $v_r < v_u$. This signal is used by the digital-number generator to arrive at the number for which $v_r = v_u$ within the value of the LSB in the DAC. At the end of this process, the register contains the desired result.

The digital number generator begins with all its bits at 0, so the output of the DAC is 0.0 V. Then the most significant bit is set to 1, which makes the DAC output voltage one-half its full-scale output. If a comparison of the DAC and input voltages indicates that the DAC voltage is higher, then input voltage is less than half scale and the bit is changed back to 0; if not, the bit is kept at 1. Then the next most significant bit is set to 1 temporarily to test in which quarter the unknown value lies. Then the third bit is tested, and so on.

A 12-bit converter requires only 12 successive bit tests to complete the conversion. Because bit tests in modern converters can be completed in a fraction of a microsecond, these converters are capable of a thousand to millions of conversions per second. Several other types of ADCs are described in Chapter 9. The ADC accuracy is generally specified by the manufacturer, often in relation to 1 least significant bit (LSB). The resolution is exactly equal to the voltage value of 1 LSB of the DAC. Therefore, to acquire voltages to an accuracy of 0.1% (1 part per 1000), one needs a converter of at least 10 bits ($2^{10} = 1024$) and an accuracy of 1 LSB or better. Note also that this accuracy is relative to a full-scale voltage. To realize an accuracy of 0.1% down to one-quarter of full-scale, one needs a 12-bit converter accurate to 1 LSB.

The application of a successive-approximation converter in computer-based data acquisition requires both analog input signal conditioning and an output interface to the central processing unit (CPU) bus. These functions are illustrated in Figure 2–16. An input selector switch selects which of the (usually 8 or 16) input signal voltages will be digitized. An amplifier is then used to bring the anticipated signal level within the range of the ADC. Optimally, the signal voltage is between 0.25 and 0.99 of the full-scale ADC input voltage. The sample-and-hold circuit (S&H) follows the variations in the input signal magnitude until the hold command is applied, after which the S&H maintains a constant output equal in magnitude to the analog signal at the time of the hold command. This operation serves two essential functions: to maintain a constant ADC input voltage during the successive-approximation conversion process and to fix precisely the instant of acquisition, that is, the time at which the ultimately stored number represented the analog signal magnitude. When the conversion is complete, an end-of-conversion (EOC) signal is generated. This signal causes the contents of the ADC register to be

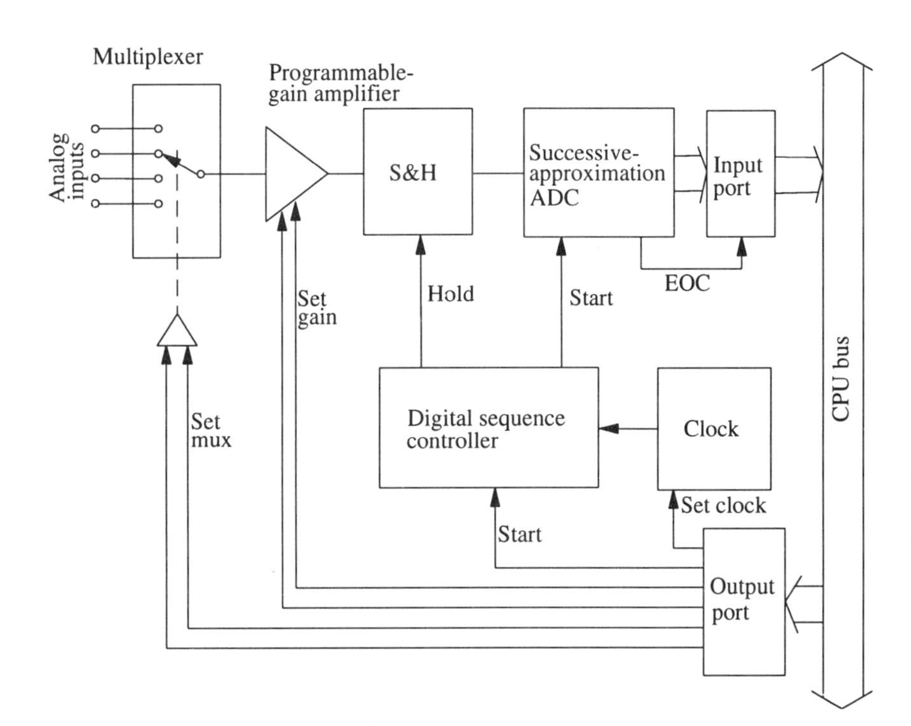

Figure 2–16. System for computer acquisition of analog data. Selected by the multiplexer (mux) and appropriately amplified, the signal voltage is sampled and held by the S&H at the desired acquisition time. The ADC then converts the sampled voltage to a proportional number that is transferred by the input port to the CPU bus. Selection of the active input, amplifier gain, and acquisition timing can be controlled by the program running in the computer. When the conversion is complete, an end-of-conversion (EOC) signal is generated.

transferred to the CPU bus input port register and a flag to be set that lets the CPU know that a new data point is ready to be sent to memory. At this time, the process can be repeated and another piece of data can be acquired.

The time at which the signal is sampled is called the **acquisition time**. In this system, the acquisition time is the instant at which the hold signal is sent to the S&H. Acquisition time can often be determined to within a few nanoseconds. The durations of the conversion and storage processes that follow do not affect the precision of the acquisition time; they affect only the maximum possible acquisition rate.

The maximum data rate (data points per second) is the reciprocal of the conversion time (seconds per data point). The **conversion time**, then, is the minimum time between successive samplings of the analog input signal. Either the analog-to-digital conversion time or the time to store the conversion results can determine the overall minimum conversion time. A computer data acquisition program cycle that checks the ADC status, moves the conversion result to a location in memory, increments the memory location for storage of the next data point, and returns to check the ADC status can take 3 to 100 μs to complete. This will be the limiting factor in a system where the time from start to EOC for the ADC is 4 μs. Thus, a system with a conversion time of 25 μs has a maximum data rate of 40 kHz. The relationship between the analog signal bandwidth and the sampling rate required for adequate digital recording is presented in Section 2–5.

2–4 Measuring Time

The analog and digital techniques of data encoding have been extensively explored in previous sections. In this section, the third form of data encoding, that of *time*, is developed. In time-encoded signals, the information is contained in the time relationship of the signal variations rather than in the amplitude of the variations.

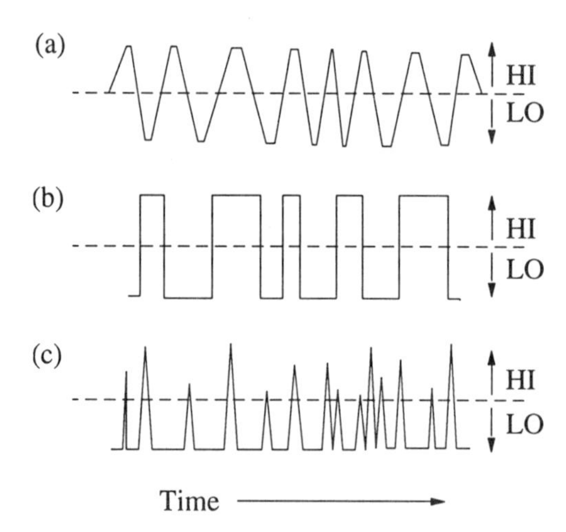

Figure 2–17. Time domain signals. The data are contained in the relative time of the crossing of the transition between HI and LO levels. Though the encoding quantity cannot generally be determined by inspection of the waveform, signal a could be encoded as the frequency of the waveform, signal b could be pulse width-encoded, and signal c could be the output of a detector of randomly timed events in which the average number of events per unit of time is sought.

Typical time-encoded signals are shown in Figure 2–17. In the signal shown in Figure 2–17a, information could be conveyed by the frequency of the oscillations. The establishment of a threshold level, shown by the dotted line, can aid in the measurement of signal frequency. The time from one crossing of the threshold level to the next (in the same direction, such as LO→HI) is the **period** of the oscillation. The number of periods (cycles) in 1 s is the frequency. The signal in Figure 2– 17b may be pulse width-modulated. The duration of the pulses may contain the information, with a narrow pulse representing a small value and a wider pulse representing a larger one. Again using the threshold level, we can define the pulse duration as the time between a LO→HI transition and the subsequent HI→LO transition. Radiation or particle detectors often produce an output signal similar to that of Figure 2–17c. Here, the threshold level is set low enough to include all true events but high enough to exclude electrical-noise pulses. The number of LO→HI transitions per second is the rate of the events.

An important feature of time-encoded signals is their insensitivity to electrical noise compared with analog signals. Because the information is in the relative times of the threshold crossings and not in the amplitudes of the signals on each side of the threshold, substantial variations in signal strength and superimposed noise can occur with little effect on the precision of the data conveyed. The effect that noise does have on data precision is illustrated in Figure 2–18. The effect of

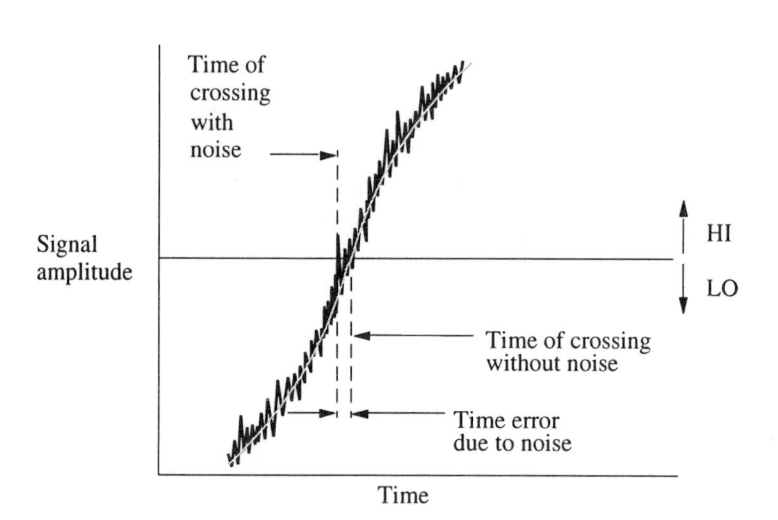

Figure 2–18. Noise-induced error in time of threshold crossing. The presence of noise can cause a premature or delayed crossing of the threshold. The error is reduced by decreasing the noise amplitude and increasing the rate of signal amplitude change through the threshold region.

a given amount of noise will decrease as the rate of change of signal level through the threshold region increases.

Time-encoded signals, like analog-encoded signals, are continuously variable, because the frequency or pulse width can be varied infinitesimally. However, the encoded variable of time-encoded signals cannot be measured continuously with time, nor can it be measured at any instant in time. The minimum time required for conversion of data encoded in time to any other encoding is necessarily at least one period or one pulse width. The remainder of this section is devoted to techniques for converting time-encoded information to digital form.

Measuring a Time Interval

The general scheme of digitizing described thus far is to count the units contained in the quantity to be digitized. Time is thus measured or digitized by counting the time units between the events that mark the beginning and the end of the interval. A system that accomplishes this is shown in Figure 2–19. Time units are generated by a precision oscillator or **clock** that produces a known number of events each second. An example is a 1-MHz crystal-controlled oscillator that produces a pulse every microsecond.

The counting gate is used to control the time interval over which the units of time will be counted. In response to a start signal, the gate is opened to allow the clock output to appear at the counter input, and when the stop signal occurs, the gate is closed. The number of counts now stored in the counter register is the number of time units (for example, microseconds) between the occurrence of the start and stop signals. The exact start and stop times are the times the start and stop input signals cross their respective threshold values.

The resolution of this time measurement is determined by the clock frequency; for a 100-Hz clock, the resolution is 0.01 s; for a 10-MHz clock, the resolution is 0.1 μs. The maximum count multiplied by the resolution determines the maximum length of time that can be measured. The precision of the measurement cannot be greater than 1 part in the total count, because 1 count is the smallest unit of measurement and because changes in the relative timing of the clock events and the counting interval can cause a variation of 1 count in repetitive measurements.

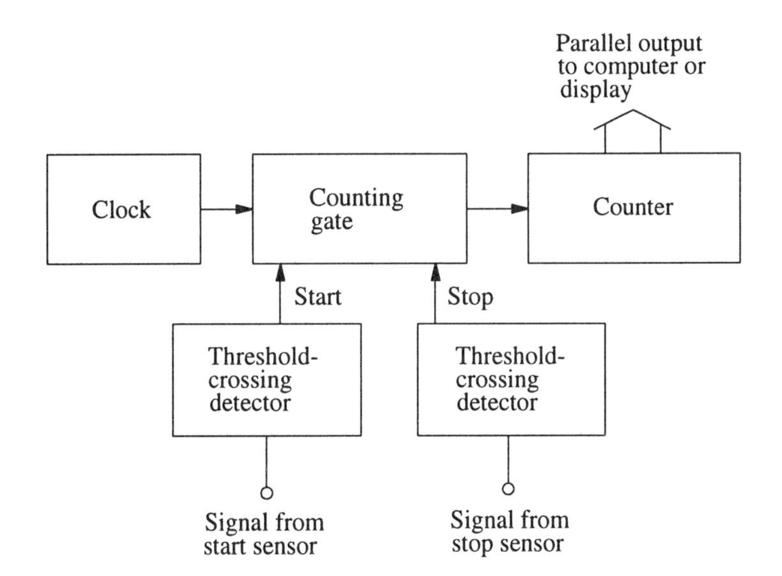

Figure 2–19. Time interval measurement. The time between the sensing of the start and stop events is measured by counting the number of clock cycles that occur in this interval.

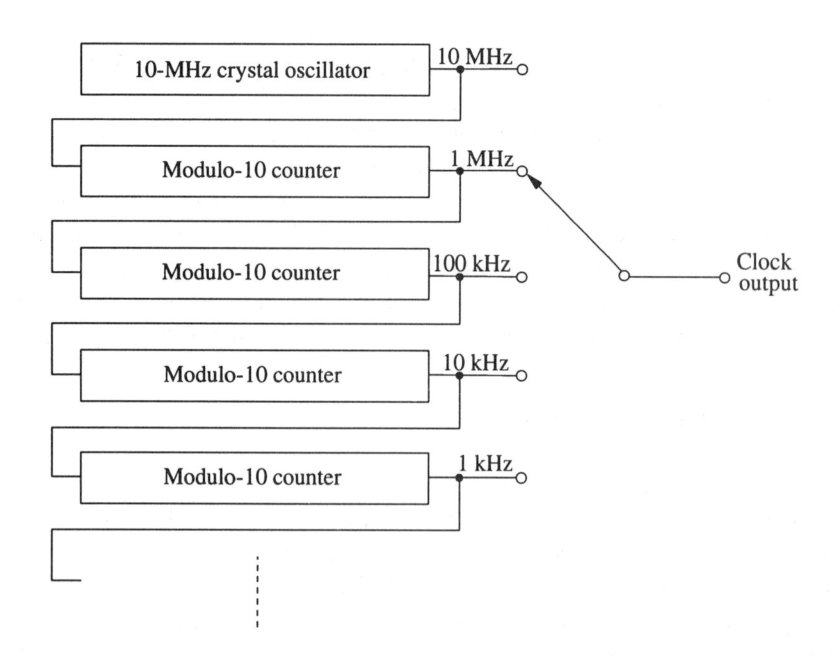

Figure 2–20. Precision clock circuit. The frequency at the output of each modulo-10 counter is exactly 1/10 the frequency at its input. The time base accuracy of each output is equal to that of the crystal oscillator.

Timing errors in the start and stop signals of the type illustrated in Figure 2–18, if larger than one period of the clock, will become the limiting factor in the precision of the measurement.

The accuracy of the time interval measurement, if not limited by the precision, depends only on the accuracy of the precision clock oscillator. Crystal oscillators used for this application have frequencies in the range of 1 to 50 MHz and can be accurate to 1 part in 10^5 to 10^8. A clock with a crystal oscillator having a frequency of 10 MHz would have a time measurement resolution of 100 ns. This measurement may be too fine and may result in too large a count for longer time measurements. Practical time measurement clocks use a series of counters called a **scaler** to obtain precise clock signals of lower frequencies. A typical clock composed of a crystal oscillator and a series of modulo-10 counting circuits is shown in Figure 2–20. The available clock frequencies are scaled down from the original 10 MHz in successive decade steps through the modulo-10 counters. A scaler composed of eight decade frequency dividers will provide a clock output as low as 0.1 Hz. Because there is no noise or variance in the counting (scaling) operation, all clock outputs are as accurate and precise as the crystal oscillator used. Time interval measurements are one of the most accurate of physical measurements. Most measurement techniques with accuracies in the range of 1 part in 10^6 or better are based on time interval measurements.

Measuring the Signal Period

The determination of the period of a repetitive waveform is accomplished in much the same way as the measurement of a time interval: the number of unit time increments is counted during one or more complete cycles (periods) of the input signal, as shown in Figure 2–21. The input signal is connected to the threshold-crossing detector, which will produce one LO→HI and one HI→LO transition each period of the input signal. A scaler is used to set the number of periods N over which the time interval is to be measured. The signal period is equal to the number of counts measured multiplied by the clock period divided by N. With a 1-MHz

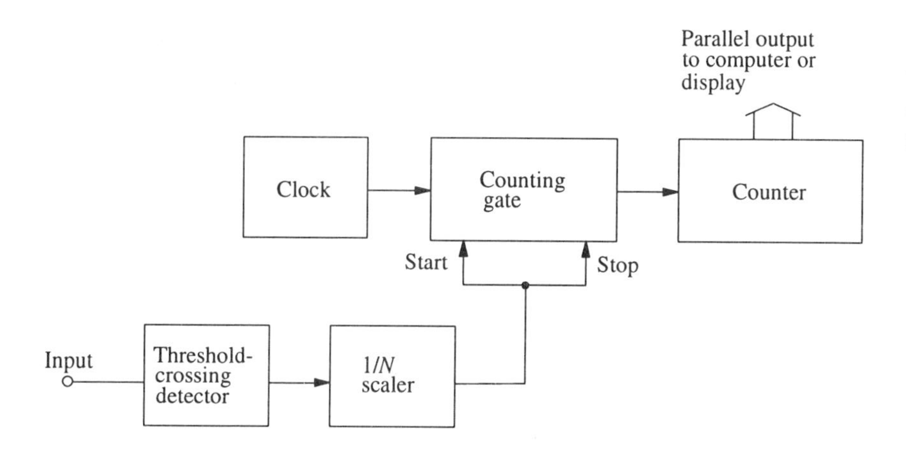

Figure 2–21. Period measurement. The counter counts the number of clock cycles per N periods of the input signal.

clock output, the resolution (time per period per unit count) is $1/N$ µs. For a 100-Hz input signal, the time interval measured is $N/100$ s, and the total count would be $N \times 10^4$ counts at 1 count per µs. If the counter is an eight-digit counter, the maximum value for N (without overflowing the count) is 10^4. Although such a measurement would provide an eight-digit result with a resolution of 0.1 ns, it would take 100 s to accomplish.

Input signal scaling is particularly valuable for determining the periods of higher-frequency signals, where the duration of a single period would provide very few counts and thus poor relative resolution. Scaling the input frequency also divides over N cycles the error in the start and stop times of the measured interval, and it averages the variations in the signal period over the measurement interval. For every period measurement situation, there will be an optimum choice for the clock frequency and the scaling factor N. In general, N should be determined by the input frequency and the longest desirable measurement time. The clock frequency should then be chosen to provide the needed resolution. As in many digital measurements, it is easy to get more figures displayed or printed than are significant.

Measuring Frequency

A frequency measurement determines the number of cycles of an input signal per unit time. The basic counting and timing functions are arranged so that each cycle of the input signal is counted over the period of one clock cycle, as shown in Figure 2–22. Generally, clock frequencies of 1 kHz or lower are used. If the clock frequency is 1 Hz, the count will be the number of input signal cycles per second. Variations in the frequency of the signal being measured are averaged over the clock period. For high-frequency input signals, the number of counts accumulated in 1 s can be quite large, but for lower-frequency measurements, only a few digits of the counter will be used. Thus, the relative resolution decreases as the frequency decreases. This can be overcome by using still longer counting intervals for low-frequency signals, but the measurement time will increase proportionately. A better solution for low frequencies is to measure the period and take the reciprocal. Very high frequency signals may exceed the maximum counting rate of the counting circuitry. For these signals, a high-speed prescaler (usually one or a few decades) is used to reduce the input frequency to a value low enough for accurate counting.

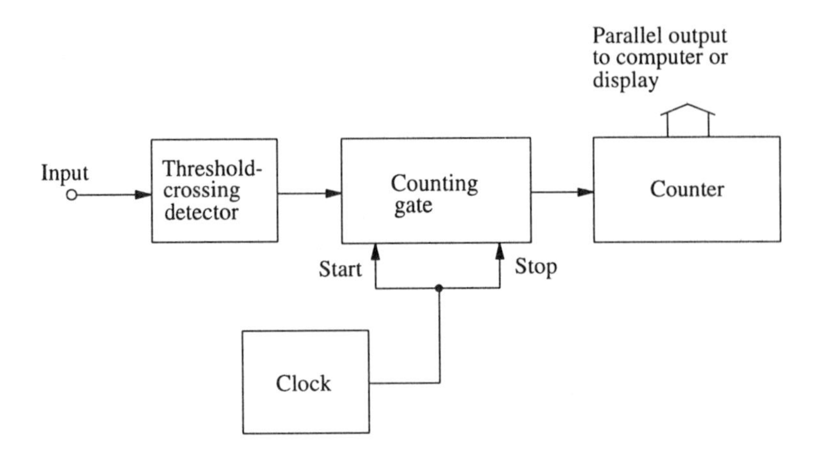

Figure 2–22. Frequency measurement. The counting gate is open for one period of the clock. The resulting count is the number of cycles of the input signal over that time interval. If the clock period is 1 ms, the counter reads directly in kilohertz.

The basic counter, gate, and scaling circuits can also be rearranged to measure the frequency ratio of input signals, as shown in Figure 2–23. The number of counts obtained is the number of cycles of the A input signal that occur over N periods of the B input signal. The counting gate will be open for N/f_B s, during which Nf_A/f_B counts will accumulate. For signals close together in frequency, the number of counts in the result will be approximately equal to N. The frequencies of both the input signals are averaged over the period of the measurement.

A **counting instrument** contains a basic counting circuit, input signal-processing circuitry (variable gain amplifier and threshold setting), a counting gate with start and stop inputs, a method for storing the resulting count while it is being displayed, and a sequencer for automatically performing repetitive measurements. The count begins with the opening of the gate and ends with its closing. At this time the counter contents are transferred to the readout register for display. At the same time, the gate is reset so that it will open on the next occurrence of the start signal. A **frequency meter** is an instrument with these basic recycling counter circuits as well as a clock system composed of a crystal oscillator and a scaler. Instruments dedicated to the frequency meter function often contain signal prescalers so that their upper input frequency limits can extend to the hundreds of megahertz. As we have seen, period and frequency ratio measurements are accomplished with the same basic circuit functions as those used in the frequency meter. Instruments that are designed to provide all modes of counting and timing measurements are called **universal counters**. A mode switch is used to select the

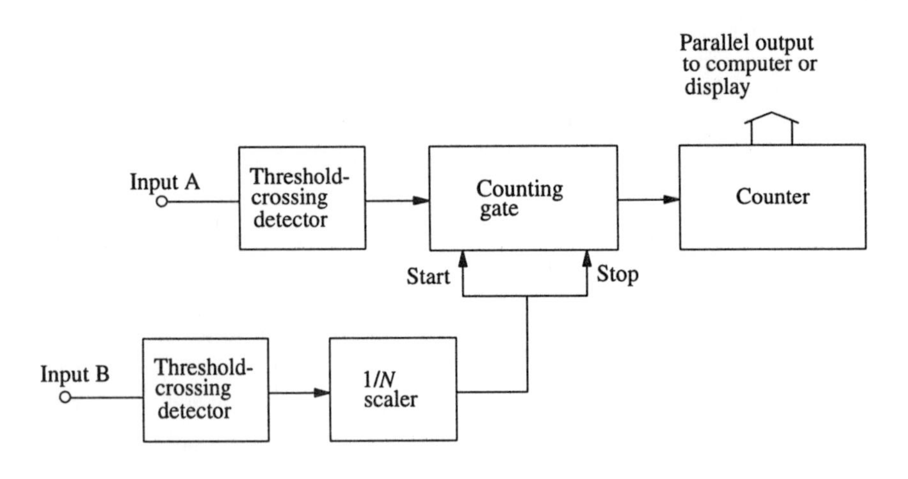

Figure 2–23. Frequency ratio measurement. The counting gate is open for N periods of the B input signal. The count is equal to the number of cycles of the A input signal that occurs over that interval.

measurement function. Other switches select the clock frequency and the scaler value. Signal-conditioning circuitry is provided for two inputs (A and B) for frequency ratio measurements.

Measuring Time in Microcomputers

If a computer is to measure or control the time of events in the real world, it needs some way to keep track of time; that is, it needs a clock. A **real-time clock** is basically a counter that is incremented or decremented at regular intervals by a time unit generator. A **time-of-day clock** is a real-time clock in which the number in the counter represents the actual time of day in hours, minutes, and seconds (and perhaps days, months, and years as well). Most computers now have battery-operated world-time clocks as part of their standard hardware. A computer can use a time-of-day clock to record the actual time of events. It does so by reading the counter contents as part of its programmed response to the event. A computer can also be programmed to turn devices on or off at particular times by jumping to the control subroutine when the clock counter reaches the appropriate state. In many instances of time measurements or timed control, only relative time, the time difference between events, is important. In such cases, a real-time clock that measures elapsed time, like a stop watch, is appropriate. In an **elapsed-time clock**, the counter contains the number of time units since the previous event (count-up) or until the next event is to occur (countdown). Computer clocks can be either software- or hardware-based. In a **software clock**, the standard oscillator generating the time units is the computer clock oscillator scaled by software instructions, and the counter is a memory register that is incremented through a software instruction. A **hardware clock** has a dedicated crystal oscillator, scaler, and counter.

A software clock uses an instruction loop as the time unit generator. The execution time of one cycle through the loop is equal to the time unit to be counted. One of the steps in the instruction loop is to increment (or decrement) a register that serves as the clock counter. The software clock is extremely simple and adequate for many purposes. However, it has serious limitations. The resolution of the time measurement is the execution time of one loop through the program. The maximum resolution thus equals the minimum loop execution time. To ensure reproducibility, no variable-length instructions can be included in the timing loop. Furthermore, programs execute at greatly different rates on various computers, so the counted time unit can be dependent on the specific computer running the program. Software clocks are normally avoided in scientific applications when accuracy in the timing function is required.

A block diagram of a hardware clock is shown in Figure 2–24. Both the time unit selection scaler and the counter are interfaced to the CPU bus. The counter can be preset to any value by the CPU and instructed to count up or down. The clock counter register contents can be transferred to another register by the CPU at any time. The programming for elapsed-time measurements becomes simple, because after the initial event, it is necessary only to clear and enable the clock, wait for the terminating event, read the clock, and calculate the time. Because program steps are at a minimum, the accuracy of the interval timer is good, even for short time intervals. In the timed-control mode, the program is also simple. The counter is preset to the desired interval and is decremented by the scaled oscillator signal until overflow occurs. The counter then automatically clears and presets itself, and the cycle is repeated.

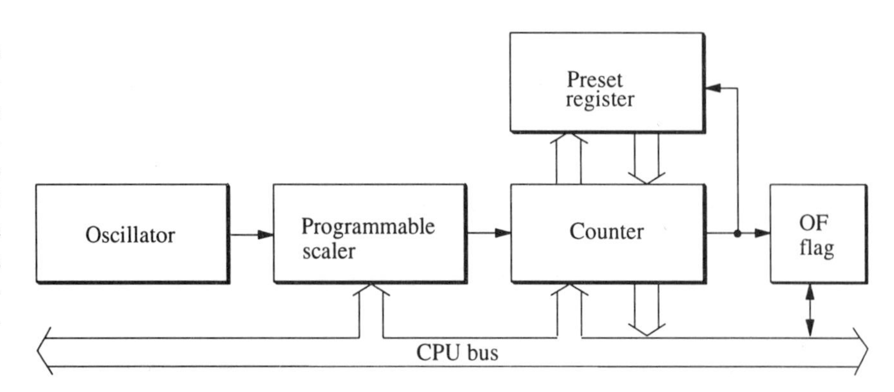

Figure 2–24. Hardware clock. The time unit is selected by programming the scaler. The counter can be cleared or preset at the beginning of the timed period and set to count either up or down. The CPU then either reads the relative time from the counter register or waits for the overflow (OF) flag to signal the end of a preset time interval.

The characteristics of timed operations are affected not only by the design of the clock but also by how interactions between the clock and world events are implemented. Two types of interactions can be distinguished: computer *control* of the time of an event and computer *measurement* of the time of an event. In the event time measurement case, the computer watches for the time of an event. In the event control case, the computer watches for the desired event time to be indicated by the clock counter and then enters a subroutine to command the event. The CPU can recognize the desired time by one of three methods: (1) comparing repetitive clock counter readings with the desired value, (2) presetting the clock counter to set its flag at the desired time and waiting in a flag check loop, or (3) presetting the clock counter to generate an interrupt at the desired time.

For timing on the microsecond time scale, all software response and control approaches are too slow. In these cases, interaction between clock and event must be direct; the clock flag output must be connected to trigger the event directly. The computer is used to preset the appropriate delay between the clock counter reset and overflow, but the timing is in the hardware, where it can be performed with nanosecond precision. An analogous set of approaches is possible for interactions between the clock and the world in event time measurement. In a programmed time measurement, the computer branches to a subroutine to read the clock counter in response to a flag or interrupt from the event to be measured. Subroutine execution time can be a factor in measurement accuracy for very short times or very accurate measurements. Direct control of a gate to the clock counter by the event detector allows timing in hardware alone. Integrated circuits containing complete real-time clock systems have been designed for direct connection to the CPU bus; they vary considerably in the programmed and direct-connection timing modes available.

2–5 Using the Laboratory-Interfaced Microcomputer

In the early days of laboratory applications of minicomputers, the interface circuits for transferring data between the measurement and control devices and the CPU bus were completely customized, that is, specially designed for each device and application. However, it soon became apparent that the same elements appeared in many applications: an ADC, one or more DACs, some digital (HI→LO) control lines, and a real-time clock.. Recognizing this commonality, a number of computer peripheral manufacturers began to market general-purpose input/output (I/O) boards (sometimes called laboratory interface boards) for specific minicomputers and microcomputers. The great virtue of these boards is that they need only be plugged into the CPU bus and one is ready to connect to the real world. Only two elements remained to be dealt with: conditioning the signal to the levels and forms required by the specific devices and choosing the appropriate data-gathering and data treatment operations (and software to implement them). Providing the background to perform these functions is a principal goal of this book. This section introduces the elements of the hardware and the processes involved in applying the laboratory-interfaced microcomputer.

Laboratory Interface Systems

There are now a great many suppliers of laboratory interface systems designed for use with microcomputers. Interface systems generally consist of various combinations of ADCs, DACs, digital I/O, and timers along with the circuitry necessary for computer interfacing. These functions may be sold as separate modular cards for user system configuration or as single cards with preconceived combinations of functions and specifications. The interface to the computer may be accomplished in two ways: to the CPU bus or to a standard communication port, which is in turn connected to the CPU bus. These options are illustrated in Figure 2–25.

In the bus-interfaced systems, the ADC, DAC, digital I/O, and timer circuits are built on cards designed to plug into and interface with the CPU bus. Obviously, these systems are specific for certain classes of computers. Separate function cards (such as DACs only) can be plugged into the CPU bus along with cards performing the desired combination of other functions for custom configurations. For simplicity and to save on the number of required bus card slots (sometimes at a premium), single cards with preselected combinations of functions are also made.

The other option for the laboratory interface (also illustrated in Figure 2–25) is connection to the computer through a standard interface port. A few types of interface connections are available for virtually any type of computer hardware. These include the parallel port (often used for the printer), the serial port (used for printers and modems), and the IEEE-488 port (most often used for instrumentation control). The specifications for the signals and the connector arrangement for these

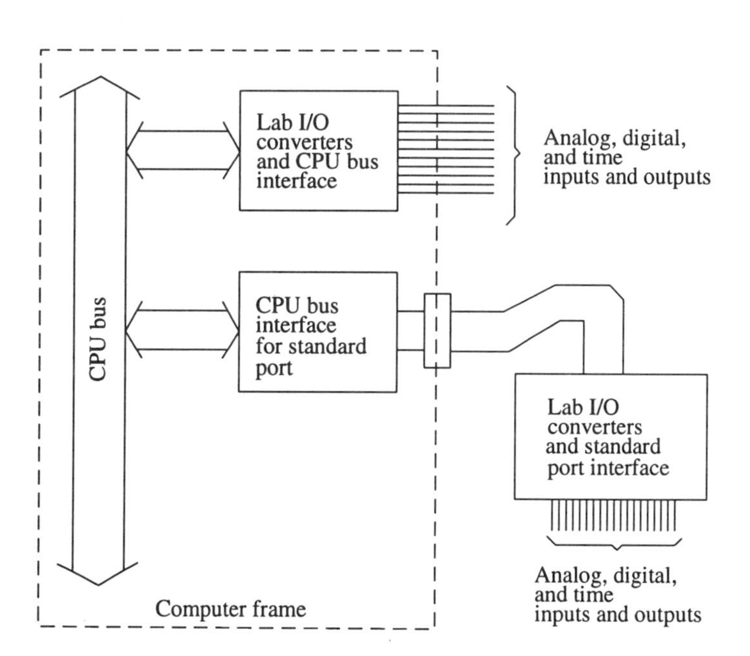

Figure 2–25. Laboratory interface options. The bus-interfaced laboratory I/O system connects directly to the CPU bus. All laboratory connections are brought to or into the computer framework. The port-interfaced I/O system is located outside the computer framework and provides an interface to a port of a type available with virtually any computer.

ports are quite carefully controlled so that designers of peripherals can make devices that are port-specific rather than bus-specific. As can be seen in the diagram, the port interface requires two levels of interfacing: one to the port standard and the other (inside the computer) between the port and the CPU bus. This added complexity may be offset by the ability to readily change computer and laboratory interface hardware independently.

Bus-interfaced systems. Many peripheral equipment suppliers now feature plug-in analog I/O boards for specific computer systems. These may be input boards, output boards, or combination boards. Most analog I/O boards include input multiplexers, an input amplifier, an S&H, an ADC, one or more output DACs, and bus-interfacing circuitry. Some include on-board random-access memory (RAM), read-only memory (ROM), and real-time clocks. A typical real-time analog I/O board is shown in block diagram form in Figure 2–26. The input multiplexer allows 16 channels of single-ended analog input or 8 differential inputs. The programmable gain amplifier often has software-selectable gains of 1, 2, 4, and 8 or 1, 10, 100, and 500. A 12- or 16-bit ADC may be provided. Though an amplifier with gains from 1 to 500 offers the widest dynamic range, some signal levels will use as little as 10% of the effective ADC input range, removing over 3 bits of its resolution. The binary gain sequence, on the other hand, can maintain a minimum resolution within 1 bit of the ADC resolution. The S&H preceding the ADC can have an uncertainty in analog sampling time as low as 5 ns.

The two 12-bit DACs may be driven by double-buffered registers. Their outputs can then be updated simultaneously for driving two-coordinate plotters or displays. A clock can provide precisely timed analog-to-digital conversions, or timed events can be externally triggered. Digital I/O occurs through two or more 8-bit ports, either of which can be dedicated to input or output. To control and

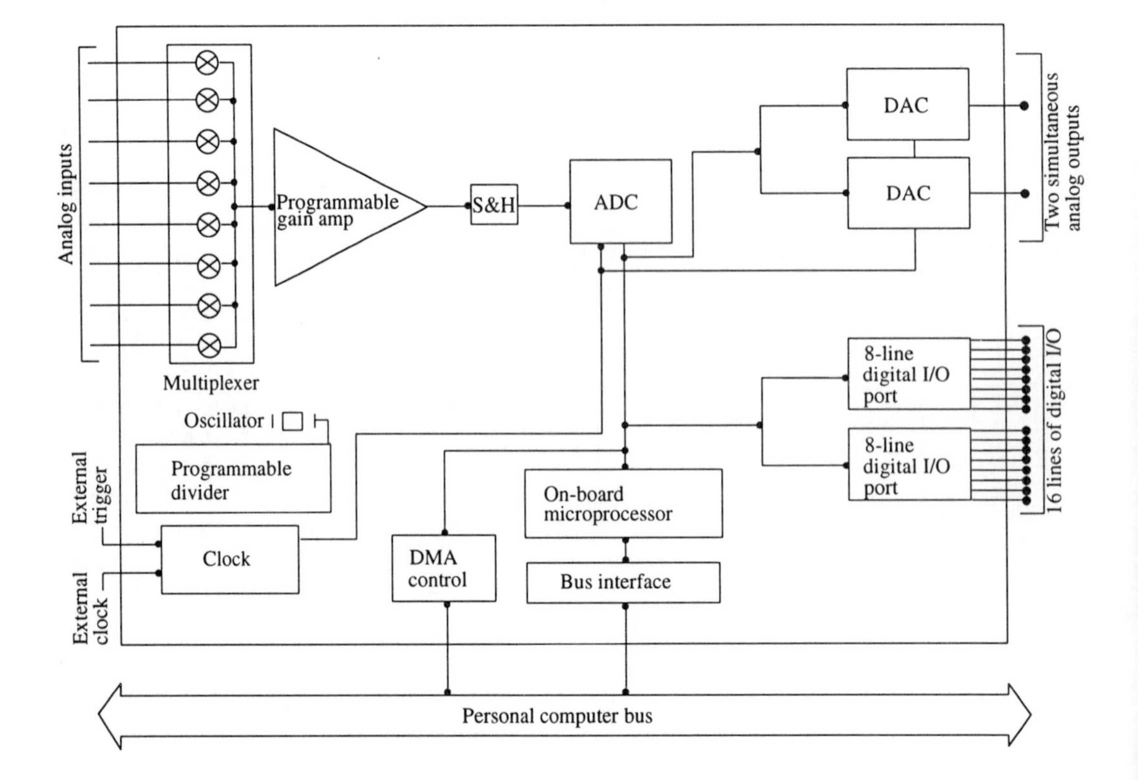

Figure 2–26. General-purpose analog and digital I/O board. Multiple analog and digital outputs are included as well as an on-board clock for timing.

coordinate all these features, an on-board microprocessor is often used. This is a growing trend among peripheral suppliers, who find microprocessor control the most effective way to achieve power and versatility in their products. The microprocessor can control the CPU bus interface functions as well as the real-world interfaces provided on the card. In such a case, instructions to the card for data acquisition and other functions cause the execution of the appropriate subroutines stored in the microprocessor ROM memory. The CPU bus of an on-board microprocessor can also provide a convenient data link among the many registers and devices on the card. Some RAM memory may also be included as a microprocessor-controlled buffer for either input or output data.

A major specification of analog interface cards is the **analog-to-digital throughput**, the maximum rate at which sustained conversions can occur. The throughput is generally limited by either the analog-to-digital conversion speed or the speed with which data can be transferred between the card and the CPU bus. Under program control, data transfer through the CPU bus is limited by the time required for the acquisition program to check the converter output flag, acquire the conversion value, and transfer the acquired value to a location in memory. Programmed throughput rates in excess of 20 kHz require fast hardware and efficient acquisition routines. A faster method of data transfer can be used when substantial blocks of data are to be moved to or from successive locations in memory. This technique is called **direct memory access** (DMA). During DMA operation, the device transferring data to or from the memory acquires control of the bus, leaving the CPU on temporary hold for the time required for the data block transfer. Systems with DMA capability can transfer data at the maximum rate allowed by the bus or the system memory, whichever is less. When dynamic RAM is used, the maximum DMA rate is limited by the memory refresh circuit requirements. Through the DMA option, data throughput rates in excess of 100 kHz are routinely accomplished.

Port-interfaced systems. Industry-standard serial and parallel data transfer ports are part of the basic configuration of virtually all modern computers. As such, they provide a nearly universal means of getting data into and out of the computer. The serial port, sometimes called the com port, is often used to connect communication devices such as telephone line modems. It is a true bidirectional communication path, highly suitable for connection of external data interfaces. The parallel port, on the other hand, was originally designed to provide data to the printer. In this application, most of the data flow is from the computer to the printer; only data on the printer status line go the other way. Thus, these parallel ports are much less suited for data acquisition than the serial port. A recent and welcome variation is the bidirectional parallel port, which, if available, can provide very rapid data transfer with a wide variety of peripheral devices. A specialized port, optimized for speed and flexibility in data acquisition and instrument control applications, is the general-purpose interface bus (GPIB or IEEE-488).

The serial port has the advantage of generally requiring no additional hardware for connection to any computer system. As the name implies, the data are transferred serially, 1 bit at a time. This has the advantage of requiring few wires for connection to the interfaced device and the disadvantage of lower transfer rates than are possible with parallel data transfers. The data transfer rate through the serial port is specified as the **baud rate** (basically, bits per second), which can be as high as 125 kilobaud. Because some of the bits must be used to indicate the start and stop of each byte, the maximum serial data transfer rate is about 10 kilobytes/s.

Some stand-alone instruments provide serial port connections for parameter setting and data transfer. Each instrument to be interfaced requires a separate port so that interface cards with banks of serial ports are needed for more complex systems.

The GPIB is more efficient than the serial system as an instrumentation interface. It uses 8-bit parallel data transfer and will accommodate up to 15 devices through the same port. Data transfer rates of up to 1 megabyte/s are now possible through the GPIB system. The advent of large-scale integrated circuits that handle the GPIB and an industry-wide agreement on the physical and protocol standards used in bus operation have spurred its widespread adoption. GPIB cards are made for virtually all scientific computers. An increasing number of instruments, including oscilloscopes, waveform generators, meters, power supplies, and timers, come with GPIB connections built in. Groups of such instruments can be readily interconnected into a sophisticated measurement and control system coordinated by a GPIB-interfaced computer.

Acquiring and Generating Analog Signals

We have seen that the digital acquisition of an analog signal is quantized, both in the amplitude of the acquired number and in the discrete times at which the analog amplitude is sampled. The optimum treatment of the analog signal prior to acquisition and the choice of acquisition rate and resolution depend on an awareness of the signal features that convey the desired information and the nature of the accompanying electrical noise. Similarly, the digital generation of an analog signal involves the creation of a time-variant amplitude from a series of discrete digital values. The bandwidth and noise components of the resulting signal depend on the resolution and update rate of the DAC. The relationships between signal and noise frequencies and the sampling process are explored next.

Sampling and bandwidth. An exact description of a signal with unrestricted variation can be obtained only when the interval between samples approaches zero. For band-limited signals, however, there is a finite sampling rate that is sufficient to include all the information in the signal. The Nyquist sampling theorem provides the quantitative basis for the rate at which samples must be taken, which is based on the bandwidth of the signal. The **sampling theorem** states that if a band-limited dc signal is sampled at a rate that is *at least twice the highest frequency component* in the signal, then the sample values exactly describe the original signal. If the sampling rate is $1/\Delta t$ (where Δt is the sampling interval), the signal must have no signal components at frequencies greater than $1/(2\Delta t)$. The critical frequency $1/(2\Delta t)$ is called the **Nyquist frequency**. A signal with Fourier components extending from 0 to 200 Hz, for example, should be sampled at a rate of at least 400 samples per s, or every 2.5 ms. To ensure accurate sampling, the signal should be band-limited by an appropriate input filter (*see* Chapter 7) prior to the sampling step. It is important to point out that sampling rates considerably higher than the Nyquist criterion are often used to ensure adequate sampling. As a rule of thumb, a sampling rate of 10 times the Nyquist frequency is often used.

If the Nyquist criterion is not followed and the sampling rate is too low, two kinds of errors result. First, the information in the signal at frequencies above half the sampling frequency is lost, and second, the undersampled high frequencies show up as spurious low frequencies. This latter error, known as **aliasing**, is illustrated in Figure 2–27. As a result of undersampling at 200 Hz, information

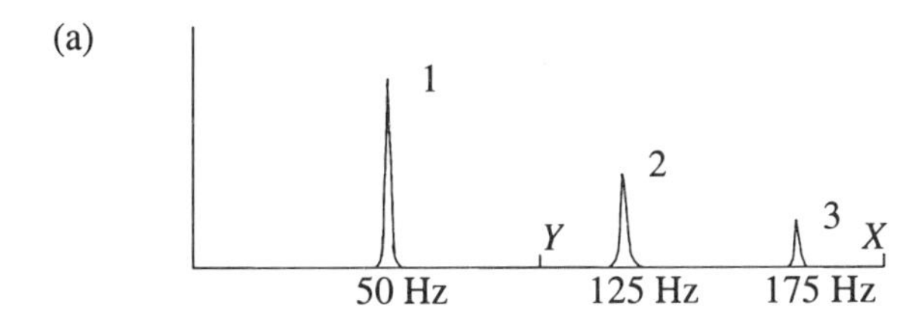

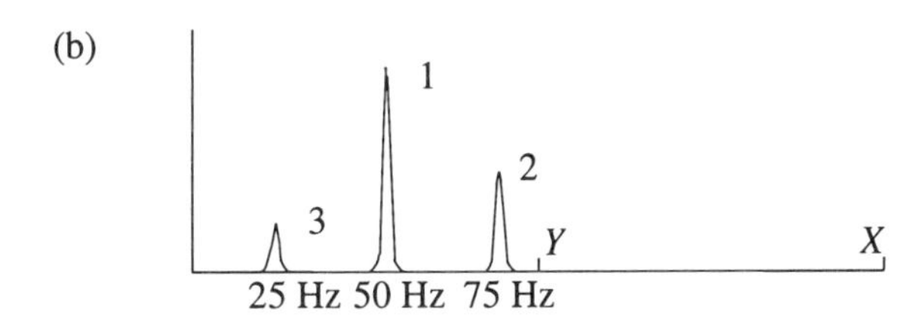

Figure 2–27. Aliasing. (a) The original analog waveform contains components at 50, 125, and 175 Hz. If the sampling rate is 400 Hz, the Nyquist frequency is 200 Hz (point *X*), and no aliasing occurs. (b) If a sampling rate of 200 Hz is chosen (point *X*), the components above 100 Hz (point *Y*) are undersampled and produce low-frequency aliases.

about frequency components above 100 Hz is lost, and low-frequency components are added to the signal as shown in Figure 2–27b. The 50-Hz component is still properly sampled, but the 125- and 175-Hz components have aliases at 75 and 25 Hz. The way in which aliasing comes about can be appreciated by the simple example shown in Figure 2–28. A familiar example of aliasing is the appearance of the slow rotation of the wheels of a rapidly moving stagecoach in Western movies, because the frequency of spoke rotation is undersampled by the framing of the movie camera. The stroboscopic effect is another example of the application of undersampling to reduce the apparent frequency of the observed motion.

It is interesting to point out with respect to Figure 2–27 that if no frequency components below 100 Hz are present in the original signal, the undersampling, or foldover, would not be serious. The aliased high frequencies would not overlap any other signal, and the position of foldover could be accurately predicted. This points out a more general statement of the Nyquist sampling theorem that applies to band-limited ac and dc signals: *a signal or waveform sampled at twice its bandwidth is adequately sampled.* Thus, if all frequency components are located in a 100-Hz bandwidth, a 200-Hz sampling rate is adequate, even if the 100-Hz bandwidth is between 1000 and 1100 Hz. This fact can considerably reduce the sampling rate for narrow-bandwidth ac signals. However, because aliases may be generated in the process, the reconstruction of the signal from its aliases may be

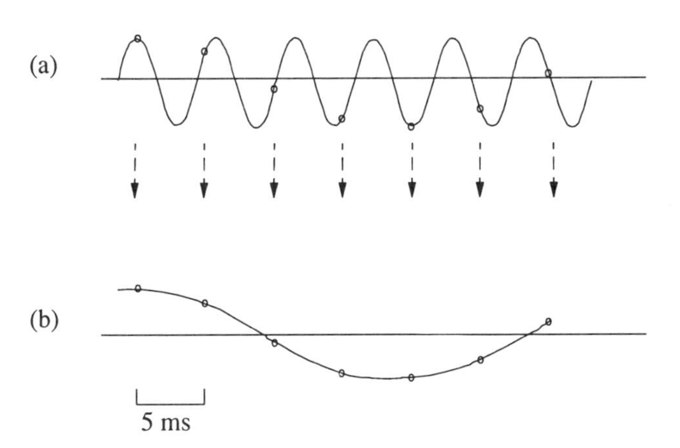

Figure 2–28. (a) Aliasing of 175 Hz to 25 Hz in a 175-Hz sine wave is shown. (b) Samples taken at 200 Hz. When the sample points are connected, the 25-Hz alias is revealed.

quite complicated. Again, in practice, sampling rates much higher than the Nyquist rate are often used.

Signals and noise. Electrical signals consist of a desirable signal component, which is related to some quantity of interest, and an undesirable component, which is termed noise. **Electrical noise** may thus be defined as any part of the observed signal that is unwanted. Implicit in this definition is the concept that what is considered noise in one situation may be useful information under other conditions. It is the goal of analog signal conditioning, digital signal acquisition, and digital signal processing to enhance the elements of the signal that convey the desired information while suppressing the deleterious effects of electrical noise.

Several types of electrical noise exist, and each type has its own characteristic distribution of frequencies. The spectrum of a composite of noise sources is shown in Figure 2–29. **White noise** has an essentially flat power spectrum and can be considered a mixture of all frequencies with random amplitudes and phase angles. Low-frequency **one-over-*f* (1/*f*) noise**, also called **flicker** or **pink noise**, has a spectrum in which the power density increases approximately with the reciprocal of the frequency at low frequencies. Such a spectrum is typical of the low-frequency drifts common in transducers, amplifiers, and analog components. Discrete-frequency **interference noise** often arises from 60-Hz power lines, radio transmitters, motors, and nearby oscillators, and the power density spectrum has peaks at all the fundamentals and harmonics of these frequencies. In a real electronic system, all three types of noise are likely to be encountered. The goal in data acquisition is to limit the bandwidth in which the signal components are to be found and then to record as little of the signal outside this bandwidth as possible. In this way, the effects of the noise components can be minimized. It is particularly helpful to obtain power density spectra of both the desirable signal components and the background electrical noise so that one knows the bandwidth that must be acquired in order to maintain signal integrity and also the nature of the noise that is present. Fixed-frequency noise sources can often be identified and shielded against (*see* Appendix A). This kind of frequency analysis is also essential in choosing the optimum sampling rate and the frequencies of band-pass filters for noise reduction and elimination of aliasing.

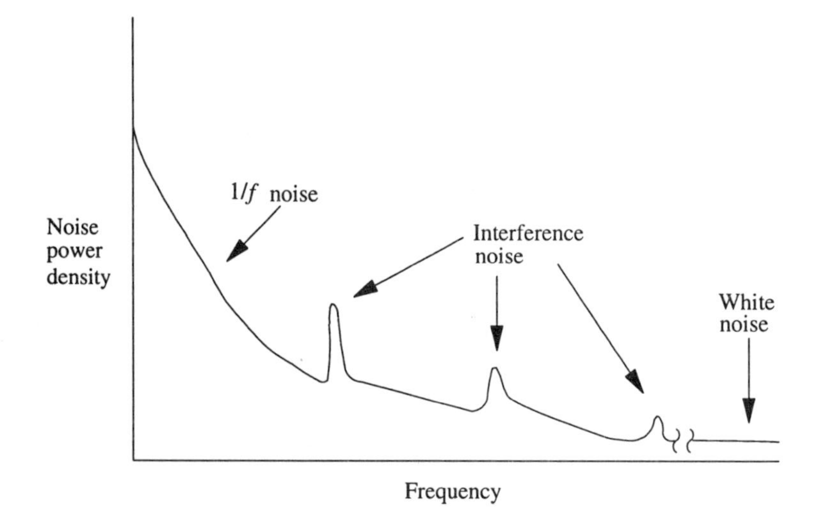

Figure 2–29. Combined noise power density spectrum. At high frequencies, white noise predominates; at low frequencies, 1/*f* noise predominates. Interference noise has components at discrete frequencies.

Averaging and smoothing. One of the simplest data-processing options available for digitally recorded data is data averaging. If the analog signal to be recorded is presumed to be constant but not free of noise, the averaging of successive samplings of the signal amplitude will improve the precision of the measurement by $\sqrt{N}$, where N is the number of samples averaged. A data acquisition routine can be set up in which N samplings of the input signal are taken and averaged (or simply summed), and the result is stored in memory. The effective sampling rate of the input signal then becomes the data conversion rate divided by N. As long as this rate satisfies the Nyquist criterion for the desirable signal components, no information is lost and the signal-to-noise ratio of the acquired data is improved. Note that to avoid aliasing the input filter need be only half the conversion frequency, not half the effective sampling rate. Thus, if the conversion speed and computer time are available, it is desirable to maintain a high conversion rate and reduce the effective sampling rate by increasing the number of successive data points summed prior to storage. Clearly, in situations where the signal bandwidth requires nearly the maximum conversion rate available, this approach to averaging will not work. In such cases, it may be possible to repeat the entire waveform to be recorded and to sum corresponding data points in each successive waveform. The repetitive signal components reinforce with each successive summing, while the random noise tends to average out. This process does not reduce the bandwidth of the recorded signal but does require a repetitive signal and a way of triggering waveform acquisition to occur at the same point in each successive waveform.

Another common signal-processing technique is data **smoothing**. This technique amounts to mathematically drawing a smooth line through successive data points that are scattered because of random noise. In this approach, it is assumed that the true data values do not change as rapidly as the noisy values. The application of smoothing routines certainly makes the resulting recorded waveform smoother, but this occurs at the expense of reducing the effective signal bandwidth. If the original signal is oversampled, smoothing is an effective method of improving the apparent signal-to-noise ratio. However, if the signal has components that are near the Nyquist limit for the sampling rate used, smoothing will cause distortion of the data elements such as peak height, peak width, and rise time. Also, because smoothed data necessarily contain more data points than are necessary for the bandwidth of the data stored, they waste data storage space.

Generating Analog Signals

A computer feeding a series of digital values to the input of a DAC will produce an analog signal at the DAC output of varying magnitude. In this way, an analog waveform of virtually any description can be created by making a table of the successive numerical values of the desired waveform magnitude and then sending these values in order to the DAC at a constant rate. This process is illustrated in Figure 2–30. The construction of an analog waveform from digital samples is the reverse of the digital acquisition of analog data, so that most of the same considerations apply. The accuracy of the resulting analog signal depends on the accuracy and resolution of the DAC. The maximum-frequency component of the analog waveform is limited to half the rate at which the DAC input value is updated. Viewed on a fine scale, the DAC output does not form a smooth curve but instead changes in a series of steps at the update rate. If a smooth output waveform is

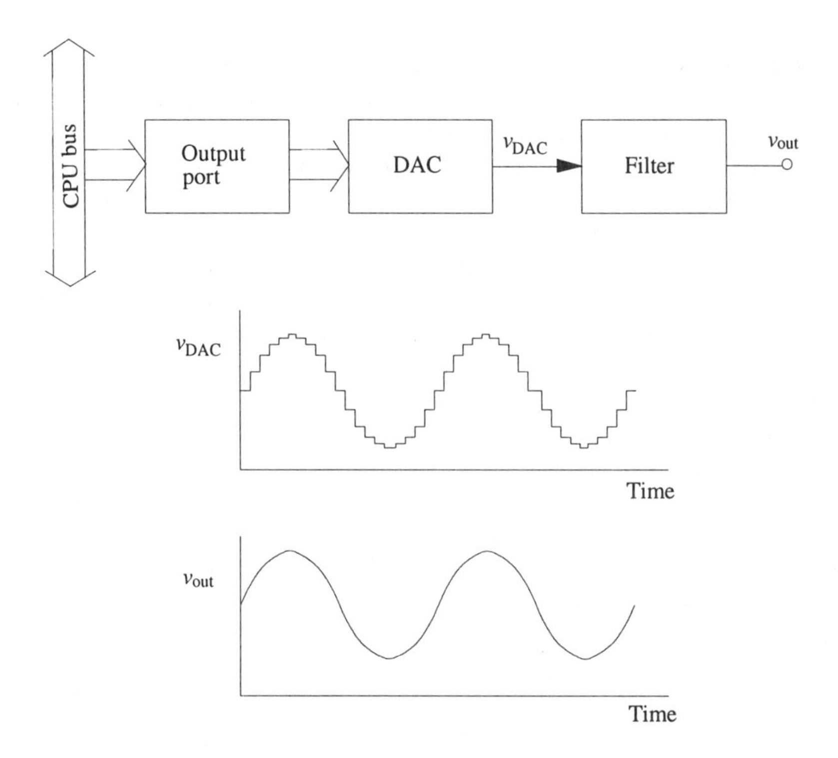

Figure 2-30. Computer generation of a varying analog signal. Digital data representing the desired output voltage values at successive time intervals are fed to the DAC at the appropriate rate. The steps in the DAC output due to the discrete time and voltage values (exaggerated in this figure) may be filtered to produce a more smoothly changing signal.

desired, the DAC output must be filtered to eliminate signal frequencies higher than half the update frequency.

The digital generation of analog waveforms is a convenient way of generating a variety of waveforms with a single system. The output waveforms can be very precise in amplitude, and the rate of generation can be readily varied by simply changing the update rate to the DAC. Such computer-generated analog signals can be connected to the analog control inputs of such devices as motor speed controllers, programmable power supplies, and servo positioners. In consumer electronics, digital generation of analog waveforms is used in compact disk players, digital tape players, and digital television.

Using the Digital I/O

Signals at the digital I/O terminals are in the form of logic levels, HI or LO. For each class of logic circuit, the terms HI and LO have a specific defined range. Most logic interfaces currently use the ranges defined by the transistor–transistor logic (TTL) family of logic devices in which the HI signal is between 2.5 and 5.0 V and the LO signal is between 0.0 and 0.4 V. In addition, the LO signal must be capable of maintaining the LO level while absorbing a positive current of several milliamperes; in other words, it must have a very low output resistance in the LO state. Digital data appear at a digital output port in response to a command to transfer a byte of data to that port. Similarly, the digital data levels at the 8-bit inputs of a digital input port can be acquired by the computer through the input port read operation. The 8 bits of input or output data can be a byte of data such as a number or a printer character, but more often, each bit represents the state of a particular part of the system, where only two states are possible (such as on or off, adequate or inadequate flow, active or inactive, enabled or disabled). In such cases the operator may often be interested in just 1 bit at the input port or may want to change just 1 bit at the output port. In performing operations on just 1 bit in a word, appropriate software is required to avoid inadvertent changes to the other bits.

Digital I/O can provide very important functions in computer control of experiments. The output port signals can be used to turn instrument functions on and off, advance stepper motors a desired number of degrees, set parameters and operating modes, and so forth. Input port connections can be used to monitor the state of instrument conditions, including the presence of appropriate power, cooling, and sample position. In an increasing number of cases, the instrument may provide control inputs and system monitor outputs that operate at TTL levels suitable for direct connection to digital I/O ports. However, it is still often the case that TTL levels at the digital output port must be converted to the levels required to perform the desired operation and that the monitor signal levels must be converted to TTL levels. Failure to make the appropriate level conversions is likely to cause damage to one or both of the devices thus connected.

Programming Experiments

The programming of laboratory experiments is becoming an increasingly simple task since general-purpose software for common laboratory operations has become available. Suppliers of such software include specialized software vendors as well as analog I/O and GPIB board manufacturers. This software must be written specifically for a particular combination of computer type and operating system. In addition, it must contain software drivers for the interface hardware to be used. Such software packages are assemblages of routines that acquire a data point, send a value to a DAC or output port, read a counter, etc. These subroutines can be linked to automate complete experiments, including parameter changing and waveform generation as well as data acquisition. Before the experiment-controlling program is run, other subroutines in the software package facilitate the setting of variables in the laboratory interface system such as the mode and timing of ADC triggering, the number of consecutive data samples to be collected, the number of input channels to be scanned, the frequency and optimum gain setting for each channel, and the destination of the collected data.

The development of the automated experimental sequence itself may be done by selecting the desired operations from menus, connecting icons representing the desired operations in sequence, or writing a program linking the needed subroutines. In most cases the assembly of the selected steps into a final executable program is transparent to the user. When the selections are complete, the experiment program is ready to run. In some of the more sophisticated laboratory software packages, the experimental program can include a variety of data processing and display operations and the course of the experiment can be made to change depending on the values of the data being acquired.

Now that the time required to program an automated experiment is often less than the time it would take to perform the experiment manually, it has become worthwhile to automate experiments that will be performed only a few times. Another advantage of programmed experimental control is that it is relatively easy to alter some aspect of the experiment and run it again. Thus, the experimenter's energy can be put into optimization of the experimental process rather than the tedium of the measurement and control sequence itself. Once the experiment is completed, the files containing the experimental data can be reviewed, a process facilitated by other routines in the same software package. Data in files can be operated on mathematically to obtain sums, differences, statistical variances, or even Fourier transforms. Raw or processed data can be viewed numerically or graphically on screen under fine or course inspection. Final results can be printed out in report or graphical format. The combination of the personal computer, the laboratory interface, and general-purpose laboratory software packages is rapidly becoming the most powerful and versatile tool in the scientific laboratory.

Related Topics in Other Media

Video, Segment 2

- Use of solderless breadboards for rapid wiring and testing of circuits
- Frequency measurements and counting
- Analog-to-digital conversion of signal
- Sampling theorem and aliasing

Laboratory Electronics Kit and Manual

- Section A
  - ¤ Measuring analog signals with a digital multimeter
  - ¤ Converting light and temperature into electrical signals by transducers
  - ¤ Troubleshooting by measurement of electrical continuity

Chapter 3

Understanding and Measuring Analog Signals

Electrical signals in our instruments and computers represent information that is used in measurement and control processes. Consider the amperometric determination of oxygen illustrated in Figure 3–1a. If the platinum electrode is polarized sufficiently negative with respect to the reference electrode, electrons in the metal reduce the oxygen to water. The current in the external circuit is then proportional to the rate of arrival of oxygen molecules at the electrode surface. The platinum electrode is an oxygen-sensitive electrode under these conditions. This arrival rate is limited by the rate of diffusion of O_2 molecules to the electrode and is directly proportional to the oxygen concentration. As shown in Figure 3–1b, the platinum electrode must be from 0.6 to 0.8 V negative with respect to the reference electrode for the current to be proportional to the oxygen concentration (Figure 3–1c). Note that the oxygen concentration is represented or encoded by the electrical current in the external circuit. Once a current signal proportional to the oxygen concentration has been obtained, it can be modified, manipulated, and processed (e.g., converted to voltage, amplified, filtered, converted to a parallel digital signal, or processed in a computer to obtain the oxygen concentration) prior to being displayed in a form that we can easily read and interpret.

In the example in Figure 3–1, the voltage v between the two electrodes is not related to the oxygen concentration. This voltage is not exactly equal to the applied voltage v_{app} because of voltage drops in the solution and the external circuit. To obtain a linear relationship between the oxygen concentration and the current, as shown in Figure 3–1c, the voltage between the electrodes must stay constant and be independent of the circuit current. Typically, a control system monitors v and adjusts the v_{app} so as to maintain v at the preset value. Here, two different electrical signals are used for two different pieces of information: one signal is related to the oxygen concentration, while the other is used to control the voltage between the electrodes.

Like the oxygen electrode, most sensors used in instruments produce analog electrical signals. The characteristics of the analog quantities of current, voltage, charge, and power are thus important to know, as are the properties of simple direct current (dc) circuits for manipulating and controlling these quantities. The basic analog electrical quantities are described in Supplement 1 and briefly summarized in this chapter.

Analog signals can also be time-variant. With such signals the amplitude at a particular time or the entire amplitude–time relationship can carry the desired information. Analog operations can change these relationships in a variety of intentional or unintentional ways and thus influence the encoded information. Periodic waveforms can also be used to encode information as the amplitude or frequency of the waveform, the width or magnitude of a pulse, or the phase angle between two time-varying signals. Thus, it is important to understand the basic properties of such periodic signals and the ways in which circuit components can alter signal levels and frequency composition. With such an understanding we can troubleshoot alternating current (ac) systems, develop ways to extract the informa-

2861-2/94/0057$09.60/1 Published 1994 American Chemical Society

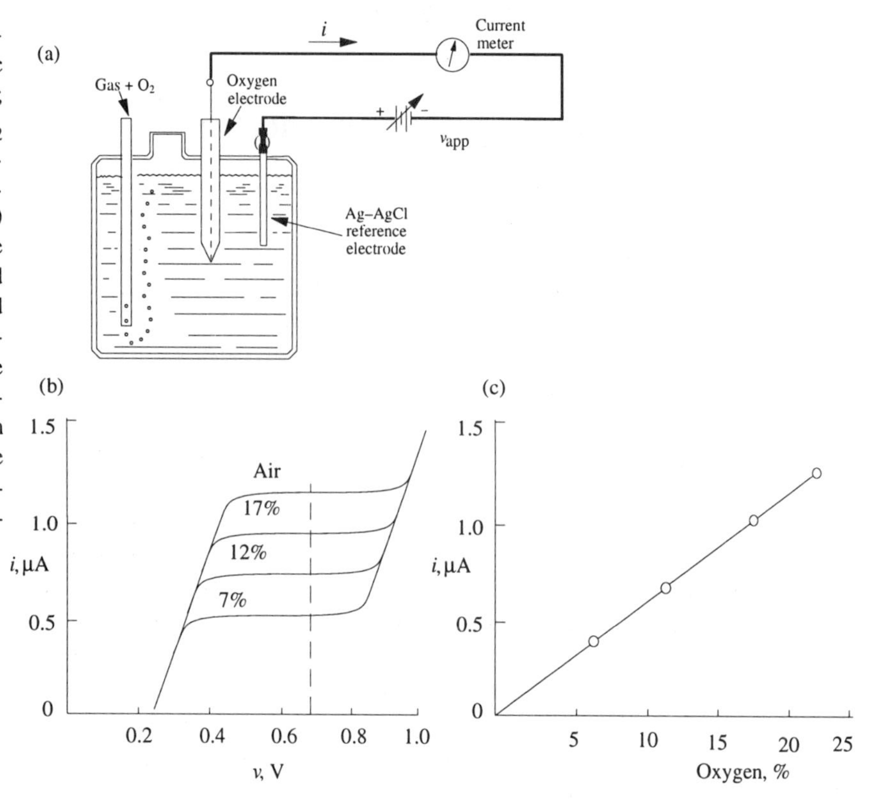

Figure 3–1. Amperometric determination of oxygen. (a) Schematic diagram of measurement system; (b) current–voltage curves for O_2 electrode; (c) output current vs. percent oxygen for v = 0.68 V. The current–voltage curves (b) allow the appropriate bias voltage to be chosen. Thereafter, a fixed value of bias (shown by the dotted line) is applied and the O_2 concentration is obtained from the measured current and the calibration curve (c). A control system monitors the voltage between the electrodes and varies v_{app} to maintain the control voltage at the preset value.

tion of interest from extraneous noise, or tailor a measurement or control system for optimum behavior. Such ac signals and systems are briefly discussed in Supplement 2.

One basic electrical circuit used time after time in modern electronics is the voltage divider. It is often used as an attenuator for signals that are too large for recorders, amplifiers, analog-to-digital converters, and many other units. Voltage dividers appear so often in electrical circuits that it is important to recognize them and understand their behavior. A related circuit for current is the current splitter, which also appears frequently in modern electronic systems.

Measurements of dc and ac quantities are basic to modern analog and digital instrumentation. This chapter introduces measurement principles as well as specific test instruments such as the digital multimeter, the analog multimeter, and various computer-based systems. Every time a measurement device is used to obtain the value of current, voltage, or power in a circuit or from a transducer, a perturbation of that circuit or transducer occurs. This perturbation or loading effect is unavoidable, but under appropriate conditions the effect can be made so small that the errors introduced are not significant.

The major tool for observing periodic, repetitive, and transient signals is the laboratory oscilloscope. This device has improved greatly in recent years, and digital storage oscilloscopes with built-in memory and many advanced features are becoming routinely available.

3–1 Starting with the Basics

The basic electrical quantities known as charge, voltage, current, and power are described in Supplement 1 and summarized in Table 3–1. Because we deal with

Table 3–1. Basic Electrical Quantities

Quantity	Symbol	Unit	Definition
Charge	q, Q	Coulomb, C	Basic electrical quantity. One electron has a charge of 1.603×10^{-19} C.
Voltage	v, V	Volts, V	Electrical potential energy difference resulting from charge separation. One volt results when 1 J is required per coulomb of charge.
Current	i, I	Amperes, A	Rate of charge flow. One ampere is 1 C/s.
Power	p, P	Watts, W	Rate of performing work: $P = IV = I^2R = V^2/R$

*It is common to use uppercase letters (Q, I, V, P,) when constant or static quantities are discussed and lowercase letters (q, i, v, p,) when varying quantities are indicated.

Table 3–2. Basic Laws of Electricity

Name	Equation	Usage
Ohm's law	$I = V/R$ or $i = v/R$	Relates current and voltage in a circuit or circuit element. An element of resistance R with current I through it has a voltage drop of IR across it in the direction of the current.
Kirchhoff's voltage law	$V = IR_1 + IR_2 + \ldots + IR_N$	For a series circuit of N resistors and a voltage source, the sum of voltage drops ($IR_1 + IR_2 + \ldots + IR_N$) equals the source voltage V.
Kirchhoff's current law	$I_t = I_1 + I_2 + \ldots + I_N$	For a parallel circuit with N conducting paths, the total current I_t into the junction equals the sum of the currents ($I_1 + I_2 + \ldots + I_N$) in each path.

current and voltage so often, it is important to differentiate between these basic quantities. Voltage is an electrical pressure or driving force. When a conducting path is present, this driving force can cause charges to flow and a current to exist in the conductor. In a flow system analogy, voltage is similar to pressure and current is similar to flow rate.

Current and voltage are related to each other in electrical circuits by the basic laws summarized in Table 3–2. The derivations and descriptions of these laws are given in Supplement 1. We will use these basic laws in discussions of the important voltage divider and current splitter circuits.

Voltage Divider

Voltage must often be attenuated before being applied to a circuit or measuring device. For example, if we want to record the time variations in the voltage of a 1-V battery on a strip chart recorder with a fixed 10-mV full-scale range, we must attenuate the battery voltage by at least a factor of 100 before applying it to the recorder. A simple series circuit, such as that shown in Figure 3–2, can function as a **voltage divider** circuit. Compare the *IR* drops across the two resistors with the source voltage. The first drop, IR_1, is 0.5 V, or 1/10 of the 5 V from the source. The second drop, IR_2, is 9/10 of the source voltage. The resistance R_1 is 1/10 [100 Ω/(900 Ω + 100 Ω)] of the total series resistance, and R_2 is 9/10 of the total. The voltage drops have thus distributed according to the ratio of the individual resistances to the total resistance, leading us to what is often called the **voltage divider theorem**. This theorem can be readily verified from Ohm's and Kirchhoff's laws. Let us call b_1 the fraction of the total source voltage that appears across R_1. This fraction is

$$b_1 = \frac{IR_1}{V} = \frac{IR_1}{IR_1 + IR_2} = \frac{R_1}{R_1 + R_2} = \frac{R_1}{R_s}$$

where R_s is the total series resistance $(R_1 + R_2)$.

Thus, to attenuate a voltage by a factor of 100 before connecting the voltage to a recorder, we need to construct a series circuit with resistances of R and $99R$ Ω. Typical values might be 1 and 99 kΩ. The voltage drop across the 1-kΩ resistor would be sent to the recorder input.

For dividers with N resistors in series, the fraction b_N of the total source voltage that appears across the Nth resistor is

$$b_N = \frac{IR_N}{V} = \frac{IR_N}{IR_1 + IR_2 + \ldots + IR_N} = \frac{R_N}{R_s} \quad (3\text{–}1)$$

Two voltage dividers used for attenuating voltages are shown in Figure 3–3. A switched-range divider is shown in Figure 3–3a. In any position, the fraction of V appearing at the output is the total resistance across which the output is taken divided by the total series resistance (1000 Ω). If the resistors used in this divider are 0.1% precision resistors, the attenuation factor (divider fraction) can be quite

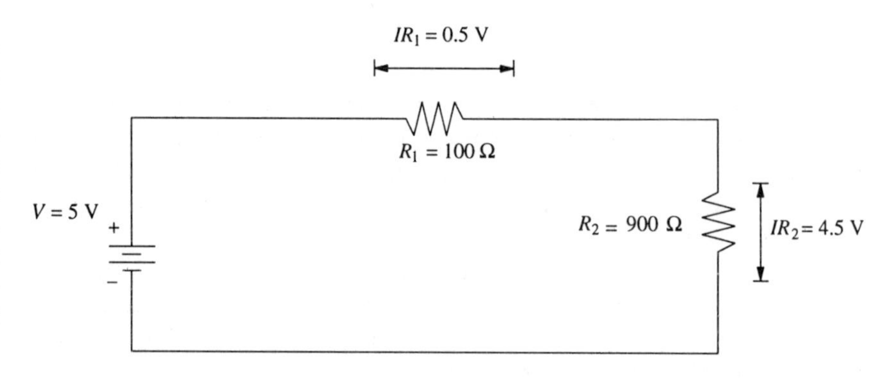

Figure 3–2. Series circuit. In a series circuit, there is only one path for current, and this path includes every circuit element in sequence. Each terminal of each element is in contact with only one other element. For example, the two resistors have only one point in common (right contact of resistor 1 is connected to top contact of resistor 2).

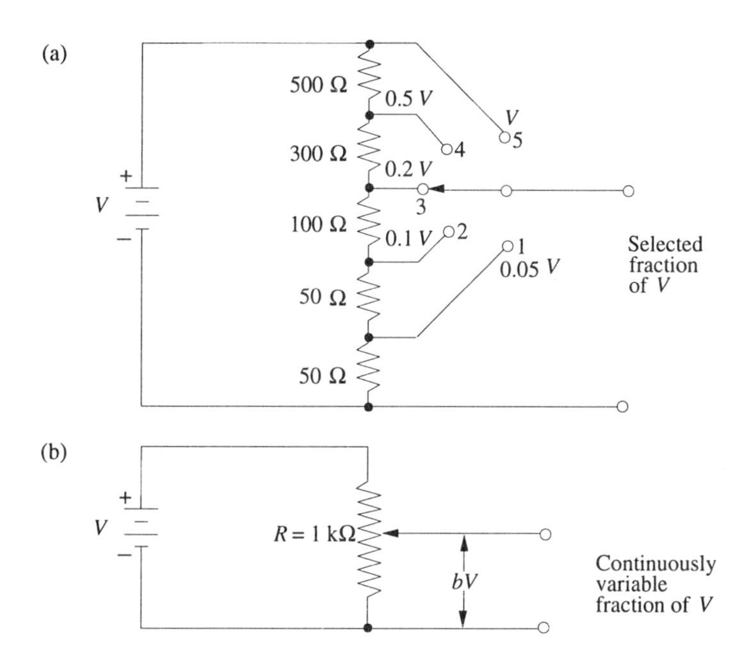

Figure 3–3. Voltage dividers. The voltage divider is a series circuit in which a fraction of the applied voltage V appears across each resistor. The fraction can be varied by fixed resistors and a selector switch (a) or by a continuously variable potentiometer (b). In part a, the total series resistance is 1000 Ω (500 Ω + 300 Ω + 100 Ω + 50 Ω + 50 Ω). When the switch is in position 1, the fraction of the source voltage V appearing at the output of the switch is 50 Ω/1000 Ω, or 0.05. When the switch is in position 2, the IR drop across the two 50-Ω resistors appears at the switch output. Hence, the fraction of V at the switch output is 0.1 in switch position 2. In part b the slider selects a continuously variable fraction of V.

precise. Such a switched-ranged voltage divider is often used with instruments such as multimeters and chart recorders to provide different ranges.

A continuously variable voltage divider is shown in Figure 3–3b. Here, a resistor with a variable slider contact called a **potentiometer** is used. Such a voltage divider is often used as a fine adjustment on oscilloscopes, chart recorders, and other instruments.

Because the voltage divider equation was based on the simple series circuit, the output voltage fraction equals the resistance fraction only if the current through all the resistors is the same. The connection of a load across the divider output terminals diverts current from the selected resistors and causes an error in the output voltage fraction. For the error to be negligible, the resistance of the load must be very large compared with the resistance of the divider. Such loading errors are discussed in greater detail in Section 3–3.

Current-Splitting Theorem

It is often desirable to attenuate currents in order to increase the current sensitivity of a meter or to avoid overranging an instrument. When two resistors with resistances R_1 and R_2 are connected in parallel, the total circuit current I_t is split between them in accordance with Kirchhoff's current law. It can be readily shown (*see* Supplement 1) that I_1/I_t, the fraction of I_t that passes through R_1, is

$$\frac{I_1}{I_t} = \frac{\frac{V}{R_1}}{\frac{V}{R_1} + \frac{V}{R_2}} \qquad (3\text{–}2)$$

Equation 3–2 can readily be rearranged to yield

$$\frac{I_1}{I_t} = \frac{R_2}{R_1 + R_2} \tag{3–3}$$

The fraction of the total current through either resistor is the resistance of the other resistor divided by the sum of the resistances of the two resistors. This "current-splitting" relationship is frequently used in calculating the desired values for parallel resistors. For example, if you want 1/10 of the total current to go through a meter of 9-Ω resistance, a parallel resistance of 1 Ω is needed. Current splitting in parallel circuits is analogous to voltage division in series circuits. If one of the parallel paths has a resistance that is much lower than that of the other path, then the first path carries nearly all the current.

3–2 Measuring Analog Electrical Quantities

Many laboratory instruments can measure analog signals. The older moving-coil meters have largely been supplanted by digital meters such as the digital multimeter (DMM) and the digital voltmeter (DVM). For slowly changing signals, strip chart recorders (*see* Chapter 5) can be used to provide a record of signal amplitude versus time. The oscilloscope is frequently used to display rapidly changing signals (*see* Section 3–4) on the screen of a cathode ray tube (CRT). In recent years computer-based measurement instruments have become widely used. These include conventional instruments with computer interfaces, instruments with integral computers, and peripheral devices that allow measurements to be made with a standard personal computer.

The DMM

In recent years the price of digital meters has dropped substantially so that they are now routinely used in troubleshooting and maintenance applications. Digital meters are based on the **digital voltmeter (DVM)**. A typical DVM consists of a dual-slope analog-to-digital converter (ADC) (described in Chapter 2, Section 2–2) and a numerical display. A **digital multimeter (DMM)** consists of a DVM with appropriate converters and/or voltage dividers (*see* Chapter 2, Figure 2–14). These circuits are used to convert current or resistance into a related voltage prior to measurement with the DVM. Here, we consider the operating principles of general-purpose DMMs, including simple hand-held and laboratory units. Computer-based multimeters are discussed later in this section.

Voltage ranges. The ADCs in most DVMs have one or two fixed full-scale ranges. For example, a dual-slope converter might have a fixed range setting of 199.9 mV full scale. To provide the necessary versatility for measuring voltages of wider dynamic ranges with sufficient resolution, an input voltage divider is normally used to scale the input voltage. With a five-position divider (dividing by 1, 10, 100, 1000, and 10 000) and a 200-mV full-scale converter, full-scale voltage settings from 200 mV to 2 kV are available. Figure 3–4 shows a typical input voltage divider that provides a high input resistance (10 MΩ).

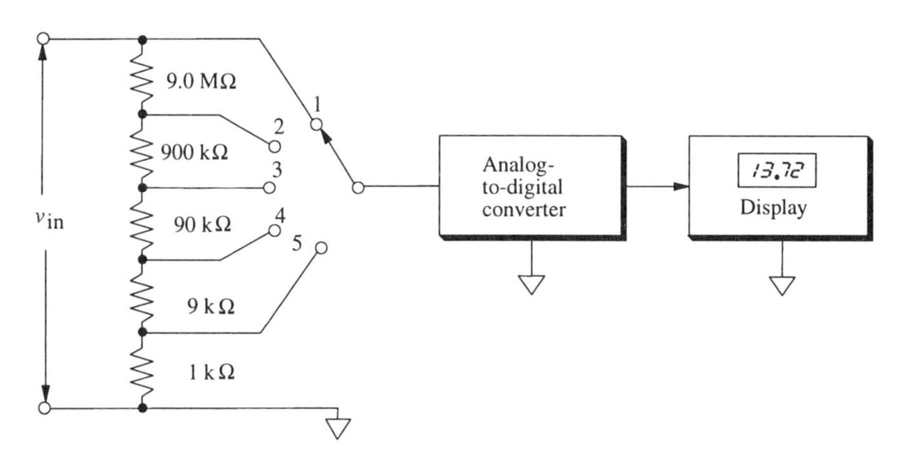

Figure 3–4. Input voltage divider for DVM. In position 2 the input voltage v_{in} is divided by 10, and in position 3 it is divided by 100, etc. Thus, five full-scale voltage ranges are available with a fixed 10-MΩ input resistance. The downward-pointing triangles indicate a connection to common, a point of reference for all voltages. The interconnection of common points is often symbolized in this way.

Some more expensive DMMs have **autoranging** capabilities. With these the full-scale range setting is changed automatically to provide the highest resolution without overranging. Most DMMs also give a polarity indication and automatically select the appropriate decimal point as different full-scale settings are selected.

Current measurements. It is possible to measure current with a DMM by inserting a current-to-voltage converter ahead of the DVM. A precision resistor network is used in inexpensive meters to perform this function. Figure 3–5 shows a typical shunt network for current measurements. Note that the current through the precision resistor produces an *IR* drop that is measured by the DVM. Also, a significant amount of resistance is introduced into the circuit being tested when the meter is in position 1 or 2. Some more expensive DMMs use operational amplifier current-to-voltage converters to avoid this problem (*see* Chapter 5).

Modern DMMs have switch-selectable full-scale current ranges from microamperes to hundreds of milliamperes or even amperes. Compared with low current ranges, higher current ranges often require a different connection to the meter to prevent damage to sensitive components. Inputs for current are also frequently fused to afford further protection in the case of current overloads.

Resistance measurements. By adding a resistance-to-voltage converter ahead of the DVM, the DMM can also measure resistance, as shown in Figure 3–6. Here, a constant current source produces a current I in the unknown resistance R_u. The unknown resistance develops a voltage drop IR_u. The resulting voltage is measured with the DVM. The two-wire resistance measurement works well with moderate resolution (3½ digit) DMMs. However, voltage drops in the test leads

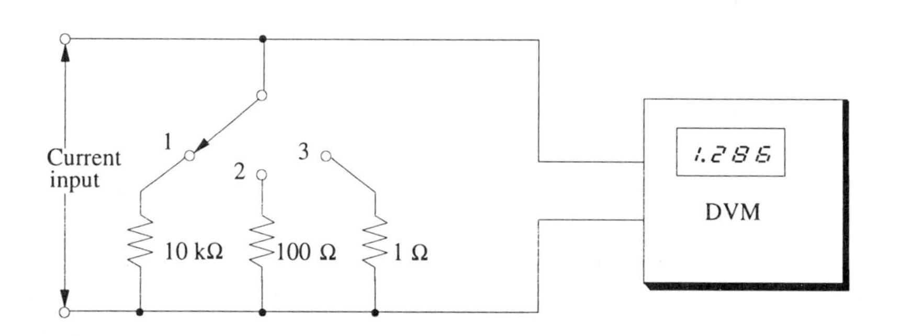

Figure 3–5. Shunt network for current-to-voltage conversion with a DVM. If the DVM is 200 mV full scale, the full-scale range is 20 µA for position 1, 2 mA for position 2, and 0.2 A for position 3.

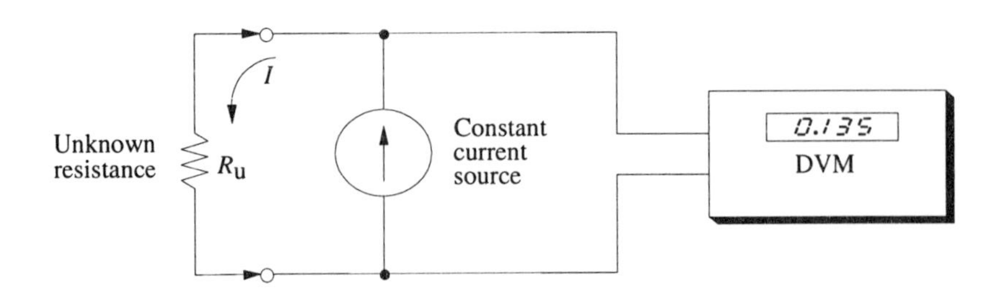

Figure 3–6. Resistance measurement circuit for DMM. The basic DVM measures the *IR* drop across the unknown resistance R_u when a known constant current *I* is applied. The readout is proportional to R_u and can be calibrated to read directly in ohms, kiloohms, or megaohms.

can lead to noticeable errors with higher resolution meters. With these meters, a four-wire measurement scheme is normally used. One pair of leads supplies current to the unknown resistor, and a second pair senses the *IR* drop across R_u. The only current through the sense leads is the input current to the ADC, which can be as small as picoamperes.

With simple hand-held DMMs, the constant current for resistance measurements is usually supplied by a battery and a large resistance in series. With such meters, errors can result when high values of resistance are measured, because the current is not determined solely by the battery voltage and the series resistor. Some more expensive DMMs use operational amplifier constant-current sources that are extremely constant even though R_u values may vary widely. The operational amplifier resistance-to-voltage converter is described in Chapter 5.

DMM measurements of ac quantities. To measure ac quantities with the DMM, an ac-to-dc converter is switched into the circuit ahead of the DVM. The output of the converter is a dc voltage suitable for measurement by the voltmeter part of the DMM. On the ac volts scale, an input divider, which must be nearly frequency-independent, is used to select the appropriate range for the ac-to-dc converter. On the ac current scale, the input current is sent through switch-selectable resistors to produce an ac *IR* drop proportional to the current.

Conversion from ac to dc generally involves sampling the maximum absolute voltage of the input waveform. Thus, the ac DMM responds directly to the peak or peak-to-peak voltage of the ac waveform. In general, the ac quantity of most interest is the root-mean-square (rms) voltage (*see* Supplement 2, Section S2–3). To obtain ac measures different from the one the meter responds to directly, the ac meter scale can be calibrated with a conversion factor included. With a DMM that responds to the peak-to-peak voltage, a conversion factor of 0.707/2 is needed to convert to rms values for sine waves. This factor is correct, however, only for sinusoidal signals; the peak or peak-to-peak responding DMM does not read the correct rms amplitude for nonsinusoidal signals.

True rms-responding meters are available for cases in which an rms measurement is needed on a waveform for which the relationship between the average or peak value and the rms value is not known. This measurement is often obtained by squaring the signal, averaging it, and taking the square root. This procedure gives an output voltage equal to the rms value of the input voltage. Some instruments use digital computation methods to produce an rms response.

Complete DMM. A block diagram of the complete DMM is shown in Chapter 2, Figure 2–14, and a typical front panel is shown in Figure 3–7. Switches or push buttons on the front panel select ohms, volts, or amperes as the measured quantity. These controls switch the *R–V* converter, the voltage divider, or the *I–V* converter into the circuit. Another front-panel switch selects ac or dc for voltage or current

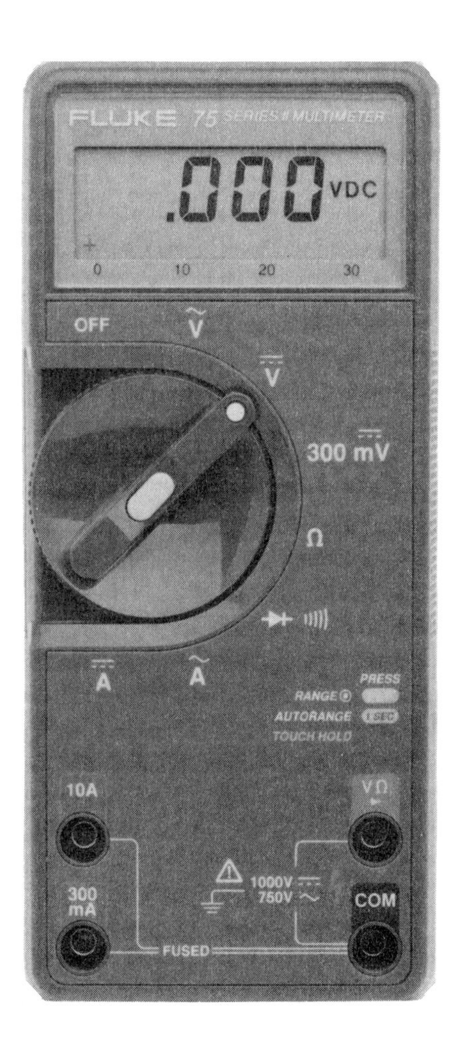

Figure 3–7. Typical front panel of DMM. The range and function controls select the appropriate input signal conversion and range circuits so that a dc voltage appears at the input of the dual-slope ADC (*see* Chapter 2, Figure 2–14). (Photo supplied by and used with permission of John Fluke Mfg. Co., Inc.)

measurements. This switch places the ac-to-dc converter in the circuit (for ac) or out of it (for dc). In addition, range selection controls select the *R–V* converter sensitivity, the voltage divider attenuation, or the *I–V* converter sensitivity. Once the quantity of interest has been converted to a dc voltage, the dual-slope ADC presents a parallel digital signal to the front-panel display circuits.

Multimeters are available that vary in readout precision (number of digits of display) and in the accuracy with which ac and dc measurements can be made. Typical hand-held 3½-digit multimeters are available, with basic dc accuracy figures ranging from about 1% to better than 0.1%. Accuracies for ac vary from about 3% for less expensive units to better than 1%. More sophisticated systems can have as many as 6½ digits of readout resolution (*see Computer-Based Measurement Systems* in this section).

A particularly useful feature of general-purpose DMMs is autoranging. After the user chooses the function (dc voltage, current, etc.), the meter automatically selects the range with the greatest accuracy and resolution. Most autoranging meters also allow this feature to be defeated for situations in which a single range is more convenient (i.e., following signal changes). Some multimeters have an analog bar graph display in addition to a digital readout. The extent and direction of changes in the length of the bar graph can be observed more readily than changes in numerical readout; it is thus more convenient when slowly changing signals are being observed.

Analog Meters

The **moving-coil current meter** is the basis for many older analog measuring devices. The position of a pointer is affected by the magnitude of an electrical current. A scale behind the pointer can be calibrated in the desired units.

Moving-coil current meter. The D'Arsonval style of moving-coil meter is illustrated in Figure 3–8. A coil of fine wire wound on a rectangular aluminum frame is mounted in the airspace between the poles of a horseshoe magnet. Hardened steel pivots attached to the coil frame fit into jeweled bearings so that the coil rotates with minimal friction. An indicating pointer is attached to the coil assembly, and springs attached to the frame return the needle (and coil) to a fixed reference point. When a current passes through the coil, the direction and magnitude of rotation depend on the direction and magnitude of the current. The pointer position is viewed against a printed scale in order to obtain a meter reading related to the current in amperes.

Suppose that the current in the light-emitting diode (LED) circuit of Figure 3–9a is to be measured with the moving-coil current meter. To measure the current, the circuit must be opened and the meter inserted as shown in Figure 3–9b. When the resistance in the circuit is small, the meter resistance R_m can alter the current from the value it had in the meter's absence (*see* next section).

Current ranges. The moving-coil meter itself has a fixed range determined by the design of the coil, springs, and magnet. The fixed-range meter has a resistance R_c given by the resistance in the meter's coil. To allow current measurements over a wide range without use of a separate meter for each range, a technique called shunting is used. A **shunt** is a parallel path around the meter that causes only a fraction of the circuit current to pass through the meter itself.

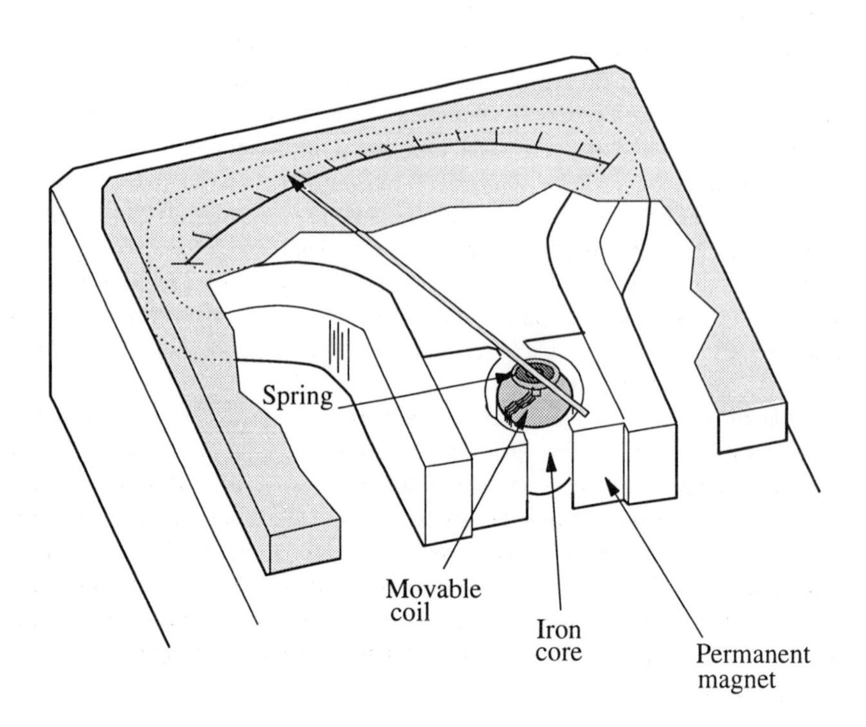

Figure 3–8. Moving-coil meter. The current to be measured passes through the movable coil. A turning force on the coil assembly results from interaction of the magnetic fields from the coil and the permanent magnet. This force acts against a restoring spring to produce a rotation of the pointer that is proportional to the current.

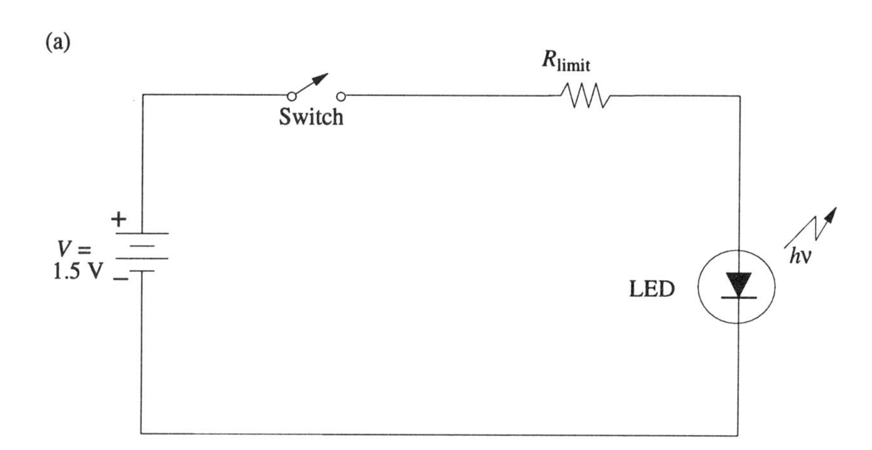

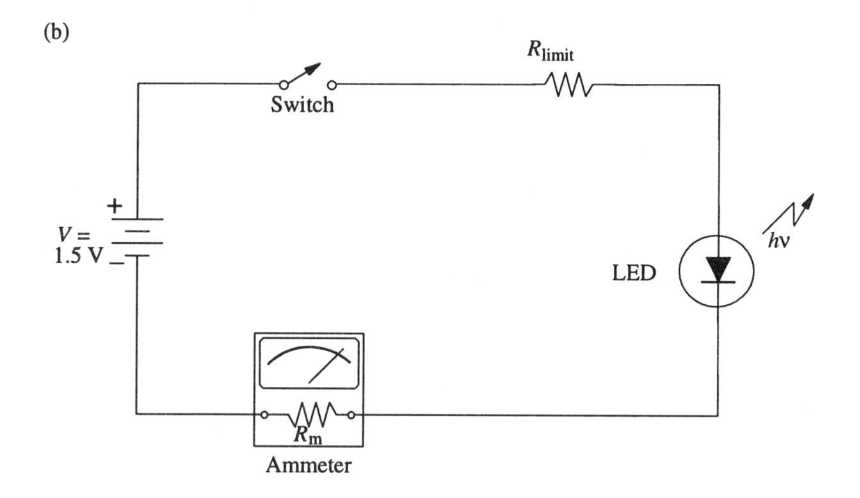

Figure 3–9. Current measurement. To measure the current through a circuit element such as the LED and a current-limiting resistor (a), the current meter must be inserted into the circuit *in series* with the element (b). In this way, the currents through the element and the meter will be identical.

A simple shunt is shown in Figure 3–10. According to the current-splitting relationship (equation 3–3), the fraction of the total current i_t that exists in the meter i_m is $i_m/i_t = R_{sh}/(R_c + R_{sh})$, where R_{sh} is the resistance of the shunt. For example, if a 1-mA meter movement with a coil resistance of 46 Ω is to read 10 mA full scale, only 1/10 of the total current should pass through the meter. With $R_c = 46$ Ω, the value of R_{sh} should be $R_c/9$, or 5.1-Ω. Because the shunt resistance and the coil resistance are in parallel, adding the shunt reduces the effective meter resistance R_m to $R_cR_{sh}/(R_c + R_{sh})$. In this example, adding a 5.1 Ω shunt resistor would decrease R_m to 4.6 Ω. For higher current ranges, the value of R_{sh} can become impractically low. For these ranges it is common practice to connect a resistor in

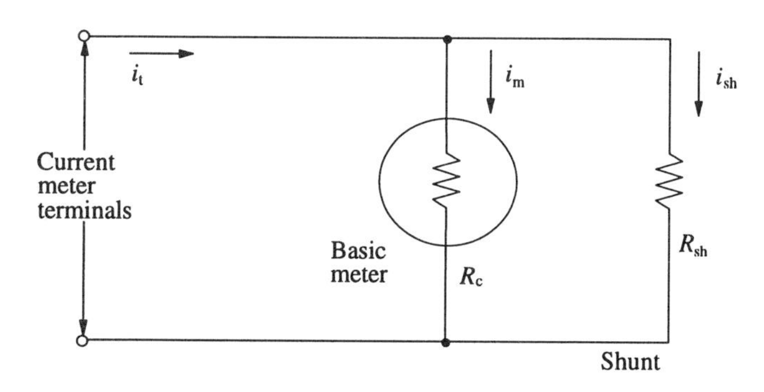

Figure 3–10. Current meter shunt for extended range. If R_{sh} is 1/9 of R_c, 9/10 of i_t will go through the shunt resistor and 1/10 will go through the meter coil. If a 1-mA meter movement has a resistance of 46 Ω, a shunt resistance R_{sh} of 5.1 Ω would allow the meter to read 10 mA full scale, a 10-fold increase in the basic range.

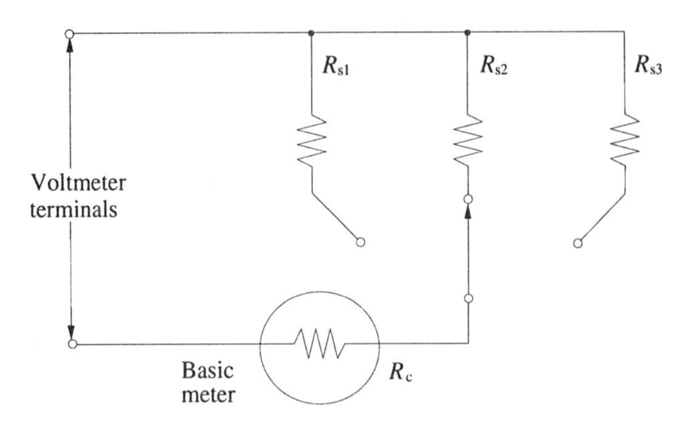

Figure 3–11. Voltage measurement with a current meter. Various values of R_s can be used to set the voltage required to produce the full-scale current in the meter. If a 50-μA meter movement has a resistance of 5000 Ω and it is desirable to measure 10 V full scale, the total meter resistance ($R_s + R_c$) would be $R_s + R_c = 10\ \text{V}/(50 \times 10^{-6}\ \text{A}) = 200\ \text{k}\Omega$. The series resistance R_s would then be 195 kΩ.

series with the meter coil to increase the effective coil resistance. A versatile current meter usually has several different full-scale current ranges accessible by switching.

Voltage measurements. The moving-coil meter is also readily adaptable for voltage measurements. Because the coil has constant resistance, the current through the meter is proportional to the voltage drop across it. The full-scale voltage sensitivity V_{fs} is the full-scale deflection current I_{fs} multiplied by the coil resistance R_c. To obtain multiple voltage ranges, it is necessary only to add series resistors to the meter circuit, as shown in Figure 3–11. With a series resistance R_s, the full-scale voltage sensitivity is $V_{fs} = I_{fs}(R_s + R_c)$.

The ratio $(R_s + R_c)/V_{fs}$ is a constant for a given meter movement. This constant is called the ohms-per-volt rating of the voltmeter. The resistance of the voltmeter on a given scale is the ohms-per-volt rating multiplied by the full-scale deflection voltage of that scale. Thus, a 20 000-Ω/V meter has a resistance of 20 kΩ on the 1-V full-scale range. The current sensitivity of a meter is the reciprocal of the ohms-per-volt rating.

Suppose that the voltage drop across the LED in the circuit of Figure 3–9a is to be measured with a voltmeter. The appropriate circuit is shown in Figure 3–12. In order that the meter not change the *IR* drop across the circuit element, the meter resistance must be much larger than the resistance of the element (*see* next section).

Analog multimeters. With the addition of a battery and range-determining resistors, the moving-coil meter can also be used to measure resistance. However, the resistance scales of such meters are nonlinear. Analog multimeters combine such measurement functions as current, voltage, and resistance in a single instrument. Many can also provide ac measurement functions. Another name for the moving-coil or analog multimeter is a **volt-ohm-milliammeter (VOM)**. DMMs

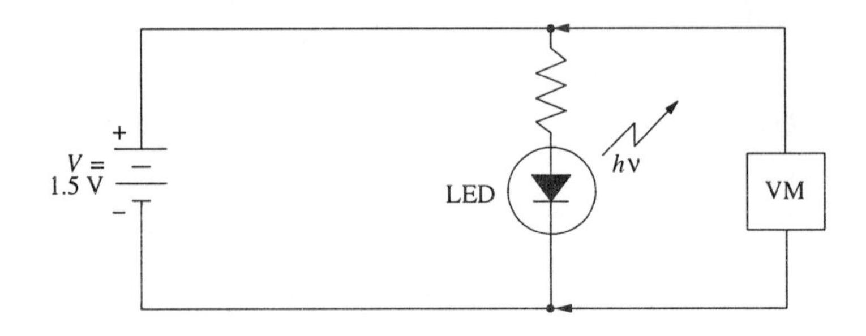

Figure 3–12. Measurement of voltage. To measure the voltage across circuit elements such as the LED and current-limiting resistance, the voltmeter (VM) is connected *in parallel* to ensure that the voltages across the meter and the circuit elements are identical.

have almost totally replaced these analog meters. The digital meters eliminate the reading error associated with observing the pointer against the calibrated scale. In addition, they can provide significant improvements in measurement accuracy.

Computer-Based Measurement Systems

Multimeter systems considerably more sophisticated than the general-purpose DMMs discussed previously are available. Bench-top units can have as many as 6½ digits of resolution and a basic accuracy of better than 0.01%. Many of the modern bench-top DMMs have an IEEE-488 interface. This allows external control of functions, ranges, and often reading rates. Calibration and diagnostic testing can also be accomplished under external control. In addition to allowing communication between the DMM and a computer, an IEEE-488 interface allows the DMM to communicate with printers, recorders, and many other devices.

Even more sophisticated DMMs have internal microprocessors and built-in mathematical capabilities. Such systems can add, subtract, multiply, and divide as well as store and compare acquired information. Measured values can be inserted into a calculation prior to the display or storage of results. With such systems, the user can preset HI- and LO-limit information for tolerance-testing purposes. All measurement results are then compared with these values, and the display indicates whether the test was passed or not. Such DMMs can also average repetitive measurements and display statistical information such as standard deviations or signal-to-noise ratios. Modern microprocessor-controlled DMMs also have an IEEE-488 interface that makes them fully programmable.

3–3 Assessing Errors in Analog Data Measurement and Transfer

Whenever a measuring device is connected to a circuit (e.g., a transducer circuit to measure light intensity) or the output of one circuit is attached to another (e.g., a transducer output is connected to an amplifier), a perturbation error is possible. That is, the connection of the measuring device or subsequent circuit perturbs the original circuit. Although this perturbation effect can never be totally eliminated without disconnecting the measurement device or ensuing circuit, it can be made negligibly small. The perturbation effects discussed here are among the most common sources of errors when instruments and computers are used.

Voltage Measurement and Transfer

An important theorem in electricity, **Thevenin's theorem**, states that *any voltage source composed of batteries and resistors that has two output terminals is equivalent to a single battery and a series resistor,* as shown in Figure 3–13a. The equivalent voltage V_e of the complex source is the voltage between the two terminals when no load is present (i.e., no device is connected between the terminals). However, to measure the voltage, the voltmeter is placed in the circuit as a load, as shown in Figure 3–13b.

Many of the voltage sources that we deal with have significant internal resistances. For example, glass–calomel electrode pairs used for measuring pH can have R_e values of 100 MΩ or more. Some voltage sources have very low

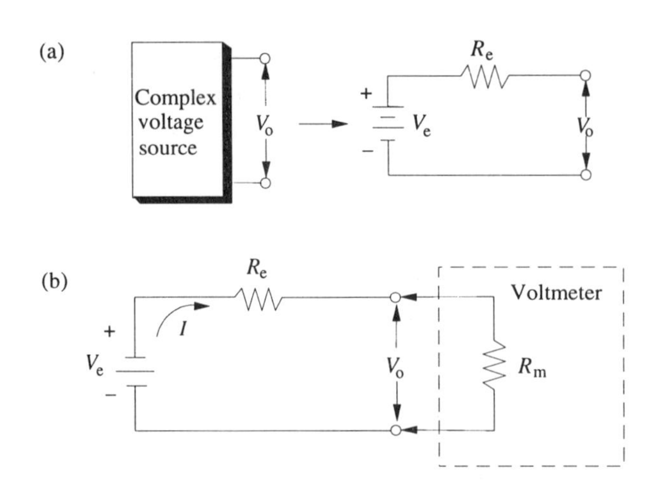

Figure 3–13. Voltage source equivalent circuit (a) and voltmeter connection (b). According to Thevenin's theorem, any complex voltage source behaves as though it were a "pure" voltage source delivering voltage V_e and a series resistor with resistance R_e. (a) In the absence of the voltmeter, the voltage at the source terminals V_o is equal to V_e, the no-load voltage. (b) When the meter is connected, the terminal voltage V_o measured by the meter is $V_e R_m/(R_e + R_m)$. The loading or perturbation error of the meter is negligible if R_m is very much larger than R_e.

internal resistances; batteries may have R_e values of fractions of ohms. When the meter is connected to the source as shown in Figure 3–13b, the circuit is completed and current is drawn. The current I in the circuit causes a voltage drop IR_e across the internal resistance of the source and a reduction in the voltage reaching the terminals from V_e to $V_e - IR_e$. The resulting circuit is a voltage divider, and the voltmeter will measure a voltage V_o equal to $V_eR_m/(R_e + R_m)$. Clearly, the meter resistance R_m must be large with respect to the internal resistance R_e if we are to read the correct no-load voltage. As a rule of thumb, to measure V_e with a maximum error of 1%, the meter resistance R_m must be $100R_e$. For a 0.1% relative error, R_m should be $1000R_e$. When the output voltage of a battery is measured, this requirement poses no problem, because even the least expensive meters have input resistances of thousands of ohms. However, to measure the output of a glass–calomel pH electrode pair, we need a meter of extraordinarily high resistance. Recall that typical DMMs have input resistances of 10 MΩ. High-input-resistance meters for pH measurements usually have an isolation or buffer amplifier to reduce the current drawn from the source. Simple operational amplifier buffers can increase the input resistance to 10^{12} Ω or more, as discussed in Chapter 5.

The equivalent voltage V_e and resistance R_e of a voltage source can often be determined from the circuit of the source, as shown for the voltage divider in Figure 3–14. The equivalent voltage V_e is the calculated no-load voltage (divider output in this case). The equivalent resistance R_e is the calculated resistance between the terminals when the voltage V is replaced by a short circuit. For the divider, this resistance is the parallel combination of R_1 and R_2.

The perturbation or loading error can also occur with a computer data acquisition system. In fact, the loading error has become more noticeable with high-resolution digital and computer readout systems. With a 4-digit display, for

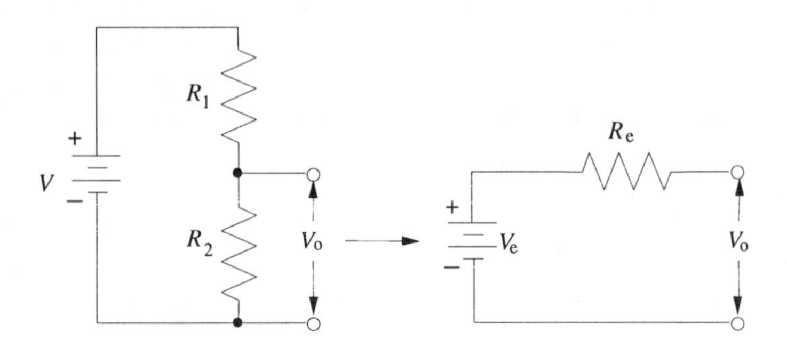

Figure 3–14. Thevenin's theorem applied to the voltage divider. V_e is the no-load divider output voltage, and R_e is the resistance between the terminals with the source V shorted. Therefore,

$$V_e = VR_2/(R_1 + R_2)$$

and

$$R_e = R_1R_2/(R_1 + R_2)$$

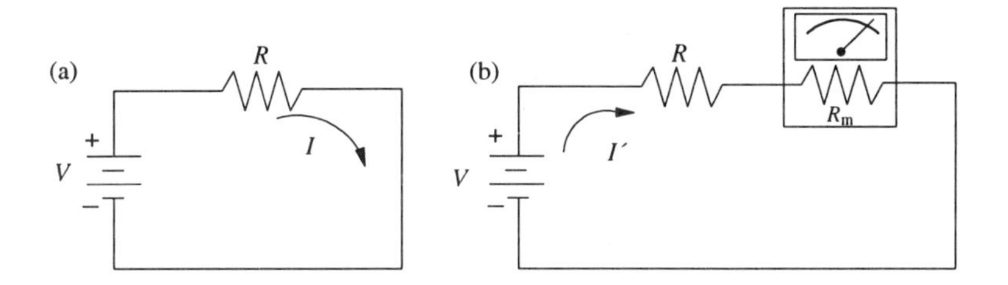

Figure 3–15. Current measurements. (a) Without the current meter, the circuit current is $I = V/R$. (b) With the current meter, the current is reduced to $I' = V/(R + R_m)$.

example, the loading error must be less than 0.01% in order not to affect the least significant digit. Analog meters, by contrast, have reading errors of about 1%.

The same considerations apply when voltage levels are being transferred from one circuit to a load. Thus, a voltage amplifier must have an input resistance much larger than the resistance of the source being amplified in order to avoid a significant perturbation error. When the signals involved are ac signals, the same considerations apply except that we must worry about the source and load *impedances* (*see* Supplement 2, Section S2–4). Again, for efficient voltage transfer, the load impedance should be large compared with the source impedance.

Current Measurement and Transfer

When currents are being measured, the considerations are quite different from those when voltage is being measured. Let us measure the current in the circuit of Figure 3–15a by inserting a current meter as shown in Figure 3–15b. The insertion of the meter changes the current from I to I' as shown. For this perturbation error to be negligibly small, the current meter resistance R_m must be much smaller than the circuit resistance R. Under these circumstances, $I' \approx I$.

This current perturbation error can be significant for multimeters based on simple current-to-voltage conversion and voltage measurement, such as the DVM with shunt network shown in Figure 3–5. With this DMM, the shunt resistance is 10 kΩ on the most sensitive current scale. Hence, a significant meter resistance would be inserted into the circuit being tested on this scale. More expensive current meters use operational amplifier current-to-voltage converters as described in Chapter 5. These can reduce the effective meter resistance to ohms or less while allowing the measurement of very small currents. A theorem analogous to Thevenin's theorem that applies to current sources is called **Norton's theorem**. It states that *any two-terminal circuit composed of batteries and resistors is equivalent to a single current source and a parallel resistance.* The Norton equivalent of a two-terminal network is shown in Figure 3–16a. This circuit is also the equivalent circuit of any real current source such as a current transducer. Just as

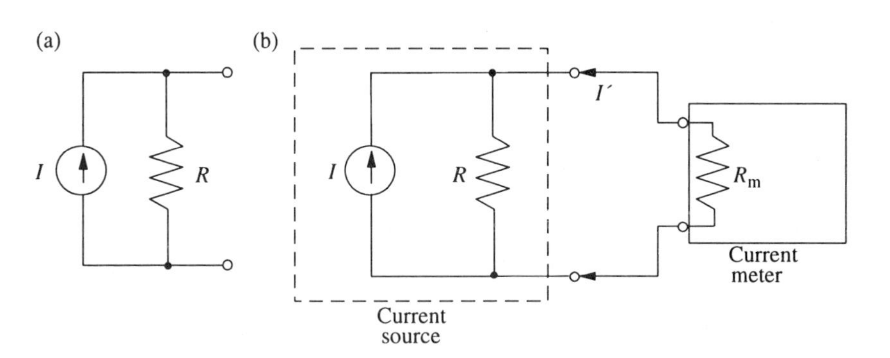

Figure 3–16. Current source equivalent circuit (a) and current meter connection (b). (a) With a short circuit at the output, the full current I is delivered between the terminals. (b) With a current meter load, the total current I is split between the internal source resistance R and the meter resistance R_m.

all real voltage sources have some internal series resistance, all real current sources have some internal parallel resistance. When a current meter is attached to a nonideal current source as shown in Figure 3–16b, only a fraction of the total current I goes through the meter, because the internal source resistance R and the meter resistance R_m form a current-splitting network. The current through the meter I' is $[R/(R + R_m)] \times I$. Hence, only when R_m is much smaller than R will I' be essentially equal to I. As a rule of thumb, R_m should be less than $0.01R$ to keep the perturbation error smaller than 1%.

Norton's theorem and Thevenin's theorem in fact provide equivalent ways of looking at two-terminal networks. The current I from the current generator in the Norton circuit is just V/R in the Thevenin circuit. Likewise, the series resistance R_e in the Thevenin circuit is just the parallel resistance R in the Norton circuit. The same network can readily be treated with either model with equal results. It is more convenient, however, to use Thevenin's theorem for voltage sources and transducers and Norton's theorem for current sources and transducers.

The same considerations apply when we are transferring current from a source to a load. To avoid attenuation of the source current, the resistance of the load must be small compared with the resistance of the current source. When the currents are ac, impedances instead of resistances must be considered. Again, we want the input impedance of the load to be much smaller than the output impedance of the current source. In ac circuits, loading can affect not only the amplitudes of signals but also their frequency characteristics. Input circuits to amplifiers and oscilloscopes, such as voltage dividers, can be made nearly frequency-independent as described in the next section.

Power Transfer

In power transfer, as opposed to current or voltage transfer, the goal is to deliver the maximum *power* to a load. It can readily be shown that maximum power transfer occurs when the source impedance is equal to the load impedance. This situation is familiar from the home stereo system, where the power amplifier output impedance is normally matched to the low-impedance load (typically 8 Ω) presented by the loudspeakers. The preamplifier, by contrast, is normally of high input impedance compared with typical source impedances (phono pickups or microphones).

Thus, for maximum power transfer efficiency, impedances must be matched, but for current or voltage transfer, large impedance mismatches lead to the highest accuracy. Because of these different characteristics, it is extremely important to know the type of source being measured, amplified, or otherwise treated.

3–4 Using the Oscilloscope To Observe Signals

The **oscilloscope** is an almost indispensable measurement and test instrument because of its ability to display rapidly changing voltage information as a function of time t or to plot one voltage x against another voltage y. Several types of oscilloscopes are available. The analog oscilloscope displays x–y or x–t plots directly on the face of a cathode ray tube (CRT). Some analog oscilloscopes are **storage oscilloscopes,** in which a special screen phosphor or a charged mesh behind the screen is used to maintain the intensity of the display. The **digital oscilloscope** is inherently a storage device in which the input signal is converted into a parallel digital signal, stored in memory, and then converted back into an analog signal for display on a conventional CRT. In addition to these, some oscilloscopes, called **sampling oscilloscopes,** can display extremely rapid, repetitive waveforms. Such oscilloscopes can extend the range of observations into the picosecond time regime. Some new oscilloscopes use a liquid crystal display instead of a CRT. These oscilloscopes will undoubtedly become widely used as their speed improves and their price decreases. Because the majority of oscilloscope types still use CRTs for display, this section begins with a brief discussion of the operating principles of the CRT.

The CRT

The CRT consists of an electron gun (Figure 3–17a) and deflection plates (Figure 3–17c) combined in a vacuum tube with a fluorescent screen on the enlarged end (Figure 3–17b). The electron gun provides a beam of electrons that is sharply focused at the fluorescent screen. The end of the tube is coated with various phosphors that emit visible radiation at the point of bombardment with electrons.

Intensity and focusing controls. The intensity of the visible light emitted by the screen phosphors is proportional to the number of electrons per unit time striking a given area of the screen. Electron flow is regulated by partially surrounding the cathode with a metal grid. A variable voltage applied between the grid and the cathode acts as the **intensity control**. As the grid–cathode potential is made

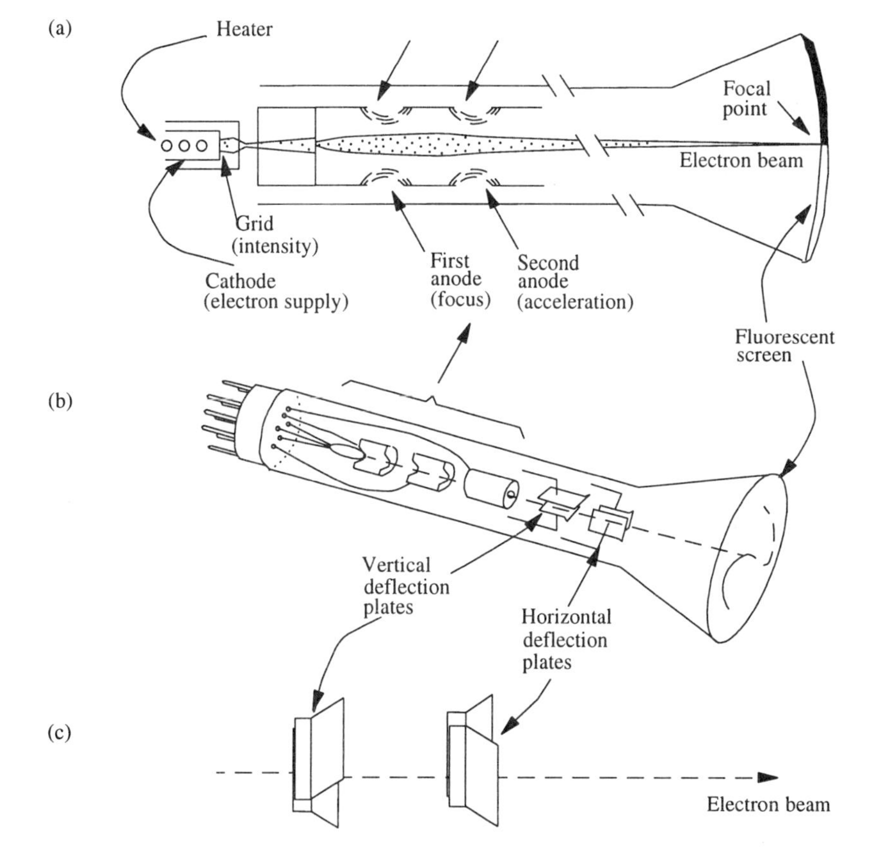

Figure 3–17. CRT. (a) Electron gun; (b) cutaway sketch; (c) deflection plates. A beam of electrons is generated by the electron gun and directed by the deflection plates to the desired positions on the fluorescent screen.

more negative, the repulsion of electrons is greater, the fraction that reaches the fluorescent screen is smaller, and the visible light spot is less intense.

To obtain a point source of electrons on the screen, it is necessary to focus and accelerate the electrons by two anodes that are at high positive voltages with respect to the cathode. For a typical 5-in. (ca. 13-cm) tube, the second anode is often 2000 to 10 000 V more positive than the cathode, and the first anode is about 350 to 750 V more positive. The electrostatic field between the two hollow cylindrical anodes provides the necessary focusing of the electron beam. The diverging electrons entering the first hollow anode are forced to converge because of the field between the first and second anodes. The desired focus point is, of course, the fluorescent screen, and the correct focus can be obtained by varying the voltage on one anode with respect to the other. The focus control is a front-panel adjustment on most oscilloscopes.

Electron beam deflection. After passing through the two anodes, the electron beam passes two sets of deflection plates, which are mounted as shown in Figure 3–18a. The first set of plates, mounted in the horizontal plane, are the **vertical deflection plates.** The plates mounted perpendicular to the first set are the **horizontal deflection plates.**

If one vertical deflection plate is made positive with respect to the other, the electron beam is deflected toward this positive plate, as illustrated in Figure 3–18b. The greater the applied voltage, the greater the deflection of the electron beam. Horizontal deflection of the beam can be accomplished similarly by using the horizontal deflection plates. If voltages are applied to both sets of plates simultaneously, the position of the electron beam on the CRT screen depends on the signs and magnitudes of the two deflection voltages. In this manner the electron beam can produce an *x–y* plot of one deflection voltage as a function of the other.

Analog Oscilloscopes

In addition to the CRT and the deflection plates, several other circuits are required to make up an oscilloscope, as shown in Figure 3–19. The voltage needed to deflect the electron beam is usually much higher than the measured voltage levels. For this reason, **deflection amplifiers** are used to increase the signal voltage before application of the voltage to the plates. The quality and type of an oscilloscope depend greatly on its amplifiers. The deflection amplifiers of modern oscilloscopes cover the range from dc to more than 100 MHz for general-purpose scopes or to beyond 1 GHz for special wide-band scopes. The internal sawtooth sweep gener-

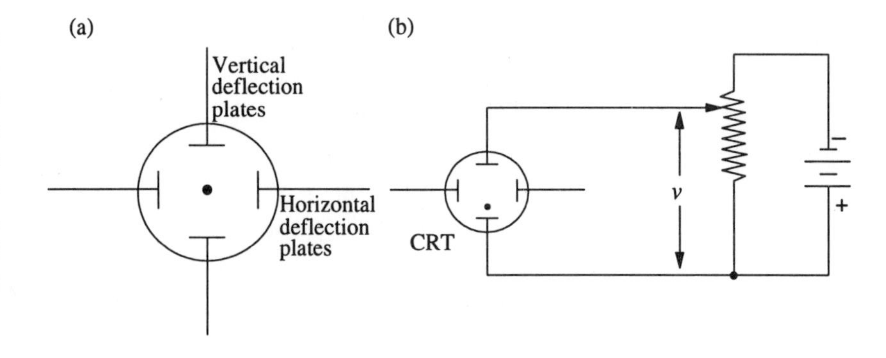

Figure 3–18. CRT deflection plates. (a) Schematic; (b) application of a deflection voltage to the vertical plates. In part a the spot in the center represents the image on the screen caused by the undeflected electron beam. In part b, as v is increased, the electron beam position is lowered on the CRT face.

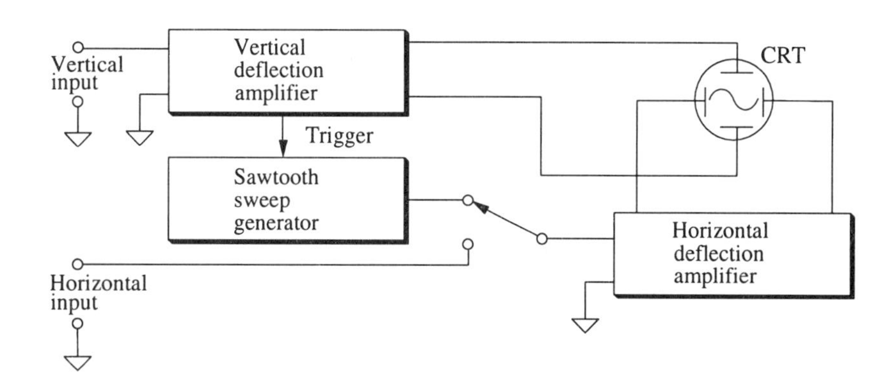

Figure 3–19. Block diagram of basic oscilloscope. The horizontal deflection amplifier can be connected to an external signal or to a sweep generator that causes the beam to move at a constant rate from left to right across the face of the oscilloscope.

ator described below is used to provide a horizontal deflection that varies linearly with time.

The input resistance of many general-purpose oscilloscopes is 1 MΩ. A probe with an internal resistance, as shown in Figure 3–20, is frequently used to increase the input resistance 10-fold. The two resistances R_p and R_m form a voltage divider, or attenuator, that decreases the maximum sensitivity of the oscilloscope 10-fold. Many oscilloscope probes have adjustable compensation capacitors built in to reduce the effect of scope input capacitance. The compensation capacitor with capacitance C is in parallel with the resistance R_p as shown in Supplement 2, Figure S2–14. The compensation network can form a frequency-independent voltage divider when the capacitor is properly adjusted. Extremely fast scopes often have 50-Ω input resistances to match the impedance of the coaxial cable connections used.

Linear sweep for horizontal deflection. For many applications of the oscilloscope, voltages are plotted on the vertical axis versus time on the horizontal axis. To provide a uniform time scale, a voltage that increases linearly with time is applied to the horizontal deflection plates. This voltage causes the light spot to move horizontally across the face of the tube at a uniform rate. Because the electron beam is "swept" across the screen at a constant rate, it is customary to refer to the waveform as the sweep. The oscillator that produces the sweep is called the sweep generator. The application of such a linear horizontal sweep is illustrated in Figure 3–21.

Good linearity of the sweep is a major requirement for a reliable oscilloscope. In most modern scopes the time base is calibrated in seconds per centimeter and is accurate to about 1 to 2%. The sweep is obtained by charging a capacitor with a constant-current source (*see* Chapter 5, Section 5–6). This process provides a

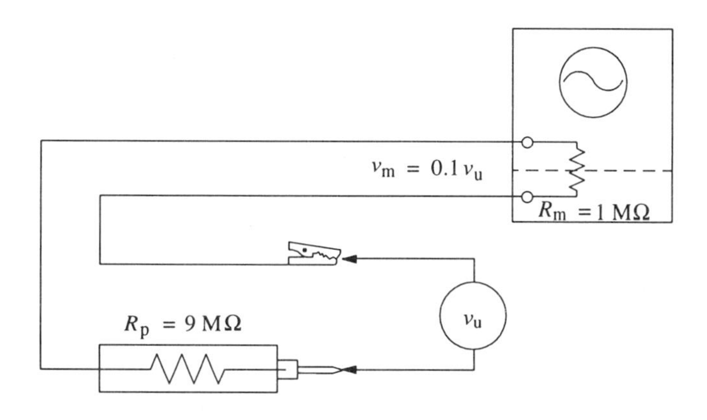

Figure 3–20. A 10× attenuator probe for an oscilloscope. Sensitivity is sacrificed for an increase in input resistance.

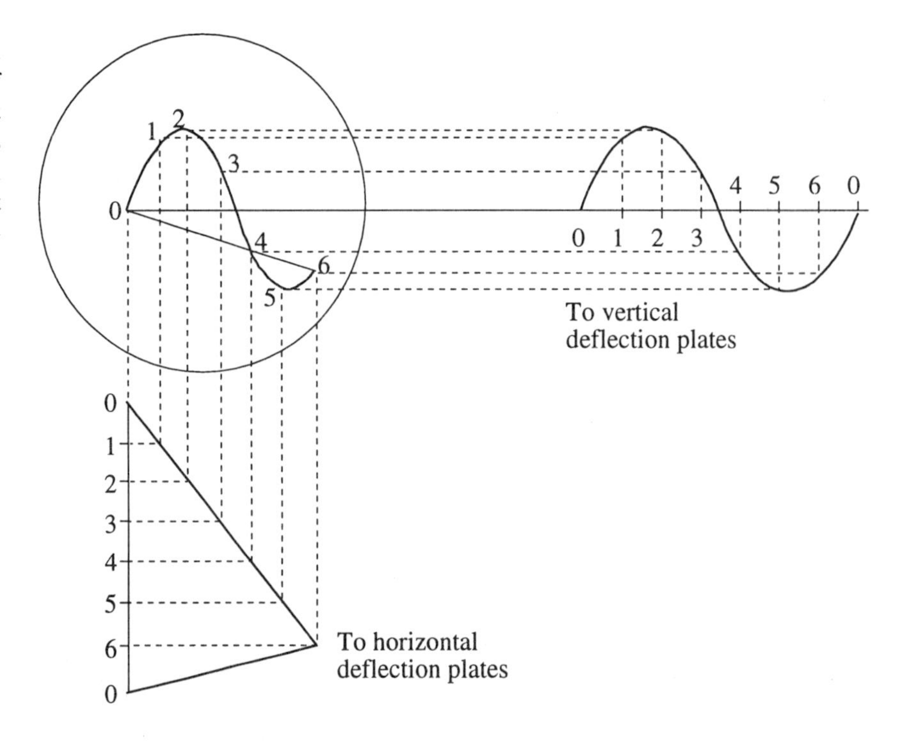

Figure 3–21. Linear horizontal sweep for displaying the plot of vertical deflection voltage against time. To observe waveforms as they are usually drawn, a signal that varies linearly with time must be applied to the horizontal deflection plates.

capacitor voltage that changes linearly with time. At the end of the sweep, the electron beam is moved back across the screen to the starting point by discharging the capacitor, which returns the sweep voltage to zero. Although this occurs very rapidly, a visible **retrace** or **flyback** trace of low intensity would be observed if it were not blanked by applying a negative grid voltage during the retrace time interval.

Triggered-sweep oscilloscope. To provide a continuous, stable display of a repetitive waveform signal on the oscilloscope screen, the starting time of the sweep must coincide with a single point on the waveform to be observed. For a triggered-sweep oscilloscope, achieving this coincidence involves using the input signal to start the sweep at the same point on the waveform for each sweep. The triggering feature allows considerable flexibility in choosing the display mode. For instance, it is possible to choose whether the time base is triggered on a positive or a negative slope of the input waveform. It is also possible to select the triggering level, that is, the sign and magnitude of the signal needed to trigger the sweep (*see* Chapter 5, Section 5–2, for a discussion of the circuits used for selecting trigger levels).

Other advantages of a triggered sweep are that the input signal need not be repetitive to be displayed (i.e., single events can be shown, because the event itself starts the sweep) and that a calibrated time base is provided, making it possible to perform time interval and frequency measurements. The scope displays for various triggering conditions are illustrated in Figure 3–22.

All trigger circuits require some time to operate. There is thus a time delay (usually nanoseconds) between the time that the input signal satisfies the trigger condition (level and slope) and the beginning of the sweep. To avoid the loss of information that could occur during this time, some oscilloscopes use a delay line between the output of the vertical amplifier and the vertical deflection plates. This

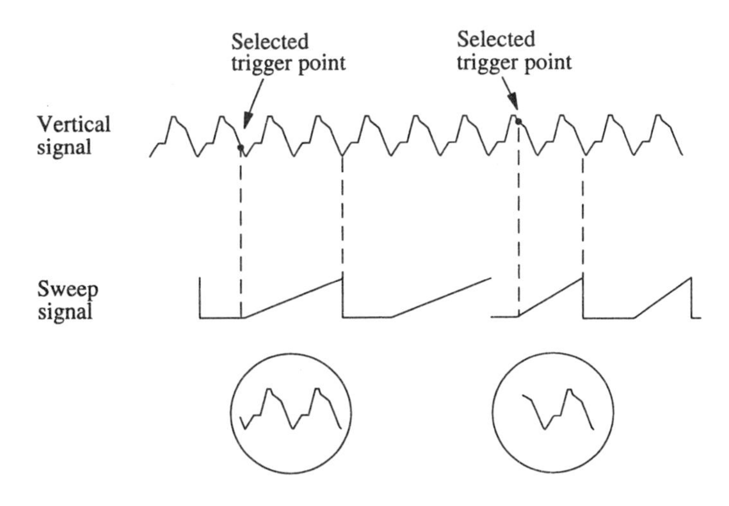

Figure 3–22. Triggered-sweep oscilloscope. Both the trigger point and the sweep rate (time window observed) can be independently selected.

line delays the application of the amplified input signal to the vertical plates until the triggering delay has been exceeded.

Some oscilloscopes have a **delayed-sweep** feature that creates a small time delay between the trigger point and the start of the horizontal sweep. This time delay can be varied by means of a front-panel control. With a delayed-sweep oscilloscope we can expand the time base to observe a small feature or section of a waveform that is of particular interest while triggering on a stable prior portion.

Dual-beam and dual-trace oscilloscopes. It is often desirable to display more than one signal versus time simultaneously. Dual-beam and dual-trace oscilloscopes allow two (or more) signals to be displayed simultaneously for comparison. The **dual-beam oscilloscope** has two separate electron guns and separate pairs of vertical deflection plates for each beam and is relatively expensive. A less expensive approach is time-sharing one electron gun and one pair of vertical deflection plates, as is done in the **dual-trace oscilloscope**.

The sharing of a single analog measurement or recording system by more than one input signal is accomplished by a single-switching method called **multiplexing** (Figure 3–23). Each input channel has its own preamplifier with separate range (gain) and offset (position) controls in order to allow the dual-trace display of widely different signal levels. The multiplexer is a two-position solid-state analog switch (*see* Chapter 5, Section 5–2) with four modes of control. In the Ch.A or Ch.B position, the multiplexer output is continuously connected to channel A or B, respectively. In the chop mode, the multiplexer switch alternates back and forth between channels A and B at a rate of 100 kHz to 1 MHz. With a horizontal sweep rate of 100 ms/cm or slower, the traces for both signals appear to be continuous. As the sweep rate is increased (less than 100 ms/cm), the chopping of each signal becomes apparent. At still faster sweeps, the chop mode becomes unusable. In the alternate mode, the channel selector is alternated after each sweep. Channel A is displayed for one entire sweep, channel B is displayed during the next sweep, and so on. This mode is practical when the observed signals are repetitive and when alternate sweeps are frequent enough to appear continuous to the eye. Sweep rates faster than 1 ms/cm generally repeat each trace 50 times or more per second. The retrace portion of the sweep signal is used to alternate the channel selector switch. With many modern oscilloscopes the position of the time base selector switch automatically governs whether the scope is in the chop or the alternate mode. For

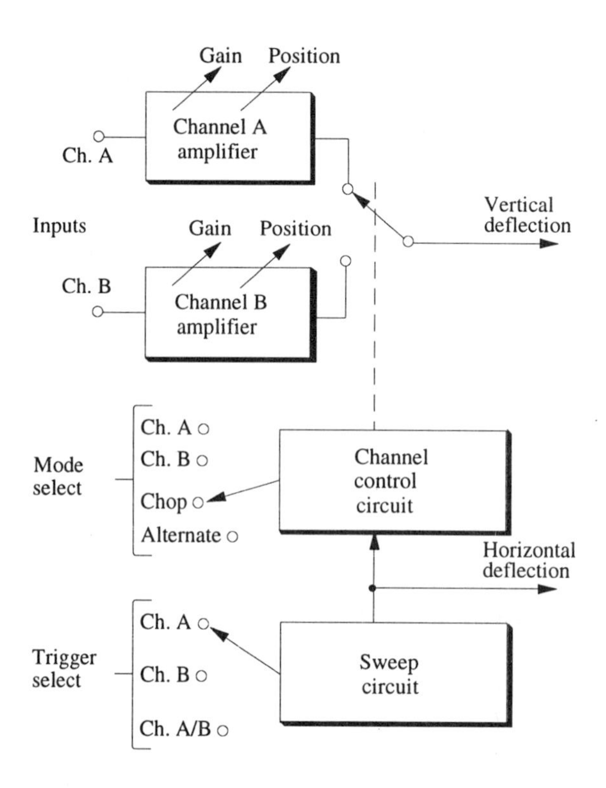

Figure 3–23. Input multiplexer for dual-trace oscilloscope. The rapid alternation of control of the vertical deflection between two signals (A and B) allows both signals to be observed on the CRT.

fast sweep rates, the alternate mode is used; for slow sweep rates, the chop mode is chosen.

The sweep trigger selector switch on a dual-trace oscilloscope provides for triggering from the channel A signal, the channel B signal, or the multiplexed signal (Ch. A/B). Selecting this last trigger source, which is used only in the alternate mode, allows each signal to determine its own trigger point, but the time correlation between the two traces is lost. Four-channel oscilloscopes have become more common in recent years, and models with eight channels are available. These multichannel models are just extensions of the two-channel multiplexer shown in Figure 3–23.

The x–y display. When two signals are connected to the horizontal and vertical inputs of an oscilloscope, the oscilloscope becomes an x–y plotter that displays the functional relationship between the two signals. Instead of plotting two signals against time, the oscilloscope plots the value of one signal against the value of the other as both signals vary with time. If two periodic signals are used and the time relationship of the two signals shifts, the pattern changes. Because the time–amplitude relationship of one signal is being plotted against that of the other, the resulting pattern must contain information about the time relationship between the two signals. These patterns, called **Lissajous figures**, are useful for phase angle and frequency ratio measurements.

The phase angle between two sine-wave signals may encode information. The form of the Lissajous display as a function of phase difference between signals is shown in Figure 3–24. The peak-to-peak amplitudes of the input signals can be obtained from the maximum horizontal and vertical excursions of the trace. For the equal-amplitude signals shown, the phase angle θ, the major measure of phase difference, can be determined as illustrated in Figure 3–25.

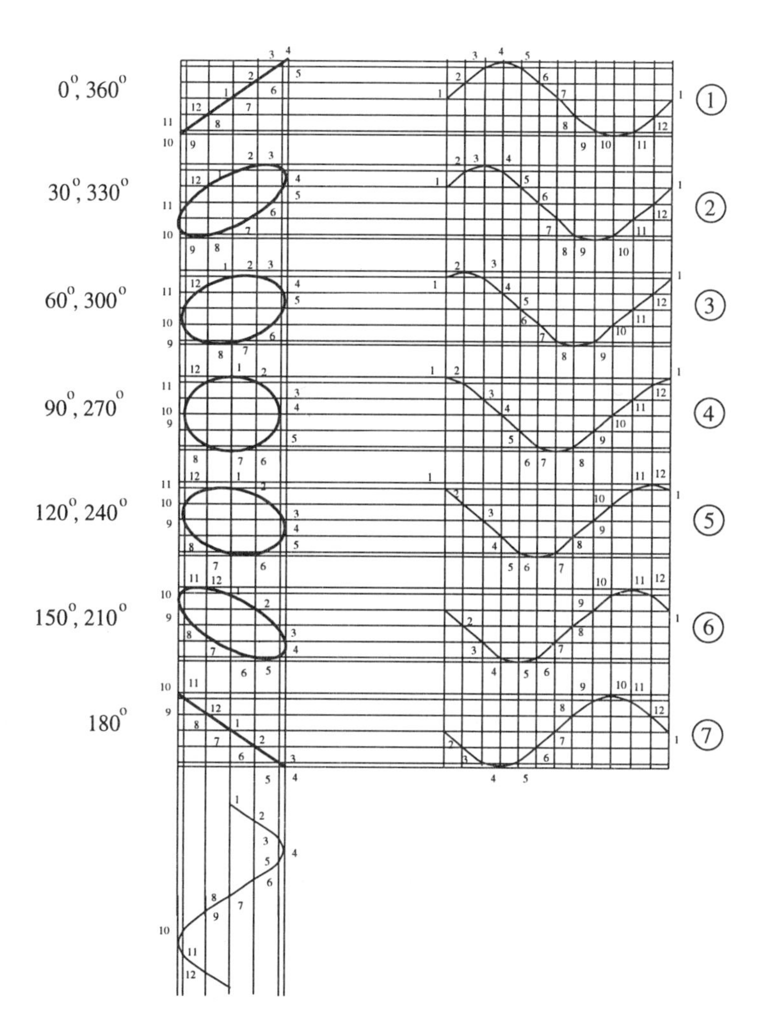

Figure 3–24. Lissajous patterns for different phase angles. The signals on both axes are sine waves of the same frequency. If the signals differ slightly in frequency, their relative phases change continuously, and the pattern seen will vary from that at the top to that at the bottom and back again.

If two signals with a frequency ratio that can be expressed as a small whole number or a simple fraction are applied to the horizontal and vertical inputs of the oscilloscope, patterns similar to those in Figure 3–26 are obtained. These patterns can be used to determine the frequency ratio of the two signals. For example, two signals of equal frequency give a circle, ellipse, or straight line, as shown in Figure 3–24. If the horizontal-to-vertical frequency ratio is 2:1, the figure eight pattern shown in Figure 3–26b results. Although the measurement of frequency ratio is readily accomplished with digital techniques, the oscilloscope method is still used

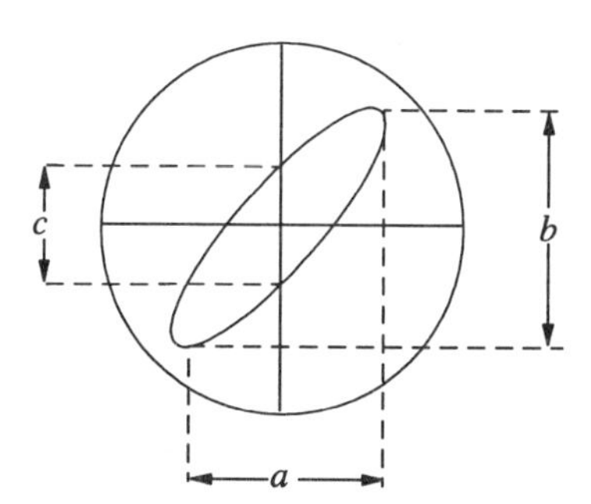

Figure 3–25. Determination of phase angle from Lissajous pattern. The values of b and c are determined from the graduations on the CRT, and the phase angle is calculated from $\sin \theta = c/b$.

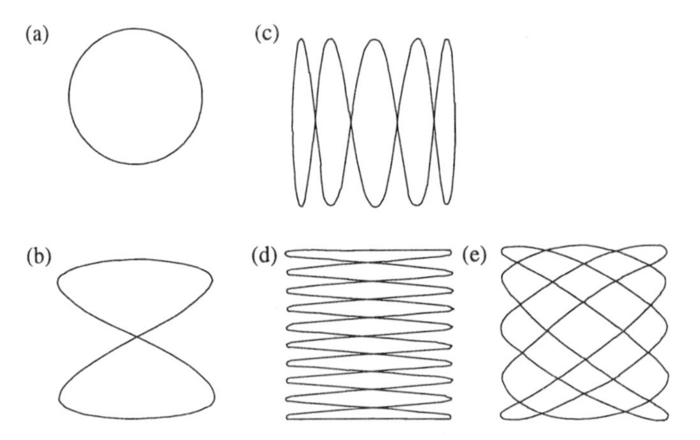

Figure 3–26. Lissajous figures for various horizontal-to-vertical frequency ratios: (a) 1:1; (b) 2:1; (c) 1:5; (d) 10:1; (e) 5:3. The frequency ratio is the ratio of the number of nodes along the horizontal extremity to the number of nodes on the vertical extremity.

because of its visual nature and because phase and amplitude information can be obtained simultaneously.

Coupling modes. Most oscilloscopes allow the input signal to be ac- or dc-coupled to the scope amplifier. In the **dc-coupled** mode, the input signal is connected directly to the vertical amplifier. In the **ac-coupled** mode, a capacitor is inserted between the scope input and the vertical amplifier. Because the capacitive reactance (*see* Supplement 2) is very high for dc components and very low for ac components, dc signals are effectively blocked by the capacitor. This ac coupling makes it possible to observe small ac fluctuations on top of a large dc level, as shown in Figure 3–27. The dc coupling mode should be used for signals with low-frequency components to avoid distortions.

Digital measurement options. Some high-quality analog oscilloscopes provide digital readout of voltage amplitudes and time values. With a DVM feature, the user can set two cursors on the screen in the horizontal plane to select the portion of the waveform to be quantified. A numerical display (often with the appropriate units) of the voltage difference between the cursors is presented. With a digital time measurement feature, the scope displays the time between two cursor positions in the vertical plane. Such measurement options eliminate the errors caused by limited screen resolution and the necessity to interpolate between screen markers.

Many of the newer oscilloscopes have IEEE-488 interfaces that allow the oscilloscope to be programmed externally. In addition, the positions of the front-panel controls and cursors can be read by a computer. Some systems allow waveforms to be transferred between oscilloscope and computer or vice versa.

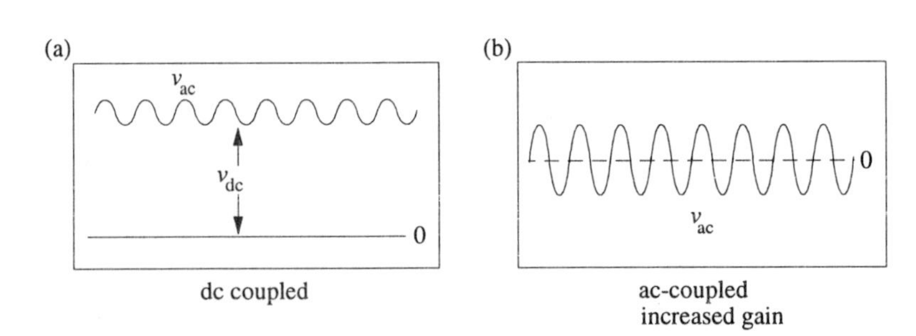

Figure 3–27. Comparison of dc and ac coupling for an oscilloscope. (a) With dc coupling, both the dc and the ac components of a signal can be observed. (b) With ac coupling, the dc component is blocked, and expanded-scale measurement of the small ac feature is possible.

With the connection of a printer or plotter on the IEEE bus, a hard copy of the oscilloscope display can be obtained.

Storage Oscilloscopes

With repetitive waveforms a stable pattern is achieved with an ordinary triggered-sweep oscilloscope, because each sweep reinforces the previous trace. Single events, however, may occur too rapidly to be viewed on the oscilloscope unless storage is provided. A storage oscilloscope retains the trace on the screen long enough to allow observation and measurements. Two types of storage oscilloscopes, the analog storage oscilloscope and the digital storage oscilloscope, are available.

Analog storage oscilloscopes. With analog storage oscilloscopes a special storage CRT is used to retain the waveform image. The bistable CRT uses a phosphor screen in which individual phosphor particles have two stable states, written or unwritten. With this type of scope, waveform images can be retained for several hours or until erased by the operator. Bistable displays are bright and long-lasting but of rather low contrast. Variable-persistence CRT displays are used when long storage periods are unnecessary. These CRTs have a storage mesh in front of the phosphor screen. The electron beam, controlled by the input signal, charges the storage mesh where it "writes." Flood guns in the CRT illuminate the phosphor where allowed by the storage mesh. With front-panel controls, the charge on the mesh can be varied to change the contrast between the trace and the background. The persistence can be adjusted so that the stored trace fades out as a new waveform is being stored, or the variable persistence can be used to provide integration so that only the coincident parts of a repetitive waveform are retained.

The fast-transfer storage CRT uses an intermediate mesh target to capture the waveform, which is then transferred to another mesh that is optimized for longer-term storage. This mode provides faster writing speeds than other types of storage CRTs. The second target can be of the bistable or variable-persistence variety.

With an analog oscilloscope, the image on the screen fades with time. Also, the brightness of the trace depends on the writing rate. These features make it difficult to display and store with uniform brightness a slow waveform with very fast transitions. The digital storage oscilloscope does not suffer from this intensity problem.

Digital storage oscilloscopes. A completely different principle is used with digital storage oscilloscopes. The input analog signal is digitized and stored in digital memory, and the contents of the memory are converted back into an analog domain for display on a conventional CRT. The digital scope is thus a combination of a digital data acquisition system and a waveform generator (*see* Chapter 2, Section 2–3). In addition, some digital scopes can computer process the acquired data to provide waveform analysis, signal averaging, frequency spectrum display, transformation of coordinates, or other functions.

Figure 3–28 is a block diagram of a basic digital storage scope. Most digital storage scopes provide a numerical display of waveform amplitudes and time values. As with an analog scope with digital measurement options, the user adjusts cursors to select the portion of the waveform to be measured. With the digital scope the data display rate is independent of the data acquisition rate. This feature allows

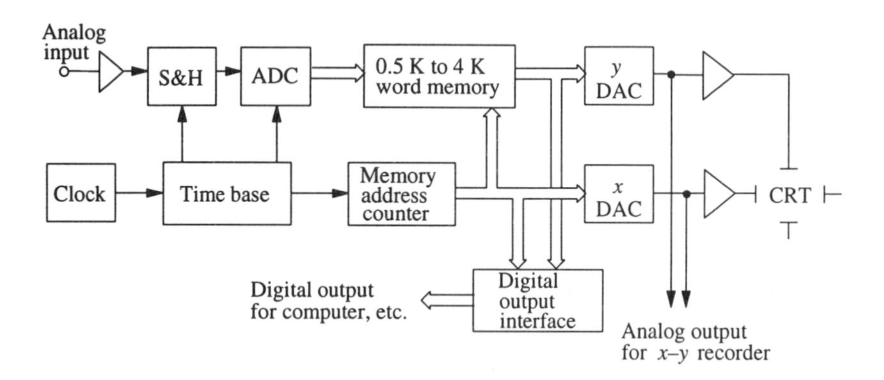

Figure 3–28. Block diagram of a digital storage oscilloscope. The analog input signal is digitized and stored in a solid-state memory at a rate determined by the time base settings. The memory contents are then read out in order to the *y* DAC (digital-to-analog converter) for display on the CRT or (more slowly) for the external recorder output. The readout can be repeated indefinitely. The *x*-axis deflection signal is obtained by analog conversion of the memory address. Data acquisition in the digital oscilloscope is synchronized with the input signal waveform by using the trigger signal to control the memory address counter. If the counter is running continuously before the trigger and proceeds less than a full count after the trigger, the memory will contain a pretrigger portion of the waveform. This unique ability to trigger *after* the event can be very useful. S&H, sample and hold.

horizontal expansion of the trace after acquisition, an operation impossible to achieve for a transient signal with an analog scope.

In contrast to the analog storage oscilloscope, digital oscilloscopes allow indefinite retention of data. The sampling theorem discussed in Chapter 2, Section 2–3, states that we must sample at twice the the rate of the highest frequency component in the signal to avoid errors. Typical digital storage scopes have bandwidths in the 50- to 100-MHz region, although more expensive units can achieve bandwidths of 500 MHz. The fastest storage scopes use a hybrid analog/digital technology. Analog scopes without storage capabilities are available with bandwidths of greater than 1 GHz. Equivalent time sampling, as discussed below, can extend the upper frequency range on a digital scope for repetitive signals.

Several manufacturers provide hardware and/or software that allows a personal computer to emulate a storage oscilloscope. The hardware consists of a data acquisition (analog input/output) board for digitizing and displaying the input waveform. Typically, the software allows display of waveforms and measurement parameters as well as disk storage of measurement information. Many packages also have such advanced features as zooming between cursor points; averaging waveforms; calculation of maximum, minimum, and average values; curve fitting; and calculation and display of Fourier transforms.

At present these personal-computer-based systems are slower and less versatile than dedicated oscilloscopes. The power of the computer, however, provides some extremely useful advanced features. In the near future we will certainly see more computing power being incorporated into digital oscilloscopes as well as new features being added to personal-computer-based systems.

Sampling Oscilloscopes

Quite often we want to measure a signal waveform whose rate of change is faster than the response speed of an ordinary oscilloscope. For example, we might want to record the shape of a pulsed-laser output that lasts less than 1 ns. If the signal is or can be made to be repetitive, data points sampled from many repetitions of the waveform can be combined to produce a reasonable recording of the waveform. The relatively simple device that accomplishes this is shown in Figure 3–29.

The repetitive input signal is applied to the input of a sample-and-hold circuit (*see* Chapter 2, Section 2–3, and Chapter 7, Section 7–2) and to the trigger input of a fast-sweep generator. The generator is adjusted to be triggered by the signal feature to be recorded and is set to complete its sweep in the period of interest after

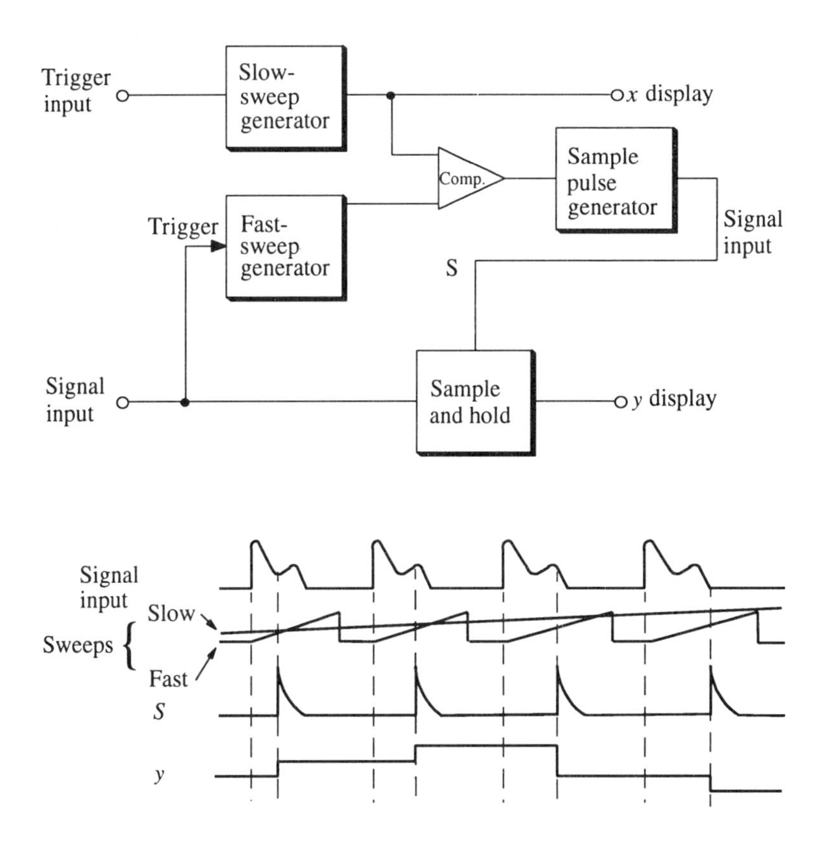

Figure 3–29. Equivalent time converter for a sampling oscilloscope. A sample is acquired whenever the slow- and fast-sweep generators are of equal voltage as determined by a comparator (Comp.) circuit (*see* Chapter 5, Section 5–2). One sample is thus obtained each time the fast sweep is triggered. Because the sampling instant occurs a little later in the fast sweep each time, the sampled point is taken at increasing times in the signal waveform. The waveforms show that plotting the sampled amplitude against the sampling delay time gives points on the original waveform.

the trigger. A sweep generator that is slow enough for the oscilloscope is also started. The time between the fast-sweep trigger and the sampling instant is called the **sampling delay time.** The amplitude of the slow sweep is proportional to the sampling delay time, and thus, an equivalent time sweep signal is provided. The ratio of equivalent time to real time is equal to the ratio of the slow-sweep rate to the fast-sweep rate. The number of data points recorded in a single slow sweep is equal to the number of fast-sweep triggers that occur in that time. The sampling operation can extend bandwidths to 15 GHz or more and thus allow observations of phenomena on the picosecond time scale.

Related Topics in Other Media

Video, Segment 3

- Use of voltage divider for signal attenuation
- Loading errors caused by meters
- Operation and use of dual-trace oscilloscope
- Use of digital oscilloscope in Fourier waveform analysis

Laboratory Electronics Kit and Manual

- Section B
    - ¤ Measuring resistances and studying series circuits and voltage dividers
    - ¤ Measuring loading errors

- Appendix 1
    - ¤ Familiarization with oscilloscopes

Chapter 4

Supplying Power to Instruments and Computers

Whether we are using a microcomputer, watching a TV program, or operating an electronic device or instrument in the home, office, or laboratory, we depend on the reliable operation of power supplies. True, most instruments that we use, such as consumer products or scientific instruments, are probably "turnkey" systems. That is, we plug them in and turn the key or flip a switch, and the systems work like the manufacturer said they would. Or do they? Are there peripheral devices you want to include in your system? Could you arrange to run the system from your car battery? Will the system work in another country? Are there modifications you'd like to make? These questions and more require at least a basic understanding of power supplies, because all electronic circuits require power to operate.

For those who build up their computer or instrument systems from functional modules, power supplies with the right characteristics must be chosen. For those who use laptop or notebook computers or other types of compact portable instruments, power supplies must be recharged occasionally and maintained in operating condition. Thus, if we're using or adapting a turn-key system, or building our own system with functional modules or maintaining a limited-life supply, we should understand power supplies. Let's see what some of the considerations might be.

4–1 Characterizing Power Supplies

In general, regulated, noise-free dc voltages (*see* Note 4–1) provided by power supplies are necessary for electronic equipment to operate. The desired design features of a computer-controlled instrument system or other electronic device determine the requirements and often the kind of power supply to be used. The types of dc power supplies used to fulfill specific design criteria for the required dc voltages include nonrechargeable and rechargeable batteries, dc voltage regulators, and various systems for converting from one dc voltage to another dc voltage (dc to dc) or from an ac voltage to a dc voltage (ac to dc). It is also necessary to invert dc voltages to ac voltages (dc to ac) in certain applications. These types of conversions are briefly described in this section, and they are illustrated in Figure 4–1. They are investigated more thoroughly in Sections 4–2 through 4–6.

Note 4–1. ac and dc Signals

In electronics it is useful to define an **ac signal** (ac current or ac voltage) as that part of the signal that varies with time. The average value about which the ac signal varies is called the **dc signal** (dc current or dc voltage; *see* Supplement 1, Figure S1–5, and Supplement 2, Figure S2–3). A regulated dc voltage remains constant as a function of time and has negligible ac voltage superimposed.

Batteries

For a small calculator or quartz watch, a small nonrechargeable battery is a logical choice. The lifetime of the battery is usually one or more years, and the infrequent replacement of the battery causes no problem. With more complex electronic systems such as laptop computers and portable field test pH meters, which require more power, it is necessary to use rechargeable battery supplies together with a power management protocol so as to conserve and recharge the batteries. These supplies must be compact, lightweight, efficient, cost-effective, and reliable.

2861-2/94/0085$11.20/1 Published 1994 American Chemical Society

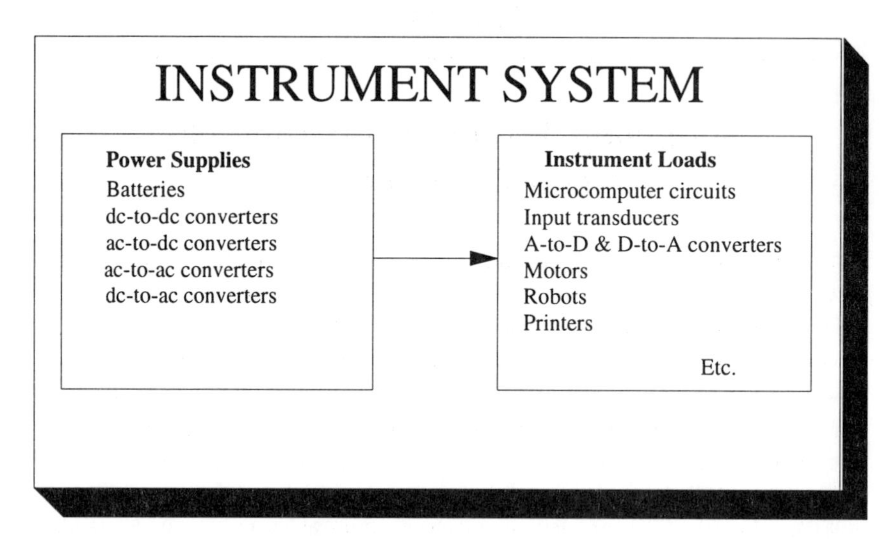

Figure 4–1. Types of power supplies used to operate various instrument loads. A, analog; D, digital.

Several types of battery power supplies meet these requirements, but none are ideal. All are subject to maintenance and require some type of power management system, but the portability justifies this inconvenience. We will therefore consider the characteristics and limitations of these types in Section 4–2.

Conversion

To obtain the regulated dc voltages required for many electronic circuits, it is usually important to use a dc-to-dc converter and regulator. These systems accept battery inputs over a wide range from peak charge to discharge and provide stable regulated output voltages. Recent advances in battery technology and associated dc-to-dc converters have increased the number of portable systems. Improved portability will greatly influence the next generation of computer-based instruments. Therefore, we will investigate the characteristics of these dc-to-dc power supplies in Section 4–3.

dc-to-ac Inversion

If the only available power source is a battery or other dc voltage source and if various devices require ac voltages for operation, then some type of dc-to-ac inverter is necessary, as discussed in Section 4–6. Also, a dc-to-ac inverter is used in the switch mode dc power supply, which is one of the most desirable supply systems for modern electronic instruments such as microcomputers (*see* Section 4–3).

ac-to-dc Conversion

At present, most electronic instruments, including measurement and control systems, are designed to operate from the conventional ac electrical power fed into outlets such as wall sockets in buildings. In the United States, power lines into buildings generally provide either 115- or 230-V, 60-Hz ac. Line voltage is the root mean square (rms) voltage available at the power socket. In many other countries the line voltage is 240 V, 50 Hz, so equipment is often not usable from one country to another without some modification. Two major types of ac-to-dc conversion systems will be discussed in Section 4–4.

Although ac-to-dc power supplies are generally reliable, we should know some things about them in order to prevent problems and to troubleshoot the problems that develop. From the time we pick up the instrument line cord to plug it into the wall socket, it helps to be aware of certain features of the power supplies. The question of the ac line voltage and frequency is usually made obvious by the type of socket and plug, but these features cannot be ignored. Some computers and other instruments have an internal switch that allows operation on either 115- or 240-V ac line voltage. If the switch is in the wrong position for ac input, there will be serious damage to the instrument. In some countries the type of socket is not an assurance of the voltage and frequency available, and serious damage results if the socket is misinterpreted.

Many instruments can also be damaged by high-voltage transients and other noise fed through the power lines, and data can be lost by a drop in supply voltage. Thus, it is important to know whether some additional protection device is necessary between the ac outlet and the instrument plug or within the instrument. From the input connection, to the connection of functional modules, to power management, to connection of test instruments for tracing of power supply troubles, there should be a concern for "making the right connections."

In this chapter, various power supply types, parameters, principles, and components are discussed to provide a guide for making the right connections.

4–2 Supplying Power with Batteries

Batteries convert chemical energy into electrical energy and consist of electrochemical cells in series to provide the necessary voltage. Each cell consists of an electrode material that readily gives up electrons (such as zinc, lead, or cadmium) and an electrode material that readily accepts electrons (such as carbon–manganese dioxide, lead dioxide, and nickel–nickel hydroxide) separated by an electrolyte that forms an ionic path to complete the electrical circuit. There are two classifications of batteries: primary (nonrechargeable) and secondary (rechargeable).

Primary Batteries

Primary batteries or cells have only a one-time use because they are nonrechargeable. They are used where the discharge rate or "power drain" by the device (load) is relatively low. The carbon–zinc cells are familiar because of their application in flashlights and toys. Because each cell has only about a 1.5-V output, several cells are frequently used in series to obtain the desired voltage and power. The most common dry batteries have voltages of 1.5, 3, 6, 9, 22.5, 45, 67.5, and 90 V. During discharge, the zinc metal of the negative electrode (in the shape of a can) is converted to a zinc salt in the electrolyte, leaving two electrons in the zinc can for each atom of zinc dissolved. Manganese dioxide (MnO_2) is reduced at the positive carbon rod. This chemical action establishes a voltage of 1.5 to 1.6 V between the two electrodes.

The service life of a battery is the number of hours that a fresh battery will satisfactorily operate the device under normal operating conditions. This life can vary widely depending on the quality of the battery, the length of time it was stored before use, the temperature during storage, the rate of discharge, the number and duration of the off periods, the temperature during discharge, and the lowest voltage for satisfactory operation of the circuit.

Secondary Batteries

The lead–acid storage battery is perhaps the most familiar example of a secondary (rechargeable) battery because of its use in automobiles. Several important features of the lead–acid (or any) rechargeable battery are readily recognized by most drivers. If you don't turn off your car lights when you park your car for a few hours, you will probably have to call for help to start the car when you return, because even with a rather large battery, the service life before discharge is relatively short when the power drain is high. However, when the battery charger is operating with the engine running, you can operate the lights, stereo, and other electrical equipment in the car without concern about loss of battery power. The decrease of battery power on a very cold day, as indicated by a reluctance of available battery power to turn on the engine, illustrates the dependence of battery operation on temperature.

Lead–acid battery. The chemical reaction for the lead–acid battery is

$$\underset{+\text{ plate}}{PbO_2} + 2H_2SO_4 + \underset{-\text{ plate}}{Pb} \rightleftarrows \underset{+\text{ plate}}{PbSO_4} + 2H_2O + \underset{-\text{ plate}}{PbSO_4}$$

Each cell consists of a lead electrode (negative plate) and a lead dioxide electrode (positive plate), each with various alloys in the grid of the plate, and sulfuric acid electrolyte. Microporous rubber separators are between the plates. The surface area and the number of parallel positive plates determine the available power. During discharge, sulfuric acid is converted to water. When a reverse, or charging, current is forced through the cell, lead sulfate is converted back to lead and lead dioxide, and the cell returns to nearly its original state. Dilution of the sulfuric acid causes the cell voltage to decrease during discharge. The decrease is slow at first but becomes quite rapid during the last one-third of the battery's service life.

Each cell has a nominal value of 2 V (fully charged, about 2.1 V). A 12-V battery has six cells in series and is about 12.6 V when fully charged. The energy available from a fully charged battery is referred to as "battery capacity" and is indicated as ampere–hours (A-h). The capacity depends on discharge current, temperature, and final acceptable cutoff voltage as well as the state of charge at the time of use.

Nickel–cadmium rechargeable battery. In recent years the nickel–cadmium (Ni–Cd) battery has come into widespread use. One form of this cell, which requires little attention other than recharging as required to maintain the necessary voltage level, is shown in Figure 4–2a. Under discharge, the cadmium is oxidized and supplies electrons at the negative electrode. At the positive electrode, nickel oxide is reduced to a lower oxidation state by accepting electrons. The elec-

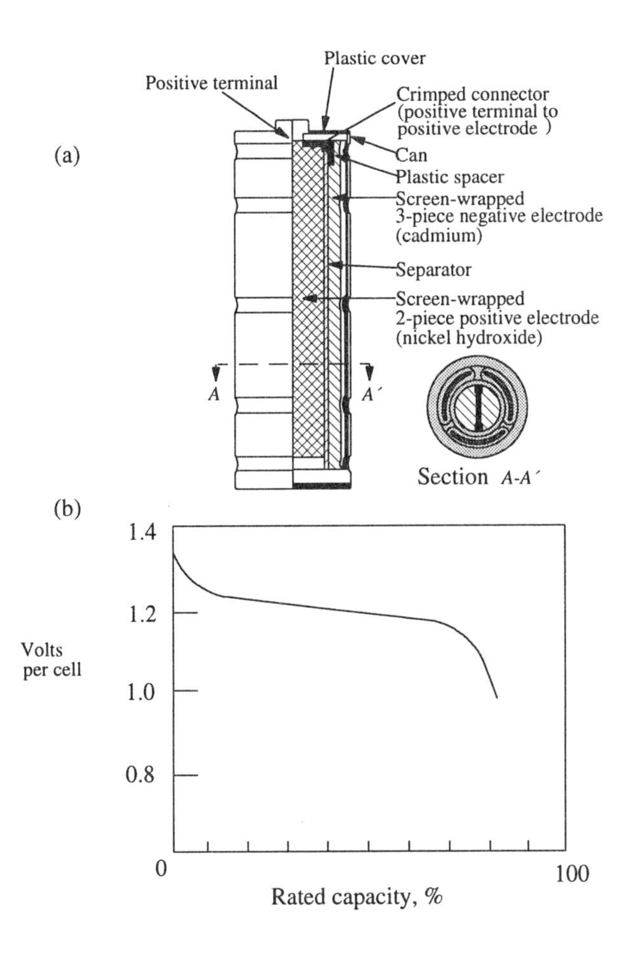

Figure 4–2. Nickel–cadmium cell (a) and typical discharge curve (b). The open-circuit voltage of this cell is about 1.3 V.

trochemical reactions provide an open-circuit voltage of about 1.3 V per cell. During the charging process the nickel oxide is reoxidized to its higher oxidation state and the cadmium oxide is reduced.

Nickel–cadmium batteries offer the advantages of high current capability, long service life, reasonably constant voltage, low internal resistance, long storage life, and wide operating temperature range (–40 to +65 °C). Ni–Cd batteries are especially suited to instruments with heavy power drain. A typical discharge curve for a Ni–Cd cell is shown in Figure 4–2b. The cell voltage is typically 1.4 to 1.5 V in the fully charged or "float" mode. The cell voltage drops from about 1.3 to 1.2 V over a long period of discharge. Then, as discharge continues, the cell voltage decreases abruptly and should be discontinued when the cell voltage reaches 1 V.

Power management. The recharging or replacement of batteries requires that specific procedures be followed so that the associated equipment operates when needed. For example, the Ni–Cd battery is used in popular laptop and notebook microcomputers, and the battery pack requires frequent recharging. It is necessary to follow the manufacturer's recharging protocol in order to realize the longest possible battery life. If the Ni–Cd battery is recharged before it is fully discharged, a memory effect shortens the battery's useful life.

For various types of pocket calculators and phone dialers with memory pads, the stored information will be lost unless directions for replacing the batteries are followed. In one type, the battery must be replaced within a time limit, such as

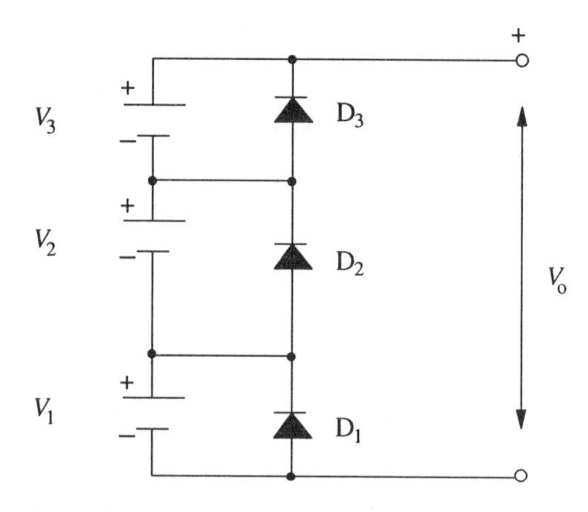

Figure 4–3. Battery replacement system for a memory calculator. Each cell may be replaced without disrupting delivery of input power from the other two. The diodes D_1, D_2, and D_3 do not conduct when the cells are in the circuit because they are reverse-biased. However, when one cell is removed, the diode across it conducts because it is now forward-biased, and it maintains a complete circuit to provide the required output for operation of the memory.

30 s. This type depends on a capacitor to hold enough charge so that the required voltage for the electronic circuits is maintained while the battery is being changed.

Another type of memory calculator battery supply has three individual cells in a circuit arrangement that allows one cell at a time to be removed and replaced. This simple circuit is illustrated in Figure 4–3. The three cells, supplying voltages V_1, V_2, and V_3, are in series, and each has a semiconductor diode (D_1, D_2, and D_3, respectively) in parallel. These diodes do not conduct when the cells are connected because they are reverse-biased. (A review of semiconductor devices is presented in Supplement 3, and a brief description of diodes is included in the next section.) When a cell is removed, the diode across it can conduct because it is now forward-biased. Thus, when one cell is removed, the output voltage is the sum of the voltages of the other two cells minus the voltage drop across the diode of the cell being replaced. The diode voltage drop is 0.6 V, but the available voltage from the two cells is sufficient to operate the electronic circuitry while each cell is replaced.

4–3 Regulating dc Voltages (dc-to-dc Conversion)

Modern integrated circuits require regulated dc voltages. These can be provided in several ways by controlling unregulated dc supplies. Any power supply can be represented by the Thevenin equivalent circuit shown in Figure 4–4, with the load resistance R_L connected at its output. It can be seen in this figure that

$$v_o = V - iR_S \tag{4–1}$$

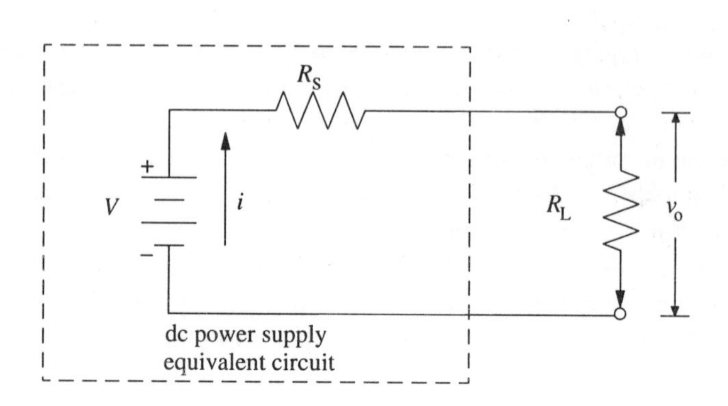

Figure 4–4. Power supply equivalent circuit. The voltage v_o across the load is $v_o = V - iR_S$. Any change in the current i or voltage V changes v_o.

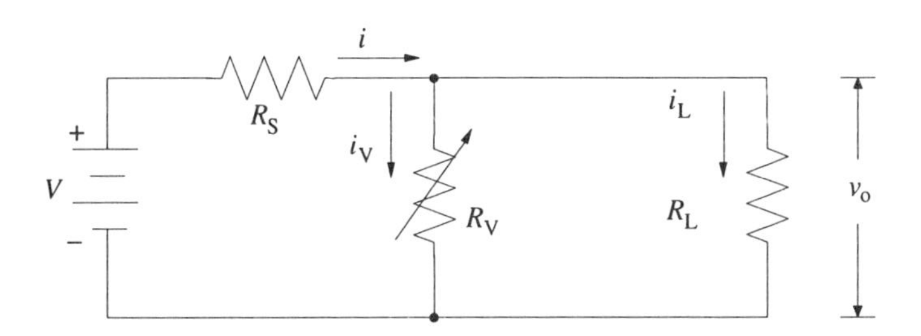

Figure 4–5. Shunt regulation of dc output voltage. In the shunt regulator, a part i_L of the total supply current i goes through the load resistance R_L, and a part i_V goes through a variable resistance R_V. Because $v_o = V - iR_S$ and because $i = i_L (R_V + R_L)/R_V$, it follows that

$$v_o = V - i_L \left(\frac{R_V + R_L}{R_V} \right) R_S$$

If the load changes, the current changes, and the effect on voltage v_o depends on the relative values of R_L and R_S. To regulate fully (to control) the dc output voltage v_o, it is necessary to compensate for changes in the unregulated V and in load current. Usually, this is accomplished by using the shunt regulator, linear (series) regulator, or switch-mode regulator. Each has certain advantages and disadvantages.

Shunt Regulator

The shunt regulator is illustrated in Figure 4–5, where it is shown that

$$v_o = V - i_L \left[\frac{R_V + R_L}{R_V} \right] R_S \tag{4–2}$$

Thus, as the load increases (i.e., R_L decreases and i_L increases), the variable resistance R_V should automatically increase so that the output voltage v_o remains constant. If the unregulated supply voltage V increases, the value of R_V should decrease so that the second term in equation 4–2 increases and the output voltage is kept constant. The reverse would be true for decreases in load or supply voltage. A device that could effectively vary R_V in such a way as to hold v_o constant at all times is a valuable regulator. This is the role of a Zener semiconductor diode.

Before introducing the Zener diode regulator, a brief discussion of semiconductor diodes and Zener diodes is necessary. Details about the pn junction diode are reviewed in Supplement 3.

Semiconductor diodes. A pn junction is formed in a semiconductor crystal at every interface between p-doped and n-doped regions. The holes from the p region and the electrons from the n region are free to cross the junction. The transport of positive charge into the n region and negative charge into the p region develops a potential difference between the n and p regions. Thus, an equilibrium condition in which there is no net flow of charge across the boundary is established. A **contact potential** results.

The depletion of holes and electrons in the junction region causes this **depletion region** to be essentially nonconducting compared with the remainder of the n- and p-doped regions. A contact potential, illustrated schematically in Figure 4–6a, appears across this insulating pn junction region and is about 0.6 V for silicon. The symbol for the pn junction diode is shown in Figure 4–6b.

When a voltage source is connected across the pn junction diode so that the negative terminal is connected to the anode and the positive terminal is connected

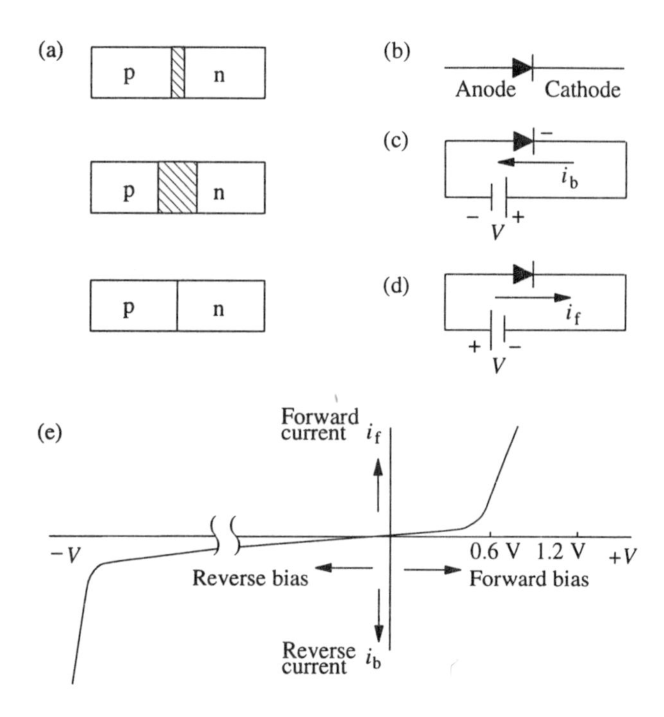

Figure 4–6. Silicon semiconductor diode. (a) pn junction schematic; (b) diode symbol; (c) reverse-biased and (d) forward-biased configurations; (e) plot of current through the diode for forward and reverse applied voltages. The forward current increases rapidly as the forward bias is increased above the 0.6-V junction potential. The reverse current is extremely small (about 10 nA) until a reverse breakdown potential is reached.

to the cathode, the diode is said to be **reverse-biased** (Figure 4–6c). Holes in the p-type material and electrons in the n-type material move away from the junction. This increases the thickness of the depletion region, which results in a very low reverse current of about 10 nA at room temperature.

When a voltage is connected across the diode so that the positive terminal is connected to the anode and the negative terminal is connected to the cathode, the diode is said to be **forward-biased** (Figure 4–6d). The depletion layer is reduced until the fringes of the n and p regions begin to overlap, and then the holes travel easily from the p to the n region and electrons travel easily from the n to the p region. This provides a forward current i_f that is the sum of the electron and hole currents. A plot of forward and reverse currents as a function of applied bias voltage is shown in Figure 4–6e.

The forward current increases rapidly with increase of forward-bias voltage after the 0.6-V junction voltage is exceeded (Figure 4–6e). The effective diode resistance is thus very low with sufficient forward bias. The reverse current is extremely small over a wide range of applied reverse bias, and the effective diode resistance is very high. However, the reverse bias can be increased so much that the insulating capacity of the junction breaks down, and the high reverse current can destroy ordinary diodes. The **reverse breakdown** occurs with tens to hundreds of volts and depends on the concentration of dopants in the semiconductor. The higher the breakdown voltage, the wider the range of reverse voltages for which the regular junction diode is useful. However, special diodes called **Zener diodes** are designed to operate with reverse bias so that they can be used in the reverse breakdown region at safe and useful currents. Zener diodes provide the important control function for the shunt regulator introduced in this section.

Zener shunt diode. Because the Zener diode is designed to operate in the reverse breakdown region, it is given a special diode symbol (Figure 4–7a). The straight line representing the cathode on the regular diode is replaced by a Z to

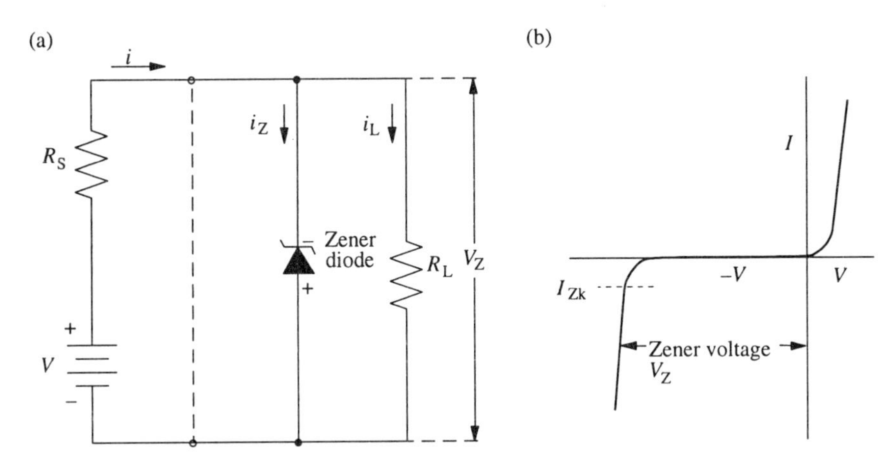

Figure 4–7. Zener-regulated supply (a) and characteristic curve (b). Because the reverse breakdown voltage of the Zener diode remains essentially constant for a wide range of reverse currents, the effective resistance of the diode, $R_z = V_z / i$, changes to compensate for changes of supply voltage V or load current i_L.

indicate that it is a Zener diode and should be operated with reverse bias (i.e., the cathode should be connected to the plus terminal of the voltage supply). The voltage across the diode is regulated at the Zener voltage V_Z as long as the Zener current is greater than I_{Zk} in the region of the knee on the diode I–V curve of Figure 4–7. The product of the Zener current and the Zener voltage should not exceed the power dissipation rating for the Zener diode [$P_{Max} = I_{Z(Max)} V_Z$].

The Zener diode shunt regulator is commonly used as a reference voltage source for comparator circuits and feedback control regulator circuits. In such applications the load is light and essentially constant. Output voltage precision of 1 part in 10 000 or better is obtainable from carefully designed Zener reference sources. Reference voltage sources are available from many manufacturers.

The important digital-to-analog converters (*see* Chapter 7) for computer interfacing require precisely regulated reference voltages, and these can be readily provided by Zener shunt regulators.

Linear (Series) Regulator

A method of providing voltage regulation of an unregulated voltage source V is to introduce the equivalent of a variable resistance R_V in **series** with the load resistance R_L, as shown in Figure 4–8. A comparator/servo system (*see* Chapter 5) monitors the output voltage v_o, compares it to a reference voltage v_r, and automatically changes the effective resistance R_V of a series element such as a transistor so that

$$v_o = V - i_L R_V \tag{4–3}$$

This type of series regulator can provide excellent voltage regulation. Because of developments in integrated circuits (ICs), excellent regulation is possible by using a compact low-cost package. Such IC regulators are now widely used in instrumentation.

A typical three-terminal IC voltage regulator is shown in Figure 4–9a. It is a small, simple-looking package with only three leads but contains dozens of components, including about 20 transistors. The functional units in the IC package are shown in Figure 4–9b. They include the Zener reference voltage source supplying v_r, the comparator control system, and the series power transistor (C, B,

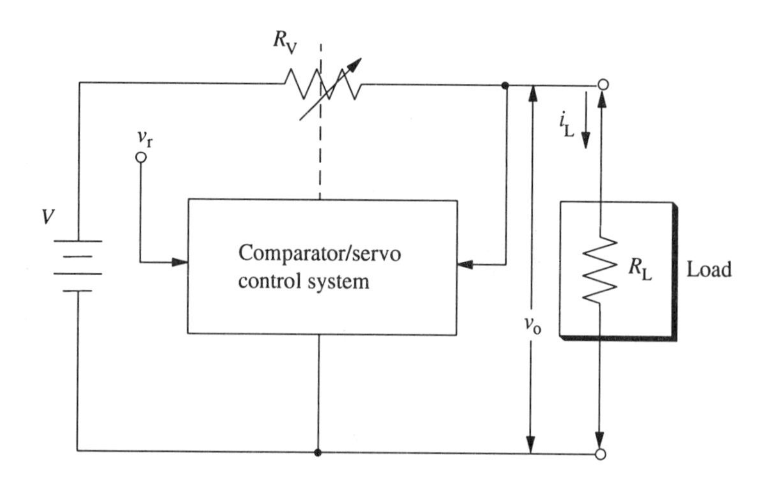

Figure 4–8. Series regulator circuit. The output voltage is $v_o = V - i_L R_V$. The resistance R_V is varied automatically by the servo system to compensate for changes in V or i_L.

E in Figure 4–9b), which acts as the automatically controlled variable resistor with resistance R_V.

Series transistor. The basis of operation for a linear voltage regulator is its ability to control automatically the effective variable resistance of a power transistor, often called the "pass transistor" in this application. Its effective resistance R_V equals the collector–emitter voltage V_{CE} divided by the collector–emitter current i_C, which equals the load current i_L (*see* Supplement 3 for the operation of transistors).

Figure 4–9b shows that on the basis of Kirchhoff's voltage law (*see* Supplement 1, Section 1–2),

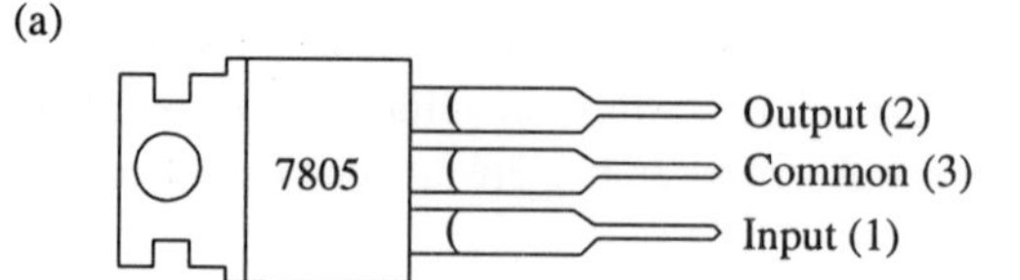

$$V = v_{CE} + v_o \quad \text{or} \quad v_o = V - v_{CE} \tag{4–4}$$

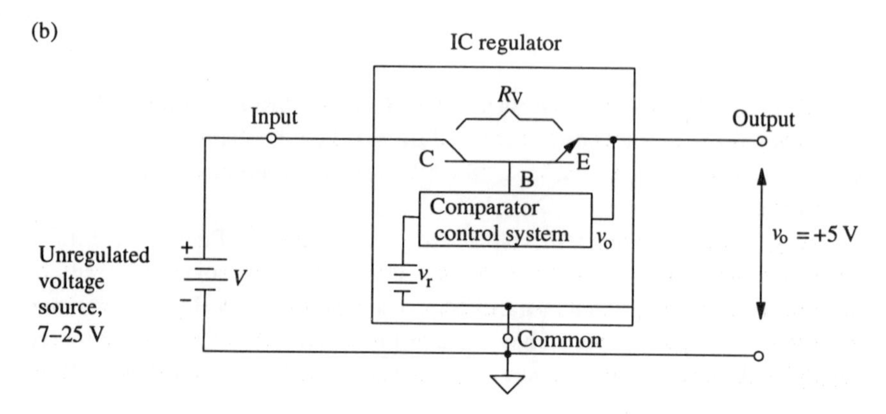

Figure 4–9. IC regulator. (a) Top view of model 7805; (b) schematic of the functional units within the IC package. C, B, and E are the terminals on the pass transistor that acts as the automatically controlled variable resistor with resistance R_V. This particular model provides regulation at +5 V with less than 0.05-V variation in output for input voltages of 7 to 25 V and output currents of up to 1.5 A. It also provides internal thermal-overload protection and internal short-circuit limiting. The numbers next to the pins are often used in schematic diagrams to avoid having to name each lead.

where V is the input voltage from the unregulated power source, v_o is the regulated output, and v_{CE} is the collector–emitter voltage drop across the series npn transistor. The output voltage v_o is continuously compared and held equal to a fixed reference voltage v_r. Therefore, a change of load or input voltage causes the comparator control system to force a change of R_V, as shown by equation 4–3. This change is accomplished by automatically controlling the base current of the pass transistor.

Efficiency. The pass transistor must be capable of dissipating the wasted power caused by the voltage regulation function. For example, if the input voltage V is 20 V and v_o is regulated at 10 V, then

$$v_{CE} = 20\text{ V} - 10\text{ V} = 10\text{ V}$$

and for a load current of 1 A, the transistor must dissipate significant power.

$$P = i_L v_{CE} = 1\text{ A} \times 10\text{ V} = 10\text{ W}$$

The efficiency is poor:

$$\%\text{ efficiency} = \frac{\text{power output}}{\text{power input}} \times 100 \tag{4–5}$$

and, for the above example,

$$\%\text{ efficiency} = \frac{10}{20} \times 100\% = 50\%$$

The low efficiency, typically 30 to 60%, is one of the disadvantages of the linear regulator. However, its excellent regulation characteristics and reliability make it the system of choice in many applications.

The minimum excess voltage necessary for a series regulator to work is typically 2 to 3 V. Thus, for a 5-V regulator with a 1-A load and a minimum input voltage of 8 V, the power dissipated by the regulator is 3 W. An external heat sink is often connected to the IC case to prevent overheating.

Protection devices. Various protection devices are incorporated within the regulator and the input to the regulator to prevent damage to both the supply and the equipment to which the supply is connected. These protection systems are described in Section 4–7.

Switch-Mode Regulator

The switch-mode regulator, illustrated in Figure 4–10, is more efficient than the linear regulator. A high-frequency switch is turned ON and OFF by an oscillator with pulsed output. The oscillator has a constant repetition rate (typically 20 kHz), but the ON time of the oscillator pulse is variable. During the pulse, the switch is

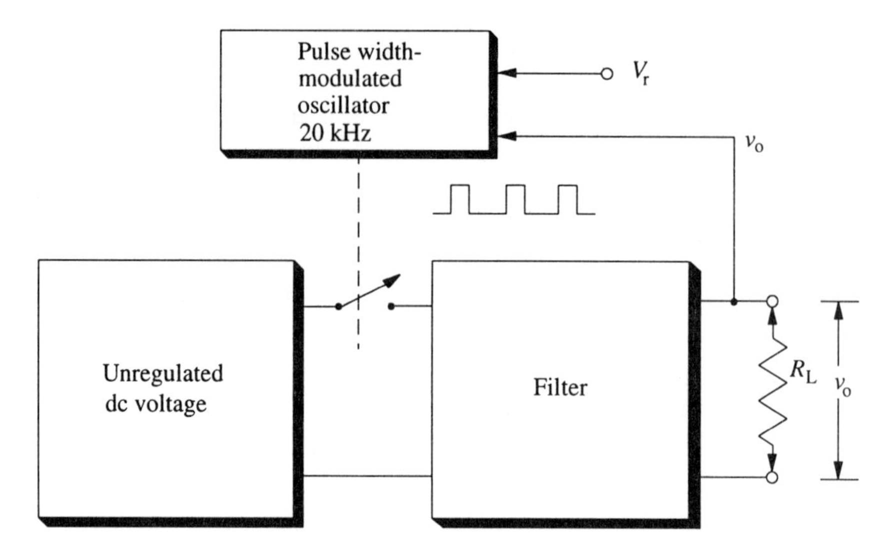

Figure 4–10. Simple switching regulator. The series switch is controlled by an oscillator of constant frequency but variable pulse width. The fraction of time that the switch is closed is controlled to maintain v_o equal to V_r.

closed and the current from the unregulated dc supply is delivered to the filter and the load. When the switch is open, the energy stored in the filter is delivered to the load. The width of the pulse is controlled by the pulse width-modulated oscillator to maintain a negligible difference between the reference voltage V_r and the regulator output v_o. In other words, the charge delivered to the filter and load by each ON time (pulse) of the switch is exactly equal to the charge consumed by the load during one period of the oscillator, and it is exactly enough to maintain v_o at the desired voltage.

Switch transistor. A transistor is used as the control switch in the switch-mode regulator. In contrast, the linear regulator uses a power transistor in the linear range (between maximum conducting and nonconducting limits). When used as a switch, the transistor is either ON or OFF (in current saturation or not conducting). This type of ON–OFF switch operation, as illustrated in Figure 4–10, provides the higher efficiency of the switch-mode regulator, which is typically 60 to 90% compared with 30 to 60% for the linear regulator.

Isolation of input voltage source. The input dc voltage source for the switch-mode regulator is often obtained from an ac line source or other source. To isolate the input source from the output voltage, an isolation system is used (Figure 4–11). The functional modules in the block that is enclosed with dotted lines in Figure 4–11 perform as a dc-to-ac converter. The ac signal is isolated from the input circuitry by the isolation system and converted to dc by an ac-to-dc converter such as that described in the next section. The principle of comparing the output voltage to a reference voltage is similar to the principle used in the simple system shown in Figure 4–10. However, the output is now isolated from the input source, as indicated by the different ground and common symbols (*see* Supplement S1–3 for a discussion of ground and common). This is usually important for safety and for noise suppression. The concepts of ac-to-dc and dc-to-ac conversions and further details on switch-mode power supplies are discussed in more detail in Sections 4–4, 4–6, and 4–7.

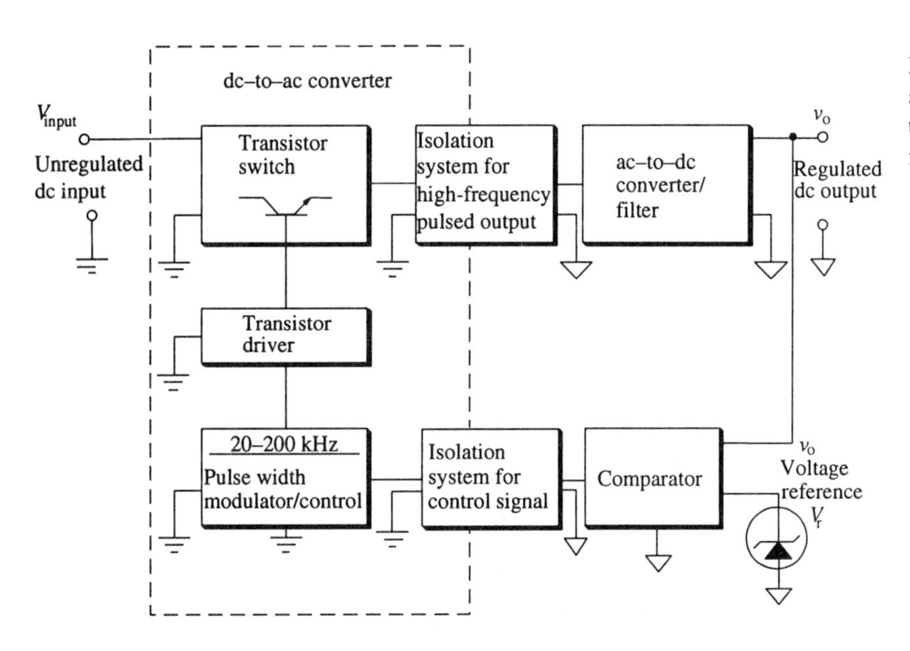

Figure 4–11. Block diagram of switch-mode regulator with isolation of the input voltage source from the output.

4–4 Converting Voltages from ac to dc

The conversion of ac to pulsating dc is accomplished with **rectifier** circuits. The basic element in these circuits is the rectifier diode. Several types of rectifier circuits are found in power supplies. They are classified according to the different configurations of their diodes as half-wave, full-wave, bridge, and voltage-multiplier rectifier circuits. These circuits are used in electronic equipment, and their characteristics are discussed in this section.

Rectifier Diodes

A **diode** conducts effectively in only one direction. As described in Section 4–3, the diode is said to be forward-biased when the voltage applied to the **anode** is positive relative to that applied to the **cathode**. The direction of forward current through a diode is from anode to cathode. When a diode is forward-biased, its effective resistance is very low, somewhat like that of a closed switch. If the diode is reverse-biased (that is, the voltage applied to the anode is negative with respect to that applied to the cathode), the effective resistance is very high, somewhat like that of an open switch. This is illustrated in Figure 4–12, where the **forward resistance** R_f is always less than the **backward resistance** R_b. From Ohm's law, i_f equals V/R_f and i_b equals V/R_b. Therefore, R_b/R_f equals i_f/i_b. The ideal diode

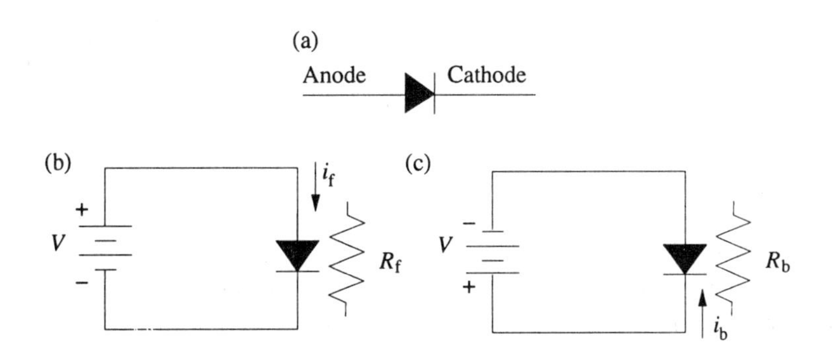

Figure 4–12. (a) Symbol of a rectifier diode. The diode is forward-biased in part b and reverse-biased in part c. The forward resistance R_f is always much less than the backward resistance R_b.

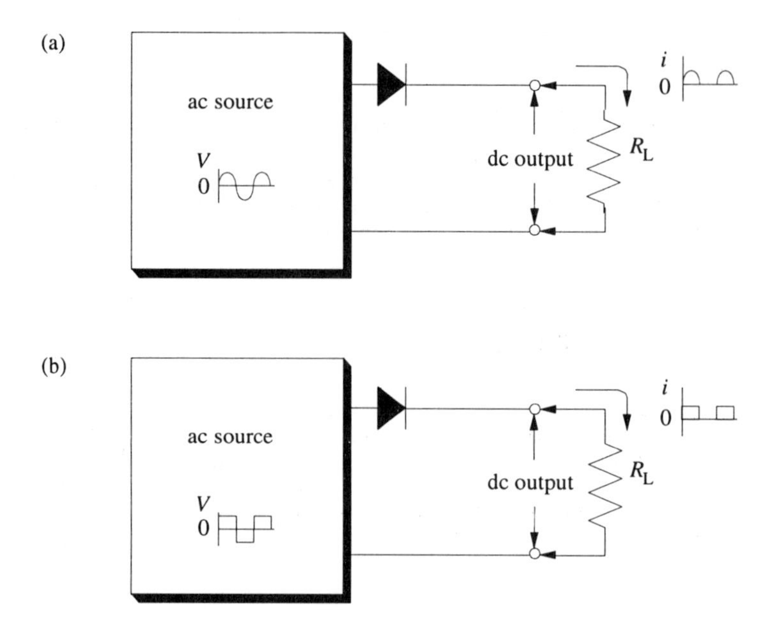

Figure 4–13. Half-wave rectifier circuit. (a) Sinusoidal ac input; (b) rectangular ac input. On the positive half-cycle of the ac voltage, the diode conducts and allows current to pass through R_L. The resistance R_L is the "load" on the circuit that is to be supplied with direct current. On the negative half-cycle, the diode is reverse-biased and therefore nonconducting.

would be a perfect conductor for forward current and a perfect insulator for reverse current. The ratio R_b/R_f approaches infinity as the diode approaches the ideal. This ratio is used as a figure of merit for rectification devices.

Half-Wave Rectifier

The simplest rectifier circuit is the **half-wave rectifier**. It is shown in Figure 4–13 with a sinusoidal ac input in part a and a square-wave ac input in part b. This rectifier circuit is so named because only half of the ac current wave is present in the output load circuit. Because only half of the input wave is used (as illustrated by the rectified outputs in Figure 4–13), the circuit is not very efficient.

In studying rectifier circuits, it is important to consider the ratings of the rectifiers. The ratings include (1) the maximum average forward current rating, which is approximately $1/2\ V_{av}/R_L$, because the diode conducts only half the time; and (2) the **peak inverse voltage** (PIV), which is the maximum voltage that should be applied to the rectifier when it is reverse-biased. For example, if 12 V_{av}(rms) is fed from a sinusoidal ac source, the diode of Figure 4–13 must withstand a peak inverse voltage of $12 \times 1.4 = 17$ V. (For a review and discussion of rms and the factor of 1.4, *see* Supplement 2, Section S2-3.) If a capacitor is connected across the output of the rectifier circuit in Figure 4–13, the diode PIV must be at least $2 \times 17\ \text{V} = 34$ V, because the capacitor holds the output at +17 V while the secondary coil goes to –17 V. The output pulse frequency is equal to the input frequency.

The effective resistance of a conducting diode is not constant but depends on the current. However, an estimate of the forward resistance R_f allows the power loss in the rectifier to be calculated as approximately I^2R_f.

Full-Wave Rectifier

Many applications require a rectifier circuit that supplies current during both half-cycles of the ac power and thus allows a more continuous current to the load. A **full-wave rectifier** circuit is shown in Figure 4–14.

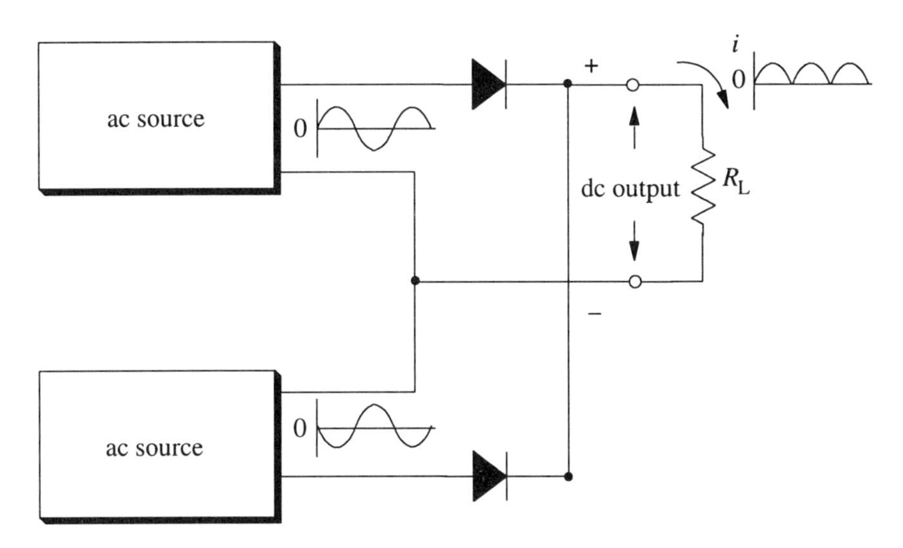

Figure 4–14. Full-wave rectifier circuit. The upper diode conducts during one half-cycle, and the lower diode conducts during the next half-cycle. Two sources 180° out of phase are required.

This circuit is essentially two half-wave rectifiers connected in parallel with their inputs out of phase by 180°. The voltage output of the full-wave rectifier depends on the voltage output of each of the ac voltage sources, which are operating 180° out of phase. For a 10-V peak output voltage from a full-wave rectifier, each of the two ac sources must provide about 7.5 V_{rms} [i.e., (0.6 V + 10 V)/1.4]. The extra 0.6 V in the calculation compensates for the voltage drop in the rectifiers. The rectifiers must withstand an inverse voltage of twice the peak value of the source. For the case given above, the peak inverse voltage is 1.4 × 15 V = 21 V. The output pulse frequency is two times the input frequency.

Bridge Rectifier

A method of obtaining full-wave rectification without the use of two ac sources that are 180° out of phase is shown in Figure 4–15a. This circuit is called the **bridge rectifier**. Because two rectifier diodes are in series with the load, the peak inverse voltage that each rectifier must withstand is equal to the peak value of the supply voltage. The output pulse frequency is two times the input frequency. In Figure 4–15b the bridge circuit is shown powered directly from a 120-V ac line. The pulsating dc is then filtered to provide about 168 V dc. An alternative way of drawing the bridge rectifier in circuit diagrams is shown in Figure 4–15c.

Voltage-Doubler Rectifier

Two rectifiers can be connected to a single ac source and wired so that their outputs are in series, as in Figure 4–16. The output voltage available from such a circuit is twice that available from the ac source with a half-wave or bridge rectifier. This kind of circuit is therefore called a **voltage-doubler rectifier**. The capacitors are essential to the operation of the circuit because they maintain the voltage developed during one half-cycle so that the voltage developed during the next half-cycle can be added to it. The capacitors with capacitances C_1 and C_2 have a filtering action that is described below. Because current is drawn from the input ac source during both half-cycles, this voltage-doubler circuit is considered to be full-wave. The peak inverse voltage applied to each rectifier is twice the peak value of the supply

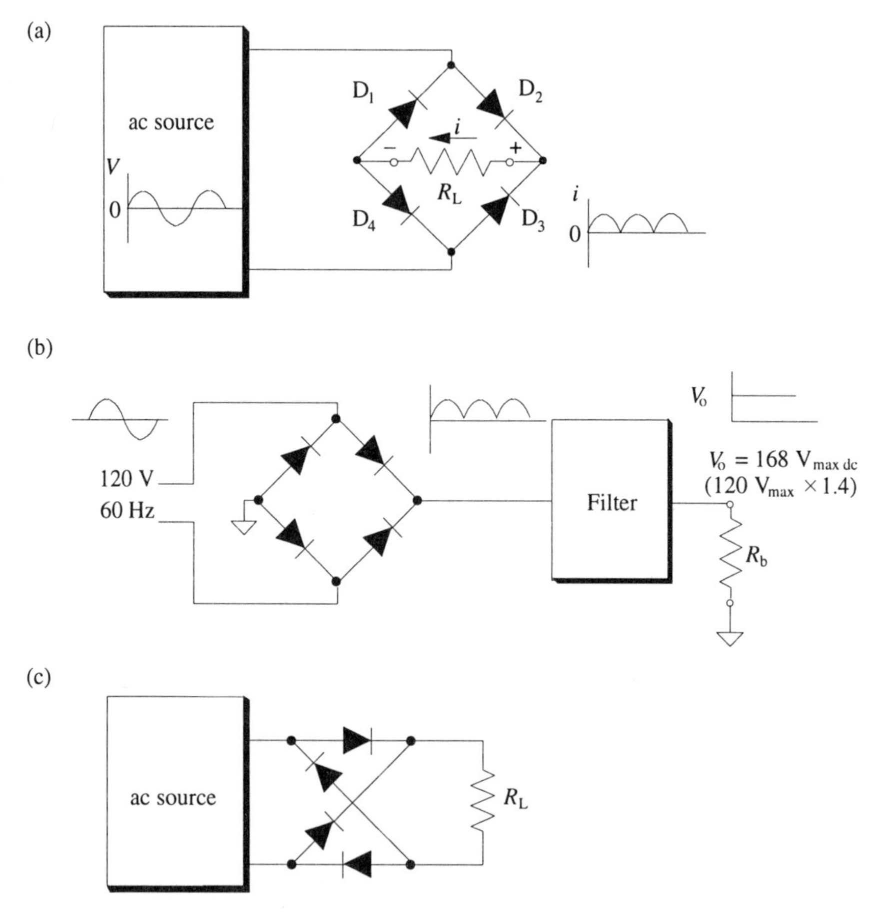

Figure 4–15. Bridge rectifier circuit. (a) Powered from any ac source. On the positive half-cycle, D_2 and D_4 conduct. On the negative half-cycle, D_1 and D_3 conduct. In each case the direction of current through the load resistance R_L is the same. (b) Powered directly from the 120-V ac power line. The output voltage is about 168 V dc maximum (1.4 × 120 V). (c) Alternative way of drawing the bridge rectifier in diagrams.

voltage. If the peak inverse voltage rating of the diode is insufficient, two diodes can be connected in series so that their peak inverse voltage ratings are additive.

Power Supply Filters

The pulsating dc voltages from the rectifier circuits described in the previous section are not useful for most electronic applications. Nearly all electronic circuits require a very smooth constant voltage. Therefore, rectifier circuits are usually

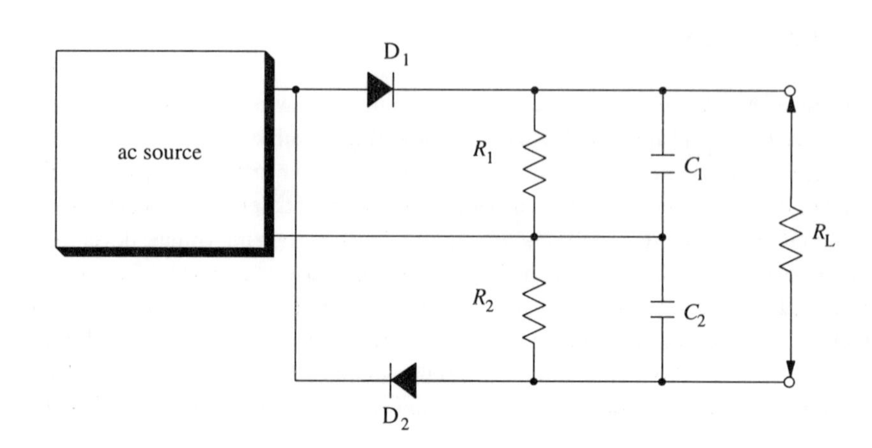

Figure 4–16. Voltage-doubler rectifier. On the positive half-cycle, capacitor with capacitance C_1 is charged to the peak value of the supply voltage. On the negative half-cycle, C_2 is charged to the same voltage. Because C_1 and C_2 are in series across the load, the output voltage is twice the peak voltage of the ac source.

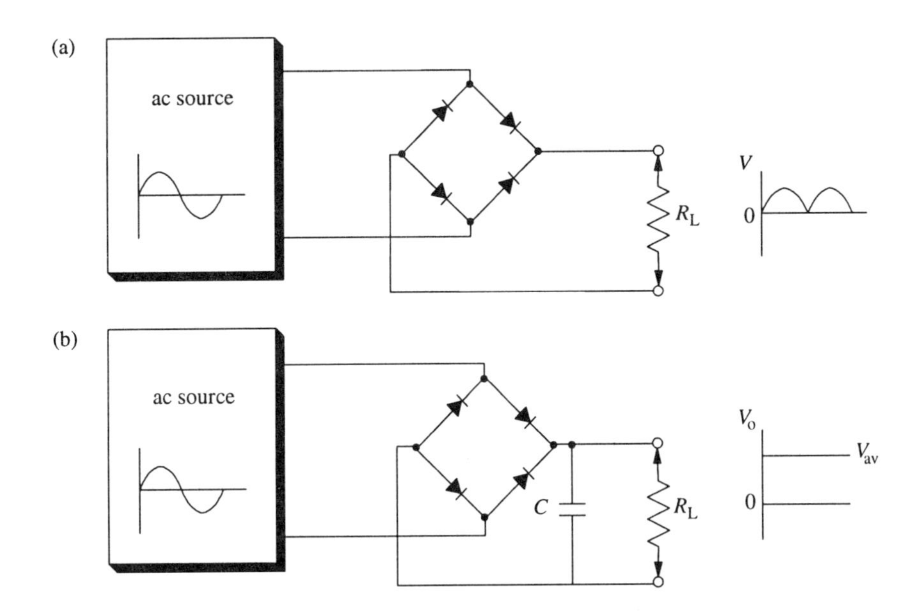

Figure 4–17. Bridge rectifier with load resistor R_L (a) and with filter capacitor across load (b).

followed by smoothing devices called **filters**, which convert the pulsating dc voltages into the required constant dc voltages.

Capacitor as filter. An effective filter is simply a capacitor connected in parallel with the load resistance R_L. A bridge rectifier circuit with and without a capacitive filter is shown in Figure 4–17. Without a filter in the circuit, the voltage across the load is always equal to the pulsating voltage from the rectifier. When the filter capacitor with capacitance C is present, the voltage across the load equals that across the capacitor, which is alternately charged by the pulsating source and discharged by the load. If the discharge between pulses is small compared with the average charge stored in the capacitor, the voltage fluctuations across the load are also relatively small. Another way to think of the capacitor filter is as a low-pass filter with an upper cutoff frequency (*see* Chapter 3, Section 3–2) of $1/(2\pi CR_L)$. To be an effective filter, the capacitor must have an upper cutoff frequency lower than the rectifier output pulse frequency.

A more detailed picture of the action of a capacitor filter on a sinusoidal input is presented in Figure 4–18. The capacitor charges toward the peak value of the input voltage. If R_L were infinite (no load), the voltage across the capacitor would quickly reach a constant value nearly equal to the peak value of the rectified ac input. In practice, R_L is not infinite, and the capacitor begins to discharge through R_L as soon as the input voltage decreases below the voltage to which the capacitor has been charged. The capacitor continues to discharge until the next pulse, when the rectifier output voltage again exceeds the voltage on the capacitor. Capacitor-charging current occurs only when the rectifier voltage exceeds the capacitor voltage.

Because of the high capacitance that can be achieved in a small volume with electrolytic and tantalum capacitors, power supply filter capacitors are almost always these types. Choosing the appropriate filter capacitor obviously involves compromises among size, expense, and maximum tolerable output fluctuations. With use and age, filter capacitors can be damaged and become ineffective. Thus, the noise or ripple in the dc output voltage increases greatly. The ripple factor is considered next.

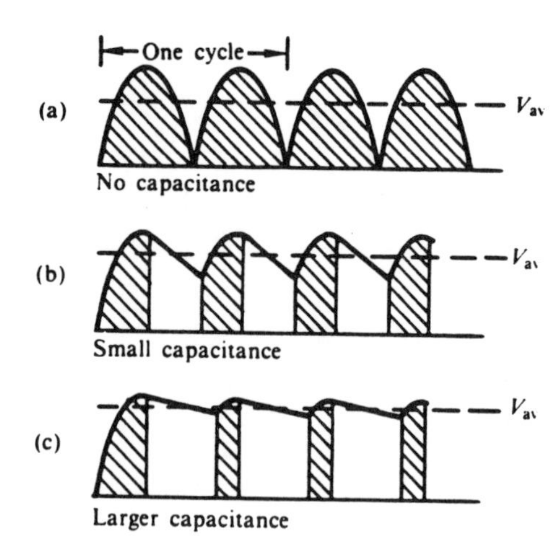

Figure 4–18. Smoothing of output voltage by a capacitor filter. (a) Output with no capacitor; (b and c) effects of increasing capacitance. Shaded areas indicate the times during which charging current is supplied by the rectifier. The dashed line indicates the average output voltage. As the filter capacitance increases from zero to a high level, the average voltage increases and the output fluctuation and charging time decrease.

Ripple factor and frequency. The effectiveness of the filter is called the **ripple factor** r and is defined as the rms value of the ac voltage component, or ripple, divided by the average dc voltage, that is, $r = I_{ac}/I_{dc} = V_{ac}/V_{dc}$. The frequency of the ripple is equal to the rectifier output pulse frequency. For a given filter the ripple factor will decrease with increasing pulse frequency of the rectifier output. This is illustrated by the comparison of full- and half-wave rectifiers shown in Figure 4–19. For a given discharge rate the ripple amplitude for the half-wave rectifier is almost twice that for the full-wave rectifier.

The ripple voltage decreases if R_L, C, or the frequency is increased (Figures 4–18 and 4–19). The approximate expression for the ripple factor, $r = 1/(2\sqrt{3}fCR_L)$, bears this out. Here, f is the frequency of the main ac component (when the rectifier circuit with capacitor filter is connected to the power lines, f is equal to the line frequency for half-wave rectifiers and twice the line frequency for full-wave rectifiers). As the ripple increases in magnitude, the average dc output decreases.

Inductor–capacitor (LC) filters. Another filter device is simply an inductor in series with the load. By opposing changes in current, the inductor tends to maintain a constant load current and thus a constant output voltage. Combinations of inductors and capacitors (LC filters) make the best passive filters. A simple LC filter is illustrated in Figure 4–20. The filter is called an L section. The inductor

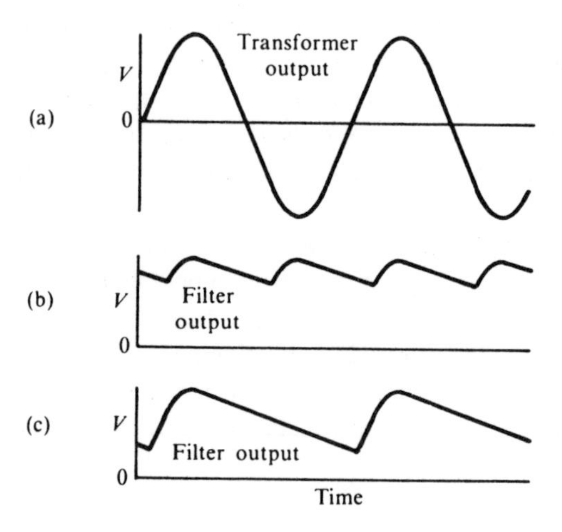

Figure 4–19. Comparison of ripple frequency and amplitude for full-wave rectifier (b) and half-wave rectifier (c) for given input frequency (a). The charging pulse frequency of the half-wave rectifier is equal to the input frequency, but that of the full-wave rectifier is twice the input frequency.

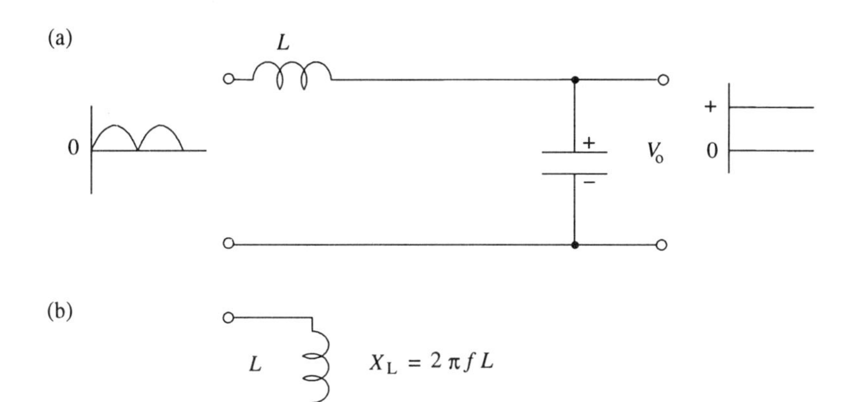

Figure 4–20. Typical schematic of inductance–capacitance (*LC*) filter (a) redrawn to illustrate ac voltage division (b).

opposes changes of current and stores energy in the form of a magnetic field. When the pulsating input voltage falls below the output voltage, a counter electromagnetic field in the inductor tends to oppose the change and maintains a rather constant current.

The L section acts as an ac voltage divider. In Figure 4–20b it is apparent that the ratio X_L/X_C should be very large for best filtering. (A review of reactance X_L and X_C is provided in Supplement 2, Section S2-4.) Also, X_C should be much less than R_L so that ac ripple will be shunted across the load. The use of high frequencies (20 kHz or more) in switch-mode regulators provides ideal conditions for use of LC filters because the ripple factor r is approximately

$$r = \frac{\sqrt{2}X_C}{3X_L} = \frac{0.47}{(2f)^2CL} \qquad (4\text{–}6)$$

4–5 Isolating and Changing ac Voltages

When ac line voltage is used as the input to power supplies, it is important to isolate the output from the input for protection against dangerous shocks. Also, as shown in Section 4–3, isolation is important in the widely used switch-mode regulators. In this case the frequency of the ac source is hundreds of times greater than that of line input voltage sources (that is, 20 to 200 kHz compared with 50 to 60 Hz). Isolation is readily accomplished with transformers. The transformer design, size, weight, and cost depend on the power supply system used and the circuit specifications.

In addition to isolation of input from the output voltage, the transformer enables output voltage to be increased or decreased compared with input voltage. That is, the transformer can step up or step down the input voltage to an output level suitable for the electronic circuits connected at the output. Supplying power to instruments and computers from ac sources requires the isolation and step-down/step-up characteristics described in this section.

Transformer

A transformer provides an effective multiplication or division of the ac input voltage. A schematic symbol and a pictorial representation of a transformer are given in Figure 4–21. The operation of the transformer is based on the principles of inductance reviewed in Supplement 2. A changing current in the **primary** winding produces a changing magnetic flux in the **secondary** coil, which induces a changing voltage across that coil. The words primary and secondary as applied to a transformer refer to the coil that is connected to the power source (the primary) and the coil that is connected to the load (the secondary).

The voltage v_s produced in the secondary winding is equal to the number of turns N_s multiplied by the rate of change of magnetic flux φ:

$$v_s = -N_s \frac{d\varphi}{dt}$$

Because the primary and secondary are often closely wound on a core of high magnetic permeability, the magnetic flux in the primary winding and the magnetic flux in the secondary winding are equal. The rate of change of flux is

$$\frac{d\varphi}{dt} = \frac{-v_p}{N_p}$$

where v_p and N_p are the voltage on the primary and the number of turns in the primary, respectively. Combining the equations to eliminate $d\varphi/dt$, we get

$$\frac{v_s}{v_p} = \frac{N_s}{N_p} \tag{4–7}$$

Only ac voltages are induced in the transformer secondary, because the transformer operation depends on a constantly changing magnetic flux. Thus, ac signals can be separated from a dc component with a transformer.

Step up/step down. The turns ratio of the power transformer determines whether the ac voltage source is increased or decreased. For example, if the turns

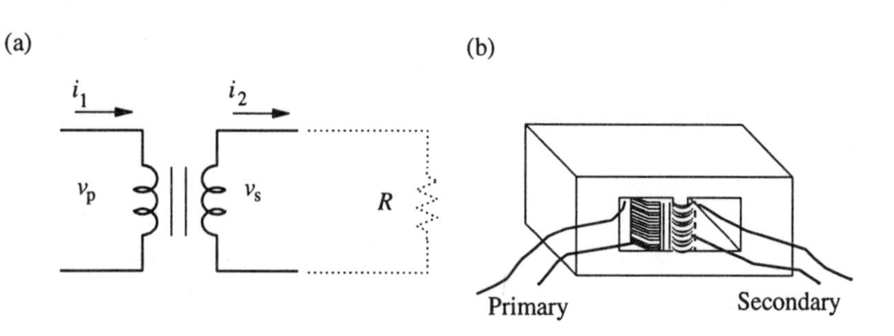

Figure 4–21. Schematic symbol (a) and pictorial representation (b) of the transformer.

ratio is 1/10 in equation 4–7, the secondary voltage will be 1/10 of the primary voltage. For a primary input voltage of 115 V ac, 60 Hz, from the line, the secondary output voltage will be stepped down to 11.5 V ac, 60 Hz. Voltages will be stepped up when the secondary windings are greater than the primary windings. For example, if the turns ratio is 5/1 and the primary voltage is 100 V ac, 20 kHz, the secondary voltage will be 500 V ac, 20 kHz.

Isolation from the ac source. Serious electrical shock can occur if line voltage (for example, 115 V ac, 60 Hz) is connected directly to rectifier circuits in two separate chassis, as illustrated in Figures 4–22a and 4–22b. A two-prong plug could be connected in either of two ways so that one chassis would be connected to the neutral line and the other chassis would be connected to the live line. By touching both chassis simultaneously, a person would be directly across the ac line, which could be fatal. Transformers provide **isolation** to prevent this problem, as shown in Figures 4–22c and 4–22d. Now the line plug can be connected either way, and both chassis will be "floating" and isolated from the power line. The power line connects only to the transformer primary winding and not to the secondary winding.

Dot convention. The use of dots on each winding of a transformer indicates the polarity of each winding relative to that of the others. That is, when a dotted end of a winding goes positive, all dotted ends go positive, and when a dotted end goes negative, all dotted ends go negative. This convention can also be stated as "All dotted ends have in-phase voltages."

Center-tap transformer. The use of the dot convention is illustrated in Figure 4–23. The secondary has a center-tap connection so that the output is equivalent to two ac sources that are 180° out of phase. The dot convention shows that when one end of the secondary is positive, the other end is negative and vice versa. This dual ac source is the type required for the full-wave rectifier described in Section 4–4 and illustrated again in Figure 4–23a. In Figure 4–23b the center-tap transformer is used with two full-wave rectifiers, each connected to linear IC regulators to provide regulated ±15 V dc.

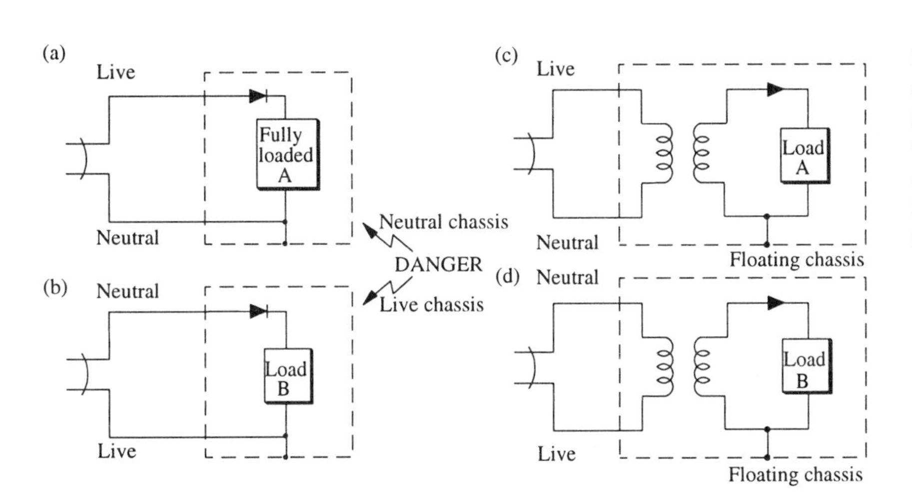

Figure 4–22. Isolation versus direct connection to ac source. (a) Line connected directly to circuit with neutral wire tied to chassis; (b) line plug reversed so chassis is live; (c and d) circuits and chassis isolated from the input line.

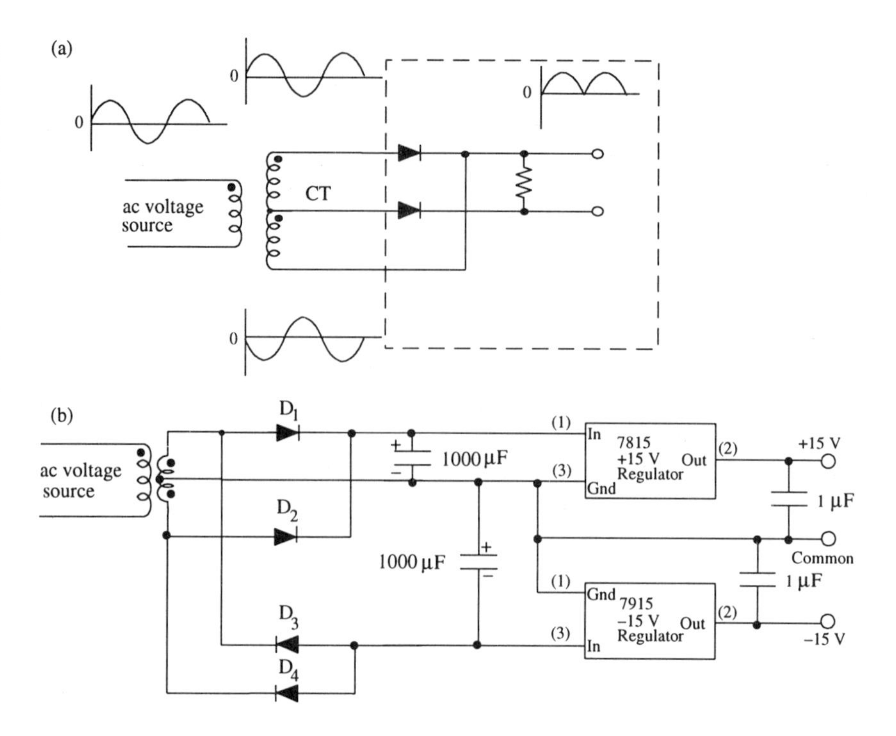

Figure 4–23. Center-tap (CT) transformer shown connected to a full-wave rectifier (a) and used in a dual-output (±15-V) power supply (b). Diodes D_1 and D_2 form a full-wave rectifier for the positive supply; D_3 and D_4 are the rectifiers for the negative supply. The filtered outputs are regulated by separate three-terminal regulators. The output 1-μF capacitors help stabilize the regulators. Gnd, ground.

Multiple Secondary and Dual (115- /230-V ac) Primary Transformers

A single transformer can have several secondary windings to satisfy several different voltage requirements. As illustrated in Figure 4–24, some transformers are also built so that dual-input primary leads can be arranged to accommodate either 115- or 230-V line voltage. In this case care must be taken to connect the primary leads correctly to avoid serious damage to the transformer, the power supply, or the equipment connected to the power supply output. If the electronic system is to operate from 115 V ac, the two primary windings are connected in parallel. If the line voltage is 230 V ac, the two primaries are connected in series.

Low- and High-Frequency Transformers

The two major types of transformers used in instrumentation power supplies are (1) low frequency, in which the primary is usually connected to the utility line (50 or 60 Hz, as illustrated in Figure 4–24), or (2) high frequency, in which the primary is connected to the switching circuit (typically, operation is at 20 to 200 kHz, as illustrated in Figure 4–25). Operation at frequencies higher than line frequencies reduces the sizes and weights of the transformers and the filter components. This reduction is especially important for compact or portable equipment. Switch-mode power supplies in the 10- to 150-W range are now manufactured with power densities of greater than 2 W/in.^{3}. For example, a 10-W (16 cm^3) model weighs just 0.5 lb (ca. 0.2 kg) and measures about 1.2 × 2.8 × 2.9 in. (3.0 × 7.1 × 7.4 cm).

Although the switch-mode power supply can be made smaller, the linear regulated power supply, which uses the low-frequency transformer off the line, is usually used for devices requiring less than 15 W. They are more cost-effective for these low-power applications.

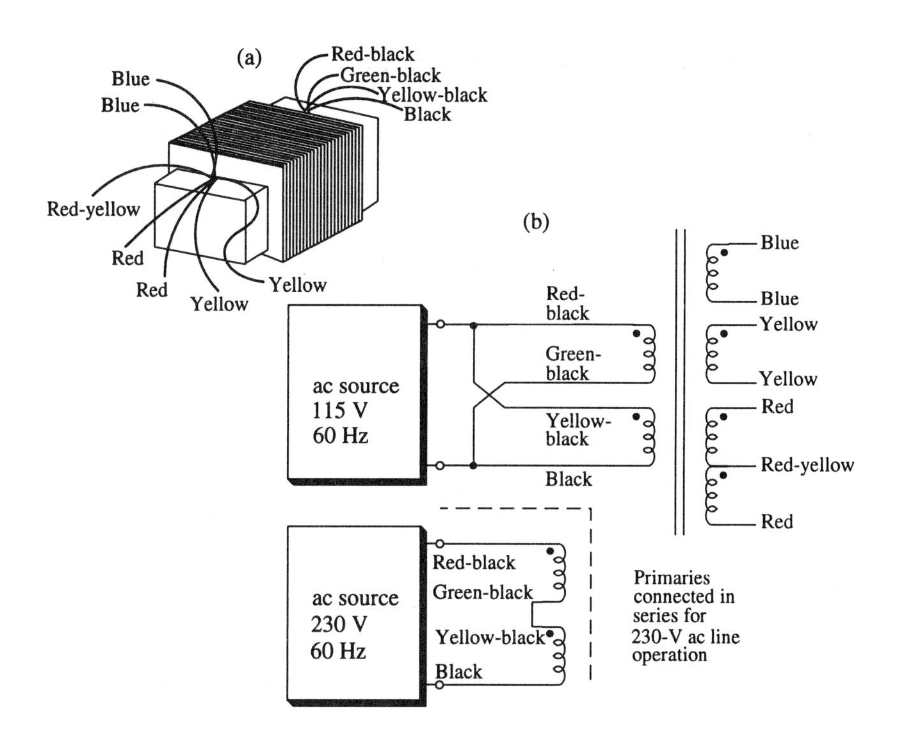

Figure 4–24. Power transformer with dual primary and multiple secondary windings. (a) Pictorial; (b) schematic.

Safety Standards

The construction of power supply transformers must comply with certain national and international safety standards. In the United States and Canada, the North American standards are set by the Underwriters Laboratories (UL) and the Canadian Standards Association (CSA), respectively. For Europe the most used (and most stringent) guidelines are those of German *Verband Deutscher Elektronotechniker* (VDE). The UL standards concentrate more on preventing fire hazards, and the VDE standards concentrate more on safety of the operator.

The standards greatly influence transformer design because they set limits for temperature rise under normal operation, winding techniques and insulation, dielectric strength of insulation, and moisture resistance. These standards affect size, weight, and cost, but when UL and VDE standards are adhered to, the transformers produced by manufacturers are safe and reliable. However, an overloaded transformer will overheat and can be damaged and cause damage, so protection devices are often included, as described in Section 4–7.

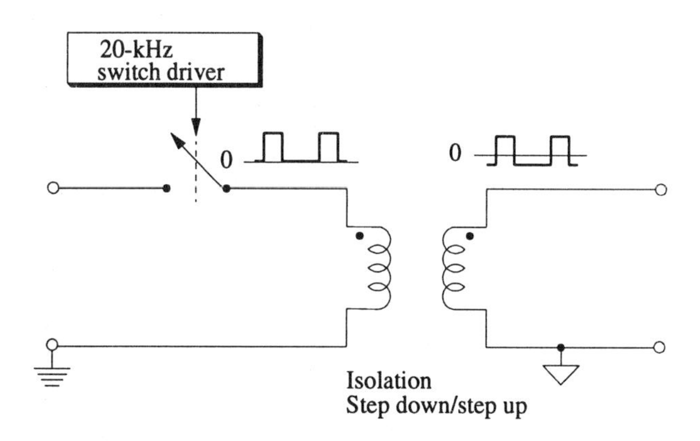

Figure 4–25. High-frequency transformer. It provides isolation and step down or step up of the 20-kHz pulse width-modulated voltage provided by the high-frequency transformer.

4–6 Inverting Voltages from dc to ac

The conversion of a dc input to an ac output is accomplished by inverters. The inverter is used in several important applications. If ac voltages are needed by the load and the only available power supply is a battery, solar cell, fuel cell, or other dc voltage source, an inverter system is necessary. Also, the inverter is an essential link in the dc-to-dc regulated power supply, where the output is isolated from the input. Likewise, it is used in conjunction with two ac-to-dc converters in the widely used ac–dc switch-mode power supply. These three applications are illustrated in Figure 4–26, and each is briefly examined in this section.

ac Power from dc Sources

Many loads that operate on ac power, such as motors, pumps, commercial appliances, and innumerable other devices and instruments, require ac of specific magnitude and frequency. Some require single-phase 120 V at 60 Hz, and others require 240 V at 50 Hz or 115 V at 400 Hz. High-power three-phase systems are operated on 120 or 208 V at 60 Hz, 220 or 380 V at 50 Hz, or 115 or 200 V at 400 Hz. If one or more of these voltages must be provided from dc power supplies, it is necessary to use a suitable inverter.

Some inverters produce an output waveform that is a square wave, as illustrated in Figure 4–27a. The switch transistors are alternately turned ON and OFF by the driver pulses obtained from the oscillator/scaler. This procedure would be suitable for some low-power applications. However, a low-distortion sine wave is desirable for medium-power applications, and it is required when the loads need high power. Therefore, the odd harmonics present in the pulsed output from a switching inverter must somehow be filtered or decreased in order to reduce load heating from the harmonics.

Many switching techniques may be used, but in all cases the inverter system must produce an output of the necessary voltage magnitude and frequency. The

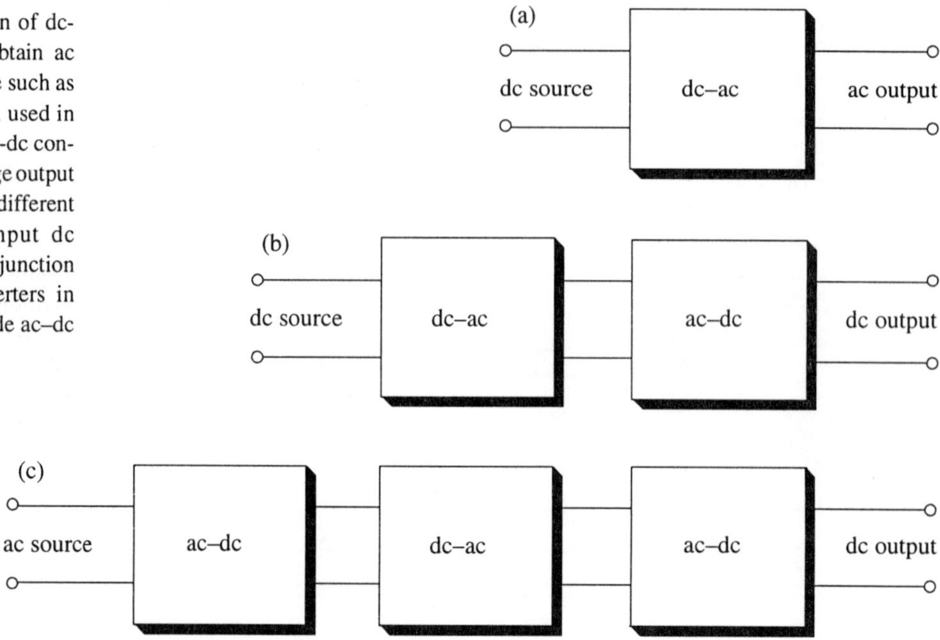

Figure 4–26. Application of dc-to-ac inverter used to obtain ac voltages from a dc source such as a battery or solar cell (a), used in conjunction with an ac-to-dc converter to obtain a dc voltage output that is isolated and of different magnitude than the input dc source (b), or used in conjunction with two ac-to-dc converters in the important switch-mode ac–dc power supply (c).

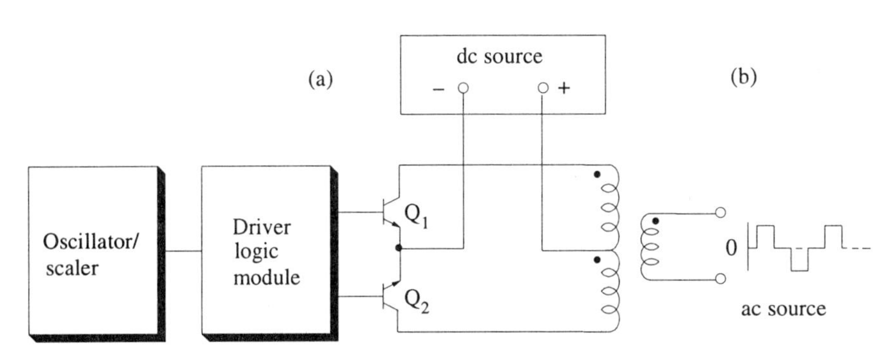

Figure 4–27. Use of an inverter for operating ac input equipment on a dc storage battery (a) to produce an ac square wave (b). Q_1 and Q_2, transistors (*see* Supplement S3–3).

master oscillator from which the desired output ac frequency is obtained may produce an output that is some multiple of the fundamental, and it is then scaled down by a counter. For example, if the switching oscillator operates at 7200 Hz, each half-cycle for 60-Hz output would require 120 pulses to be scaled by the counter to control the switch driver; at a 50-Hz output, 144 pulses would be required; and at a 400-Hz output, 18 pulses would be required. The magnitude of the output voltage can be stepped up or down by the transformers. An inverter for operating ac input equipment on a dc storage battery is shown in Figure 4–27a.

dc-to-ac Inverter in a dc-to-dc Power Supply

A dc-to-dc power supply is shown in Figure 4–28. It uses a high-frequency switching transformer to convert the dc input voltage to an ac square wave at the secondary. This ac output is isolated from the input by the transformer, which also adjusts the magnitude of the dc output to the desired value. A full-wave rectifier and LC filter provide the necessary filtering.

dc-to-ac Inverter in an ac-to-dc Power Supply

An ac-to-dc, off-the-line-regulated, switch-mode-regulated power supply is shown in Figure 4–29. All of the functional modules in this diagram have been described in previous sections. The input bridge rectifier and filter capacitor operate directly off the line to provide an ac-to-dc conversion. The 120-V, 60-Hz input is converted to 168 V dc.

The 168-V dc input is converted to ac by the transistor switches and transformer. The switching frequency is determined by the oscillator operating at 20

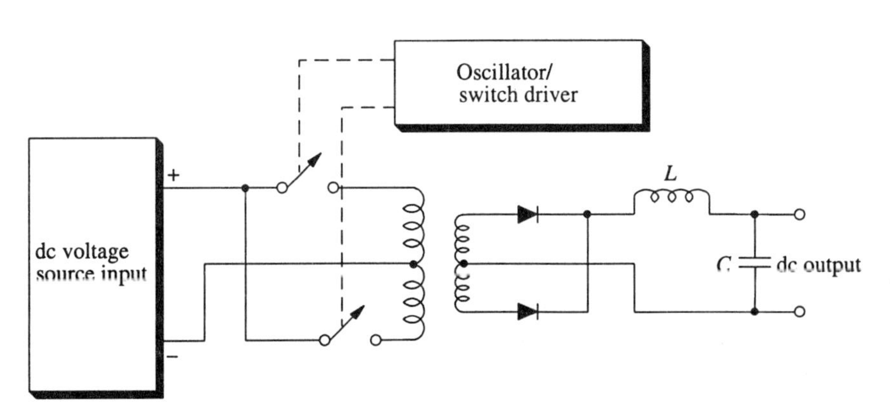

Figure 4–28. A dc-to-dc power supply. The driven switches usually operate at high frequencies (20 to 200 kHz), and the resulting pulses are converted to ac at the transformer secondary. The ac, which is isolated and adjusted in magnitude by the transformer, is rectified and filtered to give the desired dc output.

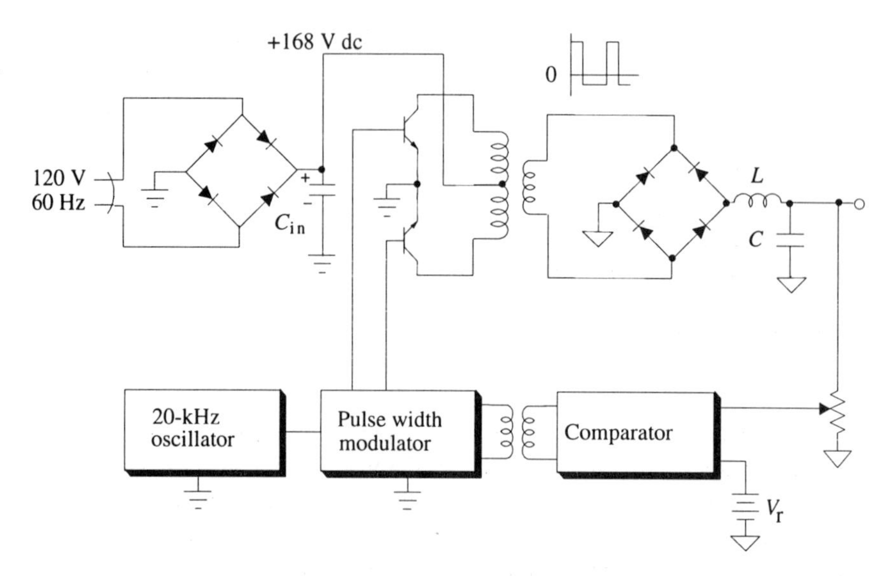

Figure 4–29. An ac-to-dc off-the-line switch-mode-regulated power supply. A dc-to-ac inverter is used to provide isolation of the output from the input and to adjust the voltage magnitude for the desired output.

kHz. The ON and OFF times of the switches are adjusted by the pulse width modulator so as to control the output at a fixed reference value, as explained in Section 4–3.

The ac output at the secondary of the high-frequency transformer is converted to dc by the bridge rectifier and LC filter, as described in Section 4–4. Thus, the off-the-line-regulated ac-to-dc supply is implemented by a sequence of conversions: ac to dc, dc to ac, and ac to dc. This series of conversions allows isolation of the output from the line and adjustment and regulation at the desired value.

The components in this power supply can be very small and lightweight compared with the transformer, capacitors, and inductors used in linear supplies. Consequently, the ac–dc switch-mode-regulated power supply is used in equipment where size, weight, and cost-effectiveness are important, such as the popular microcomputers.

4–7 Protecting Instruments and Their Power Supplies

In previous sections we considered various basic functional modules that make up computer and instrument power supplies. In this section we will look at the overall system and consider ways of protecting it from electrical damage. The emphasis will be on protecting the power supply, because various failures in or through it can also destroy other parts of the instrument.

The most frequent instrument failures are in the power supplies. This is not surprising, because components are subject to nonideal conditions such as overheating (poor ventilation, partly because of dirt buildup on parts), transient voltage spikes on the line, overloads, and large swings in the line voltage. Therefore, it is very important to use power supplies that are adequately designed and ventilated and are not overloaded. Still, even with well-designed and properly used instruments, some problems can be expected and should be protected against. Some of these problems are readily dealt with within the power supply or its input or require a power management system as with batteries and battery chargers. When loss of line voltage could cause major problems, such as in some computer systems, a more elaborate protection system is needed. These are the noninterruptible systems introduced at the end of this section.

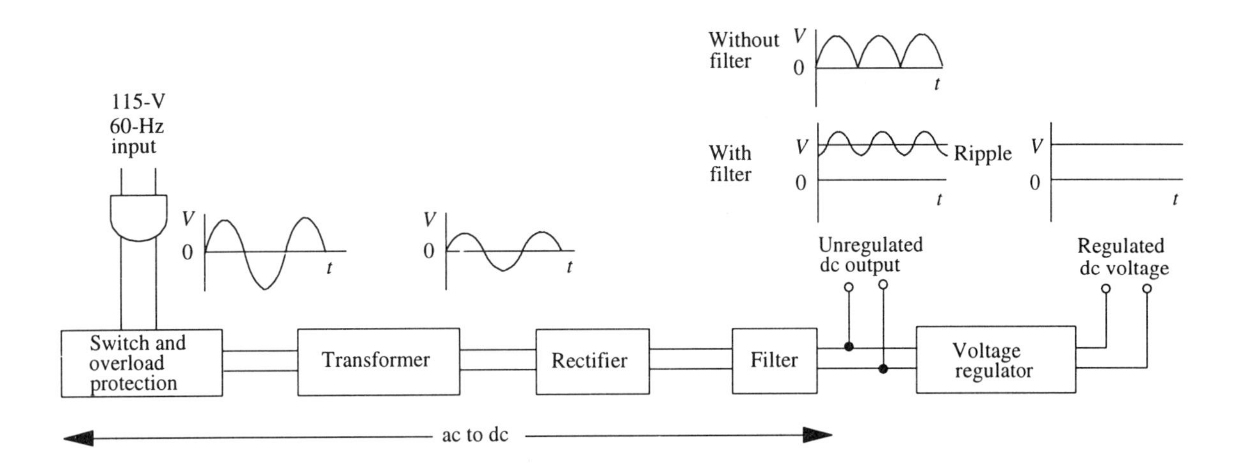

Figure 4–30. An ac-to-dc linear regulated power supply. The transformer converts the line voltage to the approximate desired voltage. A rectifier circuit converts the ac to a pulsating dc, which is smoothed by a filter network. A voltage regulator controls the output and further reduces the ripple of the output voltage. It is a dc-to-dc converter. The overall operation is ac–ac–dc–dc.

Overall Operation of Instrument Power Supply

All but one of the basic blocks in the overall block diagrams for the linear ac–dc regulated power supply in Figure 4–30 and the switch-mode ac–dc regulated supply in Figure 4–31 were introduced in previous sections of this chapter. The input and protection systems were not discussed previously, and they are the subject of this section.

Some of the protection devices for the two power supply systems in Figures 4–30 and 4–31 are similar, but others are different as a result of differences in the conversion steps shown in these figures. The line-operated linear regulated dc

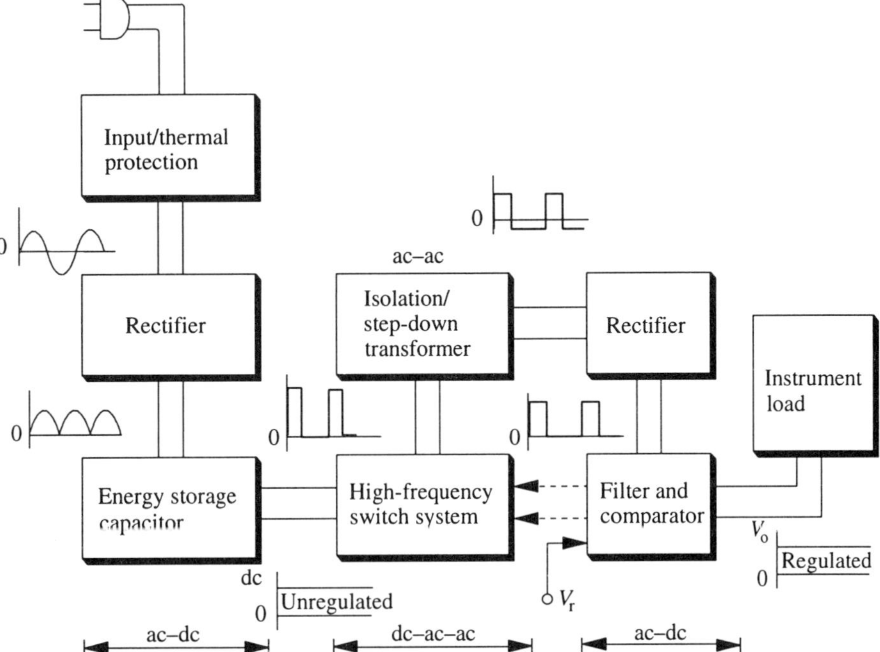

Figure 4–31. Block diagram of an off-the-line switch-mode-regulated power supply. This supply operates by a sequence of conversions: ac–dc, dc–ac, ac–ac, and ac–dc.

supply provides ac–ac voltage adjustment and isolation, ac–dc conversion with a filter, and then unregulated dc to regulated dc conversion with a linear regulator. This progression could be considered an ac–ac–dc–dc three-step process. In contrast, the line-operated switch-mode-regulated supply usually consists of an ac–dc off-the-line conversion followed by a dc-to-ac, ac-to-ac, and ac-to-dc conversion with a switch-mode regulator, high-frequency transformer, and filter. This progression could be considered an ac–dc–ac–ac–dc four-step process. The differences in techniques result in some differences in noise problems that need to be treated differently.

Input Protection

The line cord that is to be plugged into a suitable outlet is usually soldered to rugged input terminals within the power supply chassis, and the cord is made mechanically secure by some type of clamp or knot so that a jerk on the cord will not break the connection within the chassis. A poor cord connection inside the chassis could be dangerous because it could expose the user to line power on bare wires. Inside the chassis the power line is typically connected to a fuse, an ON–OFF switch with indicator light, and the input transformer (linear supply) or rectifier (switch-mode supply), as illustrated in Figure 4–32. Modern line plugs and cords have three connections: two for the power and one for a connection to **ground** as shown. If this ground wire is connected to the chassis, there is less danger of electrical shock. If a live wire touches the chassis, the short causes the protection device (fuse, etc.) to open the circuit.

Every power supply should have provision for overload protection to prevent destruction of electronic components and even severe overheating that could cause a fire. The most common devices for overload protection are fuses and circuit breakers.

Fuses. A popular type of fuse is illustrated in Figure 4–33. The fuse is in series with the primary winding of the transformer, and the fuse wire overheats and melts if excessive current is drawn by the load. Thus, the circuit is broken (forced to an open state), and the electronic equipment is disconnected from the power line until

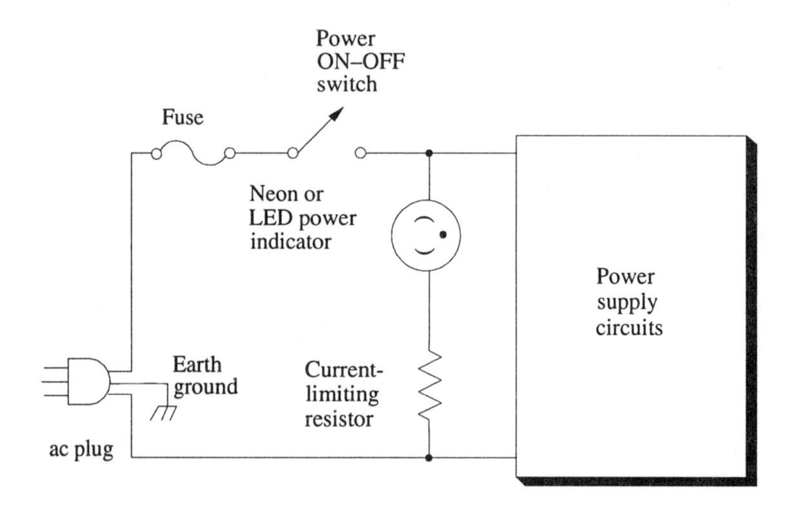

Figure 4–32. Input line connections to fuse, switch, indicator light, and power supply circuits. LED, light-emitting diode.

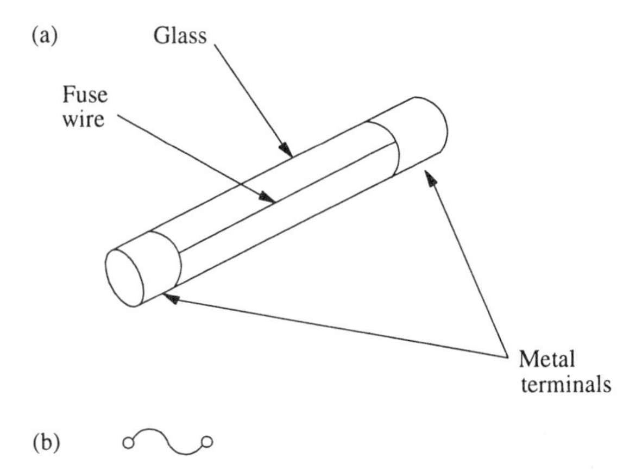

Figure 4–33. Pictorial (a) and schematic symbol (b) of typical power supply fuse. A short length of fuse wire is connected between two metal terminals. A small glass or plastic tube supports the metal terminals and protects the fuse wire.

the problem that caused excessive current can be located. The transparent glass tube makes it easy to see whether the fuse is blown.

The ratings on this type of fuse range from a fraction of an ampere to several amperes, and the fuses are made to be **slow flow** or **fast blow**. The wire in a fast-blow fuse melts if the current exceeds the rated current only briefly (for less than 0.1 s in some fuses). This fast operation is important in some applications, but in other applications, brief overloads are expected and a blown fuse would be a useless inconvenience. In these cases a slow-blow fuse with a wire that heats more slowly is used so that only continued overloads shut down the electronic system. A slow-blow fuse often contains a spring that pulls the fuse wire apart once it does begin to melt.

Circuit breakers. One type of thermal **circuit breaker** is illustrated in Figure 4–34. Current passes between points P and P′ through a strip of metal that is a thermal element. Excess current bends the metal strip, which then operates as a release spring. After the problem that caused shutdown of the electrical system is located and corrected and the metal strip has cooled, the reset button is pushed to reclose the circuit for operation. Magnetic circuit breakers are designed so that the overload current develops sufficient magnetic field in a coil of wire to activate a spring trigger mechanism and pull the contact points apart. An advantage of circuit breakers over fuses is that they can be reset rather than replaced. Fuses have the advantages of small size, lower cost, and applicability in low-power circuits.

ON–OFF switch and indicator light. Figure 4–32 shows a power ON–OFF switch and indicator light. The switch must carry the current demanded by the load

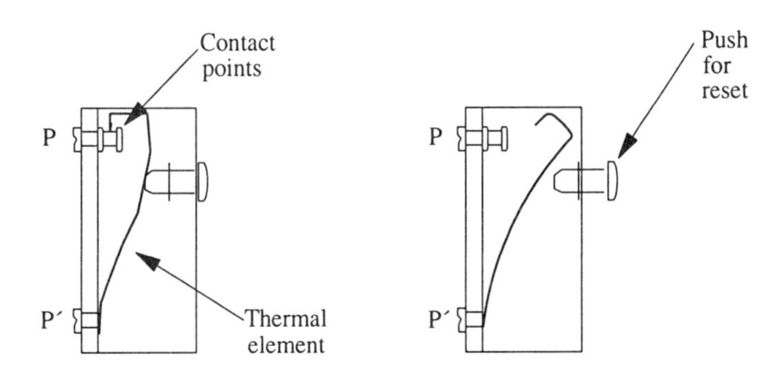

Figure 4–34. Thermal circuit breaker. An overload current through the thermal element causes it to heat and bend, releasing the latch and breaking the contact. P and P′, points of connection.

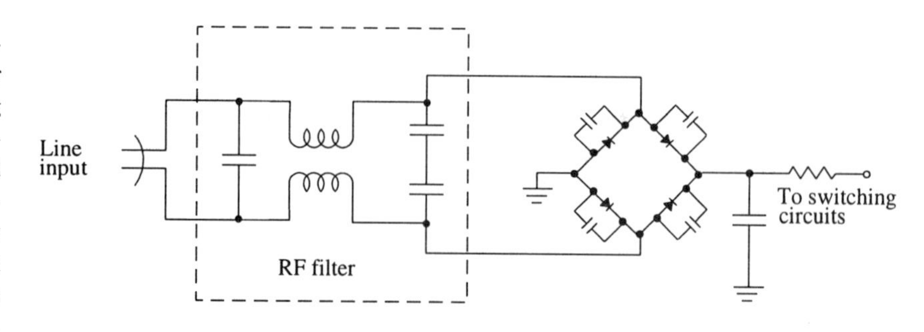

Figure 4–35. Input line protection. A radio frequency (RF) filter is connected between the line plug and the rectifier. Also, small-value capacitors are connected across the rectifier diodes to suppress high-frequency and very high frequency radiation caused by the on–off switching of the diodes at the input line frequency.

and be reliable over many ON–OFF cycles. The indicator light requires only a small current, and a current-limiting resistor allows just enough current through the light to indicate clearly when the power supply is ON.

Input line protection. Various types of voltage spikes and transient noise can come through the line and be transmitted into the instrument circuits. The instrument power supply can also generate radio frequency noise that goes into the line. These interferences can destroy semiconductor circuits and must be suppressed.

An example of line protection from noise at the input of a switch-mode power supply is given in Figure 4–35. Two protection devices are shown. One is a radio frequency (RF) filter connected between the input plug and the rectifier circuit. Another consists of small-value capacitors introduced across the rectifier diodes. These capacitors suppress high-frequency and very high frequency radiation caused by the ON–OFF switching of the diodes at the input line frequency. One way to prevent a surge current from destroying the rectifiers is to put a small feedback circuit between the RF filter and the rectifiers. Various circuit systems for so-called "soft starts" are also used. These allow the duty cycle of switching transistors to increase gradually to the normal operating point during the ON period.

It is good practice to connect a noise suppression unit between the ac power outlet and the line input to microcomputers or other equipment when line transient noise could damage the instrument or foul up circuit information.

Overload Protection

Several methods are used to ensure that electronic equipment is not subjected to excessive current that could destroy the circuits. These methods include the use of fuses, circuit breakers, and other thermal cutouts, any of which can be mounted in either the primary or the secondary side of the power supply. However, these are relatively slow acting, and it is much better to limit the current to a known safe value.

Current limiting. Current limiting can be accomplished by a variety of circuits, such as the one shown in Figure 4–36a. A resistor with a low resistance R_f is placed in series with the load to sense the output current. The low value for the resistor is important to keep power dissipation at a minimum. The voltage across the resistor is used to bias a protection transistor that reduces or turns off the conduction of the regulator. Also, it is possible to use the feedback voltage across R_f to latch the oscillator in a switch-mode supply. This latching removes the drive for the switching system, which rapidly removes the output.

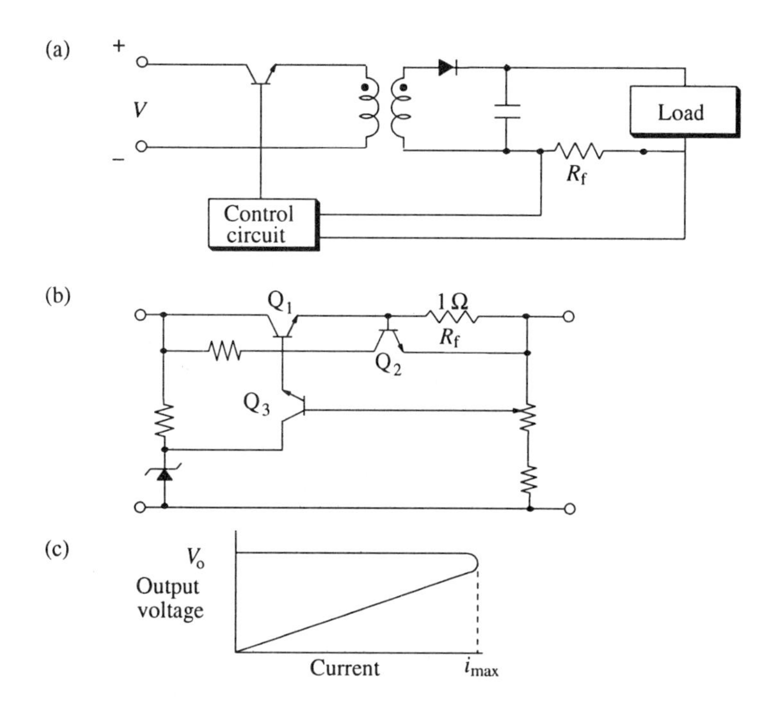

Figure 4–36. Current-limiting circuits for a switch-mode power supply (a), a linear supply (b), and a foldback of output current (c). Q, transistor.

A current-limiting circuit for a linear regulated supply is shown in Figure 4–36b. When the current is normal, the emitter base voltage of the current-limiting transistor Q_2 is insufficient to turn the transistor ON. With excess current the increased voltage drop across R_f turns Q_2 ON. Therefore, the base emitter voltage of the pass transistor Q_1 is reduced so that it conducts less and holds the output current to a safe level. (Transistor principles are discussed in Supplement 3.) When the current exceeds the preset value, it is **folded back**, as illustrated in Figure 4–36c.

Overvoltage. An overvoltage device that protects load circuits connected to the power supply is called a **crowbar** (Figure 4–37). A silicon-controlled rectifier (SCR) is a power device connected directly across the load. It fires if the output voltage goes too high, and the output is thereby brought down.

For the +5-V supply in Figure 4–37, a Zener diode of 6.2 V is chosen. Thus, the Zener diode does not conduct unless the output voltage exceeds the Zener voltage because of some failure in the power supply. As soon as the Zener diode

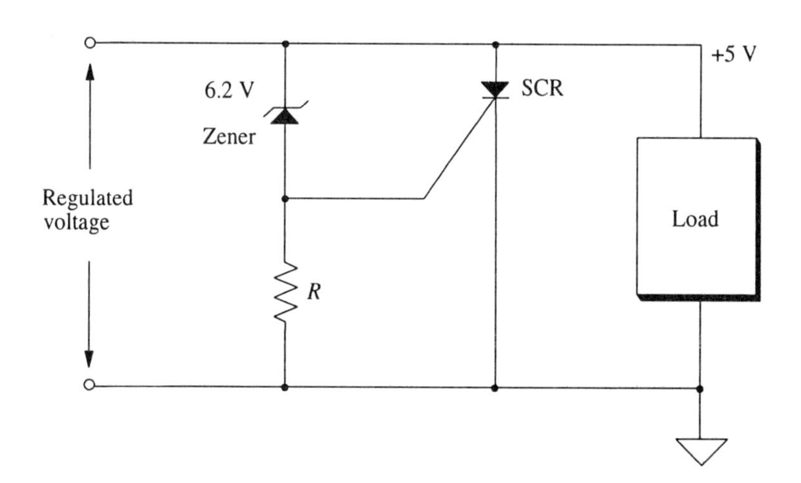

Figure 4–37. Overvoltage protection with a crowbar circuit.

conducts, there is gate current to the SCR, and the rectifier turns ON. This brings the output voltage down to a low level so that the load circuits will not be damaged.

Battery Chargers

Batteries are charged by reversing the electrode reactions that occur when the batteries are used as voltage sources. A simple battery charger is shown in Figure 4–38. A dc charging current is obtained by introducing a diode in series with the battery and applying a sufficiently high voltage at the secondary of the transformer to cause the diode to conduct during part of the ac cycle. By varying the resistance R or the sine wave amplitude, the rate of battery charging can be adjusted.

When the magnitude of v_s exceeds the battery voltage V_B, the diode conducts and produces a charging current i_c in the correct direction for recharge. When the ac voltage is less than V_B, the diode is reverse-biased to prevent discharge.

More sophisticated battery chargers utilize the regulated linear or switch-mode power supplies described in previous sections. They can be programmed with a digital logic system so that the charging follows a prescribed protocol most suitable for the type of battery being charged.

Batteries can be recharged by energy sources other than ac-powered dc supplies. Solar cells or fuel cells can be used as the primary energy source. The batteries are recharged by sunlight, and the energy stored in the batteries power the loads during the night.

Uninterruptible Power System

A power system that will provide reliable power whenever the normal source of power is outside acceptable limits and without interruption of suitable power for the load is referred to as an **uninterruptible power system** (UPS). Originally, a UPS provided backup power to loads that needed to maintain operation if the utility line power failed. At present, the quality of the utility power in many locations is so poor at certain periods that a UPS is required. Such brownout periods might occur so frequently that a UPS is essential even though costly. Computer instrumentation is a prime example of a system that would often require a UPS. In emergencies a UPS can supply power until an auxiliary generator is put into operation. Also, sensing circuitry that enables a computer to perform an orderly shutdown before the UPS runs out of power is now often included within computer systems.

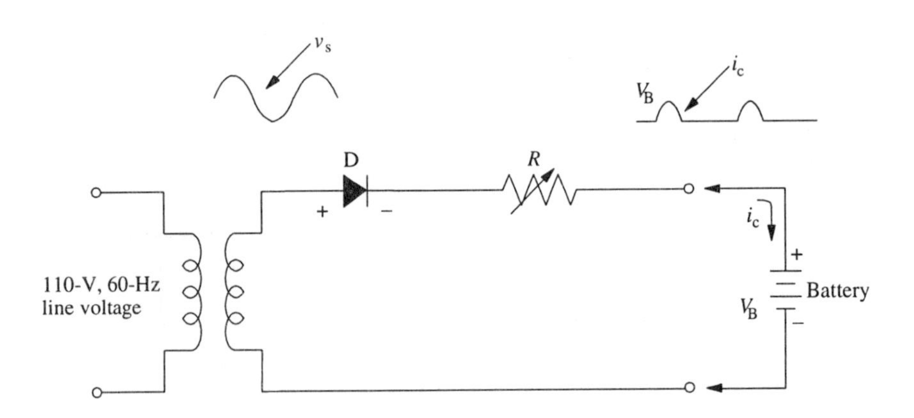

Figure 4–38. Battery charger.

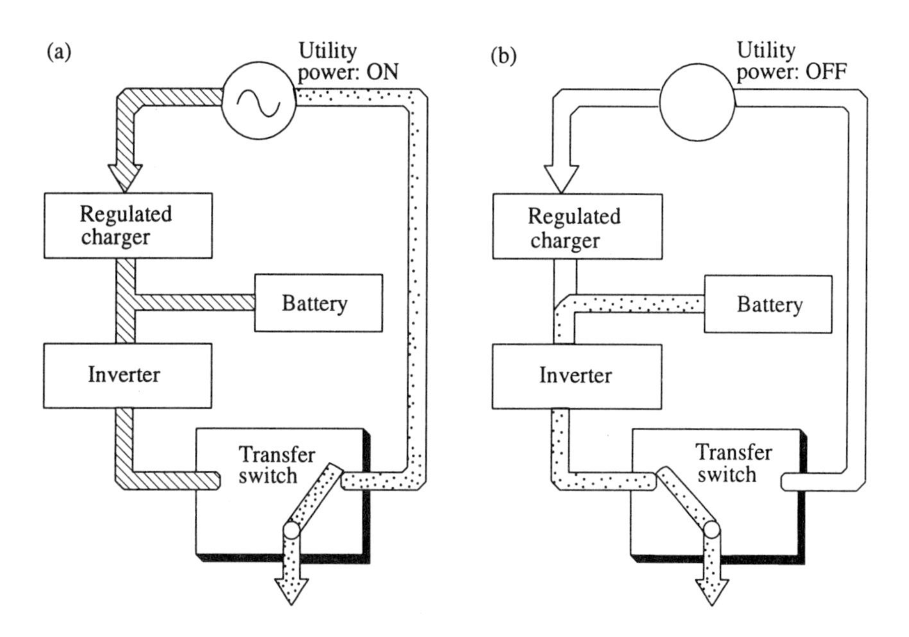

Figure 4–39. Uninterruptible power system.

The components of a solid-state UPS are an **energy storage device** (battery), which provides power during a utility power blackout or brownout; a **battery charger**, which restores the energy when utility power is restored; an **inverter**, which is powered by the battery; and an automatic transfer switch which connects the load to either the utility line or the inverter.

A UPS is shown in Figure 4–39. This is an off-line system that is suitable wherever the utility line power is normally suitable and free of transients. In the situation shown in Figure 4–39a, the utility power is usually applied to the load. The regulated charger needs only to supply recharge current to the battery and a small current to the inverter operating at no load. Figure 4–39b shows that if the utility power fails, a sensing circuit provides a control signal to the transfer switch so that the load is now powered by the inverter. When the utility power again comes on line, the transfer switch is automatically changed so that the load is again powered by the line. A more complex UPS is necessary if the utility power is normally of poor quality.

4–8 Regulating dc Current

Many transducers, light sources, detectors, and other devices require a constant current through them rather than a constant voltage across them for suitable operation. This section introduces current control.

The concept of current regulation is illustrated in Figure 4–40. The purpose of current regulation is to control the current i_c through the load resistance R_L at a desired constant value. The load resistance can vary during operation (as indicated by the arrow through R_L), and the dc voltage V from the unregulated supply can vary with time. Because the load is in a simple series circuit, the current is

$$i_C = \frac{V}{R_L + R_f + R_c}$$

Therefore, if R_L or V changes in value, the current also changes unless the value of a variable series resistance R_f is altered to compensate. In principle, whenever the

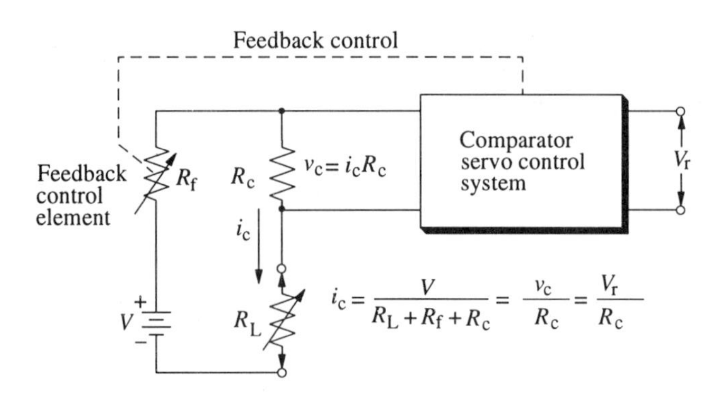

Figure 4–40. Regulated constant-current power supply. The current i_c through load resistance R_L is controlled by varying the feedback element R_f. The voltage drop $v_c = i_cR_c$ is held equal to the reference voltage V_r by the servo control system.

current tends to deviate from the desired value because of changes of R_L or V, the deviation is detected by observing the voltage v_c across a control resistance R_c that is fixed and known ($i_c = v_c/R_c$). The control voltage v_c is continuously compared by the voltage comparator with a known reference voltage V_r. If there is any difference between V_r and v_c, the electronic servo control system (described in the next chapter) immediately adjusts the resistance R_f of the feedback control element so that the desired value of i_c is maintained. The control system can be designed to respond in the microsecond-to-millisecond range, giving excellent current stability and effective suppression of ripple from the unregulated supply. Correctly choosing values of V_r and R_c allows the controlled value for i_c to be selected. This assumes, of course, that the voltage source V has sufficient voltage and current capability to supply the desired current and the IR drop across R_L, R_c, and R_f. This is also related to what is called the **compliance voltage**, which defines the load voltage range over which load current is held constant within specification. The compliance voltage determines the highest resistance that the load can have for a specified constant current.

4–9 Converting Light into Electrical Power

Solar cells are being used increasingly for the direct conversion of sunlight or even room light into electrical power. These cells are especially useful for instrumentation in remote locations. They have already found wide acceptance in very low power instrumentation such as pocket calculators.

The energy in radiation from the sun that strikes the Earth each day is many times our yearly consumption of electrical energy. On a summer day the solar radiant power is about 1 kW/m^2 at the Earth's surface. Through hydroelectric generators and fossil fuel combustion, solar radiation has long been used indirectly to obtain electrical power, but until recently there has been no efficient way of directly converting this radiant power into electrical power.

In recent years semiconductor solar cells for direct conversion of solar radiation into electrical power have been designed. These cells provide high power capacity per unit weight and have been used successfully on space vehicles and communication satellites. A representation of a silicon photovoltaic cell is shown in Figure 4–41a. Such silicon wafers, 1×2 cm in area, have been used to build up single-panel arrays of many thousands of wafers to provide a capacity of several hundred watts. A typical array is shown in Figure 4–41b.

In the silicon cells of Figure 4–41, a very thin layer of p-type semiconductor is formed on an n-type silicon strip (or vice versa) to form a pn junction and provide

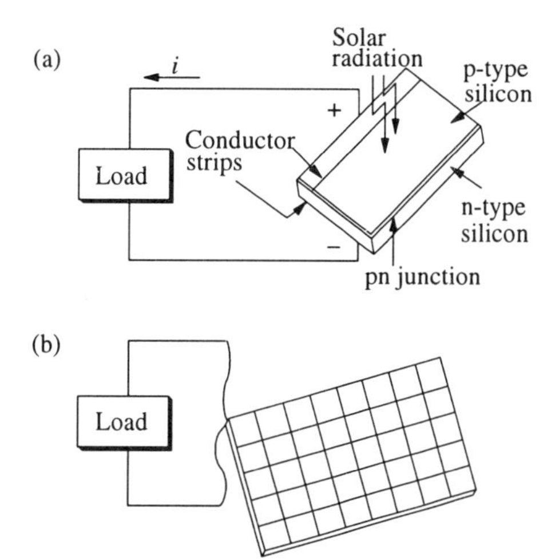

Figure 4–41. Solar cells. (a) Silicon photovoltaic cell. (b) A series–parallel array of 40 cells (each 1 × 2 cm) cells could provide about 1 W of electrical power for 10 W of incident sunlight.

the light-sensitive face. A narrow conducting strip serves as the collector terminal. The bottom of the cell is nickel-plated and tinned. The diffusion of majority carriers from both n- and p-type materials causes a combination of holes and electrons that sets up a barrier potential across the junction. The p- and n-type materials contain many valence electrons that are not affected by the impurity doping.

However, a photon can interact directly with a valence electron and provide enough energy to promote the electron to the conduction band. The free electron and the hole are now attracted by the barrier potential at the pn junction and travel in the direction opposite that of the majority semiconductor carrier. That is, the photoelectrons travel toward the positively charged n-type silicon, and the holes move toward the negatively charged p-type silicon. The result is a current *i* through a load, as shown in Figure 4–41a. The arrays used in satellites generally have an output of about 10 W/lb (ca. 0.5 kg). In view of the huge supplies of silicon and sunlight, increased application of solar cells can be expected as the technology improves.

Related Topics in Other Media

Video, Segment 4

- Different types of power supplies
- Using diodes for ac-to-dc conversion
- Filtering with capacitors
- Regulation for control of output voltage and ripple reduction

Laboratory Electronics Kit and Manual

- Section C
  - ¤ Constructing regulated power supplies
  - ¤ Testing operation of power supply and verifying regulator action

Chapter 5

Manipulating Analog Data with Operational Amplifiers and Servo Systems

One of the major advances in modern electronics is the **operational amplifier,** usually called an op amp or OA. It is the key building block in analog electronic circuits and a basic part of analog-to-digital (ADC) and digital-to-analog (DAC) converter circuits. The op amp is so widely used that many millions are sold each week. One manufacturer alone supplies nearly a quarter billion units per year.

By itself, the op amp is a conceptually simple device: a circuit that greatly amplifies the voltage difference between two input terminals. However, by connecting a few passive components in unique ways to an op amp or combination of op amps, a great variety of important control and measurement circuits can be implemented. The design of the op amp allows three distinct modes of operation: as **a comparator**, **a voltage follower**, or **a current follower**. These different modes are used individually or in various combinations to provide thousands of useful electronic circuits.

The principles of op amp circuits and many useful applications of the circuits are described in this chapter. These applications include converting from one analog signal to another (such as current to voltage or charge to voltage) and adding, integrating, differentiating, amplifying, buffering, and comparing analog signals.

This chapter starts with a general discussion of a universal concept in control and measurement systems: the principle of **null comparison with feedback control.** The null comparison concept is basic to understanding the op amp and many other circuits that are discussed here. Control by **null comparison** consists of varying the controlled quantity until there is no difference when it is compared with a reference standard. A **null comparison measurement** is based on varying a calibrated reference standard in sufficiently fine increments such that the difference between measured and reference quantities is zero (at null).

5–1 Controlling and Measuring Analog Quantities

Control and measurement are closely related. The measurement of a quantity is prerequisite to its control, and the control of a quantity implies monitoring of the quantity and application of corrective action if the quantity deviates from the desired value.

Any system that detects a difference or error between actual and desired states of a controllable quantity, such as temperature or position, and uses the error data to control a device that corrects the error is classified as a **servo system**. If the quantity to be controlled is the position of an object, an electric motor or hydraulic system might be used to control the object until the desired and actual positions coincide. Such a system is called an **electromechanical servo system** or **servo mechanism.** Such systems are essential in robots and automated industrial machinery and for positioning critical parts in scientific instrumentation.

2861-2/94/0121/$10.00/1 Published 1994 American Chemical Society

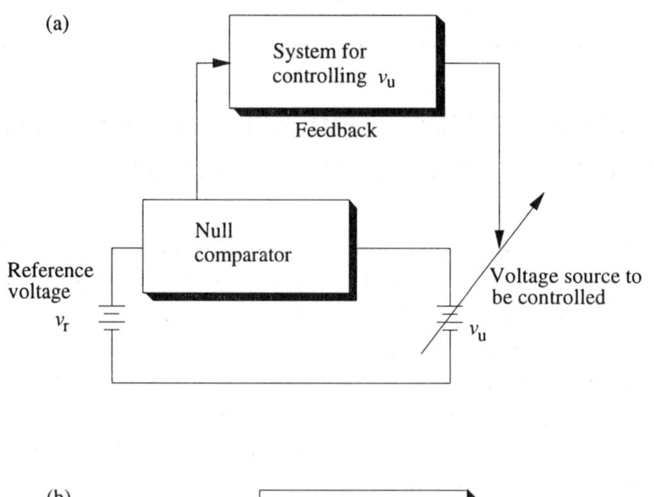

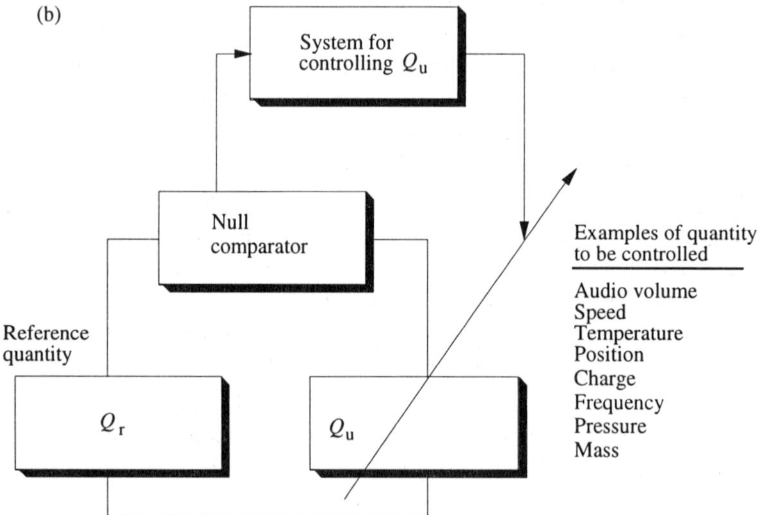

Figure 5–1. Null comparator control systems for control of a voltage source (a) and for control of any quantity, such as speed and temperature (b). The off-null signal from the null comparator activates the system for controlling the desired quantity to maintain the quantity equal to a preset reference quantity.

Servo Control Systems

The power supply regulators that were described in Chapter 4 are examples of servo control systems, as illustrated again in Figure 5–1a. The output voltage v_u of the power supply is compared with a reference voltage v_r. The off-null signal from the null comparator activates the system for controlling v_u to maintain it equal to v_r, even with changes of load or raw supply. In Figure 5–1b a more general control system is illustrated for correcting any quantity Q_u so that it is maintained equal to a reference quantity Q_r. The automatic volume control on a radio and the automatic speed control on an automobile are examples of controlling a parameter to a preset reference value.

The controlled quantity Q_u might be current, light intensity, resistance, temperature, or any of hundreds of other quantities.

Servo Measurement Systems

Servo systems that are based on the null comparison principle are also used for accurate measurements. Instead of controlling the quantity Q_u to make it equal to a preestablished reference Q_r, the reverse is done. We see in Figure 5–2 that the

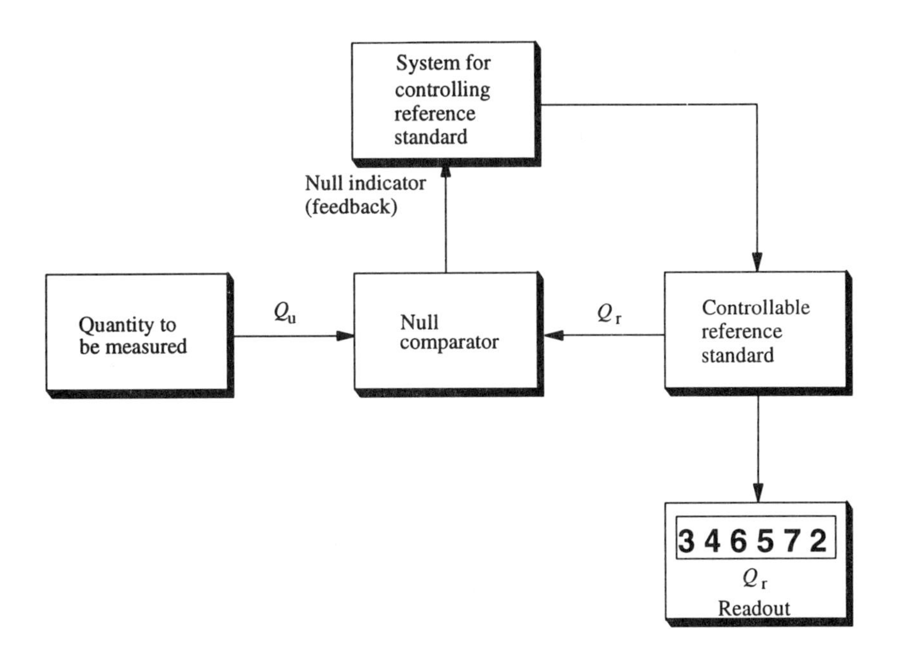

Figure 5–2. Null comparison measurement system. The off-null signal from the null comparator activates the control system, which varies the reference standard until it is equal to the quantity to to be measured. The **accuracy** is the degree of agreement between the **measured value** at null balance and the **true value.** The **precision** is the degree of repeatability and resolution of the measurement.

reference quantity Q_r is varied and maintained equal to the quantity Q_u to be measured. The quantity Q_u to be measured might be mass, length, voltage, charge, or frequency, but in any case the null comparison principle remains the same. This is an extremely important measurement concept. A study of Figure 5–2 shows that the null comparison measurement is based on a **variation of the number of units of a controllable reference standard until the reference value equals the unknown value**. Thus, the reference must be designed to vary in sufficiently small increments to satisfy the desired resolution of the measurement. Also, the null comparator must be sufficiently *sensitive* to detect the smallest difference between Q_u and Q_r that would give the desired resolution. The concept of a **null comparison measurement** is certainly simple but also profound. Given a sufficiently sensitive null comparator, the measurement accuracy does not depend on the absolute characteristics of the comparator. That is, the sensitivity of the null comparator might change by several percent with temperature, humidity, or aging of some component, but this change does not affect measurement accuracy if the reference standard remains stable.

The use of the null comparison technique for voltage measurement is shown in Figure 5–3a. The variable reference voltage v_r is manually adjusted until it equals v_u, as visually observed from the null comparator. The value of v_r is indicated by observation of the indicator scale, which is mechanically coupled to the reference voltage adjustment. If, after the balance is complete, v_u changes, then v_r must of course be reset until it again equals v_u. In Figure 5–3b, the null comparator indicates the direction of unbalance between v_u and v_r. The null comparator signal is in one direction when $v_u > v_r$ and in the opposite direction when $v_u < v_r$. At balance, the null comparator indicates zero difference between v_u and v_r. This provides another very significant advantage of using the **null comparison technique** for voltage measurements. When $v_r = v_u$, there is no potential difference to cause charge flow or current. Stated another way, at perfect null balance the measurement system has infinite input resistance. There is no "loading error." This, of course, is *ideal* for making voltage measurements. An automated electromechanical comparator system for measurement of voltage is shown in

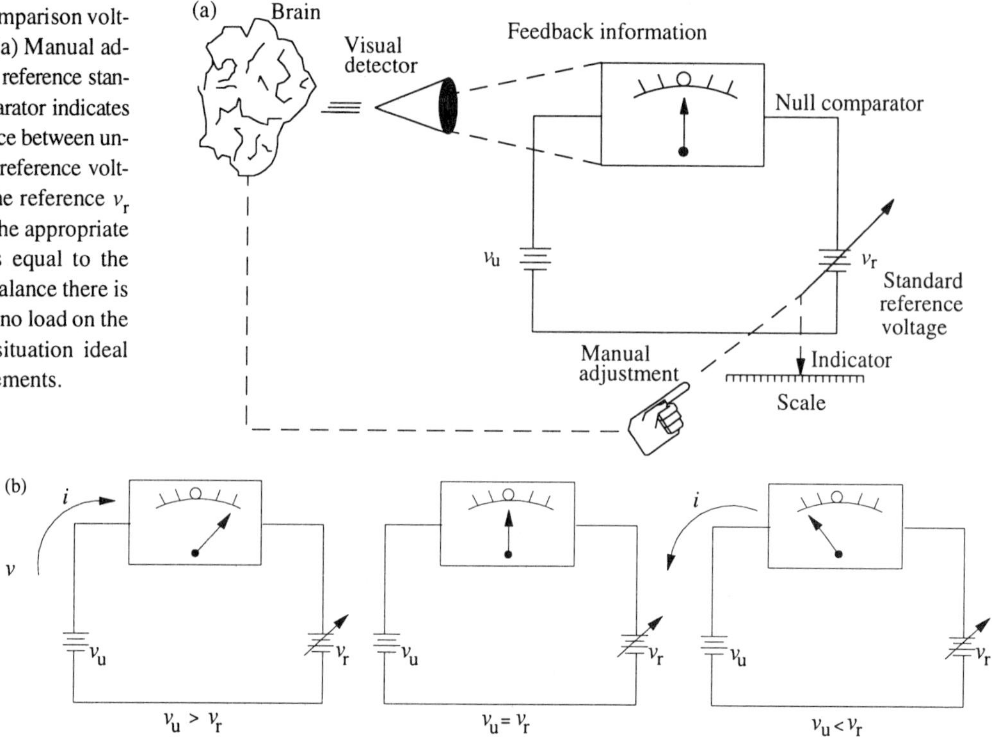

Figure 5–3. Null comparison voltage measurement. (a) Manual adjustment of variable reference standard; (b) null comparator indicates direction of unbalance between unknown voltage and reference voltage source. Thus, the reference v_r can be adjusted in the appropriate direction until it is equal to the unknown. At null balance there is no current and thus no load on the voltage source, a situation ideal for voltage measurements.

Figure 5–4. The reference voltage v_r is automatically maintained equal to v_u by a motor drive servo system and thus continuously *follows* v_u. This automated voltage measurement system will prove useful in introducing some voltage follower concepts in Section 5–3.

Figure 5–5 shows how the null comparator and the two voltage sources v_u and v_r form a simple series circuit. In practice, one point of this circuit is connected to system "common." The choice of which point is common is often an important consideration. The three possible locations of system common in the voltage comparator circuit are shown in the three figures. In Figure 5–5a, v_u and v_r are connected directly to system common. The null comparator connections are at voltages v_u and v_r with respect to common. Because a difference must be detected, the null comparator must have true difference inputs in the range for measurement of v_u. Therefore, the null comparator must be completely independent of the system common (referred to as floating), and the two inputs must respond in precisely the same way for input voltages with respect to the common (*see* Supplement S1–3 for a discussion of ground and circuit common).

In Figure 5–5b the **common** is connected between the null comparator and v_u. Because one terminal of the null comparator is always connected to common, the other terminal will also be at the common potential at the null point. That is, when v_r equals v_u, both terminals of the null comparator are at common potential even though only one terminal is physically connected to common. This condition is represented by the symbol for common surrounded by parentheses. It is called **virtual common**. Therefore, even though the reference voltage is "floating," one terminal is always at the common voltage when v_r equals v_u. Take a few moments to review the definition of virtual common given in Figure 5–5.

The third location for the system common is illustrated in Figure 5–5c. In this case v_u is floating, but at null balance one of its terminals is at virtual common,

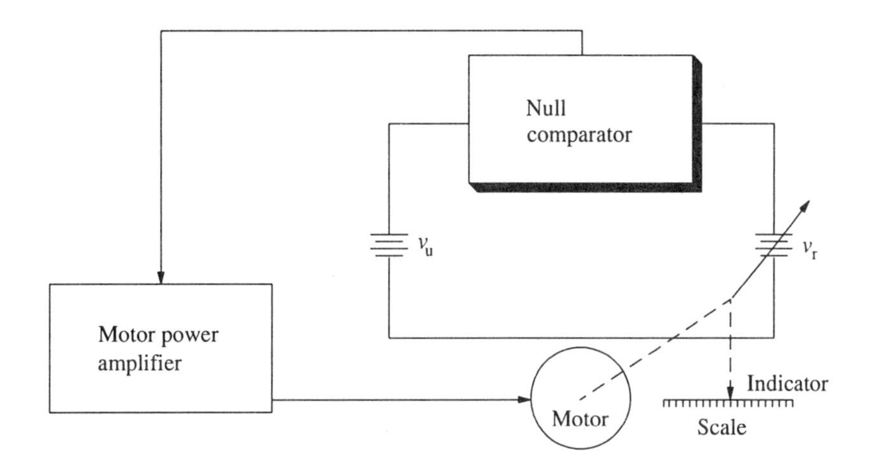

Figure 5–4. Automated electromechanical servo system for voltage measurements. The motor shaft is coupled to the variable voltage v_r, as indicated by the dashed line. The amplified null comparator output is applied to the motor, which varies v_r in the direction needed to equal v_u.

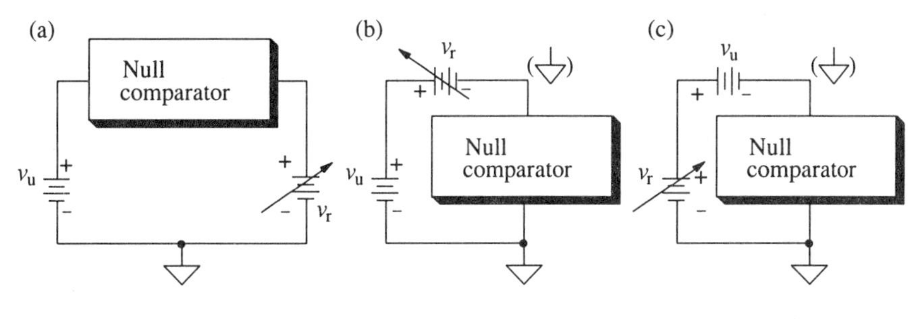

Figure 5–5. Location of system common in null comparison system. (a) Floating null comparator; (b) reference source is not connected to common but one terminal is at **virtual common** when reference and unknown voltage sources are at null balance; (c) unknown source is not connected to common but situation is similar to that in part b. In parts b and c, the null comparator is connected between **common** and **virtual common** at null balance. A **virtual common point** in a circuit is a point that is essentially at the **circuit common** potential but is not connected to the circuit common.

while both the reference source and the null comparator have one terminal physically connected to common.

The circuits shown in Figures 5–5b and 5–5c both have one end of the null comparator connected to common and the other terminal at virtual common when at null. Thus, there is no voltage drop across the null comparator. This concept of virtual common is essential for understanding the performance of operational amplifiers in many circuits.

5–2 Comparing Analog Signal Levels by Using Operational Amplifiers

Many physical and chemical quantities, such as temperature, pH, and velocity, can be converted by transducers into related voltages. Thus, the input quantities can be monitored, measured, or controlled through manipulations in the electrical domains.

For implementation of many functions and operations in the analog domain, the op amp is the key building block. Therefore, basic characteristics of the op amp are investigated first in this section. Next, the widespread use of the op amp as a **voltage comparator** is discussed and illustrated with a few applications (e.g., level detector, zero-crossing detector, and for interfacing to digital systems). The op amp voltage comparator is used extensively in circuits described in subsequent chapters. The **comparator mode** is, of course, only one way of using the op amp. It is often used in combination with either the **voltage follower** or the **current follower** modes described in Sections 5–3 and 5–4. These three modes of operation are basically different ways of using the same device. They provide the basis for the powerful array of op amp analog and hybrid analog–digital circuits that are essential to modern electronics.

The Op Amp

The integrated circuit op amp consists essentially of three interconnected basic stages, as illustrated in Figure 5–6. A high-input-impedance difference amplifier provides the two inputs. This amplifier connects to a high-gain voltage amplifier designed so that its output can swing both positive and negative. The third stage is an amplifier designed to provide very low impedance output, because this kind of output minimizes the possibility of loading. In other words, relatively high output currents can be supplied without distorting the output voltage. All of these stages are shown in Figure 5–6 enclosed within a triangle, which is the classic op amp symbol.

Typically, the overall gain of the op amp is greater than 100 000 in the **open-loop mode.** That is, when there is no external feedback from output to inputs, the operation is described as open-loop, and the gain A is maximum.

The op amp symbol as shown in Figure 5–7a is the way that it will frequently appear in schematic circuit drawings. Carefully note this symbol and the labels, which will be used in subsequent diagrams. The difference voltage between the two input terminals is labeled v_s, and the output voltage is labeled v_o. In Figure 5–7b, v_u is shown connected to one input, v_r is connected to the other input, and both inputs are shown in reference to common. In this case the op amp comparator is floating. Inside the triangle, one input is labeled with a plus sign and the other is labeled with a minus sign. These signs indicate the sign of the output voltage v_o when that input is the more positive. This can be seen from the equation for the output voltage v_o shown in Figure 5–7c. When v_+ is more positive than v_-, the output voltage v_o is *positive*. When v_- is more positive than v_+, the output voltage v_o is *negative*. Therefore, the terminal labeled with a minus sign (–) inside the

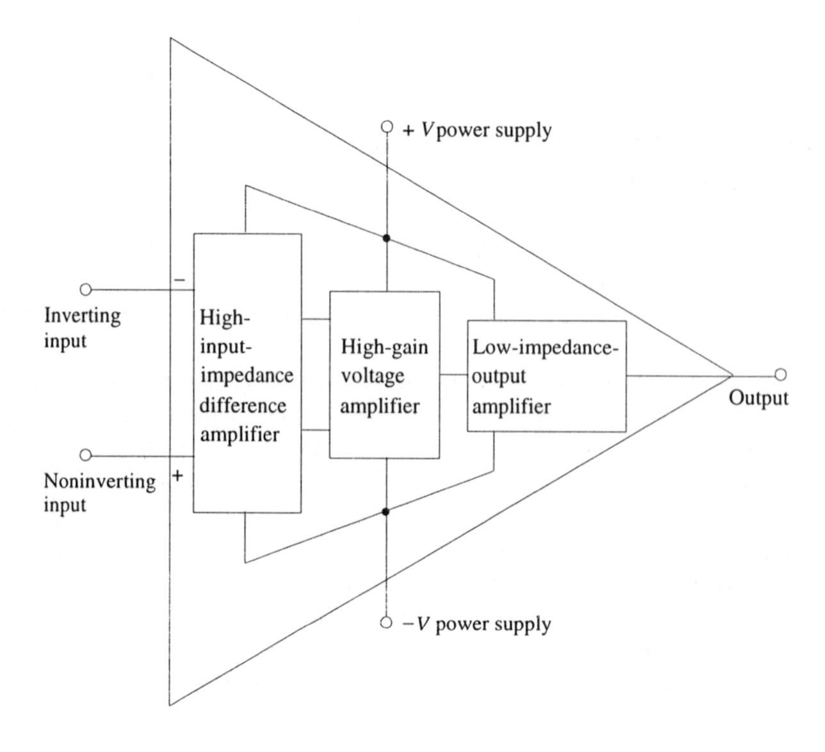

Figure 5–6. Basic circuits in the op amp. They consist of an input stage, a very high gain amplifier stage, and an output stage capable of supplying current to a load over a voltage range between the values of the plus and minus voltage power supplies.

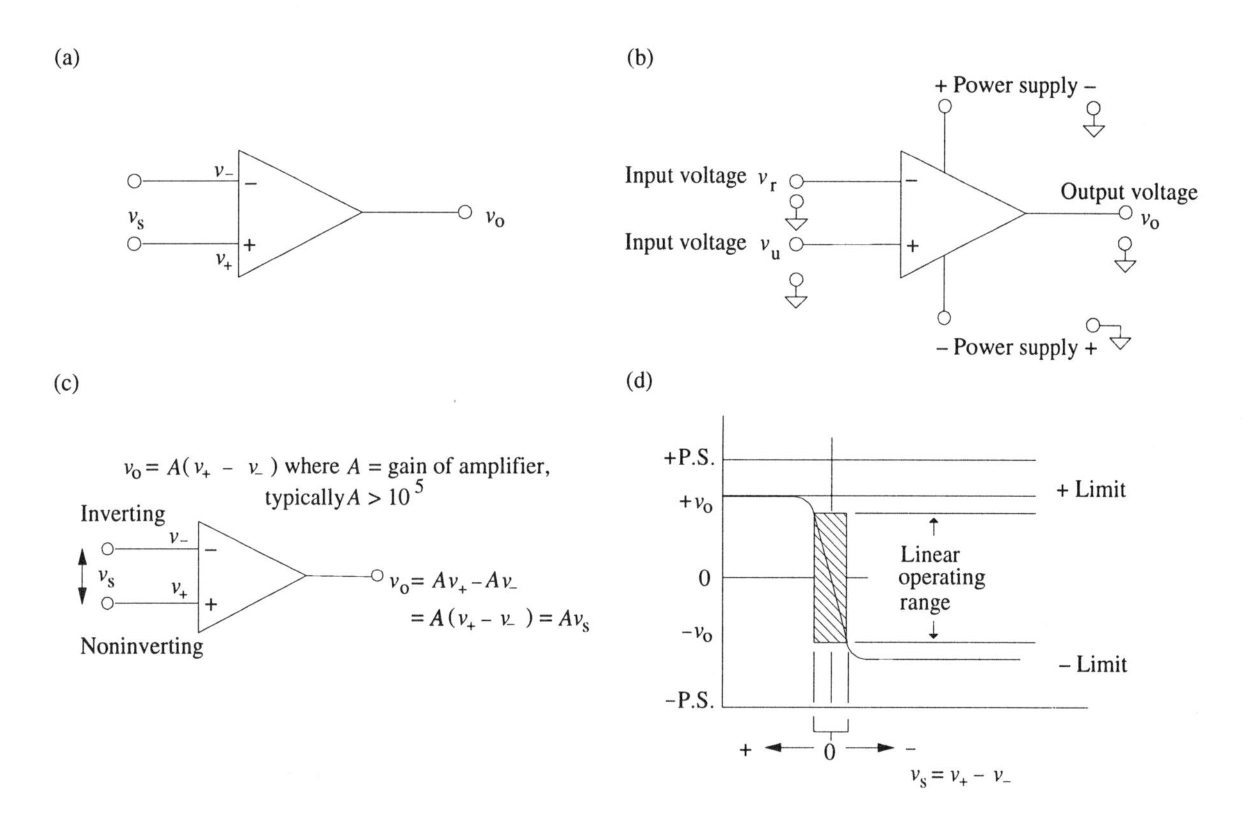

Figure 5–7. Characteristics of the op amp. (a) Symbol for op amp; (b) necessary power supply (P.S.) connections, which are typically omitted in circuit diagrams to reduce confusion, with the understanding that they are properly connected; (c) equation relating output to input and to amplifier gain; (d) transfer function relating input voltage difference to output voltage.

triangle is called the **inverting input** and the terminal labeled with a plus sign (+) is called the **noninverting input**. The plus and minus voltages provided by power supplies and shown in Figure 5–7b are necessary, of course, to operate the amplifier, but their connection is not usually included in instrument circuit diagrams.

Figure 5–7d relates the output voltage v_o of the op amp to the difference voltage v_s between the inputs. Because of the high gain (e.g., A = 100 000), only a very small input voltage difference causes the output to swing to either its plus or minus voltage limit. The output limits or limiting voltages for v_o are determined by the internal design of the op amp and are always somewhat less than the power supply voltages that are used. For example, with ±15-V supplies, the limits will be about ±13 V. Op amps can be operated over a range of power supply voltages, but typically, the range will be ±5 to ±25 V.

The Op Amp Comparator

From the equation for the output voltage (Figure 5–7c), it is apparent that if the gain A of the amplifier is very large, only a *small* voltage difference at the two inputs causes the amplifier output to go to limit. Whether it goes to plus limit or minus limit depends on which input is positive relative to the other. Thus, the output polarity indicates whether v_r must *increase* or *decrease* to equal v_u. Ideally, the output voltage would flip from plus limit to minus limit with an infinitely small difference voltage at the inputs. Because the gain A is finite, although large, there will be an operating range over which v_o will change linearly from plus limit to minus limit for a small range of input difference voltages. This is illustrated in Figure 5–7d. If the limits are ±12 V and the comparator gain is about 100 000, the output is at limit for any off-null deviation greater than about 0.12 mV.

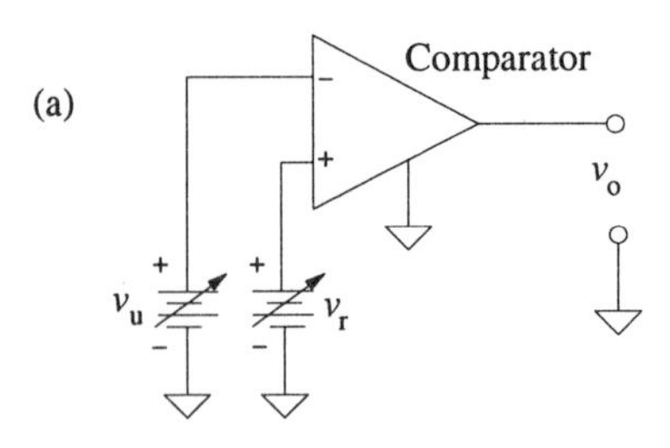

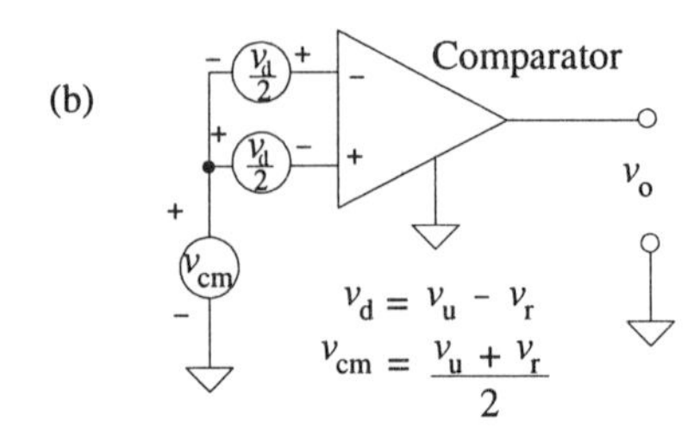

Figure 5–8. Equivalent input voltage sources that include both common mode and difference voltages. The ideal comparator response is independent of the value of the common-mode voltage.

Inputs that are perfectly balanced with respect to the common voltage respond only to a difference in voltage. The comparator shown in Figure 5–8a is in the floating null detector configuration (Figure 5–5a). To evaluate the balance of the comparator inputs, it is useful to consider that the sources v_u and v_r are a combination of a **difference voltage** $v_d = v_u - v_r$ and a **common-mode voltage** v_{cm} (Figure 5–8b). The common-mode voltage is the average of the two input voltages, $(v_u + v_r)/2$. The **difference gain** A_d is the output voltage change Δv_o for a given input voltage change Δv_d, or $A_d = \Delta v_o / \Delta v_d$. The common-mode gain A_{cm} is similarly $A_{cm} = \Delta v_o / \Delta v_{cm}$. Ideally, the common-mode response should be zero. The commonly used figure of merit for the quality of the input balance is the **common-mode rejection ratio (CMRR)**, which is the ratio of the difference gain to the common-mode gain. Therefore,

$$\text{CMRR} = \frac{A_d}{A_{cm}} \tag{5–1}$$

Of course, the better the balance, the higher the CMRR. A typical comparator CMRR is 10^4 or more. If such a comparator is used in the floating null detector configuration (Figure 5–5a), the null indication for $v_u = v_r = 1$ V could be different from that for $v_u = v_r = 0$ V by 1 V/CMRR, or about 0.1 mV. This error could be reduced by choosing a comparator with a higher CMRR, or it could be eliminated by choosing a configuration in which the comparator is not floating.

Zero-Crossing Detectors

If an input signal such as a sine wave of several volts amplitude is applied to the noninverting input of the op amp comparator shown in Figure 5–9a, the output will undergo a change from one limit to the other (about ±13 V for ±15-V power supplies) in the zero-crossing region of the input signal. Because v_r at the inverting input equals zero, the output goes to positive limit when the sine wave input goes slightly positive. When the sine wave goes slightly negative, the output goes to

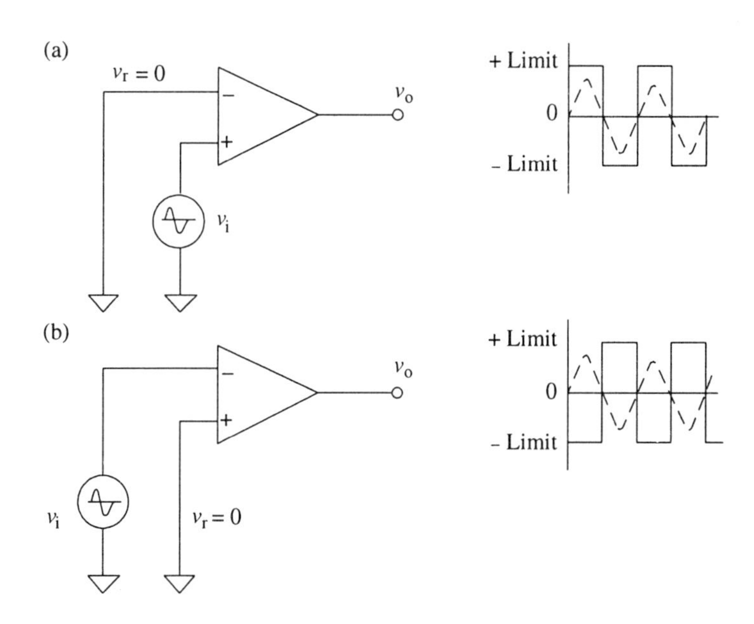

Figure 5–9. Zero-crossing detectors. (a) Noninverting, for which the polarity of the output is the same as that of the input; (b) inverting, for which the input and ouput polarities are reversed.

negative limit. Thus the comparator also acts as a sine-wave-to-square-wave converter.

If the sine wave input is applied to the inverting input and the noninverting input is connected to common, as shown in Figure 5–9b, the output will go positive when the sine wave goes slightly negative, just the opposite of the noninverting zero-crossing detector in Figure 5–9a.

Level Detector

The level detector is similar to the zero-crossing detector of Figure 5–9 except that the reference input v_r is connected to a reference voltage instead of to common, as shown in Figure 5–10. In this example the level detector is used as part of a control system. The signal voltage v_u is from a voltage transducer connected to a system that is to be controlled. Whenever the transducer voltage exceeds a preset value v_r, an output signal that initiates a correction to the system is generated and maintained until the transducer output goes below the preset reference value. The positive level detector of Figure 5–10 can be changed to a negative-level detector

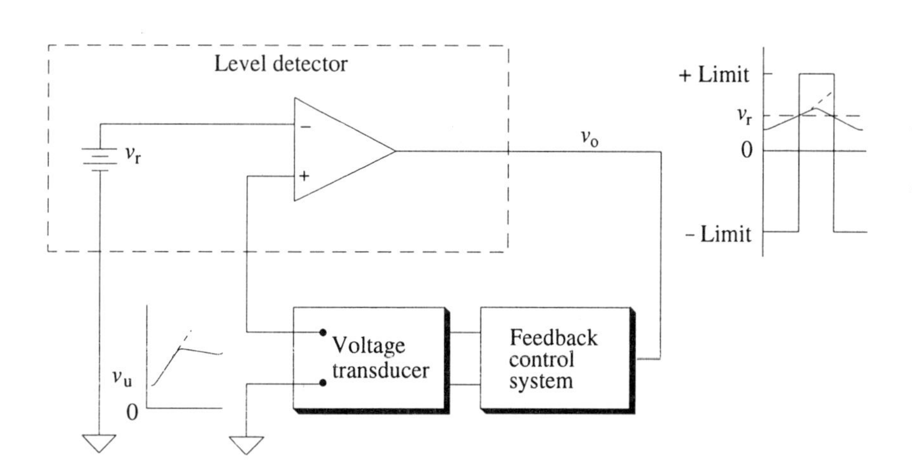

Figure 5–10. Level detector for feedback control of a system. The comparator output v_o is at + limit whenever the transducer output exceeds a set reference level v_r.

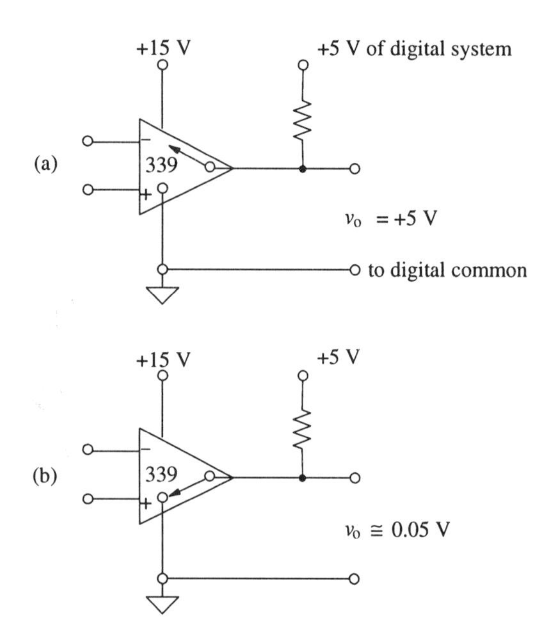

Figure 5–11. Level detector using an op amp (type 339) that provides interfacing to a digital system.

by reversing the polarity of the reference v_r. The level detectors are used in many appliances such as smoke detectors and sound-activated alarms.

Voltage Comparator for Interfacing to Digital Systems

A specially designed op amp comparator consists of an output transistor stage that has an open collector. Therefore, the output can be connected through a resistor to the digital HI-level voltage [+5 V for transistor–transistor logic (TTL) integrated circuits] as shown in Figure 5–11. If the noninverting input is more positive than the inverting input, the output switch is open and the output is digital HI (+5 V); if the inverting input is more positive, the output switch is closed and the output is LO (about 50 mV). Therefore, the output can be connected directly to a digital system.

5–3 Creating Ideal Voltage Sources

The problem of loading a voltage source so as to distort its output was emphasized in Chapter 3. It was shown that the inherent internal resistance of the source must be very small relative to the input resistance of the circuit or instrument to which it is connected. This is not feasible when the inherent resistance of the voltage source is extremely high, especially for many voltage transducers. However, the op amp can be connected so as to create nearly ideal voltage sources from those that would otherwise have loading problems. The circuit is known as an **op amp voltage follower**, and it has many applications.

The Op Amp Voltage Follower

Modern op amps are designed to have a very high input impedance (10^{12} to 10^{15} Ω) and very low output impedance (about 0.01 Ω). Therefore, if the op amp is

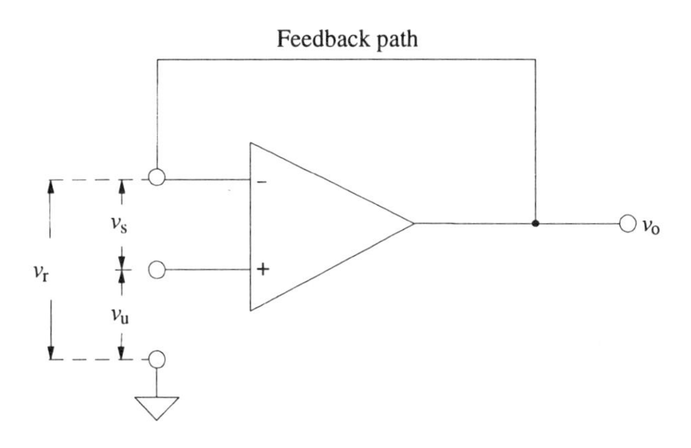

Figure 5–12. Op amp voltage follower. Because of the feedback path, the output voltage is equal to the sum of the input voltage to the noninverting input and the difference voltage between the two inputs. For all values of output voltage less than the limit voltages, the input difference voltage is very small. Therefore, the output voltage is essentially equal to and **follows** the input voltage v_u.

connected so that its output voltage equals the input source voltage, then the output is a clone of the input but protected from loading. The output voltage "follows" the input voltage but is isolated from it. Thus, the op amp voltage follower is an ideal buffer between high-impedance voltage sources and other circuits or instruments. It is often used as the input stage in pH meters and other measurement instruments.

The basic voltage follower is implemented simply by connecting the input voltage v_u to the noninverting input and a wire from the op amp output to the inverting input, as shown in Figure 5–12. The reference voltage v_r is thus obtained from the output v_o of the op amp. Because of the feedback path, $v_o = v_u + v_s$ when the op amp is operating in its linear region. Also, when the output voltage is not at limit, the input difference voltage v_s is negligibly small. Therefore, $v_o \approx v_u$. For example, if the gain A of the op amp is 100 000, then $v_s = v_o/100\,000$, and the assumption that v_s is negligible introduces an error of only 1 part in 100 000. Again, the absolute value of gain A of the op amp is unimportant if it is large. Of course, there must always be the small error v_s, because this difference input is necessary to maintain a finite output voltage. The exact relation between v_o and v_u is

$$v_o = \frac{v_u A}{A + 1} \tag{5–2}$$

This equation confirms our approximate analysis that $v_o = v_u$ with an error of 1 part in A. The actual value of A is relatively unimportant as long as it is large.

Voltage Follower with Gain

In some cases it is desirable to amplify a voltage while retaining most of the desirable characteristics of the voltage follower mode. Figure 5–13 shows a circuit that is appropriately called a **follower with gain**. The comparison voltage v_r is only a fraction b of the op amp output voltage v_o. This is accomplished by connecting a simple voltage divider at the output, which is made from two precision resistors with resistances R_1 and R_2. The redrawn follower-with-gain circuit shown in Figure 5–13b makes it easy to see that a reference voltage equal to bv_o is compared with v_u. Again, starting with the assumption that $v_u = bv_o$ and rearranging, $v_o = v_u/b$, and because the fraction $b = R_1/(R_1 + R_2)$, the output voltage becomes

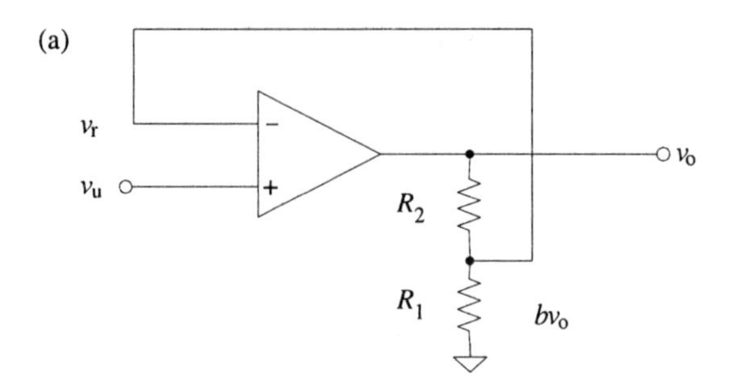

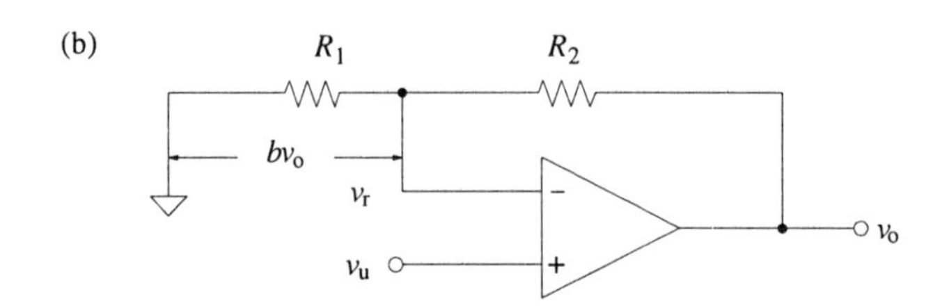

Figure 5–13. Op amp voltage follower with gain. Circuit is drawn to show voltage divider across the output (a) and to show the fraction b of the output voltage fed back to the inverting input (b).

$$v_o = v_u \left(\frac{R_1 + R_2}{R_1} \right) \tag{5–3}$$

Because of these near-ideal characteristics of accurate gain and high input resistance, the follower-with-gain circuit is used as the input amplifier for a variety of readout devices such as chart recorders and digital voltmeters. The exact gain equation is

$$v_o = \frac{v_u A}{bA + 1} \tag{5–4}$$

Equation 5–4 indicates that the error is 1 part in bA. If b is 1/100 (closed-loop gain of 100) and the op amp open-loop gain A is 100 000, the error is 1 part in 1000. As the desired follower gain increases, one must either use an op amp with higher amplification or make further sacrifices in accuracy.

Electromechanical Voltage Follower

It is interesting to compare the op amp voltage follower with the electromechanical voltage follower shown in Figure 5–14, which also serves as a measurement system. This servo chart recorder uses an op amp as an input comparator. The voltage to be measured, v_u, is connected to the noninverting input, and the variable reference voltage v_r is connected to the other input.

The variable reference voltage is produced by a motor-driven wiper contact on a slide wire connected to a selected full-scale span voltage labeled V. The pen moves across the chart because it is mechanically linked to the wiper or slider on the slide wire. Thus, the position of the pen on the chart is related directly to the reference voltage. At one end v_r equals 0 V, and at the other end it equals the full-scale voltage V applied to the slide wire.

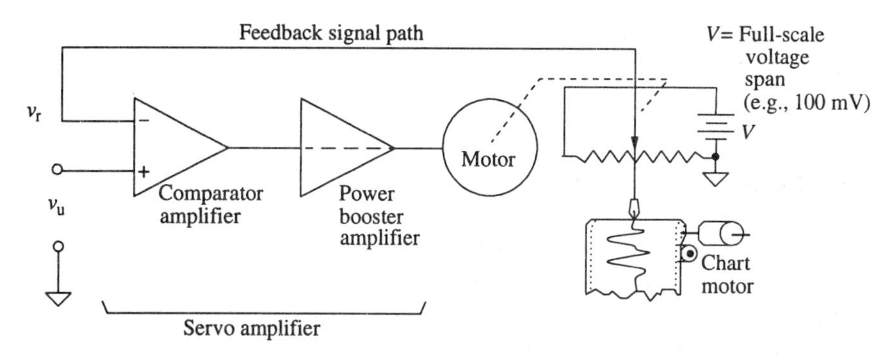

Figure 5–14. Potentiometric strip chart recorder (servo electromechanical voltage follower with strip chart display). The reference voltage automatically **follows** the input voltage. Because the chart pen position is directly proportional to the value of the reference voltage and because the chart moves at a constant rate, the chart displays input voltage versus time. An accurate recording of input voltage versus time assumes that the speed of response of the servo system enables the reference voltage (and thus the pen position) to **follow** the input voltage.

The output of the comparator is connected to a power booster amplifier that supplies sufficient power for the motor shaft to turn so as to vary v_r and move the pen. The polarity of the comparator output depends on whether v_r is smaller or larger than v_u. If v_u is greater than v_r, the motor turns in the direction that will increase v_r. If v_u is less than v_r, the reverse is true. At null there is insufficient power for the motor shaft to turn, and v_r equals v_u.

Figure 5–14 also shows that for the strip chart recorder, a roll of graph paper is moved by a chart motor at a constant rate and at right angles to the pen motion. The chart provides a recording of the reference voltage v_r versus time, and because v_r **follows** v_u (i.e., v_r changes value to remain equal to v_u) the chart is also a plot of input voltage v_u versus time. The response time of the servo system in adjusting v_r determines the rate at which v_u may change and still provide an accurate plot of v_u versus time. Typically, the response time for full-scale travel of the pen is 0.5 to 1 s.

The electromechanical servo system is inherently slow because of the mechanical positioning elements. Therefore, if the only function of the servo system is generation of a signal-following voltage v_r, rather than mechanical positioning, the same voltage follower function can be accomplished much more simply by the all-electronic op amp connected in the follower mode, which responds rapidly to changes of input voltage.

5–4 Processing Analog Signals with Operational Amplifier Current Followers

In the previous two sections, the op amp was described in two of its basic modes of operation: a **voltage comparator** and a **voltage follower.** The third mode, frequently called the **operational mode** or the **current follower mode,** is presented in this section. It is in the current follower mode that the op amp can perform mathematical **operations** such as summation, integration, and differentiation (from which it received the name "operational amplifier") as well as many other functions such as current-to-voltage or resistance-to-voltage conversion.

The basic concept of this third op amp mode is best understood by the concept of null comparison applied to the measurement of current (from which the name "current follower" is derived). Therefore, this current follower concept is introduced and then applied to the op amp. Several applications are used to illustrate the operation and the versatility of the op amp current follower. The applications included in this section are the current-to-voltage converter, voltage-inverting amplifier, resistance-to-voltage converter, and log converter. Several major applications of the current follower mode, including the summing of currents and voltages and the integrating and differentiating of analog signals, are described in Sections 5–5 and 5–6.

Null Comparison Current Measurement

The concept of null comparison can be applied to the measurement of current as well as voltage. A variable reference current source is used to offset the unknown current source. The reference current is then adjusted to bring the null detector to a zero difference indication. The same null detectors that are used for null comparison voltage measurements can be used for null comparison measurements of current, and the servo systems that are used to automate potentiometric measurements can be used to automate the null comparison current measurements. Automation is achieved through a current source that is continuously adjusted to follow the measured current.

The basic null comparison current circuit is shown in Figure 5–15. The null detector is connected as a current meter would be for measuring the magnitude of the unknown current i_u. A reference current source is then connected to the same null detector in such a way that its current through the null detector is opposite that of the unknown current. The null detector current i_{nd} is thus equal to the difference between the unknown and reference source currents. The sign of the null detector current indicates whether i_r is larger or smaller than i_u.

When the current through the null detector is zero, the voltage across the null detector must also be zero. Therefore, at null, the potential difference between points J_1 and J_2 is zero. This means that *at balance the output voltage of each current source is zero.* Both sources are under the ideal condition of a short-circuited output and are thus delivering the full values of i_u and i_r to the comparison circuit. The current comparison at null **idealizes** the load of the current sources just as the null voltage comparison **unloads** the voltage sources. If point J_2 is connected to the circuit common, point J_1 is also at the common potential at null; that is, J_1 is a **virtual common**. As with null comparison voltage measurement, either a high-resistance voltage-sensitive or a low-resistance current-sensitive null detector can be used.

For the basic voltage null comparison circuit of Figure 5–5, three positions for the connection of the circuit common are possible because of the series arrangement of the two sources and the null detector. In the basic current null comparison circuit of Figure 5–15, all components of the circuit are connected in parallel. Only two positions are possible for the connection of the circuit common: J_1 and J_2. Brief inspection shows that these two positions are equivalent. Therefore, there is only

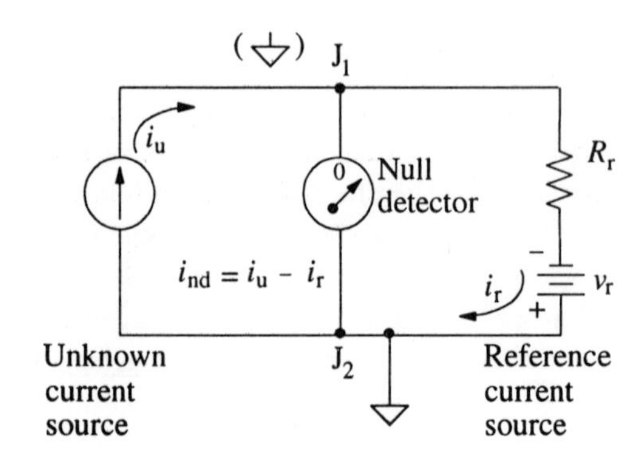

Figure 5–15. Basic current comparison circuit. Any difference in the currents i_u and i_r must pass through the null detector. Either v_r or R_r may be adjusted so that $i_u = i_r$. At null, the voltage across the null detector is zero, as though there is a short circuit between J_1 and J_2. Thus, the comparison circuit presents an ideal load to the unknown current source.

one common point configuration for the current comparison circuit, and all components have one connection to the circuit common. The other connection to each of the components is at the common potential when the system is at null. This common potential is indicated by the virtual common symbol next to J_1. The connection eliminates the need for floating sources and the concern about common mode error at the input to the null detector.

The null comparison technique for current measurement is far superior to the measurement of *IR* drop with a voltmeter because it requires no assumption about the ideality of the unknown current source. Even though manual null comparison current measurements are rarely used, a number of circuits that are very commonly used with servomechanical systems and op amps are automatic null comparison current measurement systems. The current comparison concept helps greatly in understanding the operation and characteristics of this family of circuits.

Op Amp Current Follower

The null comparison current measurement shown in Figure 5–15 is automated by providing a reference current source that is controlled by the null detector output in such a way as to maintain a null condition. When an op amp is used as the servo system, its differential inputs are the null detector, and its output voltage produces a proportional reference current through a series resistance. The resulting circuit is shown in Figure 5–16. Comparison with Figure 5–15 reveals the analogous components.

The op amp current follower mode is nearly ideal for current measurements because it introduces almost zero resistance into a circuit. Referring to the diagram for the **current follower mode** in Figure 5–16, we see again the now-familiar triangle representing the same high-gain op amp with inverting (–) and noninverting (+) inputs. However, the external connections that we make are quite different from those made in the voltage follower mode. A feedback resistor with resistance R_f is connected from the output to the inverting (–) input. An input current source supplying i_u is also connected to the inverting input. The noninverting input is connected to common. The input current i_u equals the feedback current i_f plus the op amp input bias current i_b (a nonideal leakage current that is described in more detail in Chapter 7, Section 7–4). That is,

$$i_u = i_f + i_b \tag{5–5}$$

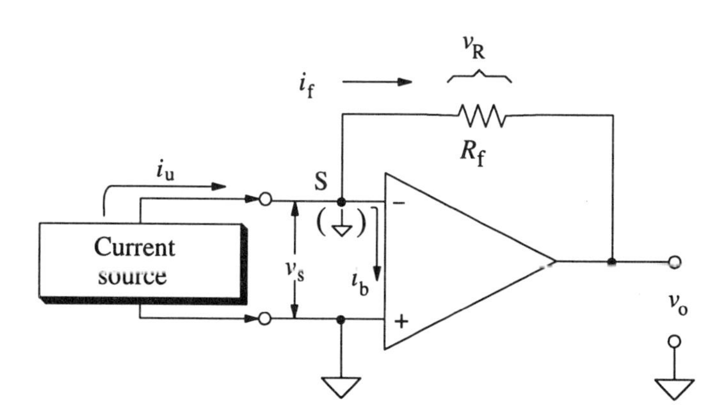

Figure 5–16. Op amp current follower. If i_b is negligible, then the feedback current i_f is essentially equal to the input current i_u. If v_s is negligible, then v_o equals $-i_f R_f$. Therefore, v_o equals $-i_u R_f$. If the direction of input current is into the op amp summing junction, the current is positive, and if it is out of the junction, the current is negative.

but i_b is typically in the 10^{-10}- to 10^{-15}-A range and thus is usually negligible. Therefore, i_u is essentially equal to i_f. Also, if the op amp is operating within limits, v_s is usually negligible. Thus, the inverting input, labeled S, is held at virtual common potential, and the output voltage v_o must then equal the voltage drop v_R across the feedback resistor but be of opposite polarity. That is, v_o is essentially equal to $-v_R$. Because the voltage drop across the feedback resistor is v_R and is equal to $i_f R_f$, then v_o is essentially equal to $-i_f R_f$. Also, because i_u is essentially equal to i_f, it follows that

$$v_o = -i_u R_f \tag{5–6}$$

In other words, the op amp circuit of Figure 5–16 generates a feedback current i_f that follows the input current i_u, and an output voltage that is proportional to the input current is produced. For example, if the feedback resistor has a resistance of $10^8\ \Omega$ and the input current is 1 nA (10^{-9}A), then the output voltage would be $10^{-9} \times 10^8 = 0.1$ V, which can be easily measured.

The current-to-voltage converter circuit is very useful as a current-measuring system. The input resistance of the current follower circuit is virtually zero, because one terminal is connected to common and the other terminal is held at virtual common by operation of the op amp circuit. The effective input resistance R_{in} of the current follower circuit as seen by the unknown current source is the input voltage v_s divided by the input current i_u; that is, $R_{in} = v_s / i_u$. Because $v_s = -v_o /A$ and from equation 5–6 $i_u = -v_o / R_f$, then $R_{in} = R_f / A$. The loading effect of R_f on the input current i_u is thus improved A times by using the current follower. Because A is generally 10^5 to 10^7, the use of the current follower provides a dramatic improvement in current measurements.

The limitations in the application of the current follower are seen in the more exact relation between v_o and $(i_u - i_b)$:

$$v_o = -R_f(i_u - i_b)\left(\frac{A}{1 + A}\right) \tag{5–7}$$

$$v_o = -i_u R_f + i_b R_f \frac{v_o}{A}$$

Either form of equation 5–7 shows that when A is very large and i_b is much less than i_u, the output voltage v_o is proportional to i_u, as expected from equation 5–6. The range of currents that can be measured by a current follower and readout is limited on the low end by the input bias current i_b of the op amp and on the high end by the op amp's output current capability. Op amp input bias currents typically vary from 10^{-11} to 10^{-15} A, and the current follower is useful in the picoampere current range with careful choice of the op amp. The output current capability of an op amp is generally 2 to 100 mA. The op amp output must supply both the feedback current i_f and the output current to the load.

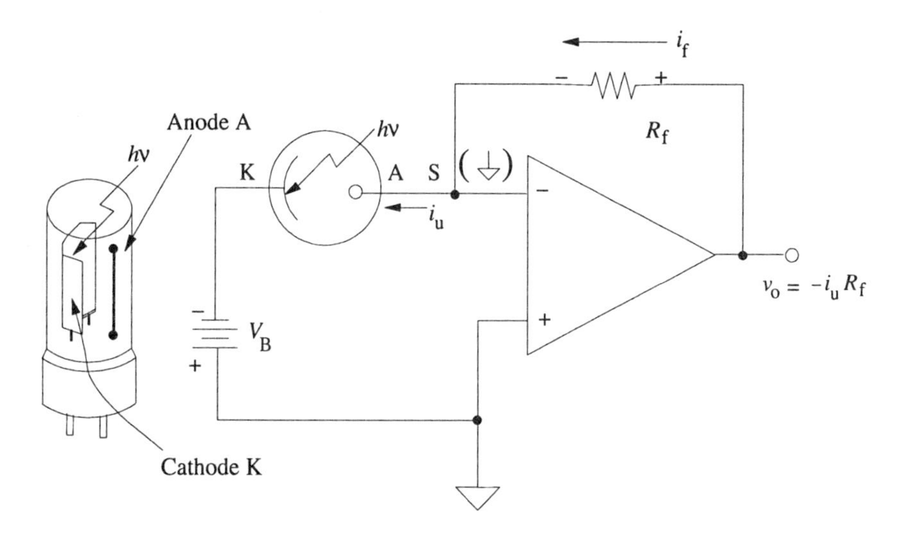

Figure 5–17. Connection of a phototube (light-to-current) transducer to an op amp current follower. Because the anode of the phototube is held at virtual common by the current follower, the bias supply is effectively connected between the cathode and the anode of the phototube. The input current from the transducer is directly proportional to the intensity of the incident light. Thus, the output voltage of the current follower is directly proportional to light intensity because it is proportional to feedback current, which is essentially equal to input current. The input current direction is out of the junction, and so current is negative and output voltage is positive.

Current-to-Voltage Converter

From the discussion of the basic op amp current follower in the previous section, it is apparent that this device is a nearly ideal current-to-voltage converter. It is widely used for measurement of the output from current transducers such as phototubes, photomultipliers, electron capture detectors, and flame ionization detectors.

The connection of a typical current transducer to an op amp current follower is shown in Figure 5–17. The ordinary two-electrode phototube has an anode and a photosensitive cathode in an evacuated glass or quartz envelope. When light, $h\nu$, strikes the photocathode, photoelectrons are attracted to the anode if it is held at a positive potential relative to the cathode. This provides an input current i_u to the op amp current follower that is proportional to the light intensity on the photocathode. The cathode is connected to the negative terminal of a source supplying voltage V_B, which polarizes the transducer terminals. The anode is connected to the inverting input of the op amp, which is held at virtual common by the op amp circuit; the plus terminal of the source supplying voltage V_B is connected to common. Therefore, the photodiode always has the bias voltage V_B between anode and cathode. When V_B is sufficiently large, all photoelectrons resulting from incident photons of light on the cathode are attracted to the anode. Therefore, the transducer current is proportional to the light intensity on the photocathode. The output voltage v_o of the current follower is equal to the input current i_u multiplied by the feedback resistance R_f and is thus proportional to the light intensity.

The flame ionization detector illustrated in Figure 5–18 is another current transducer that is connected to an op amp current-to-voltage converter. It consists of two metal electrodes opposite each other and just above a small hydrogen–air flame. If molecules enter the flame and are readily ionized by the flame energy, there will be a flow of charge carriers (i.e., electrons and positive ions) between the electrodes. The rate of flow of electrons in the external circuit depends on the rate of formation of charge carriers, which is determined by the concentration of

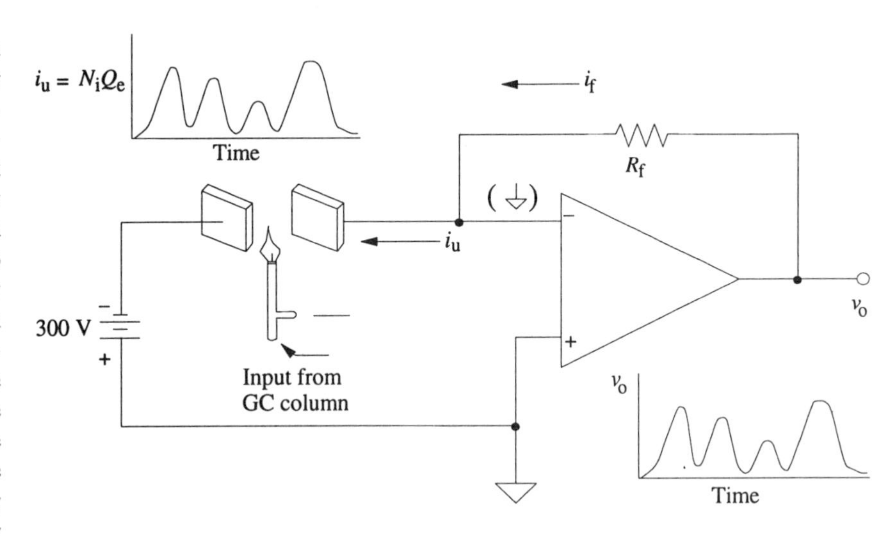

Figure 5–18. Connection of a flame ionization detector (molecular concentration-to-current) transducer to an op amp current follower. Several current peaks indicate the sequential entry of molecular constituents into a flame, where they are ionized to provide currents that are directly proportional to the concentration of ionized particles. These currents are followed ideally by the current-to-voltage converter. The resulting output voltages are readily measured. Because the output voltages are directly proportional to the currents, they are also directly proportional to the concentrations of the various constituents.

ionizable molecules in the flame at any time. Therefore, the transducer output current is related to the molecular concentration. When used with a gas chromatographic (GC) column, the molecular constituents in a complex mixture are sequentially separated as they pass through the column and enter the flame detector. The current i_u is equal to the number of charge carriers N_i produced per second that reach the detector plates multiplied by the unit charge Q_e. For example, if the number of electrons produced equals 10^{11}/s at the peak and if $Q_e = 1.6 \times 10^{-19}$ C, then $i_u = 1.6 \times 10^{-8}$ A. With a feedback resistance of 10^8 Ω, the output voltage of the current follower would be $1.6 \times 10^{-8} \times 10^8 = 1.6$ V. Typical currents from a flame ionization transducer might be in the range of 10^{-6} to 10^{-11}A. The op amp voltage peaks at the output are proportional to the concentration of specific constituents that sequentially enter the flame detector. This is illustrated in Figure 5–18 by the input and output signals.

Op Amp Inverting Amplifier

The inverting amplifier is a simple variation of the current follower circuit in which the unknown current source is a voltage source and a resistor in series. The resulting circuit is shown in Figure 5–19. Because $v_s \approx 0$, the source current i_u equals v_u/R_{in}. If v_u/R_{in} is used for i_u in equation 5–6 for the current follower, then

$$v_o = -v_u \frac{R_f}{R_{in}} \tag{5–8}$$

Thus, the output voltage is a constant multiplied by the input voltage, and the constant $-R_f/R_{in}$ depends only upon the values of R_f and R_{in}. This provides another possibility for precision amplification of a voltage signal.

In contrast to the voltage follower precision amplifier of Figure 5–13, the inverting amplifier inverts the input signal, as indicated by the minus sign in equation 5–8. The inverting amplifier also differs from the voltage follower in that it has one input connected to the circuit common, eliminating the common-mode

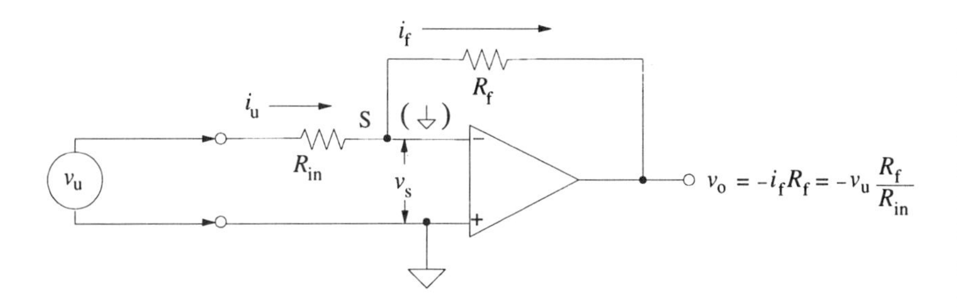

Figure 5–19. Op amp inverting amplifiers. The input current is essentially equal to the input voltage divided by the input resistance, because v_s is essentially zero. The current follower equation $v_o = -v_u(R_f/R_{in})$ shows that accurate amplifier gains can be obtained by selecting precision resistors with resistances R_s and R_{in}.

rejection error, and a simpler relationship between gain and resistance, allowing whole-decade values of resistors to produce whole-decade values of system gain. For example, if R_f is 1 MΩ and R_{in} is 10 kΩ, a gain of 100 is obtained. On the other hand, the source supplying input v_u is loaded by the resistance R_{in} of the input resistor in the inverting amplifier, but the input voltage v_u is measured potentiometrically in the follower with gain. In both inverting and follower amplifiers, higher gain is achieved with some loss in gain accuracy. Each amplifier circuit has merits depending upon the desirability of inversion and the problems of source loading.

Linear Resistance-to-Voltage Converters

According to Ohm's law, $R = V/I$, the voltage across a resistor is proportional to the resistance for a constant value of current. Similarly, the conductance, $G = i/V$, is proportional to the current through a resistor for a constant applied voltage. To take advantage of these linear relationships requires a constant current source in the first case and a current measurement with negligible input resistance in the second case. The current follower circuit introduced in this section provides a convenient and accurate means of fulfilling either of these requirements.

Recall that the current in the feedback resistor follows the current applied to the summing point and that the output voltage is proportional to R_f and the input current. In the circuit of Figure 5–20a, V and R_{in} produce a constant current i_{in}, which appears in the resistor with unknown resistance R_u in the feedback path. The output voltage is then

$$v_o = \frac{-VR_u}{R_{in}} \quad \text{or} \tag{5–9}$$

$$v_o = R_u\left(\frac{-V}{R_{in}}\right)$$

showing that v_o is proportional to R_u.

To obtain an output proportional to conductance, the position of the resistor with unknown resistance R_u is changed to the input as in Figure 5–20b. The combination of V and R_u produces a current proportional to the conductance G_u of the resistor with resistance R_u. Because the output voltage v_o is equal to $-i_u R_f$,

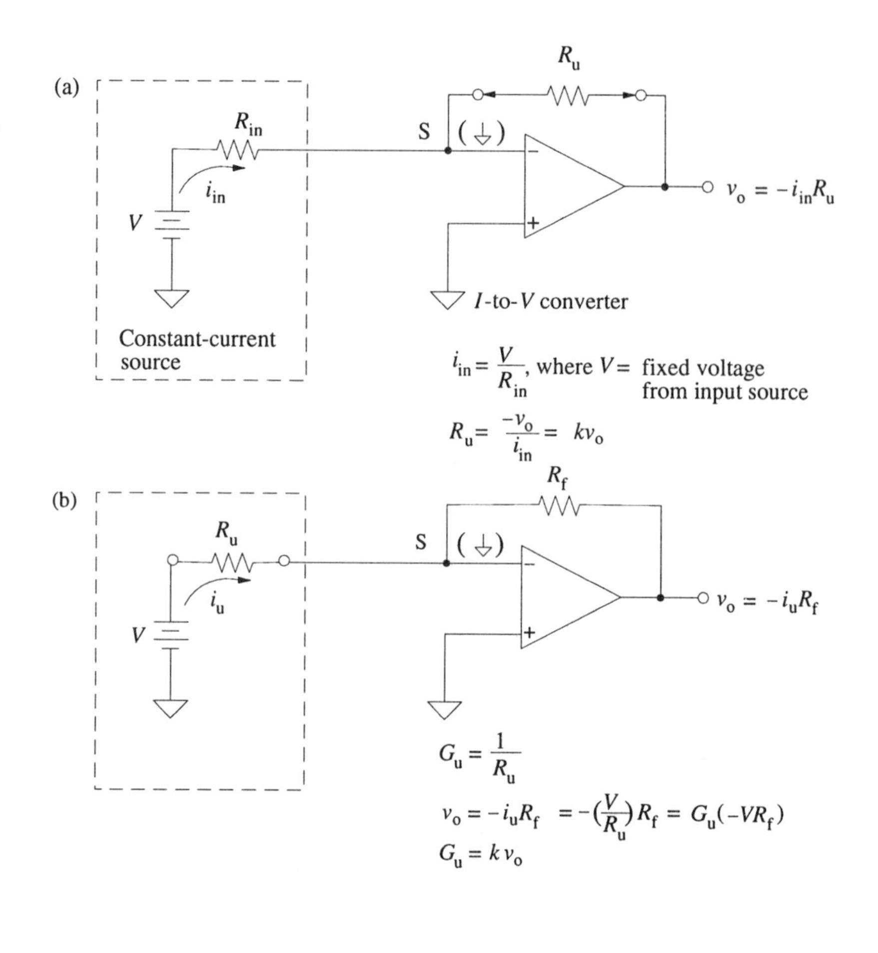

Figure 5–20. Op amp resistance-to-voltage converters. (a) Resistance is directly proportional to output voltage; (b) conductance is directly proportional to output voltage.

$$v_o = \left(\frac{-V}{R_u}\right)R_f \text{ or} \tag{5–10}$$

$$v_o = G_u(-VR_f)$$

Of course, both circuits are simply the op amp inverting amplifier, for which $v_o = -v_{in}R_f/R_{in}$, with $v_{in} = V$, and R_u taking the place of R_f in the first case and of R_{in} in the second. This simple circuit is an extremely convenient method for converting the resistance or conductance of various resistive transducers to a proportional voltage. It is also the circuit used for the resistance scales in many digital multimeters.

Logarithmic Current-to-Voltage Converter

The logarithmic transfer function is one that has many uses, including compensation for logarithmic function transducers, compression of signals with especially wide dynamic ranges, and implementation of nonlinear arithmetic operations such as multiplication and division. The logarithmic transfer function can be achieved by taking advantage of the approximately logarithmic relationship between current and voltage in a semiconductor pn junction. This provides a smooth transfer function over a wide dynamic range of about five decades.

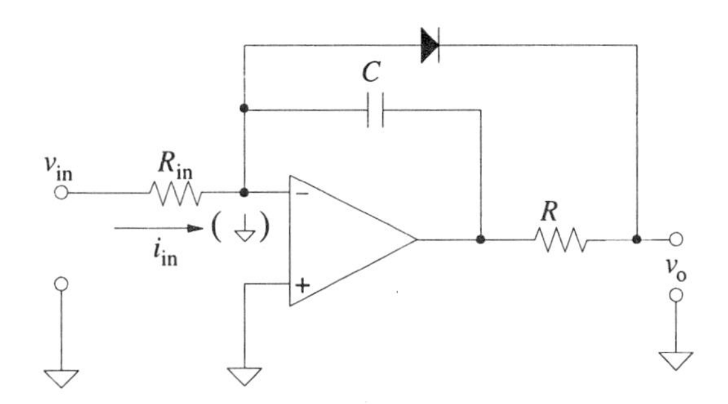

Figure 5–21. Logarithmic amplifier. An output voltage related to the logarithm of the input voltage is obtained in this circuit as a result of the logarithmic relationship between current and voltage in the diode. The circuit is basically a current follower in which the input current appears in the diode in the feedback loop. The resistor with resistance R and the capacitor with capacitance C (typically 1 kΩ and 0.01 μF, respectively) stabilize the amplifier but do not affect the output value. This simple circuit is useful, but it has a large temperature dependence of slope and offset. A transistor may be used in place of the diode.

As shown in Figure 5–21, a diode replaces the feedback resistor in the op amp current follower circuit. The voltage across the diode changes by 59z mV for each 10-fold change in current through it, where z is an empirical temperature-dependent parameter. Thus, the output voltage v_o equals $59z(\log i_{in} - I)$, where I is a term related to diode reverse-bias current. In Figure 5–21 the input current i_{in} is provided by an input voltage v_{in} connected through a resistor with resistance R_{in} to the inverting input. Therefore, v_o equals $59z[\log (v_{in}/R_{in}) - I]$ and logarithmic voltage-to-voltage conversion is achieved.

The next sections of this chapter will continue to illustrate the versatility and general applicability of the op amp current follower mode.

5–5 Adding and Subtracting Currents and Voltages

Because the current follower amplifier provides an input that is maintained at the virtual common, a current source connected to the inverting input, point S in Figure 5–22, has the same output that it would have if it were connected to the common. Therefore, multiple current sources connected to point S do not interfere with each other. The result is the **current-summing amplifier** shown in Figure 5–22. Each current source applies its current to the virtual common point S. Because the sum of all currents to point S must be equal to i_f, then $i_f = i_1 + i_2 + i_3$. For this reason, point S is often called the **summing point** of the op amp in current follower applications. Because the output voltage v_o essentially equals $-i_f R_f$, it follows that

$$v_o = -R_f(i_1 + i_2 + i_3) \tag{5–11}$$

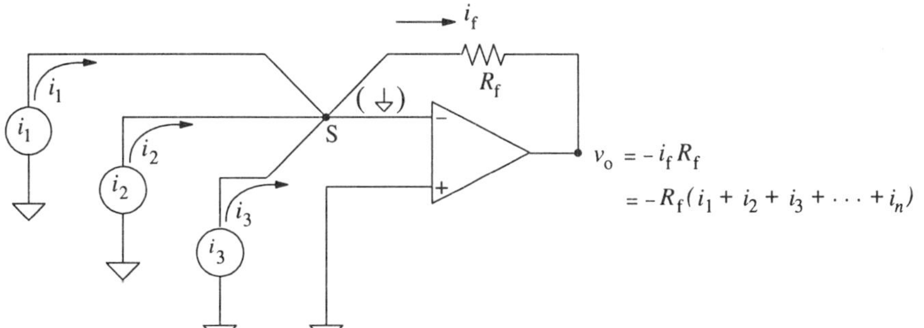

Figure 5–22. Summing-current follower. All current sources to point S are ideally loaded and do not affect each other. The feedback current is the simple sum of all the input currents, and the output voltage is proportional to that sum.

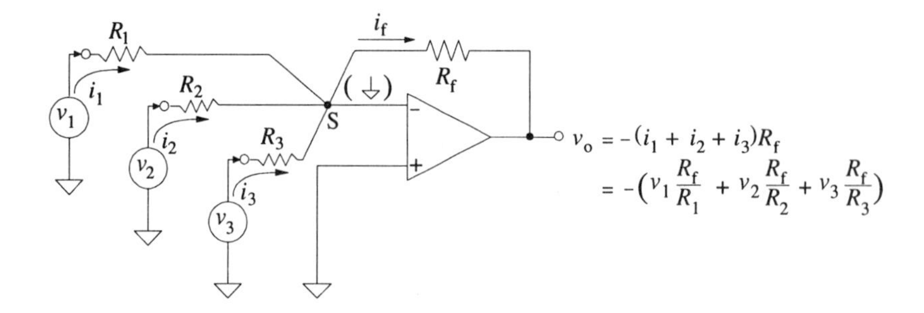

Figure 5–23. Summing amplifier. In this summing-current follower, input resistors are used to obtain input currents proportional to the input voltages. The separate input resistors allow a different gain for each input connection.

Therefore, the output voltage is proportional to the sum of the input currents. The exact equation 5–7 can be applied to determine the limitations of this circuit.

The current sources in a current-summing amplifier can be voltage sources with series resistors, as shown in Figure 5–23. Assuming point S to be at the common potential, $i_1 = v_1/R_1$, $i_2 = v_2/R_2$, and $i_3 = v_3/R_3$. Substituting these relations for the current in equation 5–11, we obtain

$$v_o = -\left(v_1 \frac{R_f}{R_1} + v_2 \frac{R_f}{R_2} + v_3 \frac{R_f}{R_3}\right) \tag{5–12}$$

Equation 5–12 shows that the output voltage is the sum of the input voltages, each multiplied by its own R_f/R_{in} ratio. This circuit is called the **weighted summing amplifier** because the contribution of each input voltage to the sum is weighted by its individual gain factor. Frequently, a simple sum is desired, so that R_1, R_2, and R_3 are all equal to R_{in}. Then

$$v_o = -\frac{R_f}{R_{in}}(v_1 + v_2 + v_3) \tag{5–13}$$

This circuit has the same output and gain error characteristics as the inverting amplifier. Summing amplifiers are used in instruments and in analog computers to perform the addition function on data in the voltage or current domains. A summing amplifier can also be used to add a constant to a signal voltage or current and thus to introduce (or eliminate) an offset in the transfer function. The input resistance of the summing amplifier is R_{in} at each input. If this presents an excessive load on the voltage source, a voltage follower circuit is used between the voltage source and the summing amplifier input. Subtraction can be achieved by inverting the appropriate inputs. Another type of algebraic subtractor is the difference amplifier described in Chapter 7, Section 7–2.

An interesting and important application of the summing op amp is for digital-to-analog conversions, as shown in Figure 5–24. The input resistors in this example follow the binary coded decimal (BCD) weights for digital data encoding. The digital data word is used to operate the gain control switches to produce an output voltage that is directly related to the input decimal number.

Each resistor produces at the amplifier summing point S a current that is proportional to the coded weight of the bit when the switch for that bit is closed, as shown in Figure 5–24b. The simultaneous closure of all the switches for which the bits in the digital word are HI produces an output voltage that is proportional

(a)

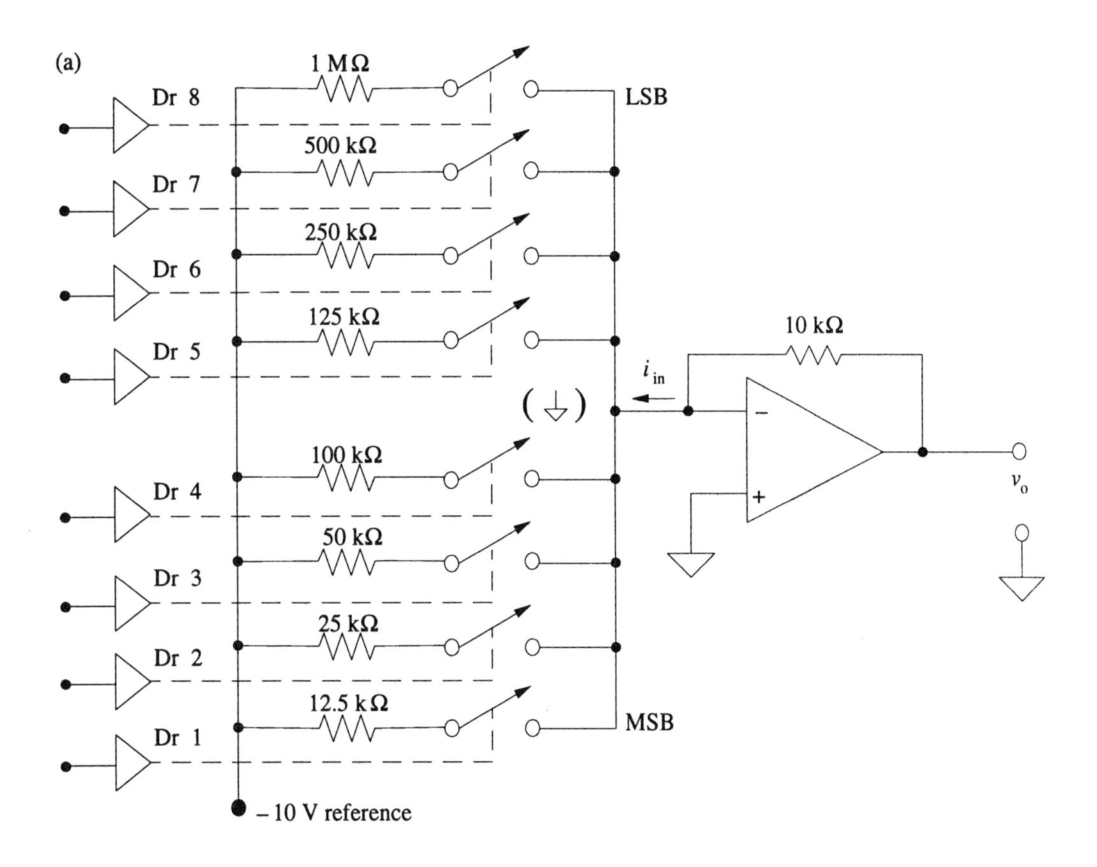

(b)

BCD Input	Decimal Equivalent	I_{in}, μA	v_o, V
0000 0000	0	0	0
0000 0001	1	−10 V/1 MΩ = −10	0.1
0000 0010	2	−20	0.2
.	.	.	.
.	.	.	.
0000 1001	9	−90	0.9
0001 0000	10	−100	1.0
0001 0001	11	−110	1.1
.	.	.	.
.	.	.	.
1001 1001	99	−990	9.9

Figure 5–24. A two-digit BCD DAC. (a) Summing amplifier circuit. The DAC function is obtained by the appropriate choice of summing resistors and the use of a reference voltage as the analog input. The encoded digital input data are applied to the switch driver inputs. (b) Relationships between BCD input, decimal equivalents, input currents generated, and resulting output voltages. Dr, driver; LSB, least significant bit; MSB, most significant bit.

to the digital number. For example, with the decimal number 65 (BCD encoded as 0110 0101) and a –10-V analog reference source, the output voltage will be 6.5V.

The value of the input resistances shown in Figure 5–24 must include the ON resistance of the series switch. Errors due to variation or unpredictability in switch resistance are greatest for the most significant bits of the digital signal.

If the –10-V reference voltage for the input resistors is replaced with a variable input voltage v_{in}, the output voltage v_o will be equal to $-v_{in}(nn/100)$, where *nn* is the two-digit BCD number (0 to 99 decimal). The circuit is now a programmable gain amplifier with gain adjustable in integer steps from 1 to 99% of v_{in}. It is called a multiplying digital-to-analog converter (MDAC). Other DAC circuits, discussed in Chapter 7, Section 7–5, use the concept of the summing op amp described in this section.

5–6 Integrating and Differentiating Analog Signals

The use of a resistor in the feedback path of the current follower amplifier results in an output voltage proportional to the feedback current. The linear current–voltage relationship is due to the ohmic nature of the feedback resistor. Devices with other current-to-voltage or charge-to-voltage relationships (transfer functions) could be used in the feedback or input paths of the basic current follower to produce a circuit that implements the transfer function of the components used. In this section we shall see how a combination of capacitors and resistors can be used with an op amp to produce charge-to-voltage converters, current or voltage integrators, and differentiators, which produce output voltages proportional to the rate of change of input voltages. These operations on information in the charge domain are very useful with certain types of transducers, in waveform generators, in analog-to-digital converters, and for other applications.

Charge-to-Voltage Conversion

Because the voltage across a capacitor is directly proportional to the charge on it ($v = q/C$), the capacitor is the basic device used in **charge-to-voltage converter** circuits. The measurements of charge and current are related because current is the rate of charge transfer. The basic op amp charge-to-voltage converter is the integrator shown in Figure 5–25. A positive charge q_{in} applied to the input accumulates on a capacitor with capacitance *C* because there is no other path for it to take. The op amp establishes an output voltage v_o that maintains point S at virtual common, and the total voltage across the capacitor appears at the output.

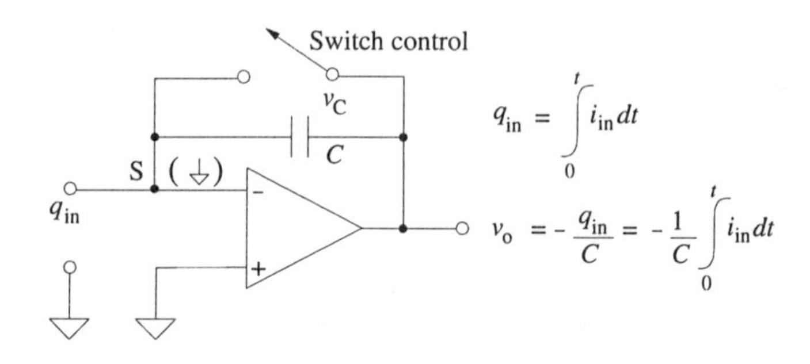

Figure 5–25. Charge-to-voltage converter. The input charge appears on the feedback capacitor and produces a proportional voltage. Because the output voltage is essentially equal to the voltage across the capacitor, the output v_o is equal to $-v_c = q_{in}/C$.

The voltage v_o is thus

$$v_o = -\frac{q_{in}}{C} \tag{5–14}$$

The output voltage v_o is thus proportional to q_{in}. The value of C is chosen for the range of charge values to be measured.

When the switch across the capacitor is closed, the capacitor is discharged and the output voltage v_o is zero. When the switch is open, the capacitor can be charged, and the output voltage **follows** the voltage across the capacitor. The capacitor **integrates** (sums the total input charges) to provide the output voltage as expressed in equation 5–14.

Equation 5–14 would be perfectly accurate for any amount of charge over any length of time (within op amp output voltage limits) if it were not for leakage of charge to point S from sources other than the input. The major source of leakage current is often the amplifier input bias current i_b. The total charge measured should be much larger than the charge leaked by i_b during the measurement time.

Current Integrator

The total charge passing a point over a given time is the integral or summation of the instantaneous currents during that time. Therefore, the charge-to-voltage converter can be used to perform the mathematical function of integration on a signal in the current domain. If q_{in} is applied to the **integrator** of Figure 5–25 as an input current i_{in} over time t, then

$$q_{in} = \int i_{in} dt \tag{5–15}$$

Combining equations 5–14 and 5–15 yields

$$v_o = -\frac{1}{C}\int i_{in} dt \tag{5–16}$$

An example of current integration is shown in Figure 5–26. The total quantity of light striking a light-sensitive phototube (light-to-current transducer) is to be measured during a light burst (pulse). The phototube is connected to the input of the integrator, and it is assumed here that when there is no light burst, the phototube current is negligible. When the light strikes the phototube, a current pulse is obtained as illustrated in Figure 5–26. The output voltage is equal to the integral of the current during the pulse. Thus, the output is proportional to the quantity of light.

Measurement of Capacitance

The op amp current integrator can be used to measure capacitance. If a known constant current i is used to charge a capacitor of capacitance C for a known time

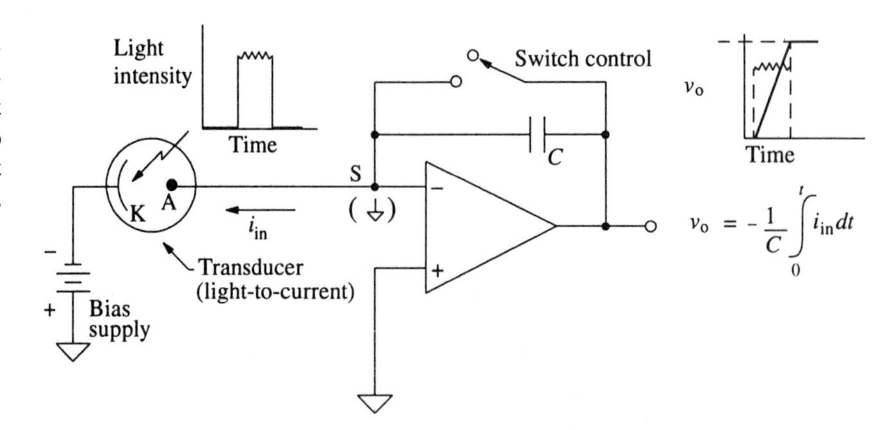

Figure 5–26. Integration of output current pulse from a light-to-current transducer. The output voltage is directly proportional to the total quantity (integral) of light during the pulse. K, cathode; A, anode.

t, then charge $q = it$ and the voltage across the capacitor is $v_c = q/C$. Therefore, capacitance can be determined by measuring the voltage output of the integrator after establishing a known charge on the capacitor. The output voltage v_o of the op amp is equal to the voltage v_c across the capacitor, and it is simple to measure v_c without disturbing the charge on the capacitor. The precision of the method depends on type and capacitance of the capacitor and the ability of the charge source to deliver the charge increment $q = it$.

Voltage Integrator

A signal encoded in the voltage domain can be integrated if a resistor of resistance R is used to convert the input voltage v_{in} to a proportional current i_{in}, as shown in Figure 5– 27a (below). Because the op amp maintains point S at virtual common, $i_{in} = v_{in}/R$. Substituting v_{in}/R for i_{in} in equation 5–16 gives

$$v_o = -\frac{1}{RC}\int v_{in}dt \qquad (5\text{–}17)$$

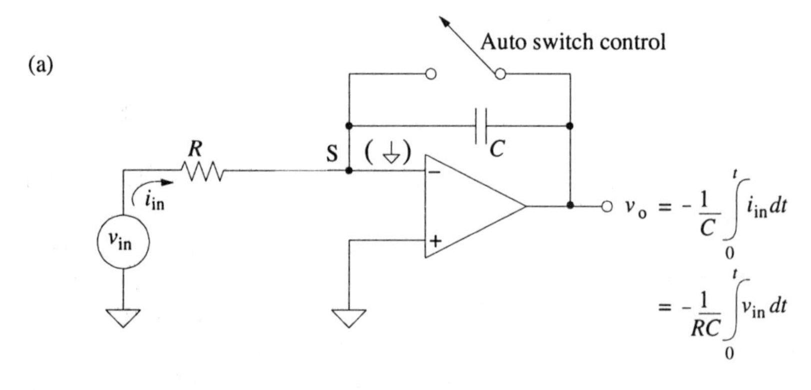

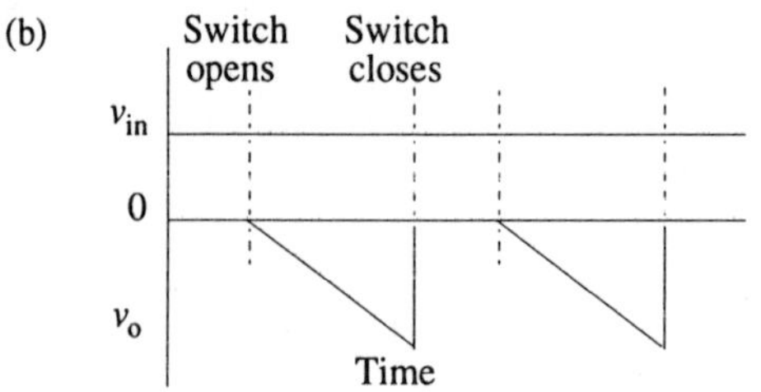

Figure 5–27. Op amp voltage integrator. (a) The resistor with resistance R in series with the input voltage source provides an input current to point S proportional to the input voltage. Thus, the output voltage is equal to $-1/C$ multiplied by the integral of the input current ($i_{in} = v_{in}/R$) from the time of opening the integrate switch (time zero) to the time t. (b) Linear voltage ramp (sawtooth signal).

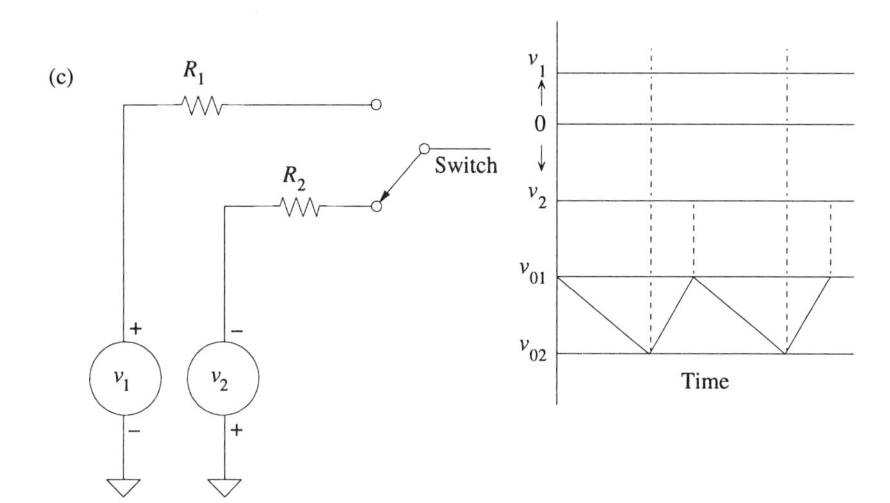

Figures 5–27—*Continued.* (c) Triangular wave obtained by switching input circuit polarity and values.

A switch is used to discharge the capacitor and reset the integrator to begin another measurement.

One of the many useful applications of the op amp voltage integrator is to generate a *linear sweep voltage* or *voltage ramp*, as shown in Figure 5–27b (on previous page). The rate of change of output voltage is proportional to the magnitude of the input voltage. This rate can be used as the linear time base in oscilloscopes, electrochemical systems, and many other applications. When the switch opens, the capacitor charges at a constant rate to provide the linear sweep voltage at the output. The capacitor discharges when the switch closes. Shown in Figure 5–27c (above) is an input current circuit where voltage sources supplying v_1 and v_2 of opposite polarity are alternately connected to the input of the voltage integrator by a switch. When the source supplying v_1 is connected, the output voltage v_o ramps down; when the source supplying v_2, which is of opposite polarity to v_1, is connected, v_o ramps up. By using the output of voltage comparators to control the switch, the output voltage will ramp up and down between the two limit voltages v_{01} and v_{02}. This type of operation is useful in certain types of analog-to-digital converters and will be discussed in Chapter 9.

Op Amp Differentiator

The op amp current follower can be connected as a differentiator to measure the rate of change of voltage inputs simply by connecting a capacitor to the input, as shown in Figure 5–28. The basic relationship for the capacitor is $q = Cv$, and the time rate of change of charge on a capacitor is

$$\frac{dq}{dt} = C\frac{dv}{dt}$$

Current is the time rate of flow of charge dq/dt. It follows that

$$i_{\text{in}} = \frac{dq}{dt} = C\frac{dv}{dt}$$

Thus, a change of input voltage dv/dt produces a proportional current, and a measurement of i_{in} becomes a measure of the time derivative of the voltage signal.

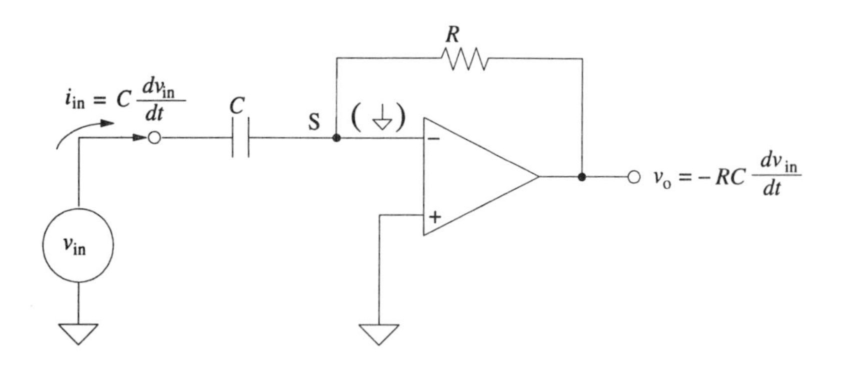

Figure 5–28. Op amp differentiator (rate-of-voltage-change-to-voltage converter).

For the current follower the output voltage v_o equals $-Ri_{in}$, and substituting for i_{in} gives

$$v_o = -RC\frac{dv_{in}}{dt} \tag{5–18}$$

The waveforms in Figure 5–29a illustrate that the differentiator is a rate-of-voltage-change-to-voltage converter. For the **linear ramp** input voltage, the op amp output voltage is constant and proportional to the rate of input voltage change. For the **sigmoid-shaped input voltage signals**, the first-derivative output signal is a voltage peak. By adding a second differentiator to the output of the first, a second-derivative signal can be obtained. In Figure 5–29b the waveforms for the second derivative of the sigmoid curve are also shown. This second-derivative output voltage is very useful in locating the inflection point of the sigmoid input. The second-derivative signal crosses zero at the inflection point regardless of the magnitude of the change or the absolute value of the input signal.

The op amp circuit in Figure 5–28 is inherently sensitive to noise voltage at the input because the output is directly proportional to the rate of change of the input signal. By connecting a low-capacitance capacitor across the feedback resistor, noise problems are minimized, because the upper-frequency response of the differentiator is reduced (*see* Supplement 2, Section S2–4).

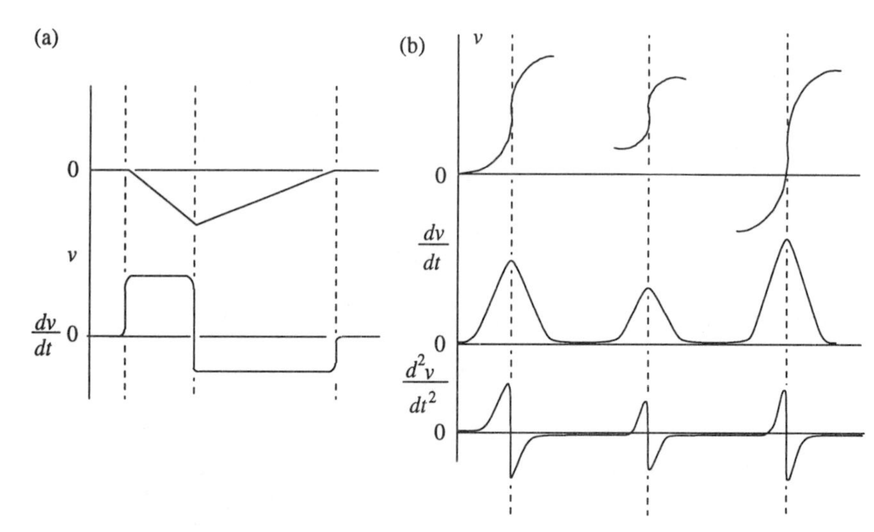

Figure 5–29. Derivative outputs from op amp differentiators. (a) For a linear ramp voltage, the output derivative voltage is proportional to the positive or negative slope of the input. (b) For sigmoid-shaped input voltages, the first and second derivatives are helpful in locating the inflection points of the input signals independent of the direct current input level.

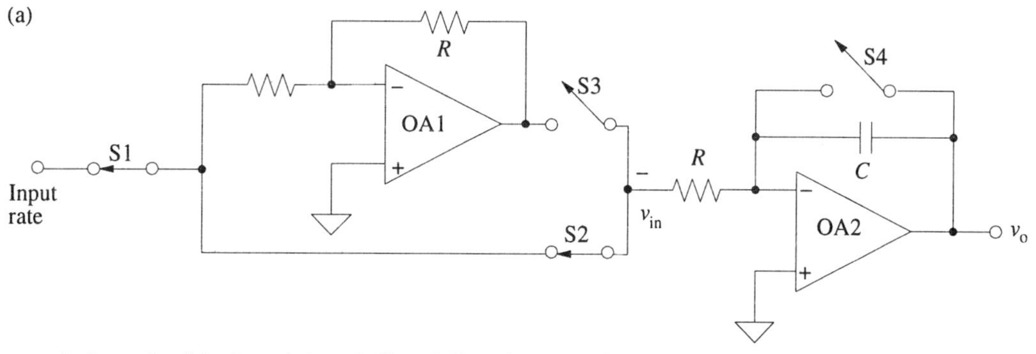

1. Input signal is directed through S1 and S2 to integrator OA2.
2. Capacitor with capacitance C charges and the integrator output changes by v_1 volts during time Δt.
3. Switch S2 opens and S3 closes; inverted input is directed to OA2.
4. Capacitor charging reverses, and the net change in integrator output is $v_2 - v_1 = v_o$.
5. Switch S3 opens; output voltage v_o is equal to ΔA and proportional to the input rate or slope.
6. Switch S4 closes to discharge capacitor so circuit is ready for next slope measurement.

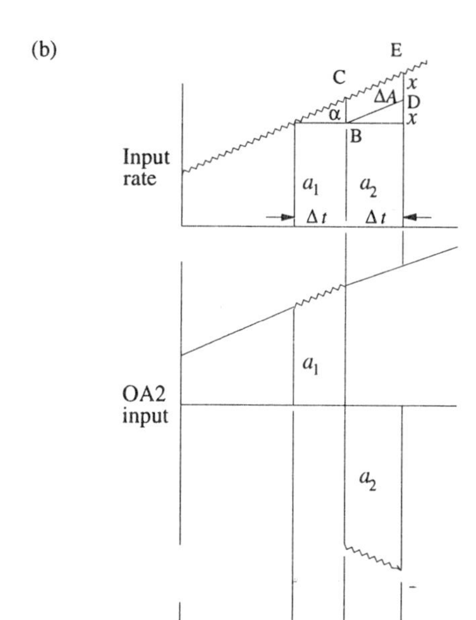

1. Input rate (slope) = tan $\alpha = x/\Delta t$; $x = \Delta t$ (slope).
2. $\Delta A = x\Delta t$; therefore, slope $= \Delta A/(\Delta t)^2$.
3. Slope $= k \cdot \Delta A$.
4. Op amp circuit measures ΔA by the following sequence: $\Delta A = a_2 - a_1$.

$v_o = \Delta A = k^*$ input rate (slope)

Figure 5–30. Integrating-type differentiator. (a) Circuit for integrating rate meter; (b) rate curve and figures showing the measurement and subtraction of two sequential areas (a_1 and a_2) to provide an output voltage directly proportional to the average rate over each increment of measurement time.

An integrating type of differentiator that is exceptionally good at discriminating against noise is shown in Figure 5–30a. Also, the circuit combines two types of op amp circuits described in this chapter. OA1 is an inverting voltage amplifier, and OA2 is an integrator. Four rapid-response electronic switches are used to perform a sequence of operations necessary to obtain an output voltage proportional to the rate of change of the input voltage.

In Figure 5–30b the principle of the technique is apparent by observing the input waveform. The integration is made over two equal time increments Δt that are sequential. The integration is performed so as to obtain an output equal to the difference ΔA between the resultant areas a_1 and a_2. The area ΔA is the parallelogram BCED, which is equal to $(\Delta t)x$. The average slope S of the input signal over the measurement period $2\Delta t$ is $S = \tan \alpha = x/\Delta t$. Because $\Delta A = (\Delta t)x$, x equals $\Delta A/\Delta t$ and the slope S equals $\Delta A/(\Delta t)^2$. The time increment Δt is a constant, so the slope is directly proportional to the area ΔA measured by the op amp circuit of Figure 5–30a. It accomplishes this by the sequence illustrated in Figure 5–30.

The integrator averages the noise on the input rate signal. The time Δt from an accurate time generator is chosen so that the average slope does not change during the measurement period, which is equal to $2\Delta t$. In general, it is desirable to make the time $2\Delta t$ as long as possible to provide better averaging of noise or noise reduction.

Related Topics in Other Media

Video, Segment 5

- Op amp as voltage comparator
- Sine wave to square wave conversion by a comparator circuit
- Effect of noise on comparison process
- Op amp as voltage follower to buffer a pH meter
- Current follower mode for measurement of light intensity
- Subtraction of background from background plus signal by op amp current summing
- Op amp integrator

Laboratory Electronics Kit and Manual

- Section D
  - ¤ Connecting an op amp voltage comparator and making null comparison measurements
  - ¤ Constructing an op amp unity-gain voltage follower and a current-summing amplifier
- Section E
  - ¤ Connecting an op amp current follower and a current-summing amplifier
  - ¤ Measuring nanoampere currents
  - ¤ Wiring an op amp inverting amplifier and measuring gain
  - ¤ Constructing a voltage-summing amplifier

Chapter 6

Thinking Digital: Logic Gates, Flip-Flops, and Counters

The impact of microcomputers and other digital instrumentation in education, business, government, industry, science, technology, and nearly all other areas of society is well documented and has been referred to in previous chapters. Digital systems are helping shape a new culture, with new ways of thinking, working, training, and entertaining. It is not surprising, then, that people from all walks of life and students from early childhood through the university and beyond are confronted by and fascinated with microcomputers and other digital systems. Most of us are now "connected" to digital instrumentation in one way or another, either through direct use or through indirect influence on our lives. This chapter introduces digital logic concepts, functions, operations, and applications that will help us in "thinking digital." In turn, we will be better prepared to make the "right connections" with the myriad digital systems, perhaps through understanding the concepts and digital vocabulary, connecting a new or modified system, or ordering the appropriate parts or system.

6–1 Implementing Basic Logic with Digital Gates

The microcomputer performs so many tasks in so many areas of our lives that its internal operations might be expected to be exceptionally complex. Yet the microcomputer needs to perform only a few basic operations, such as encoding, adding, subtracting, storing data, transferring, decoding, and reading out. The power of the microcomputer comes from its ability to accept program instructions, store and transfer data, and perform its basic logic operations over and over again with fantastic speed. By performing millions of simple logic functions each second, the microcomputer can perform very complex data manipulations and computations.

In this section the basic logic concepts and the functions and operations of digital gates are presented in the context of "thinking digital."

Two-Valued Information

As we saw in Chapter 1, a digital signal has only two defined levels of magnitude, called HI and LO. A digital signal can convey numerical information by using the HI or LO level to represent the 1 or 0 of a binary digit. As shown in Figure 6–1, 1 bit of a binary number is only one of several kinds of information that can be encoded by a two-valued signal. Another common two-valued signal is one that indicates whether a device is **on** or **off** or one that causes a device to be **on** or **off**. The signal connecting a doorbell and a doorbell button is a good example of an **on–off** signal. In a more general way, a two-valued signal can indicate the existence or nonexistence of any particular condition. The HI signal activating a doorbell indicates the presence of a caller. The temperature indicator light on the automobile

2861-2/94/0151$12.00/1

Figure 6–1. Two-valued information indicated by two-valued signals (HI or LO). Examples are binary numbers (1 or 0), on–off indicators or controls, and presence or absence of any given testable condition (True or False).

Signal Level	Number	Function	Logical Statement
HI	1	On	True
LO	0	Off	False

instrument panel indicates whether the logical statement "The engine temperature is too high for safe operation" is true or not. The use of two-valued signals is very common in scientific instrumentation for both numerical and logical information. The circuits that were designed particularly for generating and responding to digital signals, called "digital logic gates," produce digital signals that are a logical response to a **combination** of digital input signals. Like operational amplifiers, these circuits, originally developed for use in digital computers, have found widespread use in many types of instrumentation.

The AND Function: Operation and Application

Suppose, for example, that an intense light source will quickly burn out if operated without cooling water flowing through its jacket. It is desirable for the lamp switch to activate the lamp only when the coolant is flowing. Figure 6–2 shows each combination of conditions and the desired result for each. Note that the lamp should be powered only when the lamp switch is ON and the coolant is flowing. This kind of table is more often written as a truth table, as shown in Figure 6–3. Here, each column is associated with a statement, and the entry T or F indicates whether the statement is true or false. Take a moment to compare Figures 6–2 and 6–3. Again, the statement about the lamp being powered is true only when the switch ON AND the coolant flowing statements are both true. A further simplification of the truth table is achieved if symbols are substituted for the statements and **1** and **0** are used instead of **true** and **false**. This form is shown in Figure 6–4. Here, A is used for the lamp switch ON statement, B is used for the coolant flowing statement, and L is used for the condition of the lamp.

The combination of conditions and results given by the truth tables of Figures 6–2, 6–3, and 6–4 can be written as a single if–then sentence, as shown in Figure 6–5. The first sentence states all combinations of conditions under which the lamp is powered. For sentence 2, A, B, and L are used for the conditions and result. The statement in sentence 2 is reduced to an equation in line 3 which can be read "L is true when and only when A is true AND B is true." The AND operation can also

Figure 6–2. Summary of conditions and desired result for operation of cooled lamp.

Conditions		Result (Lamp)
Lamp Switch	Coolant	
Off	Not flowing	Not powered
Off	Flowing	Not powered
On	Not flowing	Not powered
On	Flowing	Powered

Lamp switch is on	Coolant is flowing	Lamp is powered
F	F	F
F	T	F
T	F	F
T	T	T

Figure 6–3. Conversion of conditions–result table to a truth table for cooled-lamp example.

A	B	L
0	0	0
0	1	0
1	0	0
1	1	1

A = Lamp switch is on.
B = Coolant is flowing.
L = Lamp is powered.

Figure 6–4. Simplification of truth table for cooled-lamp example.

1. If the lamp switch is on and the coolant is flowing, then the lamp is powered.
2. If A is true and B is true, then L is true.
   A = Lamp switch is on.
   B = Coolant is flowing.
   L = Lamp is powered
3. L = A AND B
4. $L = A \cdot B$
5. L = AB

Figure 6–5. Equivalent logic statements for cooled-lamp example.

X	Y	R
0	0	0
0	1	0
1	0	0
1	1	1

X = Truth of any condition statement
Y = Truth of any other condition statement
$R = X \cdot Y$

Figure 6–6. The AND function expressed in a truth table.

be symbolized by the dot, as in statement 4, or simply by joining the symbols, as in statement 5.

As the representation of this cooled-lamp example progresses through the truth tables of Figures 6–2, 6–3, and 6–4, the tables become less specific to the particular application until in Figure 6–4 the truth table is completely symbolic and the symbols are defined independently. Thus Figure 6–4 is a table for the *function* performed, namely, the AND function, and the symbols are defined independently. The truth table in Figure 6–6 for the AND function on input conditions X and Y and result R is the same as that in Figure 6–4. The truth table in Figure 6–6 is a general truth table for any statements X and Y where both X and Y must be true for the result R to be true.

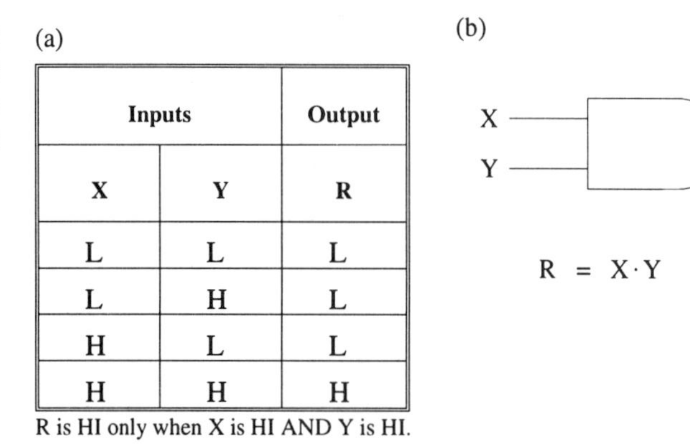

(a)

Inputs		Output
X	**Y**	**R**
L	L	L
L	H	L
H	L	L
H	H	H

R is HI only when X is HI AND Y is HI.

Figure 6–7. The AND gate. (a) Table of states for the two-valued information (HI/LO or H/L); (b) gate symbol and logic statement (R = X · Y).

A circuit that performs the AND function produces a HI or LO output signal depending on the combination of HI and LO signals at its inputs. Such a circuit is called an AND gate, and its table of states and symbols is shown in Figure 6–7. Its table of states is expressed by the symbols H and L because the table is to describe the way in which the circuit responds to actual signal levels. If a HI (H) signal is equivalent to 1 or true, the AND gate table of states and the AND function truth table are identical. The symbol in Figure 6–7b is used to represent the AND gate in circuit diagrams. This circuit is called a **gate** because a LO at one input prevents the signal at the other input from affecting the output level. Note that for the first two entries in the table of states, where X is LO (L), the output R is LO even though Y is changing. However, when X is HI, the changes in Y appear at the output R. Thus the level at X affects whether or not the information at Y is transmitted through the gate.

Now let's return to the problem of the lamp that requires cooling. Considering the equation in Figure 6–5, we see that to protect the lamp we would need a signal that is HI when the cooling water is flowing, another signal that is HI when the lamp switch is ON, and an AND gate. Figure 6–8 shows the complete implementation. A flow sensor with an output voltage proportional to the water flow is connected to one input of a comparator. The other comparator input is connected to an adjustable voltage source set to the voltage the flow sensor will produce when the coolant flow is the minimum acceptable rate. The comparator output will thus be HI when the flow is sufficient to cool the lamp, that is, when B is true. The lamp switch controls the level of signal A, producing a HI level when the operator desires

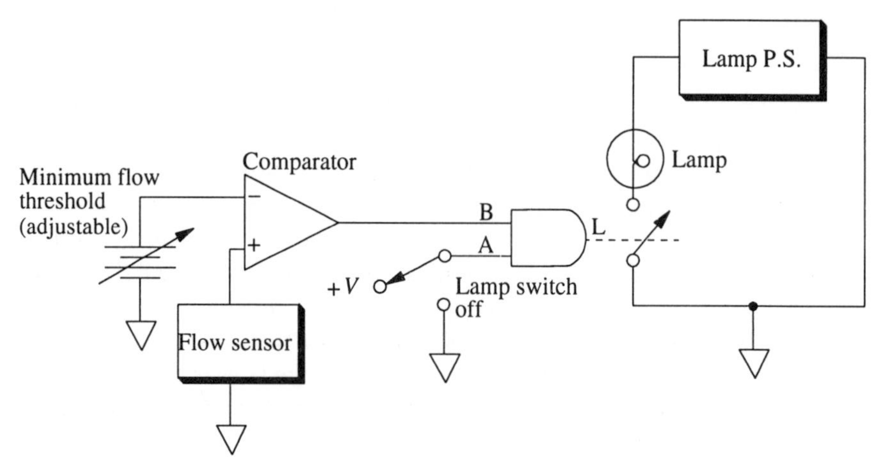

Figure 6–8. Circuit for implementation of cooled-lamp example. P.S., power supply.

the lamp to be ON. A HI at AND gate output L, which will occur only when signals A and B are both HI, will close the lamp power switch and turn ON the lamp.

This example illustrates how an operational amplifier comparator (level detector described in Chapter 5) is used to detect when an analog quantity is more or less than a particular value and how the two-valued comparator output is a digital signal encoding the truth of the comparison statement. In this case the comparator acts as a kind of "two-level" analog-to-digital converter. As we shall see in Chapter 9, the comparator is an essential component in analog-to-digital converters of all types.

The OR Function: Operation and Application

Suppose that a chemical reaction will be spoiled or will run out of control if the temperature exceeds 110 °C or the pH is greater than 8.3. We want a warning buzzer to sound if either of these conditions is approached. Sensors and comparators can be used to produce signals representing the truth of the statements "The temperature is greater than 110 °C" and "The pH is greater than 8.3." The truth table in Figure 6–9 shows the desired state of the buzzer for each possible combination of the temperature and pH conditions. W is true if A is true or B is true or both are true. The function shown by the truth table of Figure 6–9 can be expressed in any of the ways given in Figure 6–10. The first method is a logical if–then sentence; the second substitutes the symbols A, B, and W for the temperature, pH, and buzzer signals; the third expresses the relationship as an equation, where the operation is clearly an OR function; and the fourth shows that the algebraic symbol for the OR function is the plus sign.

A circuit that can implement the truth table in Figure 6–9 and carry out the OR operation for digital signals is called an OR gate. The table of states and the symbol for the OR gate are shown in Figure 6–11. As the OR function implies, the output R is HI when X is HI or when Y is HI or when both X and Y are HI. The symbol for the OR gate is different from the D shape of the AND gate. The symbol, as shown in Figure 6–11b, is used for OR gates in electronic circuits.

A	B	W
0	0	0
0	1	1
1	0	1
1	1	1

A = Temperature > 110 °C
B = pH > 8.3
W = Warning buzzer is on.

Figure 6–9. Desired state of a warning signal for dangerous conditions of pH and temperature as expressed by a truth table.

1. If the temperature is greater than 110 °C or if the pH is greater than 8.3, then the warning buzzer is on.
2. If A is true or B is true, then W is true. A, B, and W are defined in Figure 6–9.
3. W = A OR B
4. W = A + B

Figure 6–10. Equivalent logic statements for desired warning signal if pH or temperature exceeds limits for safe operation.

Figure 6–11. The OR gate. (a) Table of states for the two-valued (H/L) input and output information; (b) gate symbol and logic statement (R = X + Y)

(a)

Input		Output
X	**Y**	**R**
L	L	L
L	H	H
H	L	H
H	H	H

R is HI when X is HI OR Y is HI OR both are HI.

(b)

X
Y
R

R = X + Y

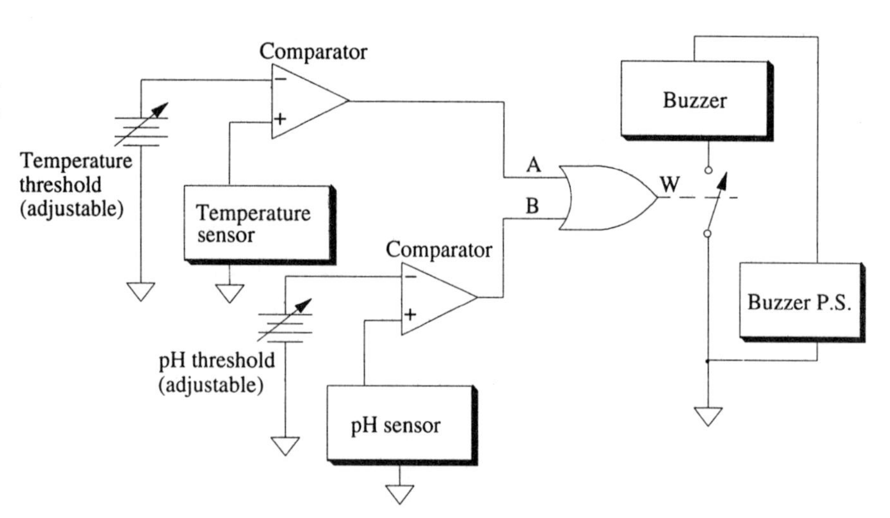

Figure 6–12. Circuit for providing a warning signal when pH and temperature exceed the safe limits. P.S., power supply.

The complete pH and temperature warning circuit is shown in Figure 6–12. Analog voltage sensors with operational amplifier comparators and adjustable threshold sources are used to produce digital signals A and B. If either sensor output exceeds the preset threshold value, the comparator output will be HI. If either A or B is high, the W output signal will be HI, thus closing the switch to sound the warning buzzer. This warning will persist until the excessive quantity is reduced to below the threshold value.

This example illustrates how digital signals can be obtained for more than one analog level by using a separate comparator and threshold source for each level detected. This concept can be extended to any number of levels and is the basis for one form of analog-to-digital converter.

The NOT Function: Operation and Application

Interlock systems in instruments are often designed to shut down the instrument (remove power) when a condition exists that is potentially damaging or unsafe. For example, damage could occur if a diffusion pump heater is on but the coolant

Function to be implemented:

Shutdown = Heater on AND coolant not flowing
OR high voltage on AND lid open

Signals available:
H = Heater on
C = Coolant flowing
V = High voltage on
L = Lid closed

Function in terms of signals H, C, V, and L:

S = H AND NOT C OR V AND NOT L

S = H AND $\overline{C}$ OR V AND $\overline{L}$

$S = H \cdot \overline{C} + V \cdot \overline{L}$

Figure 6–13. Logic statements for protection of a system with suitable interlock.

is **not** flowing. An unsafe condition would exist if the instrument lid is removed while the high-voltage power supply is **on**. Suppose we wish to generate a shutdown signal when either of these two conditions exists, as shown in Figure 6–13. Suppose also that we can easily obtain the signals H, C, V, and L as shown. Notice that the damaging condition exists when H is true and C is *not* true and the dangerous condition exists when V is true and L is *not* true. The expression for the shutdown signal is then H AND NOT C or V AND NOT L. The NOT, or the opposite of a statement, is expressed by a bar over the symbol for the statements so that $\overline{C}$ is read NOT C. This NOT symbol is then used in the increasingly symbolic forms of the equations in Figure 6–13, which are all read as "S equals H AND NOT C OR V AND NOT L."

Functionally, the NOT operation is performed by a circuit called an inverter. As shown in Figure 6–14, this circuit converts a LO input signal into a HI output signal and vice versa. Thus if we have a signal C that is HI when the coolant is flowing, after inversion the signal NOT C will be HI when the coolant is not flowing. This, then, is the appropriate signal to apply with signal H to an AND gate in order to obtain the output signal indicating potential damage. The complete implementation of the interlock example, using inverters, is shown in Figure 6–15. Note the exact correspondence between the functions implemented by the circuits and the operations indicated by the equation. Note also that the functions are performed by the gates in the same order you would follow if you were solving S

(a)

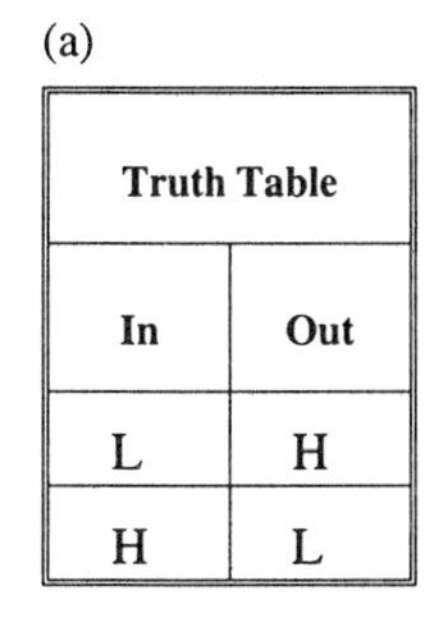

Truth Table	
In	Out
L	H
H	L

(b) Symbol

C → $\overline{C}$

C → $\overline{C}$

Figure 6–14. Describing the NOT (inverter) gate. (a) Table of states for the two-valued data; (b) gate symbol. In the symbol, the small circle indicates the inversion function. This circle is often attached to the triangular symbol of a buffer or driver amplifier, but it can also be used at the input or output of other logic functions to indicate inversion. A bar over the top of a letter indicates the inversion operation.

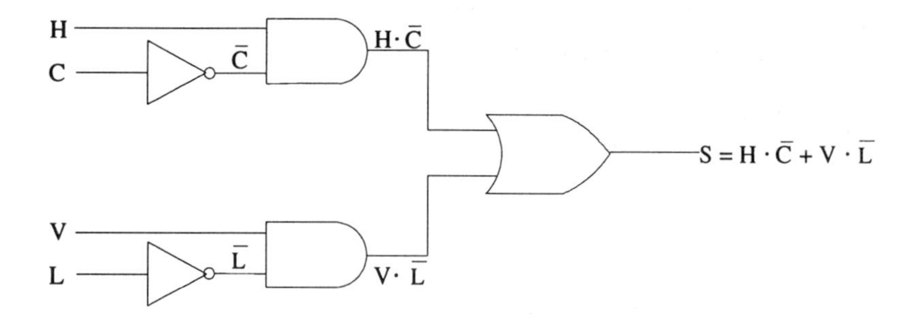

Figure 6–15. Circuit for implementation of the interlock logic statement of Figure 6–13. The signals $H \cdot \bar{C}$ and $V \cdot \bar{L}$ are obtained and then sent to the OR gate to perform the OR operation.

as an algebraic equation. That is, the dot operation is done first, and then the plus operations are done.

These three operations—AND, OR, and NOT—are the three fundamental logic operations. All logic functions, no matter how complex, can be implemented by combinations of AND, OR, and NOT circuits.

Gate circuits can be derived from the logic statements, and the same function can be implemented by different but completely equivalent gate circuits. Two logic circuits are considered completely equivalent if they have identical truth tables. The study of equivalent logic statements is part of the field of Boolean algebra. From a few simple algebraic rules, the fact that an AND gate performs the OR function on inverted inputs and outputs can be proven.

Integrated Circuit (IC) Logic Packages

The widespread application of digital logic circuits was stimulated by the development of the inexpensive integrated circuit (IC) logic package. The most common form of the IC is shown in Figure 6–16a (below). The package has 14 or 16 pins and contains a small chip of silicon on and in which the functional circuits have been fabricated. Because the silicon and its circuit constitute only a small fraction of the package cost, an effort has been made to include in the package the maximum amount of circuitry the number of connectors can support. The 14-pin package can contain four gates with three pins per gate and two pins for power connections to all. AND gates can also be made with three, four, or more inputs, and with more pins needed per gate, fewer gates fit in the package. Integrated circuit packages for OR gates and NOT (INVERT) gates are shown in Figures 6–16c and 6–16d (on the opposite page).

The truth table for a logic circuit must include all possible combinations of the states of the input signals. Thus, for a three-input gate, the truth table has eight

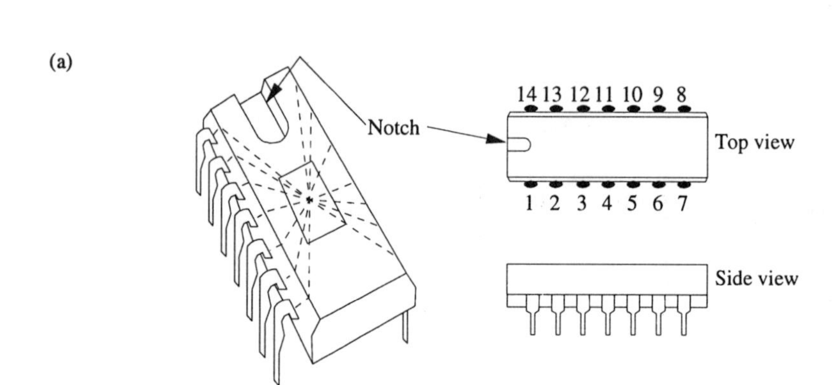

Figure 6–16. IC logic gate packages. (a) Typical physical construction; (b, on opposite page) multiple AND gates in one package; (c, on opposite page) OR gates; (d, on opposite page) NOT (inverter) gates.

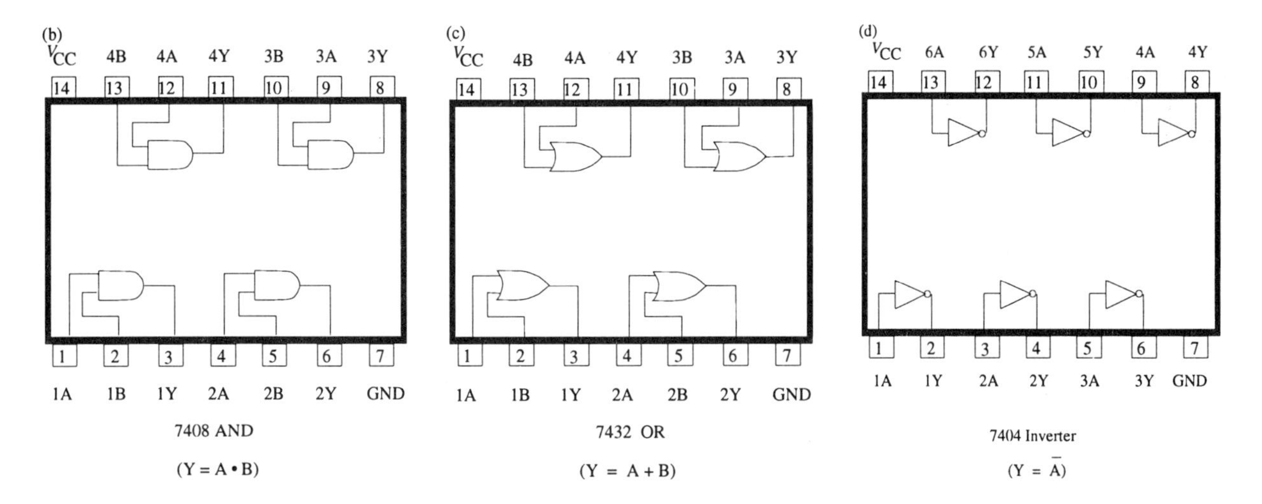

entries, as shown in Figure 6–17. The number of entries for an n-input circuit will be 2^n, that is, 8 for a three-input circuit, 16 for a four-input circuit, and so on. Some three-input and four-input IC packages are shown in Figure 6–18, on the next page.

Inputs			Output
A	B	C	R
L	L	L	L
L	L	H	L
L	H	L	L
L	H	H	L
H	L	L	L
H	L	H	L
H	H	L	L
H	H	H	H

Number of entries in table = 2^n, where n = number of inputs.

Figure 6–17. All possible combinations of input states for any three-input gate. The result is shown for an AND gate.

Gate Combinations (NAND, NOR, EXCLUSIVE-OR)

All logic functions can be performed by simple combinations of AND, OR, and NOT circuits, but some combinations are so commonly used that IC manufacturers have put these combinations in a single IC package to save IC packages, sockets, and circuit board space, which are the most expensive parts of modern electronic units. Three common prepackaged logic circuits are shown in Figure 6–18a. The first is an AND gate with an inverted output. The circle at the output indicates inversion of the output quantity. Its truth table in Figure 6–18b shows an output

Figure 6–18. Combinations of gates in IC packages. (a) Symbols; (b) truth tables; and IC TTL packages for (c) two-input NAND, NOR, and EXCLUSIVE-OR and (d) three-input NAND, four-input NAND, and four-input AND. Fewer gates fit in the 14-pin IC package.

(a)

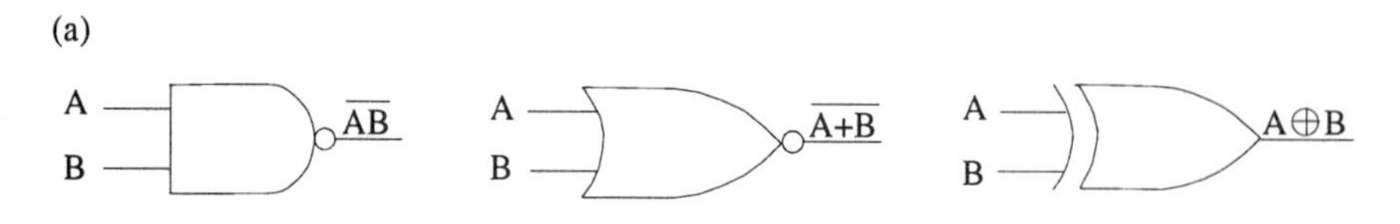

(b)

NAND (not AND)			NOR (not OR)			EXCLUSIVE - OR		
A	B	$\overline{AB}$	A	B	$\overline{A+B}$	A	B	$A \oplus B$
L	L	H	L	L	H	L	L	L
L	H	H	L	H	L	L	H	H
H	L	H	H	L	L	H	L	H
H	H	L	H	H	L	H	H	L

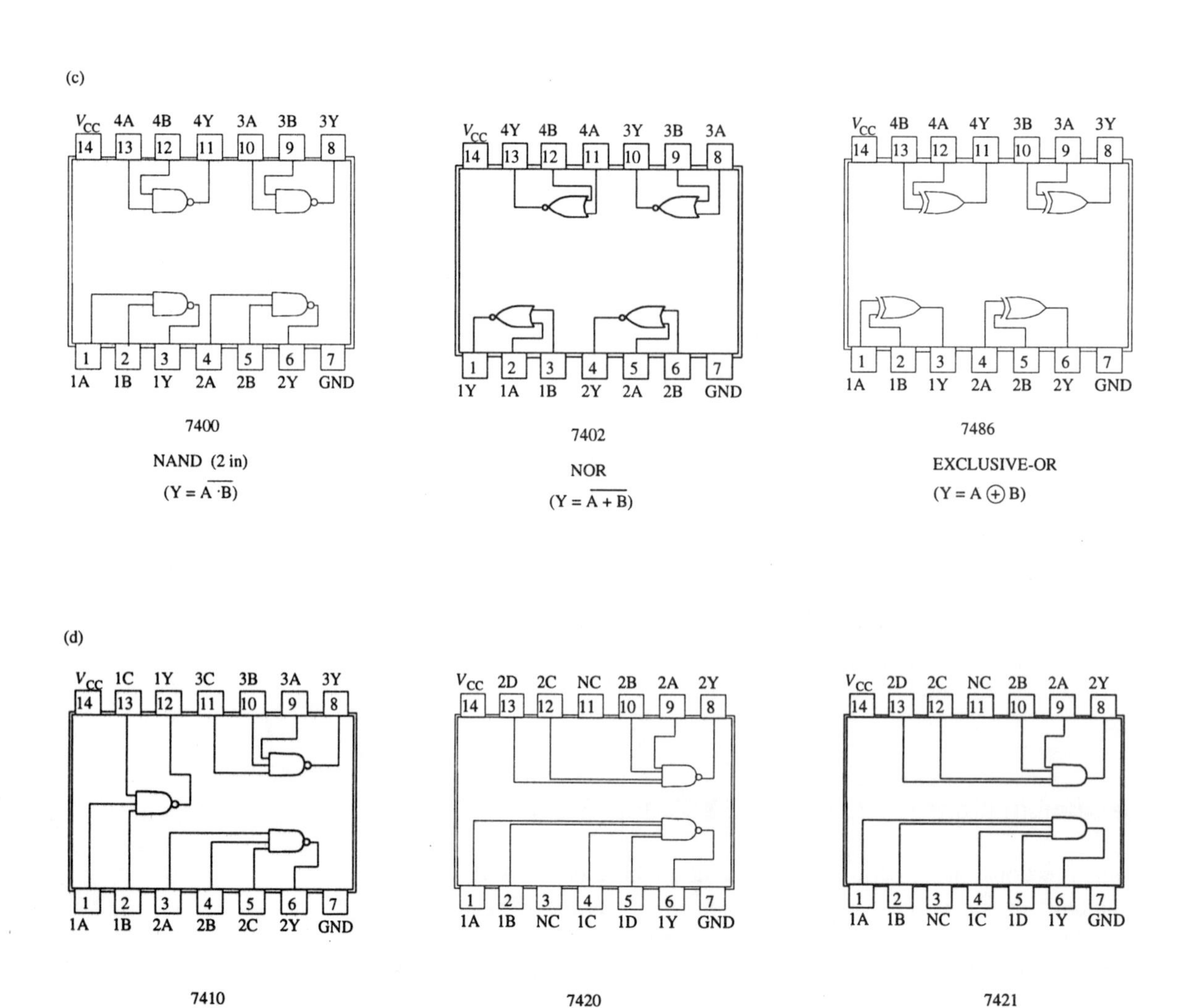

function that is the inverse of the AND gate. When an inverter is needed and a four-NAND gate package has an unused NAND gate, the two inputs can be connected to each other to provide the needed inverter. Another inverting gate is the NOR gate. As shown, its output is the complement (or inverse) of the OR gate.

The EXCLUSIVE-OR function is "A OR B *but not both.*" Only when A and B are opposite (HI–LO or LO–HI) will the output be HI. Note that the EXCLUSIVE-OR operation in Figure 6–18 is indicated by the plus (OR) sign with a circle around it. The circuit symbol is similar to the OR symbol with an added curved line at the input. The EXCLUSIVE-OR statement is written in algebraic form as

$$A \oplus B = \overline{A}B + A\overline{B}$$

Thus it could be readily implemented by gates performing the AND operation on $\overline{A}$ and B and on $\overline{B}$ and A and then the OR operation on $\overline{A}B$ and $\overline{B}A$. The EXCLUSIVE-OR function is commonly required, and prepackaged ICs simplify connections and save many gate packages.

Many other combined logic functions are available in single packages. The complexity of the logic functions varies from the equivalent of a few gates to thousands.

Families of ICs

In the design of ICs there are trade-offs between the speed of response and the power required per gate. Because different applications favor different characteristics, not all achievable at once, several different types or families of integrated logic circuits have evolved.

One family of logic circuits is called TTL, an abbreviation for transistor–transistor logic. This type of circuit is based on a technology that offers good performance for the cost and power required and has evolved well to provide improved performance over the years. It is very commonly used in electronic instruments. Over a dozen technologies have been advanced over the last 20 years, but many have not survived. Two other types, or families, of logic ICs are also widely used today. The emitter-coupled logic (ECL) family is used where speed of operation is critical. Its relatively high power dissipation, however, prevents packing a large number of gates on a chip, and the circuits are limited to relatively simple functions. The complementary metal oxide silicon (C-MOS) family, on the other hand, has extremely low power dissipation and a very small area per gate. This technology allows the use of thousands of gates per chip and is the basis of the current microcircuit revolution. Some characteristics of the TTL, ECL, and C-MOS logic families are compared in Figure 6–19.

6–2 Comparing Digital Data

The comparison of two binary words is an important function in digital circuits. Digital comparators are used extensively as decision-making elements in computers and other digital systems. When comparing two binary numbers, it might be necessary to know only that A equals B or that A does not equal B. In other cases it is important to know the relative magnitudes of A and B, that is, whether

Figure 6–19. Popular IC logic families.

Name (Abbreviation)	LO	HI	Speed, ns[a]	Power[a]
Transistor–transistor logic (TTL)	0–0.4 V	2.4–4 V	10	2 mW
Emitter-coupled logic (ECL)	−1.5 V	−0.75 V	2	25 mW
Complementary metal oxide silicon field-effect transistor logic (C-MOS)	$<V_{DD}/2$	$>V_{DD}/2$	50	10 μW[b]

[a] Values are per gate.
[b] C-MOS power depends on frequency of state because most of the power goes to charging and discharging device capacitance.

A is larger or smaller than B or equal to B. The types of logic gates that can provide this information are discussed in this section.

EXCLUSIVE-OR and EQUALITY Functions

The EXCLUSIVE-OR gate was introduced in the previous section. It is contrasted in Figure 6–20 with the EQUALITY gate. Sometimes the EXCLUSIVE-OR gate is referred to as an "inequality comparator" because it provides a HI or TRUE output only if A and B are not equal. It is thus useful in anticoincidence detection. The EXCLUSIVE-OR can also be thought of as a controllable inverter. If one input is used as a control, the other input is inverted when the control is HI and not inverted when the control is LO.

By inserting an inverter at the output of the EXCLUSIVE-OR gate, an equality operation is performed, as illustrated by the truth tables in Figure 6–20. From the truth tables it is clear that the output functions of the two gates are inversely related. The EQUALITY gate is also called the comparator gate or coincidence gate because it produces a HI (TRUE) output whenever the two input signals have the same logic level (A and B inputs both equal either HI or LO).

Figure 6–20. Comparison of EXCLUSIVE-OR and EQUALITY logic gates. (a) Truth tables; (b) symbols and logic expression for logic operation.

(a)

Variables		Result	
		EXCLUSIVE-OR	EQUALITY
A	B	$A \oplus B$	$\overline{A \oplus B}$
L	L	L	H
L	H	H	L
H	L	H	L
H	H	L	H

(b)

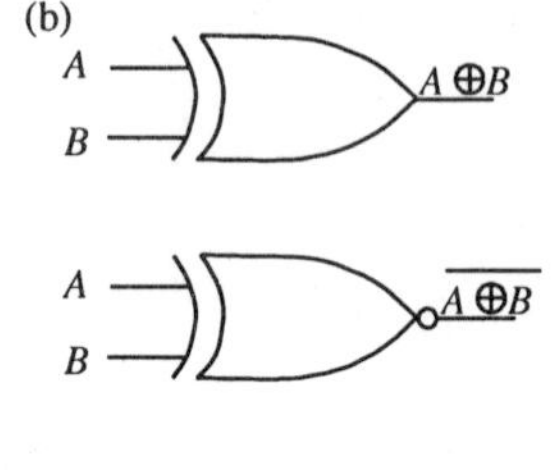

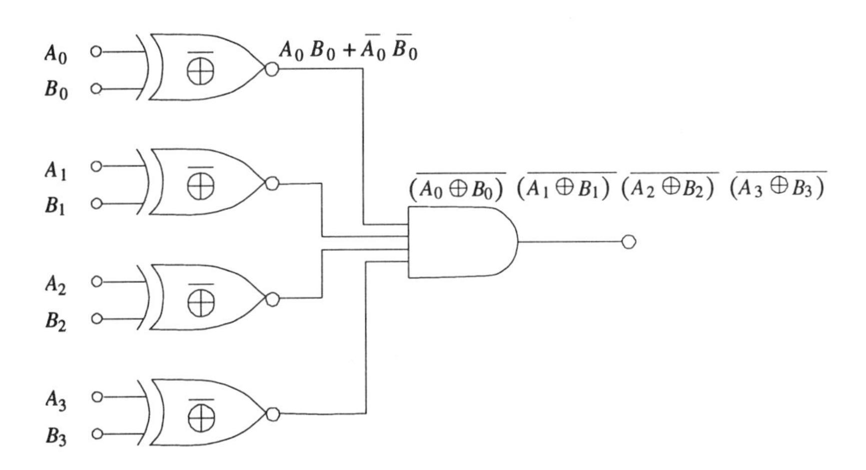

Figure 6–21. Four-bit equality detector. Adding an AND gate to the outputs of four EQUALITY gates yields a 4-bit equality detector.

Equality Detector

A 4-bit equality detector is shown in Figure 6–21. The outputs of four EQUALITY gates are connected to the inputs of a four-input AND gate. The output of the AND gate will be true only when the 2 bits of each pair are equal. The word represented by bits $A_3 A_2 A_1 A_0$ must be identical with the word represented by $B_3 B_2 B_1 B_0$ to have a HI (TRUE) output from the detector.

Magnitude Comparator

When two binary numbers are compared, there are three possible results: $A > B$, $A < B$, or $A = B$. To distinguish the three conditions, outputs are required for two of the conditions; the third can be assumed when neither of the others is TRUE. The equality condition is readily obtained by combining EQUALITY gates, as shown in Figure 6–21, and the $A > B$ condition can be obtained by comparing the magnitude of the most significant nonequal bits in the two binary words.

The magnitude comparators in computers and digital control systems enable program statements, such as "If $A > B$, do step 35 next; if $A < B$, do step 23 next; if $A = B$, stop," to provide directions for the system.

The TTL 7485 magnitude comparator is a medium-scale IC (MSI chip). Its pin configuration is shown in Figure 6–22a (below), and its functional block diagram is shown in Figure 6–22b (on the next page). It performs magnitude

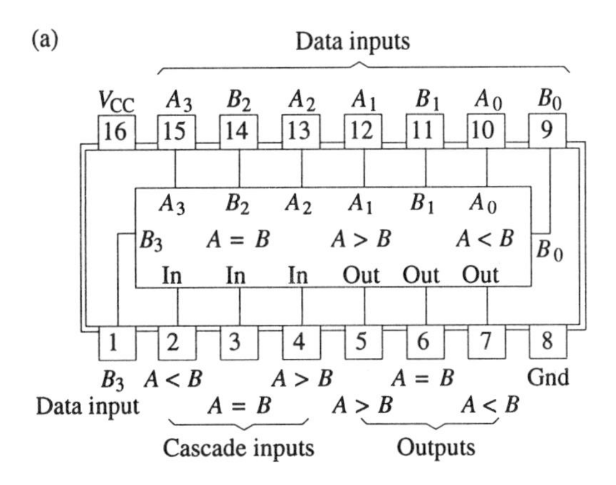

Figure 6–22. TTL IC 7485 magnitude comparator. (a) Pin configuration.

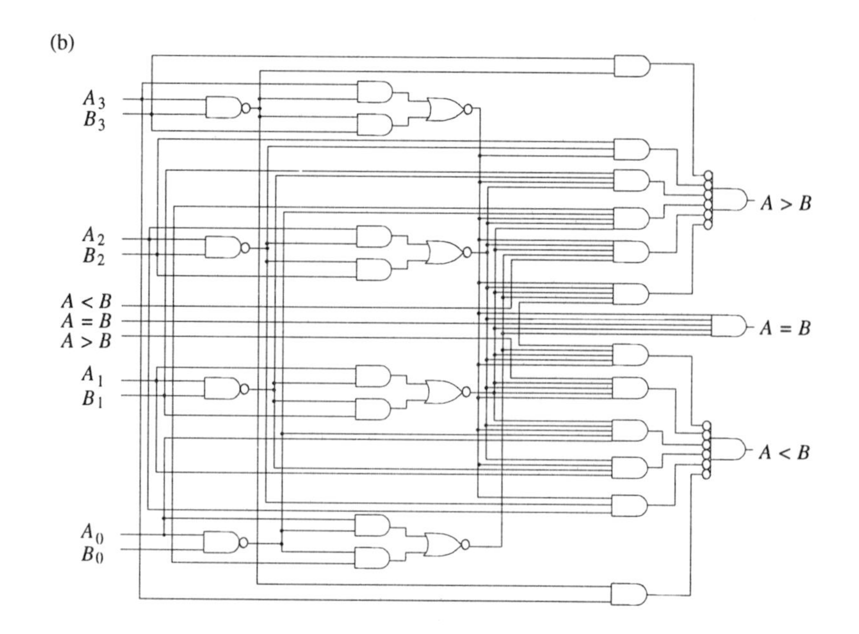

Figure 6–22—*Continued.* (b) Diagram showing gates on IC chip for implementation of functions; (c) function tables of states.

(c)

Comparing Inputs				Cascading Inputs			Outputs		
A_3, B_3	A_2, B_2	A_1, B_1	A_0, B_0	$A > B$	$A < B$	$A = B$	$A > B$	$A < B$	$A = B$
$A_3 > B_3$	X	X	X	X	X	X	H	L	L
$A_3 < B_3$	X	X	X	X	X	X	L	H	L
$A_3 = B_3$	$A_2 > B_2$	X	X	X	X	X	H	L	L
$A_3 = B_3$	$A_2 < B_2$	X	X	X	X	X	L	H	L
$A_3 = B_3$	$A_2 = B_2$	$A_1 > B_1$	X	X	X	X	H	L	L
$A_3 = B_3$	$A_2 = B_2$	$A_1 < B_1$	X	X	X	X	L	H	L
$A_3 = B_3$	$A_2 = B_2$	$A_1 = B_1$	$A_0 > B_0$	X	X	X	H	L	L
$A_3 = B_3$	$A_2 = B_2$	$A_1 = B_1$	$A_0 < B_0$	X	X	X	L	H	L
$A_3 = B_3$	$A_2 = B_2$	$A_1 = B_1$	$A_0 = B_0$	H	L	L	H	L	L
$A_3 = B_3$	$A_2 = B_2$	$A_1 = B_1$	$A_0 = B_0$	L	H	L	L	H	L
$A_3 = B_3$	$A_2 = B_2$	$A_1 = B_1$	$A_0 = B_0$	L	L	H	L	L	H

comparison of straight binary and straight binary coded decimal (BCD) (8421) codes. Three fully decoded decisions about two 4-bit words (*A*, *B*) are made and are externally available at three outputs. These devices are fully expandable to any number of bits without external gates. When cascaded, the total time for comparison is the function of the word length; however, only a two-gate-level delay (12 ns) is added for each 4-bit expansion.

Note that the $A = B$ condition is a 4-bit equality detector similar to the detector in Figure 6–21, wherein the EQUALITY gates are each shown implemented by the basic AND–OR–INVERT gates and a NAND gate. The $A > B$ output is HI whenever $A_3 > B_3$, OR $A_3 = B_3$ AND $A_2 > B_2$, OR $A_3 = B_3$ AND $A_2 = B_2$ AND $A_1 > B_1$, etc. Similarly, the $A < B$ output is HI when $A < B$. These latter conditions are decoded by six-wide AND–OR–INVERT gates.

The complete function table for a HI–TRUE magnitude comparator is shown in Figure 6–22c. X in the table indicates that the output shown is obtained whether that particular input is HI or LO. This indifference to input is referred to as the "don't-care" condition.

6–3 Selecting/Multiplexing Digital Data

When multiple digital signals need to be input sequentially into the same circuits of a digital system, it is necessary to use a gate that is the digital equivalent of a multiple-throw switch. Thus any one of the input signals can be connected to a specific circuit. Gates for controlling multiple digital signals are usually called multiplexers. They are used in computers and other digital instruments to control the flow of digital signals from one section of the system to another. A few examples of digital multiplexers are presented in this section.

Four-Line to One-Line Multiplexer

A four-wide AND–OR gate is equivalent to a four-position multiple-throw switch, as shown in Figure 6–23. If the A_c control input is HI, then the output will be the logic level at A. Only one control input is to be HI at any one time. According to the output function, the state of the control inputs A_c, B_c, C_c, and D_c determines which of the input signals, A, B, C, or D, appears at the output. This circuit can be extended to any number of inputs.

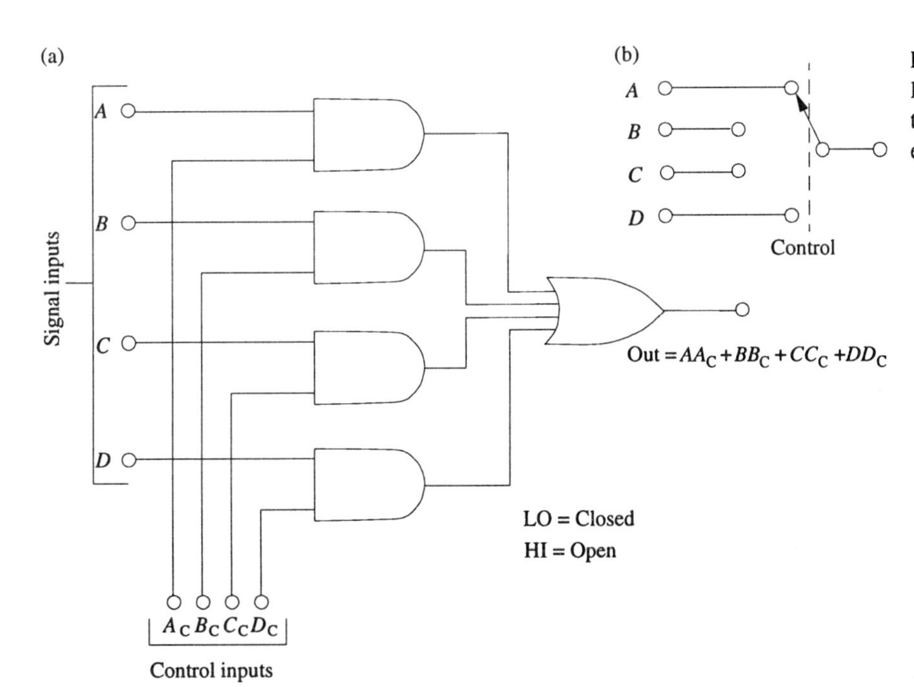

Figure 6–23. Four-line to one-line multiplexer. (a) Implementation with the AND/OR gates; (b) equivalent four-throw switch.

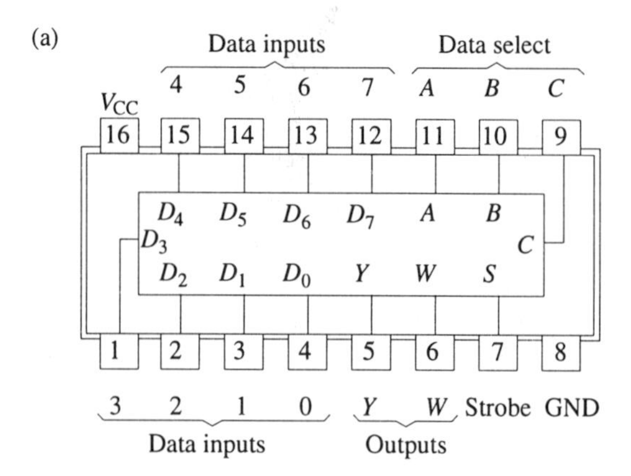

Figure 6–24. Eight-line to one-line data selector/multiplexer. (a) Pin configuration of TTL IC 74151; (b, on opposite page) functional table of states. It gives the states (1 for HI and 0 for LO) for all inputs and outputs.

Eight-Line to One-Line Selector/Multiplexer

IC multiplexers contain as many as 16 input lines switched to a single output line. An eight-line to one-line IC selector/multiplexer is shown in Figure 6–24. The eight data inputs are selected by the data select lines *C*, *B*, and *A*. Thus the specific data input connected to the output depends on the selector binary word *CBA* from 000 to 111. Two outputs *Y* and *W* are available, where *W* is the inverse of *Y*.

Two-Line to One-Line Selector/Multiplexer

A simple digital selector/multiplexer that is equivalent to a single-pole double-throw switch is shown in Figure 6–25. When the control input signal is HI, the output is *A*; when the control input is LO, the output is *B*.

6–4 Decoding Binary Data

For many applications it is necessary to decode (convert) each relevant combination of input logic levels from a digital word into a separate logic-level output. For example, to light the correct digit on a seven-segment panel display, the binary-coded decimal number must be decoded to provide decimal output signals that are used to operate the correct segments on the display. This requires a BCD-to-decimal decoder of the type described in this section. Other decoder examples are also presented.

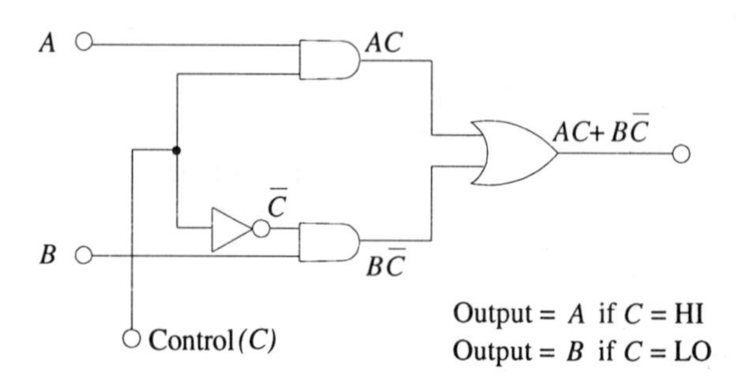

Figure 6–25. Two-line to one-line selector/multiplexer.

(b)

Inputs[a]												Outputs	
C	B	A	Strobe	D_0	D_1	D_2	D_3	D_4	D_5	D_6	D_7	Y	W
X	X	X	1	X	X	X	X	X	X	X	X	0	1
0	0	0	0	0	X	X	X	X	X	X	X	0	1
0	0	0	0	1	X	X	X	X	X	X	X	1	0
0	0	1	0	X	0	X	X	X	X	X	X	0	1
0	0	1	0	X	1	X	X	X	X	X	X	1	0
0	1	0	0	X	X	0	X	X	X	X	X	0	1
0	1	0	0	X	X	1	X	X	X	X	X	1	0
0	1	1	0	X	X	X	0	X	X	X	X	0	1
0	1	1	0	X	X	X	1	X	X	X	X	1	0
1	0	0	0	X	X	X	X	0	X	X	X	0	1
1	0	0	0	X	X	X	X	1	X	X	X	1	0
1	0	1	0	X	X	X	X	X	0	X	X	0	1
1	0	1	0	X	X	X	X	X	1	X	X	1	0
1	1	0	0	X	X	X	X	X	X	0	X	0	1
1	1	0	0	X	X	X	X	X	X	1	X	1	0
1	1	1	0	X	X	X	X	X	X	X	0	0	1
1	1	1	0	X	X	X	X	X	X	X	1	1	0

[a]When used to indicate an input, X = irrelevant.

Binary Decoder

A decoder that converts the multiple states of a binary word input into the allowed control signals can eliminate unwanted combinations and reduce circuit connections. For example, the decoder in Figure 6–26 converts the four states of a 2-bit binary word to the allowed four-line control signals. These can be used to provide the control inputs with the four-line to one-line multiplexer of Figure 6–23 described in the previous section.

BCD-to-Decimal Decoder

It was shown in Chapter 1 how a decimal digit can be encoded as a digital word that consists of 4 binary bits. The code follows the normal binary count sequence

Figure 6–26. Binary decoder. (a) Gate circuit; (b) truth table.

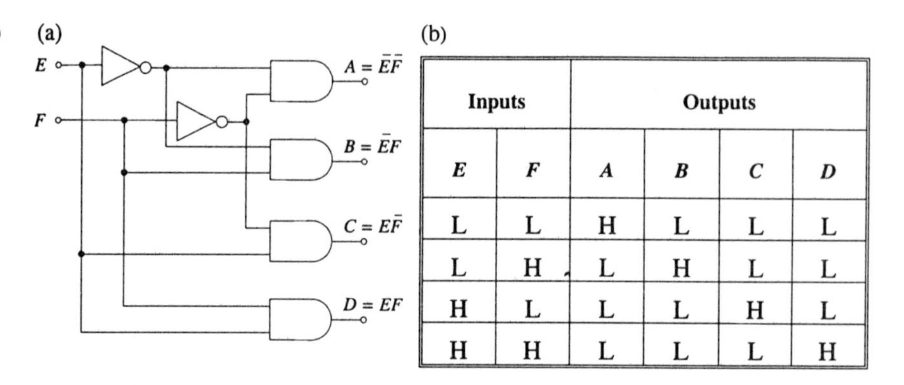

Inputs		Outputs			
E	*F*	*A*	*B*	*C*	*D*
L	L	H	L	L	L
L	H	L	H	L	L
H	L	L	L	H	L
H	H	L	L	L	H

(0000 to 1001) for decimal digits 0 to 9, as shown in Figure 6– 27a. A BCD-to-decimal decoder can convert the 4 bits of binary-encoded data into a signal representing any one of the 10 decimal digits from 0 to 9. An IC to accomplish this is shown in Figure 6–27b, and its gate circuit is shown in Figure 6–27c. Inverters are used to generate the NOT of each input, and a four-input AND function combines these as appropriate for the logic statement given in the table.

Figure 6–27. BCD-to-decimal decoder. (a) Binary to decimal decoding table; (b) TTL IC 7442 pin configuration. GND, ground.

(a)

Decimal	BCD Signals				Logic Statements			
	D(8)	*C*(4)	*B*(2)	*A*(1)				
0	0	0	0	0	$\bar{A}$	$\bar{B}$	$\bar{C}$	$\bar{D}$
1	0	0	0	1	A	$\bar{B}$	$\bar{C}$	$\bar{D}$
2	0	0	1	0	$\bar{A}$	B	$\bar{C}$	$\bar{D}$
3	0	0	1	1	A	B	$\bar{C}$	$\bar{D}$
4	0	1	0	0	$\bar{A}$	$\bar{B}$	C	$\bar{D}$
5	0	1	0	1	A	$\bar{B}$	C	$\bar{D}$
6	0	1	1	0	$\bar{A}$	B	C	$\bar{D}$
7	0	1	1	1	A	B	C	$\bar{D}$
8	1	0	0	0	$\bar{A}$	$\bar{B}$	$\bar{C}$	D
9	1	0	0	1	A	$\bar{B}$	$\bar{C}$	D

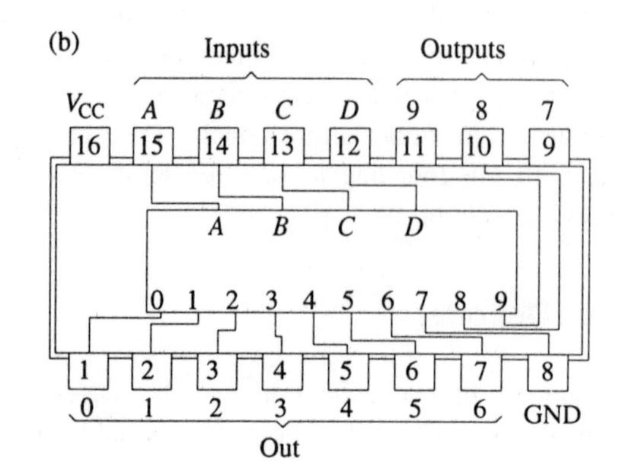

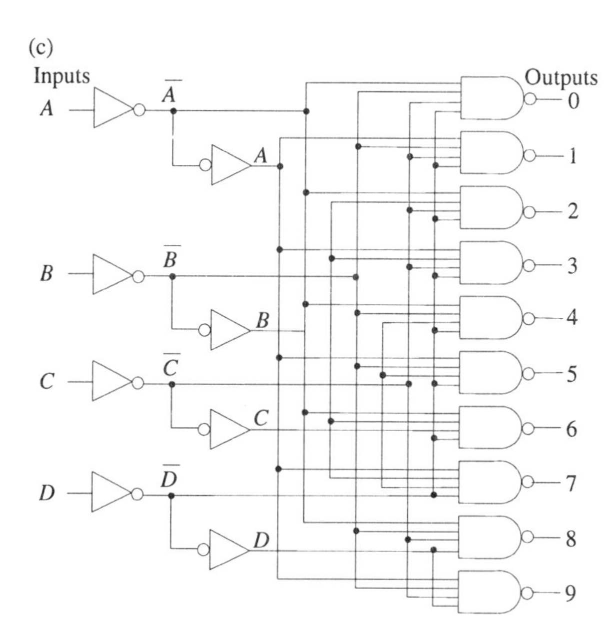

Figure 6–27—*Continued.* (c) Gates on IC chip for decoding operation.

Notice that NAND gates are actually used. This means that the output line of the encoded numeral will be LO and the others will be HI. This reversal of the more usual HI = TRUE condition is called "active LO." The convenience of connection to various circuits determines when active LO is used.

BCD-to-Seven-Segment Decoder/Driver

Often, a decoded BCD signal is used to drive a numerical display of the type shown in Figure 6–28a. In such a display, each segment is activated when it is needed in the number to be displayed. The conditions to light segment a, for example, are when the numeral is 0 or 2 or 3 or 5 or 6 or 7 or 8 or 9. This is expressed by the first equation in Figure 6–28a (below). The second equation for segment a is obtained by substituting the logic statement in Figure 6–27a for each number. If the outputs of the BCD-to-decimal decoder for these numbers are combined in an OR gate and connected to a light-emitting diode driver, segment a will light when any of the above numerals is decoded. The other segments are driven in a similar manner. A similar equation could be developed for each of the other segments. The entire combination of these operations is contained in a single BCD/seven-segment 7447 IC shown in Figure 6–28b (on the next page).

The table of states in Figure 6–28c (on the next page) completely defines the functional operation of this IC. It gives the state (1 for HI and 0 for LO) of each

(a)

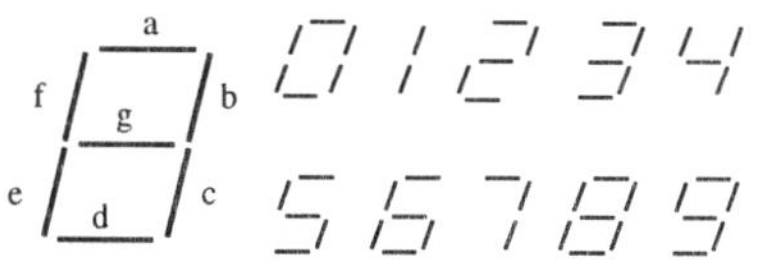

a b c d e f g

0 1 2 3 4 5 6 7 8 9

$$a = 0 + 2 + 3 + 5 + 6 + 7 + 8 + 9$$

$$a = \bar{A}\cdot\bar{B}\cdot\bar{C}\cdot\bar{D} + \bar{A}\cdot B\cdot\bar{C}\cdot\bar{D} + A\cdot B\cdot\bar{C}\cdot\bar{D} + A\cdot\bar{B}\cdot C\cdot\bar{D}$$
$$\bar{A}\cdot B\cdot C\cdot\bar{D} + A\cdot B\cdot C\cdot\bar{D} + \bar{A}\cdot\bar{B}\cdot\bar{C}\cdot D + A\cdot\bar{B}\cdot\bar{C}\cdot D$$

Figure 6–28. BCD-to-seven-segment decoder/driver. (a) Numerical display.

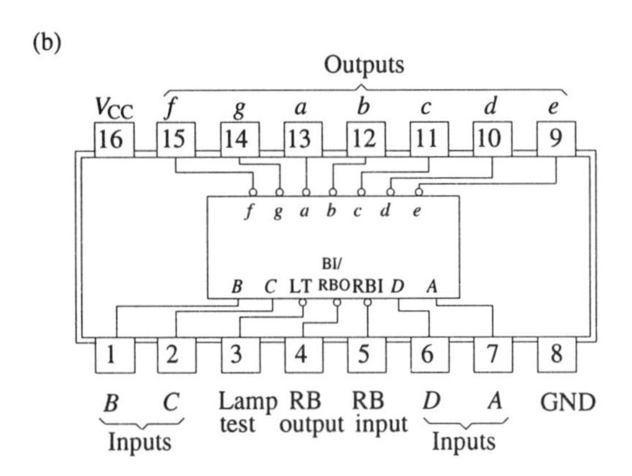

(c)

Decimal or Function	Inputs						BI/RBO	Outputs							Note
	LT	RBI	*D*	*C*	*B*	*A*		*a*	*b*	*c*	*d*	*e*	*f*	*g*	
0	1	1	0	0	0	0	1	0	0	0	0	0	0	1	1
1	1	X	0	0	0	1	1	1	0	0	1	1	1	1	1
2	1	X	0	0	1	0	1	0	0	1	0	0	1	0	
3	1	X	0	0	1	1	1	0	0	0	0	1	1	0	
4	1	X	0	1	0	0	1	1	0	0	1	1	0	0	
5	1	X	0	1	0	1	1	0	1	0	0	1	0	0	
6	1	X	0	1	1	0	1	1	1	0	0	0	0	0	
7	1	X	0	1	1	1	1	0	0	0	1	1	1	1	
8	1	X	1	0	0	0	1	0	0	0	0	0	0	0	
9	1	X	1	0	0	1	1	0	0	0	1	1	0	0	
10	1	X	1	0	1	0	1	1	1	1	0	0	1	0	
11	1	X	1	0	1	1	1	1	1	0	0	1	1	0	
12	1	X	1	1	0	0	1	1	0	1	1	1	0	0	
13	1	X	1	1	0	1	1	0	1	1	0	1	0	0	
14	1	X	1	1	1	0	1	1	1	1	0	0	0	0	
15	1	X	1	1	1	1	1	1	1	1	1	1	1	1	
BI	X	X	X	X	X	X	0	1	1	1	1	1	1	1	2
RBI	1	0	0	0	0	0	0	1	1	1	1	1	1	1	3
LT	0	X	X	X	X	X	1	0	0	0	0	0	0	0	4

Figure 6–28—*Continued* (on opposite page). (b) TTL IC 7447 BCD/seven-segment package; (c) table of states.
Notes:
(1) BI/RBO is wired-OR logic serving as blanking input (BI) and/or ripple-blanking output (RBO). BI must be open or held at a logical 1 when output functions 0 through 15 are desired and ripple-blanking input (RBI) must be open or held at a logical 1 during the decimal 0 output. X means that input may be high or low.
(2) When a logical 0 is applied to the BI (forced condition), all segment outputs go to a logical 1 regardless of the state of any other input.
(3) When RBI is at a logical 0 and $A = B = C = D =$ logical 0, all segment outputs go to a logical 1 and the RBO goes to a logical 0 (response condition).
(4) When BI/RBO is open or held at a logical 1 and a logical 0 is applied to lamp test (LT) input, all segment outputs go to a logical 0.

output for all possible states of the input. Note that some of the input entries are X rather than 1 or 0. The X means that when the other inputs are in the states given, the level at the input with the X entry has no effect on the outputs. It is useful to see how this type of table communicates the characteristics of the IC. Note, for example, for which decimal numbers a segment is 1 and for which it is 0. Compare these with the digital displays in Figure 6–28a. Note that a 0 (active LO) output lights each readout segment. This IC also converts and displays the non-BCD codes from 10 through 15. In summary, a reading of the output logic levels of the segments for any of these codes would enable you to predict the display pattern.

The complete circuit is rarely given for a function with the complexity of the BCD/seven-segment decoder. In fact, the circuit is not usually important to the user, but it is important to know exactly what output states are produced for each combination of input states. This information is provided definitively by the table of states.

6–5 Connecting Gates to the Computer Bus

In digital systems, it is often desirable to connect gate outputs to each other to allow them to share a common data line. This could, of course, be done with AND–OR–INVERT gates or multiplexers, but it is often more convenient and economical to directly interconnect all the gate outputs (and inputs) that are to share a particular line. In a bus-oriented system, the same data lines and control lines are common to each subsystem module. With normal TTL logic gates, the totem pole output (driven HI and LO levels) does not allow gate outputs to be tied together. The open collector TTL gate and the tristate logic gate, introduced in this section, have outputs that allow such wired connections. The logical consequence of connecting gate outputs together is discussed, and examples of bus-oriented operations are presented.

Open Collector Gates

The push–pull, or totem pole, arrangement can be visualized by simple switch equivalent circuits such as those shown in Figure 6–29 (refer to Supplement 3 for basic principles of the transistor). It is this push–pull arrangement that gives TTL gates their high switching speeds and good fan-out characteristics. Because the

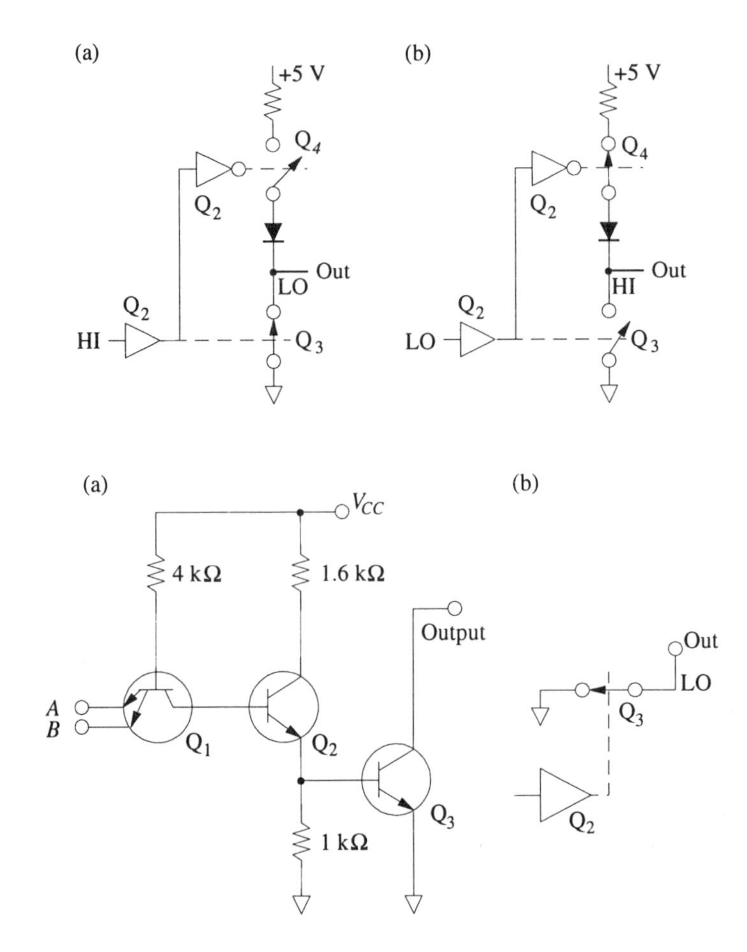

Figure 6–29. Switch equivalent circuit for TTL totem pole output. In the LO output state (a), transistor Q_2 drives Q_3 to be closed and Q_4 to be open. In the HI output state (b), Q_4 is closed and Q_3 is open. In both states the output is derived from an ON transistor. Thus both states are low-impedance outputs.

Figure 6–30. Open collector TTL NAND gate. The actual gate circuit is shown in part a. The output transistor acts simply as a switch to common, as shown by the switch equivalents in part b. An external pull-up to V_{CC} is necessary to establish the HI output level when Q_3 is off.

output impedance is low in both states, capacitive loads can be driven without serious degradation of switching times. However, if two gate outputs are connected together and one gate output attempts to go HI and the other attempts to go LO, the gate outputs work against one another.

The **open collector** gate is shown in Figure 6–30. Here, transistor Q_4 of the normal TTL gate is missing, and so only the LO output state is actively driven. To provide the HI output level when Q_3 is not conducting, the collector of Q_3 is connected to the supply voltage V_{CC} through an external **pull-up resistor**. When the gate output should be HI, Q_3 turns OFF, and the external resistive connection to V_{CC} establishes the HI output level. In the LO output state, Q_3 is conducting and the pull-up resistor limits the current through Q_3.

Because the output of an open collector gate is simply a switch to common, as shown in Figure 6–30b, the outputs of several gates can be safely connected together with a single pull-up resistor, as shown in Figure 6–31. If either of the

Open collector gates

+5 V

R_L

$M = X \cdot Y = \overline{AB} \cdot \overline{CD}$

Figure 6–31. Wired-AND connection of open collector NAND gates. If either gate output goes to LO, the output M is LO. Only when both gate outputs X and Y are HI is the output M HI. The wire connection thus performs the AND logic function on gate outputs X and Y. The overall operation is the NAND–AND function shown. R_L, pull-up resistance.

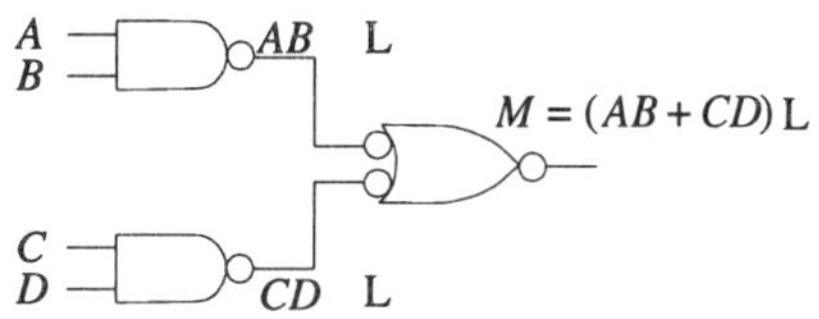

Figure 6–32. Functional equivalent of wired-AND connected NAND gates for LO–TRUE signals. The wired-AND connection performs the OR operation for LO–TRUE signals (L).

gate outputs X or Y goes LO as a result of both inputs to that gate being HI, the output of the gate array is LO because the LO output gate has an ON output transistor that sinks the current through R_L by connecting it to common. Only when both gate outputs are HI is the array output HI. Thus the HI–TRUE logic condition of an AND gate is performed on the open collector gate outputs, as shown in the Boolean expression $M = XY$. This implementation of the AND function is called wired-AND logic. The dotted AND gate in Figure 6–31 is sometimes used to show the logic function performed by the connected outputs even though an AND gate is not physically present.

According to DeMorgan's theorems, an AND gate for HI–TRUE signals performs the OR function on LO–TRUE signals. This function is shown by the equivalent gate circuit of Figure 6–32. Thus the connected open collector gates are sometimes referred to as wired-OR logic when LO–TRUE signals are used. Because gates are generally named for their HI–TRUE function, the term wired-OR is somewhat misleading.

Open collector outputs permit several devices to share a bus line, as shown in Figure 6–33. Control inputs to the transmitting gates keep all but one gate in their nondriven state. This puts the bus line under the control of the enabled gate. With some open collector gates, it is possible to tie the outputs to a higher supply voltage than the +5 V used to power transistors Q_1 and Q_2. This allows the open collector gate to drive lamps, light-emitting diodes, and C-MOS loads that require voltages higher than +5 V.

Because of the passive pull-up and the size of the pull-up resistor (usually greater than 1 kΩ), open collector gates can have substantially slower switching speeds than normal TTL gates, particularly on the LO–HI transition. The open collector gate delay is typically about 30 ns, with a nominal load capacitance of 15 pF and a pull-up resistance of 4 kΩ. The capacitance to be driven, particularly in a bused system, could be much higher than 15 pF. Thus typical bus delays due to RC time constants are often greater than 100 ns. In some logic families, such as ECL, the outputs are driven in only one state so that it is possible to connect the normal gate outputs in parallel.

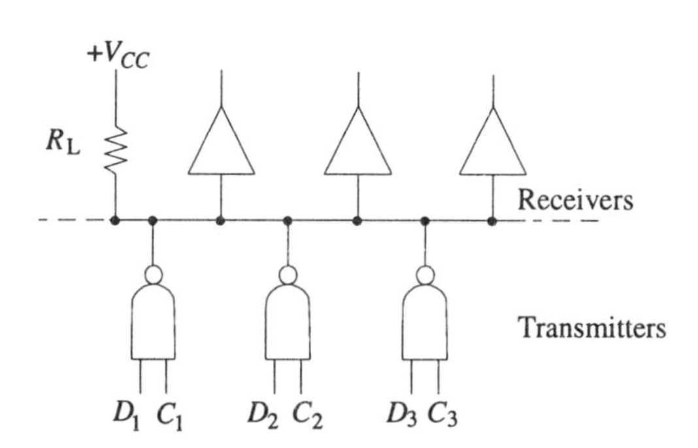

Figure 6–33. Bus line with multiple transmitters and receivers. Multiple data sources can be connected to a single line as long as only one transmitter at a time is enabled to drive the line. The others are held in their nondriven state by the corresponding gate control inputs. The pull-up resistance R_L holds the line HI when none of the transmitters is LO. The receivers present no data conflict, but each transmitter must be able to handle the total load.

Tristate Gates

The tristate gate was developed specifically for bus-oriented operations. **Tristate logic** is fully TTL compatible and is actively driven in both HI and LO states. However, in contrast to normal TTL, the tristate gate can be turned completely OFF so that it exhibits a third high-impedance state that effectively disconnects the gate output from any subsequent circuits. It thus retains the speed and drive capabilities of normal TTL and has the added flexibility of being bus compatible.

The tristate gate is similar to the totem pole TTL gate except that both output transistors can be turned OFF to give the high-impedance third state. Figure 6–34 shows a typical tristate gate structure and its output switch equivalent circuit. A HI logic level at the Disable input turns both output transistors OFF and forces the high-impedance third state no matter what the logic level at the data input. When Disable input is LO, the output is normal for a TTL gate.

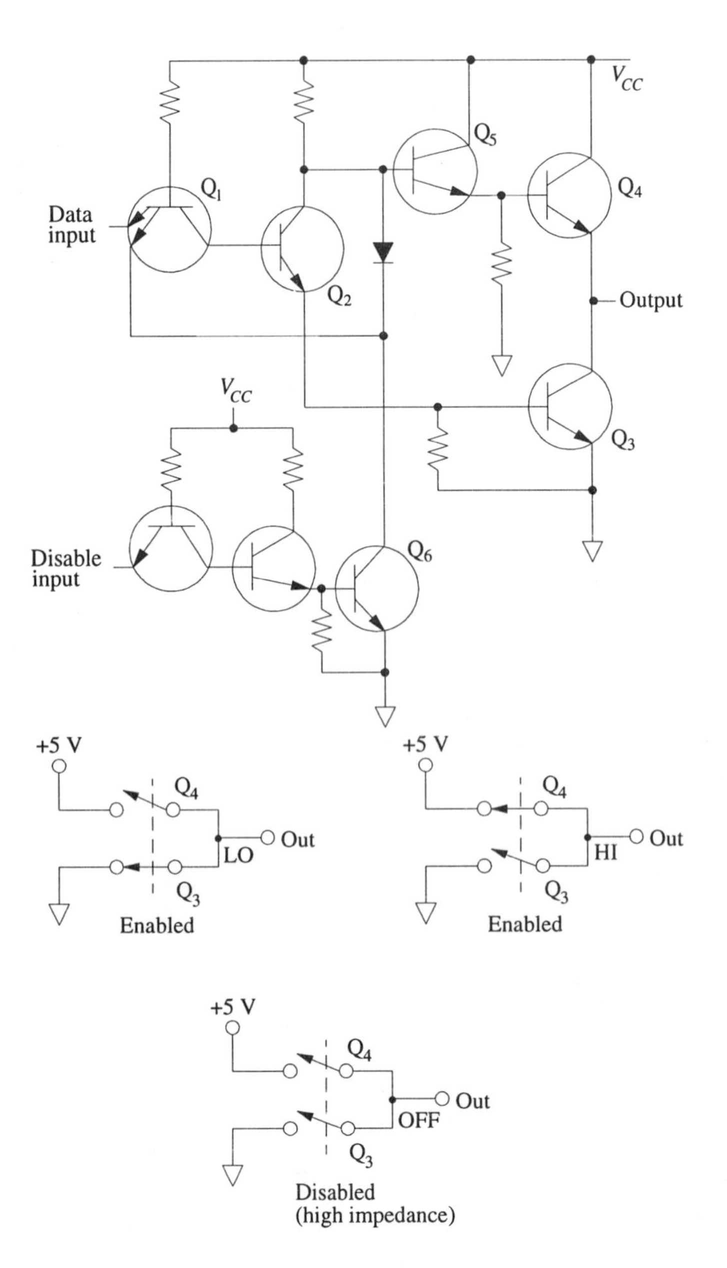

Figure 6–34. Tristate gate structure and output switch equivalents. A HI logic level at the Disable input turns Q_6 ON. This turns both output transistors OFF and effectively disables the gate. When Q_6 is ON, a LO applied to input transistor Q_1 turns Q_2 and Q_3 OFF. Transistors Q_5 and Q_4 are also turned OFF when the base of Q_5 is pulled LO by the Q_6 output. When the Disable input is LO, Q_6 does not conduct and normal TTL behavior is exhibited.

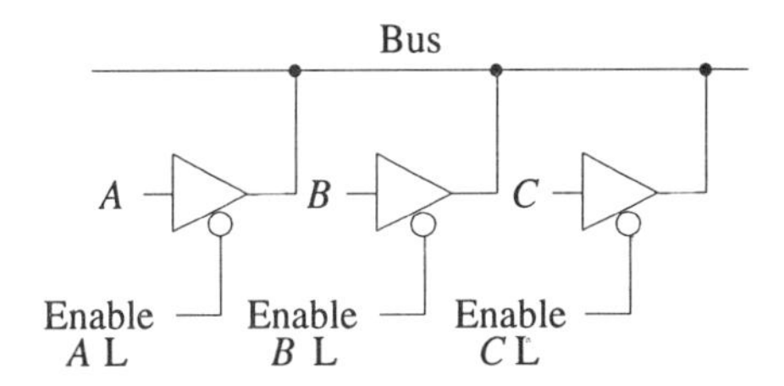

Figure 6–35. Bus connections with tristate bus drivers. The signal at A is driven onto the bus when a LO logic level is applied to Enable A, as indicated by L in the figure. Only one bus driver should be enabled at any one time.

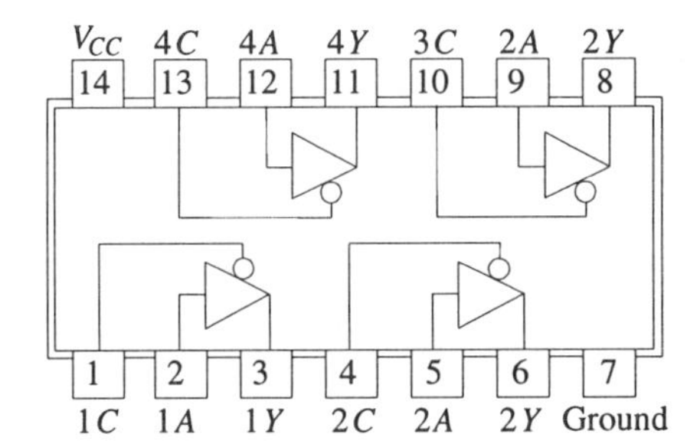

Figure 6–36. Pin configuration of 74125 quad buffer–driver with tristate output. The inputs 1C, 2C, 3C, and 4C are LO–TRUE enables. When 1C is LO, the output at 1Y is 1A. When 1C is HI, the output is in the high-impedance (disabled) state.

The tristate gate eliminates the need for pull-up resistors and significantly increases switching speeds over those of open collector gates. Typical bus delays with tristate gates are less than 10 ns, since RC time constants are smaller when both HI and LO states are actively driven.

A single bus line with tristate **bus drivers** is shown in Figure 6–35. Here the Enable input is used to place the gate in either the active state or the high-impedance disabled state. Tristate gates with either HI or LO Enable inputs are available. Normally, only one driver at a time is enabled by external control logic. One feature built into all tristate devices is a longer time delay in switching from the disabled to the enabled state than in switching from the enabled to the disabled state. This prevents data interference by ensuring that a previously enabled device will be disabled a few nanoseconds before a newly selected device is enabled.

Tristate devices are frequently found on ICs that are likely to be used with data buses. For example, Figure 6–36 shows the pin configuration of a quad tristate buffer and driver chip. Such ICs have adequate drive capability for the connection of up to 128 devices to a common bus line.

A 4-bit parallel bidirectional bus driver is illustrated in Figure 6–37a. Each buffered line of the 4-bit driver contains two separate tristate buffers in order to provide simple bus interfacing and bidirectional operation. The chip select signal $\overline{\text{CS}}$ enables the chip when the signal is LO and forces the high-impedance state when the signal is HI. The data in enable signal, DIEN, controls the direction of data flow. The complete truth table is shown in Figure 6–37b.

In addition to the examples given here, tristate outputs appear on random-access memory chips, data selector/multiplexer chips, counterlatch chips, and many microprocessor support chips.

6–6 Storing Digital Data in Flip-Flops

Digital systems need devices that can remember or store digitally encoded information. These devices are required for processing digital information, which is not all available simultaneously. This need is similar to our need for a scratch pad on which to store numbers and partial solutions when we perform a calculation. The

Figure 6–37. Four-bit parallel bidirectional bus driver. (a) Logic diagram; (b) truth table. On the bus (*DB*) side of the driver, the output of one buffer and the input of another are tied together. On the other side, the driver inputs (*DI*) and driver outputs (*DO*) are separated to permit flexibility. The control signals $\overline{\text{CS}}$ and $\overline{\text{DIEN}}$ are device select and direction controls, respectively, according to the truth table.

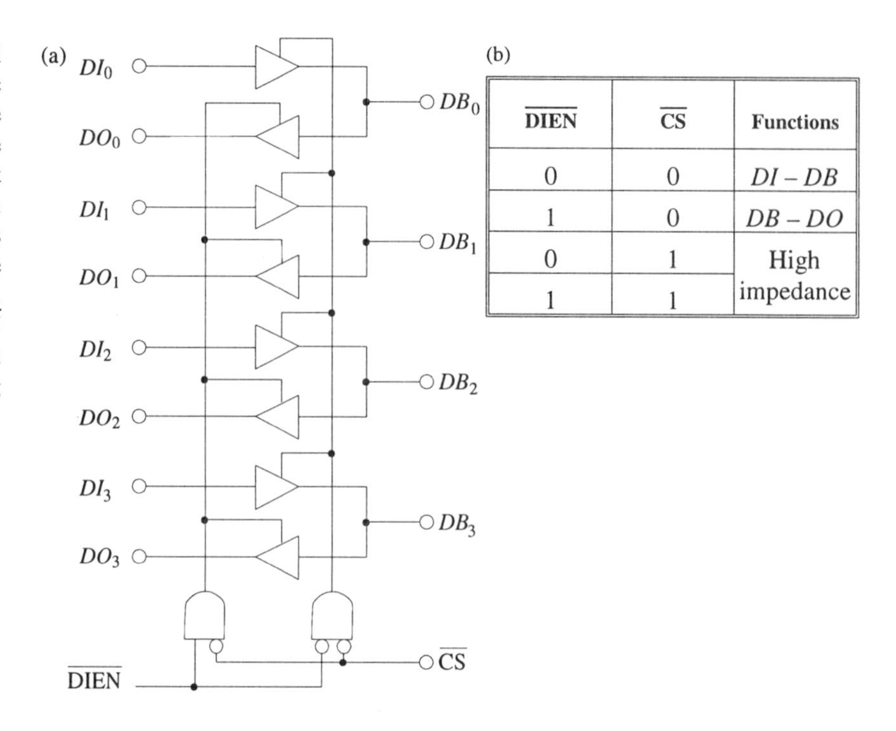

$\overline{\text{DIEN}}$	$\overline{\text{CS}}$	Functions
0	0	*DI – DB*
1	0	*DB – DO*
0	1	High impedance
1	1	

basic unit of digital information is the bit. Practical digital memories are assemblies or combinations of individual memory units, each of which can store 1 bit of digital information. Memories can be based on magnetic, electronic, or other principles.

The basic unit of electronic memory is the **flip-flop**. It is a bistable circuit that can be set to provide a HI output or a LO output and will remain in that state unattended until we want to change the value of the stored bit. An analogy for the flip-flop is a coin that can be set to either heads or tails. The coin will stay in either state until physically changed.

Cross-Coupled NAND Gate Bistable Device

The heart of the electronic bistable device is the cross-coupled NAND gate shown in Figure 6–38. Remember that for a NAND gate, a LO at either input produces a HI at the output. The Preset (Pr) and Clear (Clr) inputs are normally HI. When Pr goes LO, the output at gate 1, called *Q*, must go HI. Gate 2 now has both inputs HI, so its output, $\overline{Q}$, is LO. The LO signal from $\overline{Q}$ will keep the gate 1 output HI after the LO pulse at Pr is over. Thus the LO at the Pr input presets the flip-flop to the *Q* = HI state. The fact that this preset action is caused by a LO signal is indicated by the "not bar" on the Pr signal label. This flip-flop can be turned over, or "cleared," by applying a momentary LO at Clr to force $\overline{Q}$ HI. The two HI inputs to gate 1 make *Q* LO, which will keep $\overline{Q}$ HI. The levels at *Q* and $\overline{Q}$ are always

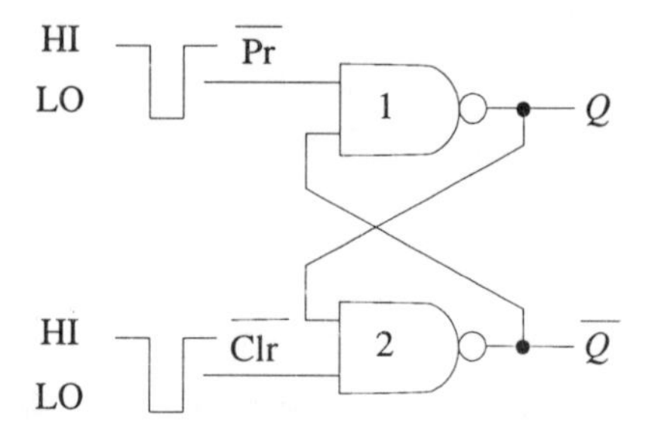

Figure 6–38. Basic bistable circuit. Preset and clear are represented by Pr and $\overline{\text{Clr}}$. A LO at Pr makes *Q* HI and $\overline{Q}$ LO. The LO at $\overline{Q}$ keeps *Q* HI, even when Pr returns to HI. The LO at Clr makes $\overline{Q}$ HI and *Q* LO. This state will remain until a LO appears at Pr.

(a)

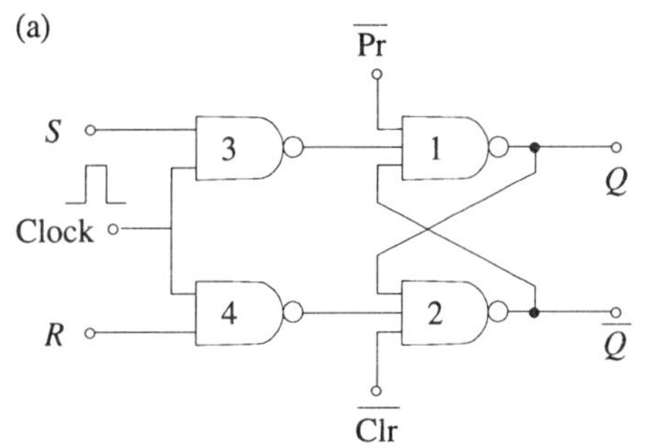

(b)

S	R[a]	Q[b]
L	L	No change
L	H	L
H	L	H
H	H	NA[c]

[a] Values of S and R during Clock = HI.
[b] Value of Q during and after Clock = HI.
[c] Not applicable
$\overline{\text{Pr}}$ and $\overline{\text{Clr}}$ are normally HI.
A LO at $\overline{\text{Pr}}$ forces Q HI and $\overline{\text{Q}}$ LO instantly.
A LO at $\overline{\text{Clr}}$ forces $\overline{\text{Q}}$ HI and Q LO instantly.

Figure 6–39. *R–S* flip-flop circuit (a) and table of states (b). The flip-flop is an electronic bistable circuit that can be placed in either state (HI or LO) by a combination of digital data and clock signals. Its state at any time is indicated by a continuously available output signal level at Q or $\overline{Q}$.

opposite, as their symbols imply. The level at Q depends upon whether Pr or Clr had the most recent LO level signal.

R–S Flip-Flop

It is often the case that the signals connected to flip-flop inputs for potential storage have many level changes, but only at particular times are they to be stored. Thus a method is required to connect the bistable circuit to the gate input signals only at the appropriate times. To accomplish this, the gating action of the gate is used. The result is the *R–S* flip-flop shown in Figure 6–39a. Signals *S* and *R* are isolated from gates 1 and 2 by gates 3 and 4 except when the clock signal is HI. A HI *S* when the clock goes HI applies a LO at gate 1, setting the Q output HI. This, of course, sets $\overline{Q}$ LO. Oppositely, a HI *R* will set $\overline{Q}$ HI, and Q will be LO. If neither *S* nor *R* is HI during the HI clock signal, neither 1 nor 2 has a LO input, so the flip-flop remains unchanged. The condition in which both *S* and *R* are HI should be avoided, because the final state of the flip-flop would be uncertain. Notice also in Figure 6–39a that gates 1 and 2 have the same Pr and Clr inputs as the basic flip-flop in Figure 6–38. Thus two methods of turning over the flip-flop are provided: gated inputs and direct inputs, Pr and Clr.

From the table of states for the *R–S* flip-flop in Figure 6–39b, you can see that the value of *S* will be stored in the flip-flop and appear at Q if the value of *R* is the complement of *S*. Thus automatic storage of a single data input signal can be achieved by inverting *S* and applying it to *R*. A circuit that does this internally is the data latch flip-flop.

Data Latch

In the data latch, shown in Figure 6–40, the *S* input is called *D* for data. It has a single data input. Gate 3 serves both to gate the input to gate 1 and to provide a $\overline{D}$ signal to gate 4. The *S* and *R* inputs of the gated latch are thus always *D* and $\overline{D}$, respectively. The table of states is limited to those states in which the *S* and *R* inputs are complementary and the Q output follows the *D* input when the load input is 1.

When more than 1 bit of information is to be stored at a time, as for a parallel digital signal, a separate data latch flip-flop is required for each bit. Thus IC packages with multiple data latches have been produced. Combinations of flip-

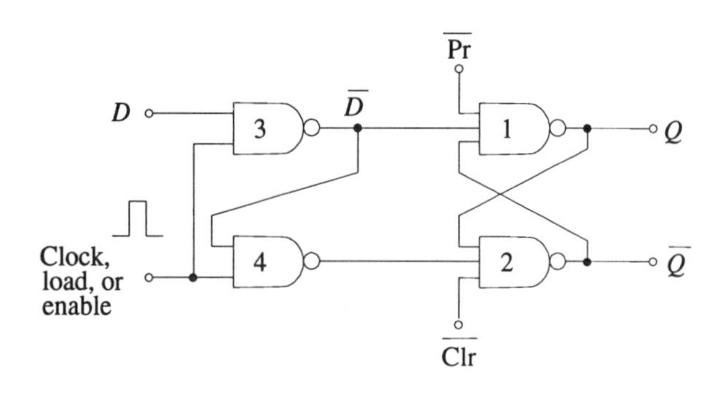

Figure 6–40. Data latch. When Clock = HI, the level at D appears at Q and is stored in the flip-flop. Pr and Clr inputs are normally HI.

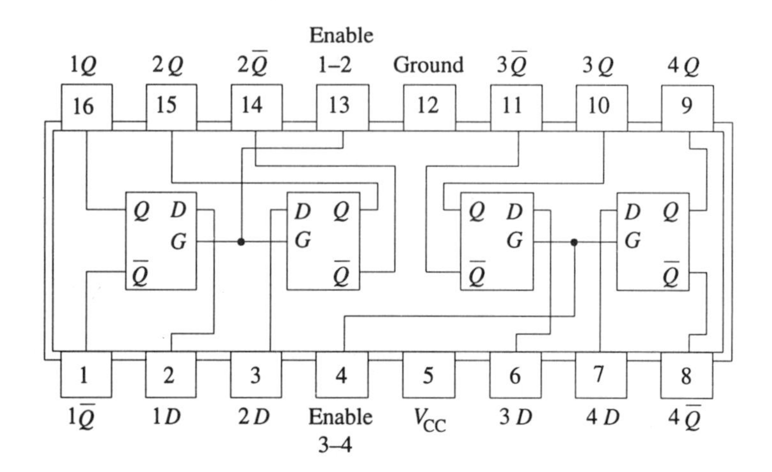

Figure 6–41. IC quad data latch. Pin connections are for TTL IC 7475. Note that each Enable (G) operates two latches in loading input data (D).

flops organized to store related bits of information are called registers. The pin diagram of an IC data latch package is shown in Figure 6–41. This package stores 4 bits at a time, so it can be used as a 4-bit register. Each pair of flip-flops is separately enabled. While both Q and $\overline{Q}$ are available for each flip-flop, no Pr or Clr is provided. Various packages give combinations of connections to meet the needs of specific applications. You can see from this example that as the on-chip circuit complexity increases, we save dramatically in the number of packages needed, but we lose access to many points in the circuit because of the limited number of pins.

J–K Flip-Flop

Another very important form of the flip-flop is the toggle flip-flop. With a toggle flip-flop, the clock signal can cause the flip-flop to change to the opposite state. That is, if Q is HI before the clock pulse, it will be LO afterward, and vice versa. An analogy here is the pull chain light switch that reverses state each time the chain is pulled. The *J–K* form of the toggling flip-flop is shown in Figure 6–42. Two *R–S* flip-flops are linked together. The clock signal to the second flip-flop (FF 2) is inverted. When the clock goes HI, the data at the outputs of gates 1 and 2 are loaded into FF 1. When the clock returns to LO, the S and R inputs of FF 1 are disabled and the data bit just latched into FF 1 is transferred into FF 2 through its now-active S and R inputs. Thus the data bit latched into FF 1 on the rising edge of the clock signal does not appear at the Q and $\overline{Q}$ outputs until the falling edge of the clock signal.

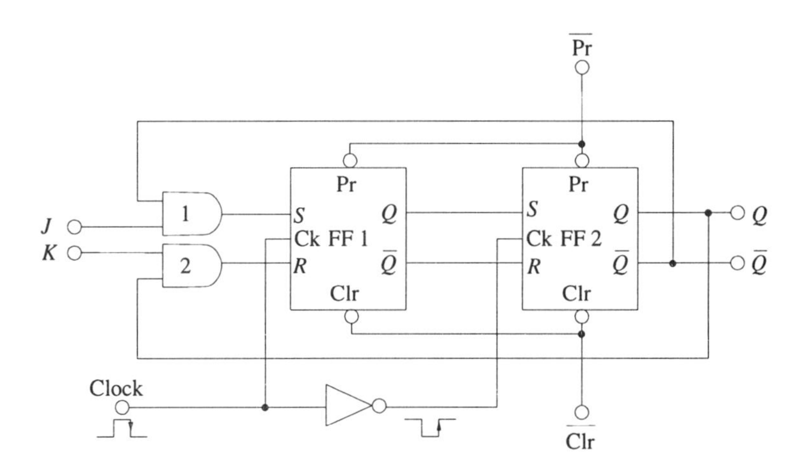

Figure 6–42. *J–K* flip-flop. (1) Toggling action. If *J* and *K* are HI, the levels of *S* and *R* of flip-flop 1 (FF 1) are the levels at $\overline{Q}$ and Q of FF 2. When clock (Ck) goes to HI, the *S* and *R* inputs of FF 1 are active but those of FF 2 are not. FF 1 is thus loaded with the opposite of the state of FF 2. When clock returns to LO, the *S* and *R* inputs of FF 1 are no longer active but those of FF 2 are, so the state of FF 1 is loaded into FF 2 and the new state (reverse of the previous one) appears at the outputs. (2) Loading action. A LO at the *J* input will cause Q of FF 1 to be LO when the clock is HI and LO to appear at the Q output when the clock returns to LO. A LO at the *K* input will cause Q to be HI at the end of the clock pulse. A LO at both *J* and *K* inputs prevents FF 2 from changing state. A LO at $\overline{\text{Pr}}$ will cause Q to become HI instantly. No other inputs can have an effect on this value until $\overline{\text{Pr}}$ returns to HI. The active low nature of the Pr input is indicated by the circle at the Pr inputs of FF 1 and FF 2 and by the bar in the $\overline{\text{Pr}}$ input label. Similarly, a LO at $\overline{\text{Clr}}$ will cause an immediate change of $\overline{Q}$ to HI.

The disabling of FF 1 inputs before the Q and $\overline{Q}$ outputs can change is essential for many flip-flop applications. One of these is toggling. This can be explained as follows. Assume that the clock is LO, the Q output is HI, and both *J* and *K* are HI. The LO clock signal to FF 1 makes the clock at FF 2 HI, so FF 2 has the same state as FF 1. The Q output presents a LO at gate 1, which will cause FF 1 to clear when the clock goes HI. This change in FF 1 does not affect FF 2, however, because its clock is now LO and its inputs are disabled. Thus FF 1 has changed state, but the outputs from FF 2 are not yet affected. When the clock goes LO again, the changed state of FF 1 is passed on to FF 2. FF 1, however, will not be affected by this change in Q and $\overline{Q}$ outputs, because its inputs are no longer enabled. The connection of the Q and $\overline{Q}$ outputs back to the opposite input gates causes the data at the FF 1 input to be always opposite the current state of the output, and it is this automatic reversal of the input levels that causes the toggling. The dual flip-flop arrangement allows this toggling to be triggered by and thus synchronized with the clock signal.

Other *J* and *K* inputs can be used to force the flip-flop to either state when the clock signal returns to LO or to prevent the clock signal from causing any change of state. The action of these inputs is explained in the loading action part of Figure 6–42. The actions of the *J*, *K*, and clock inputs are summarized in the customary way by the table of states in Figure 6–43. From this we see that the four possible

(a)

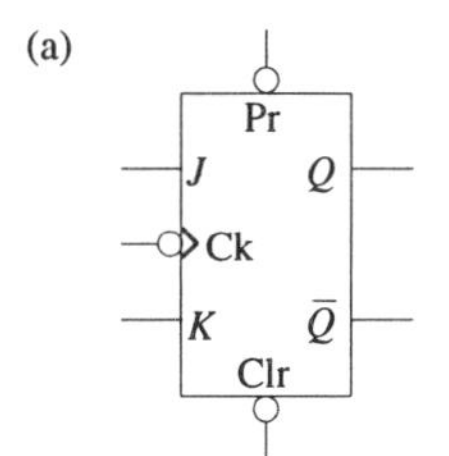

(b)

Inputs[a]		**Output**[b]
J	***K***	***Q***
L	L	No change
L	H	L
H	L	H
H	H	Change state

[a] While Clock is HI.
[b] After Clock returns to LO.

Figure 6–43. Symbol (a) and table of states (b) for *J–K* flip-flop.

combinations of J and K inputs while the clock is HI produce four different Q responses when the clock returns to LO. These combinations are the following: Q remains what it was, Q becomes LO, Q becomes HI, and Q changes to the opposite state. The choice of four modes makes the J–K flip-flop an extremely versatile device.

In Figure 6–43a, the circles at the Pr and Clr inputs have the usual inverting significance. The Pr and Clr functions are activated by LO signals at those inputs. The triangle at the clock input indicates that the output change occurs on an edge of the clock signal (rather than on a level, as with the data latch). The circle at the clock input indicates that the output changes on the HI-to-LO edge of the signal.

6–7 Encoding and Storing Digital Data in Registers

The concepts and importance of digital counting systems in instrumentation were presented in Chapter 2. We observed that counting is used extensively in making measurements. In general, a number N of events, items, or objects is determined per set of boundary conditions B. These boundary conditions B are the specified limits during which specific digital pulses N are encoded in a counting register. The encoded count information is equal to the ratio N/B (e.g., number of heart beats/second, number of seconds/100 m, number of particles/unit area, or units of charge/discharge period). Since all quantitative information is determined in terms of number of standard units, counting these units is indeed a universal measurement procedure. Therefore it is not surprising that accurate high-speed electronic counting registers are of major importance in microcomputer instrumentation and other digital measurement and control systems.

Basic registers that encode the digital data in binary or BCD numbers are presented in this section. In subsequent sections some applications of count registers are described.

Binary Counting Register

The J–K flip-flops described in the previous section are combined to form binary counters as illustrated in Figure 6–44a. Assume that the four flip-flops have been cleared ($Q = 0$) by a 0 pulse on the Clr line. When the first pulse appears at the count input, flip-flop A (FF A) toggles and A becomes 1 ($2^0 = 1$). On the next pulse, A returns to logic 0, which toggles FF B. Now B is 1, and A is 0 ($2^1 = 2$). After the next pulse, A is 1 again and B is unchanged ($2^0 + 2^1 = 3$), and so on. The waveforms in Figure 6–44b shows how FF A is set after every odd input pulse to represent the number 1. The FF B output represents the number 2, and the FF C and FF D outputs represent 4 and 8, respectively. The sum of the values of the set flip-flops represents the cumulative count at any instant. The outputs could be connected to indicators, and the instantaneous count could then be read out in binary form. The counter of Figure 6–44 can be extended to any desired number of flip-flops by adding flip-flops E, F, G, and so on in like manner. Since each successive flip-flop output represents another binary digit (power of 2), n flip-flops have 2^n states and can count from zero to $2^n - 1$. The waveforms of the four-flip-flop (4-bit) counter show the progression from 0 to 15. Then, on pulse 16, the register returns to the zero state.

It is also possible to make a counter that counts down, or subtracts 1 from the count for each input pulse. A **down-counter** and its output waveforms are shown

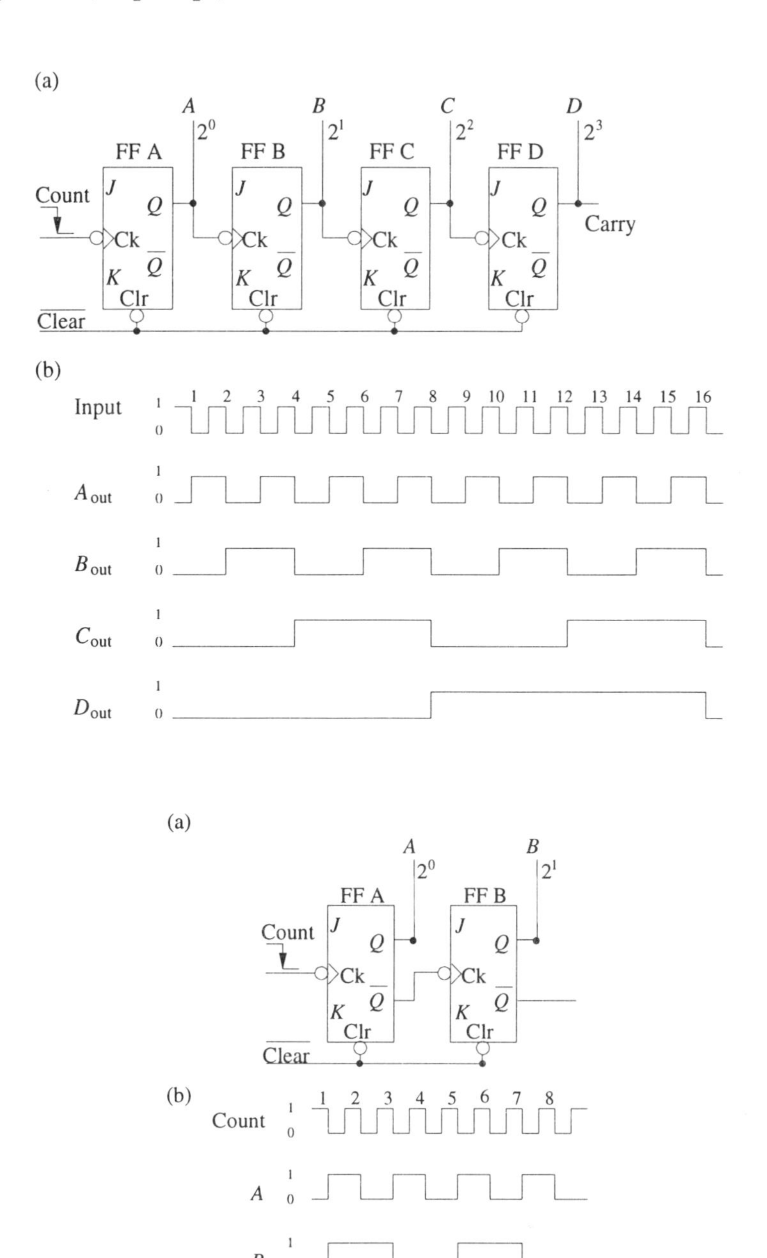

Figure 6–44. Binary counter (a) and waveforms (b). The 1→0 (HI→LO) transitions at the clock (Ck) input of each flip-flop produce first a 0→1 and then a 1→0 transition at the output. Thus the output alternates level on successive 1→0 edges at the input. Using the first flip-flop output to clock the second flip-flop produces transitions that occur every fourth transition of the input, and for the third flip-flop, every eighth transition, and so on. The result is a counter for binary-encoded numbers.

Figure 6–45. Binary down-counter (a) and waveforms (b). When the first HI→ LO transition occurs at the Count input, Q of flip-flop A (FF A) changes from 1 (HI) to 0 (LO). This toggles FF B and fills the counter. The next input pulses reduce the count until the register is cleared. Ck, clock.

in Figure 6–45. The circuit is identical to that of the up-counter except that $\overline{Q}$ instead of Q is connected to the following clock input. Additional flip-flops can be added to produce a down-counter of any desired capacity.

Synchronous Binary Counter

The counting circuits of Figures 6–44 and 6–45 are not synchronous, because each flip-flop is clocked at a different time. In each flip-flop there is a delay between the triggering edge of the clock pulse and the logic-level change at the output due to the propagation caused by the response time of the flip-flop gates. If the output of either of these asynchronous counters is sampled while the new count information is "rippling through" the counter chain, serious errors in the apparent count

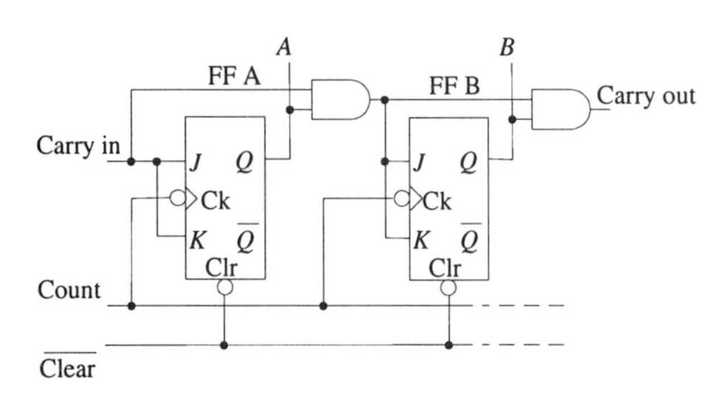

Figure 6–46. Synchronous binary counter. Flip-flop A (FF A) is gated to alternate states when clocked if the Carry in is 1. FF B should alternate only when Carry in AND *A* (the output of FF A) are 1. This is accomplished by the gate. The Carry out will be 1 only when the counter is full. Ck, clock.

could result. For instance, with the up-counter, as count pulse 8 is applied, the waveforms of Figure 6–44 show that the counter output will be 0111 (7), then 0110 (6), then 0100 (4), and finally 1000 (8). If the propagation delay is 50 ns through each flip-flop, a 10-bit counter will have error states for as long as 400 ns. This is a problem only if we need to be able to sample the parallel output at any time during the counting process. In such cases **synchronous counters**, in which all the flip-flops are clocked simultaneously and all therefore change output state at essentially the same time, are used.

A 2-bit section of a synchronous binary counter is shown in Figure 6–46. All flip-flops are clocked together, and the *J* and *K* inputs are used to inhibit all inappropriate transitions. This inhibit signal is obtained from the AND of the previous flip-flop output and its inhibit. In this way synchronous counting registers of any bit size can be made. The availability of 4-bit or larger synchronous counters in an IC package makes such counters very easy to build.

BCD Counter

Counters that store numbers in the BCD code must automatically increment following that code. Each group of four flip-flops in a BCD counter is arranged to store one decimal digit. Up to 16 combinations of output states are available from four binary circuits, but for storing a decimal number, only 10 of these states are used for the decimal numerals 0 to 9. A group of four flip-flops connected as a counting register with 10 states is called a **decade counting unit (DCU)**. To store numbers up to 9999, four DCUs are required. Because the decoding for each DCU is identical, more DCUs and decoding circuits can readily be added to store and read out as large a decimal number as necessary.

The most common storage code for decimal numbers follows the first 10 states of the binary counter. It is called the 1 2 4 8, or natural, code because the values of 1, 2, 4, and 8 can be assigned to flip-flops A, B, C, and D, respectively, and the stored decimal number can be obtained by adding the values of the set (1-level) flip-flops. The waveforms and circuit for a 1 2 4 8 BCD counter are shown in Figure 6–47. This counter can be obtained with maximum count rates of over 50 MHz. Generally, the connection between the *Q* output of flip-flop A and the clock input of *B* (and *D*) is completed by interconnecting two pins of the IC package. If unconnected, flip-flop A can then be used independently of the circuit composed of flip-flops B, C, and D, which has five stable states. Because this BCD counter is made up of a two counter followed by a five counter, it is sometimes referred to as a **biquinary counting circuit.**

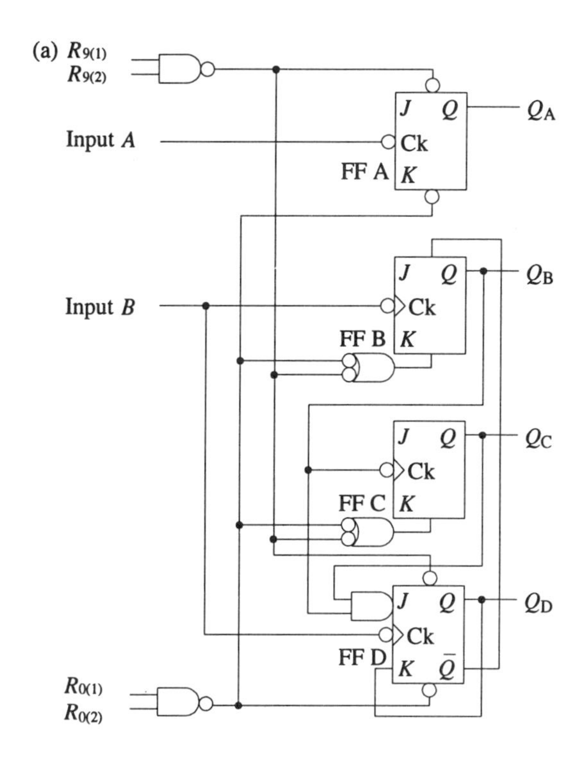

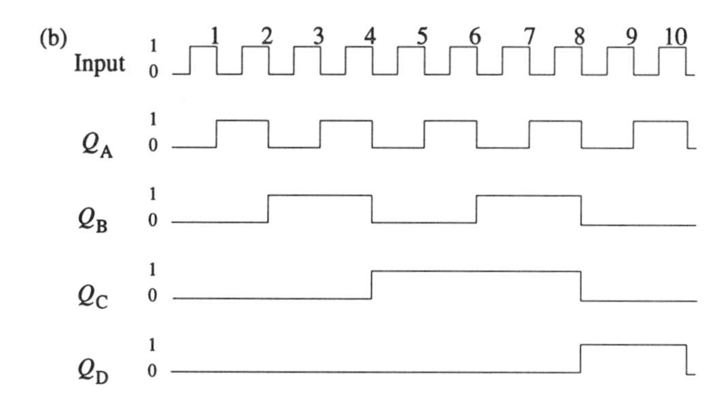

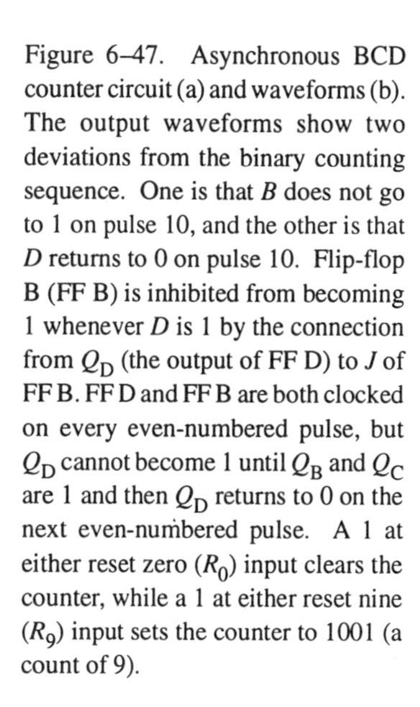
Figure 6–47. Asynchronous BCD counter circuit (a) and waveforms (b). The output waveforms show two deviations from the binary counting sequence. One is that *B* does not go to 1 on pulse 10, and the other is that *D* returns to 0 on pulse 10. Flip-flop B (FF B) is inhibited from becoming 1 whenever *D* is 1 by the connection from Q_D (the output of FF D) to *J* of FF B. FF D and FF B are both clocked on every even-numbered pulse, but Q_D cannot become 1 until Q_B and Q_C are 1 and then Q_D returns to 0 on the next even-numbered pulse. A 1 at either reset zero (R_0) input clears the counter, while a 1 at either reset nine (R_9) input sets the counter to 1001 (a count of 9).

Presettable and Up–Down Counters

A counter that can be set to any desired value before the count begins is called a **presettable counter**. A common application for presettable counters is the determination of the time required for a particular number of counts to occur. The counter is preset, and the time required to achieve maximum count (in an up-counter) or zero count (in a down-counter) is measured.

In many applications it is desirable to have a counter that can either increment or decrement the count value. Such a counter is called an **up–down counter**. In practical ICs the up–down counting capability is combined with presettability. Up–down counters are available as binary or decade counters and with either synchronous or asynchronous load. Each combination of features has advantages for particular applications.

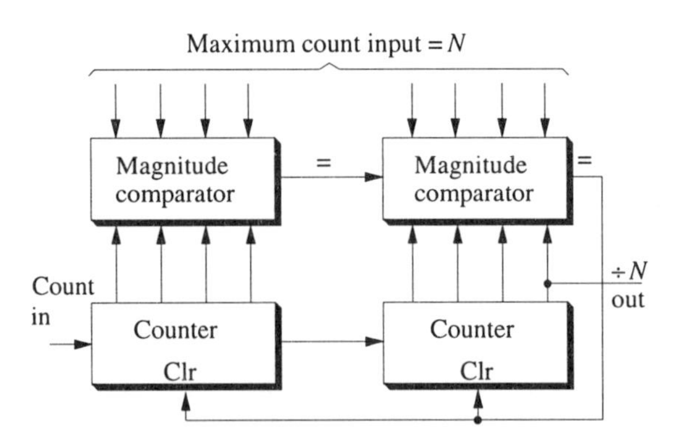

Figure 6–48. Divide-by-N or variable-modulus counter. When the counter reaches the value N at the maximum count input, the comparator-equals-output state applies a clear (Clr) signal to the counter. Because the cleared counter is no longer equal to the maximum count input, the clear signal is removed and the counter counts again from zero. The output frequency of the most significant bit flip-flop is the input frequency divided by N.

Odd or Variable-Modulus Counters

The **modulus** of a counter is the number of count input pulses for a complete count cycle. For instance, the four-flip-flop binary counter of Figure 6–44 has a modulus of 16. If it starts at 0000, in 16 pulses it is at 0000 again. Similarly, a three-flip-flop binary counter is a modulo-8 counter, and the 1 2 4 8 BCD counter is a modulo-10 counter. Many counting and scaling applications require counters with a modulus other than 2^n or 10^n. Furthermore, it is sometimes desirable to be able to select or vary the modulus of a counter as needed.

The counter can be made to have a particular modulus in several ways: the flip-flops can be wired to repeat the output cycle every N counts (fixed-modulus counter), a decoding circuit can be used to detect the Nth count and reset the counter, or the counter can be preset to a value from which it will proceed until it is full or clear. An example of the fixed-modulus counter is the BCD counter. The necessary number of flip-flops is n, where n is the smallest integer for which 2^n is larger than N. Then the clock and J–K inputs are connected to achieve the desired modulus and count sequence.

For some counting and scaling applications it is desirable to be able to change the modulus of the counter quickly and easily. This is generally done by allowing an ordinary binary or BCD counter to advance until the desired maximum count is detected and then stopping or clearing the counter. A general block diagram of such a variable-modulus counter is shown in Figure 6–48. A circuit is used to compare the outputs of the counting register with inputs that represent the desired maximum count N. When a count of N is reached, the digital comparison circuit output resets the counter. If an immediate repeat of the counting cycle is not desired, the comparator output could instead be used to close the counting gate or to preset the counter to a different number. Another, more common approach is to detect the zero state of a presettable down-counter and connect it to the preset control input. The counter then presets to the parallel data input automatically upon reaching zero and begins to count down again. The modulus of this counter is equal to the preset number.

Count Measurement Sequence

The typical sequence in making count measurements is shown in Figure 6–49. Input information flow from each flip-flop in the BCD counter registers and the latch registers into the decoder and finally into a display readout is shown. Only two digits are shown, but the same type of circuitry is required for each digit.

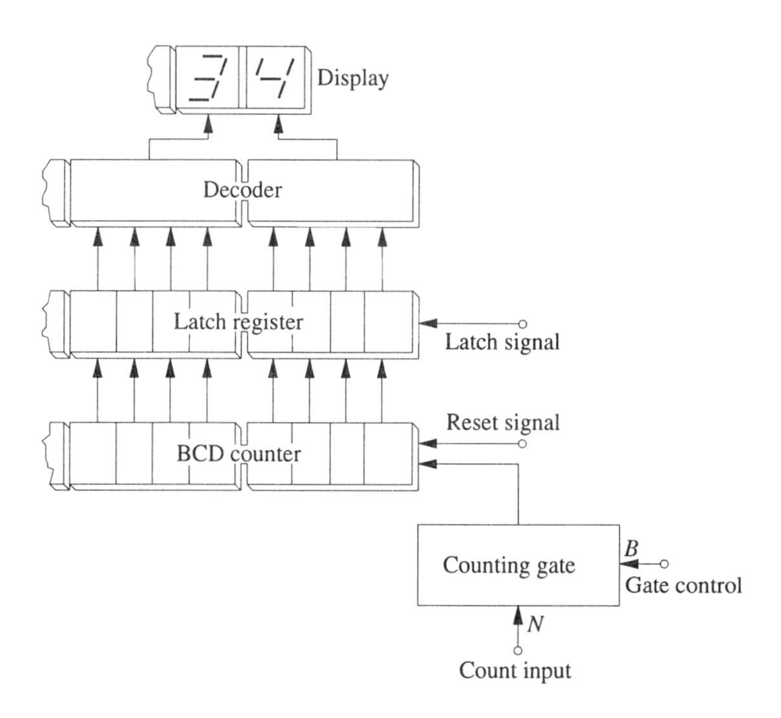

Figure 6–49. Count measurement sequence. Information flow is shown from the gate to the flip-flops in the BCD counter registers for two decimal digits. Boundary conditions *B* provide the gate control for accumulation of *N* counts in the counter registers when the gate is open. When the gate closes, a latch signal causes the encoded data in each flip-flop in the counter registers to be transferred to a data latch in the latch registers. The data in the latch registers can now be decoded and read out, while the counter registers are reset and are now able to accept new data as determined by the gate control.

The count input data *N* are made available at the counting gate input. Boundary conditions *B* provide the gate control. The gate opens as specified by *B*, and the counter accumulates digital pulses representative of *N*. The gate closes and a latch signal causes information from the counter to be transferred to the latch register, decoder, and display. The latch will hold this information in its memory flip-flops until the next latch signal requests the new information from the counter flip-flops. The counter is cleared to zero by a reset signal, so it can now accept new information. The gate control is reactivated by boundary conditions *B* to allow another measurement.

The sequence of operations that has been outlined for Figure 6–49 can be made automatically repetitive. An internal sequencer generates pulses to operate the gate, the latch, and the counter reset.

6–8 Shifting Digital Data in Registers

The movement of digital data from one bit position to the next in registers is called **shifting**. A **shift register** is designed so that the binary digit stored in each flip-flop in the register can be transferred to the next flip-flop in the series on command.

The shift register has many applications in computers and interfacing circuits. One application is as a delay element, because data shifted in does not appear at the output until *n* shifts later. Another frequent application of shift registers is as serial-to-parallel digital converters, or vice versa. An *n*-bit serial word shifted into the register appears at the *n* flip-flop outputs in parallel digital form. Similarly, parallel digital information set directly into the flip-flops can be obtained from the output flip-flop in serial form. These basic applications of the shift register are illustrated in this section.

As illustrated by the block diagram in Figure 6–50, the flip-flops in the shift register are connected so that the content of each one is transferred to the next one in line upon a shift command. In Figure 6–50a, serial data appearing at the bit 0

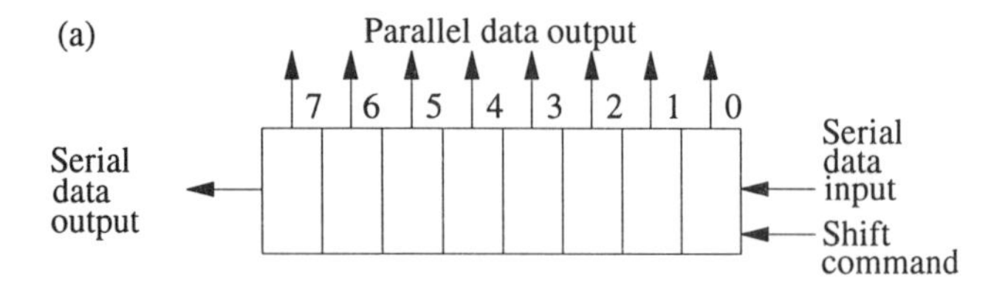

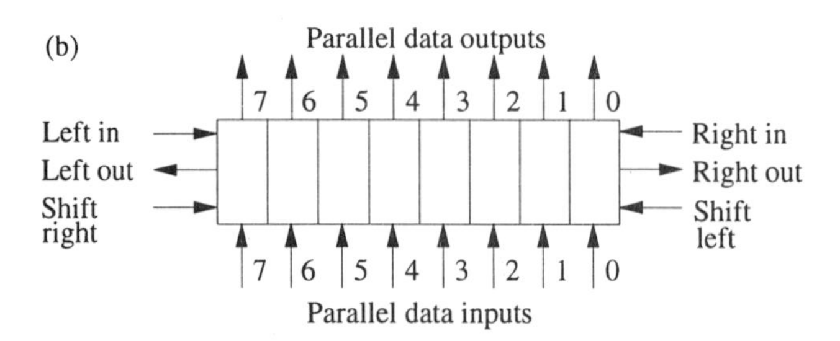

Figure 6–50. Shift registers. The level at the serial data input is transferred to flip-flop 0 on each shift command. Its level is in turn transferred to flip-flop 1, and each stored level is shifted to the next more significant bit position. Parallel output can be provided as in part a. There are also shift registers that can be parallel loaded and that can shift in either direction as in part b.

input are shifted into the register by a shift command. After eight shifts, the 8-bit register is completely filled with new data, the first bit of which is now at the serial output.

One function performed by the shift register is serial-to-parallel conversion, since the most recent eight serial bits are available at the parallel output. Another function is as a serial delay element. The input serial data appear at the serial output delayed by exactly 8 bit times. Different word lengths or delay times are obtained by using shift registers of various lengths. A parallel-to-serial conversion can be performed by loading a parallel-input shift register with the parallel word. As the shift command is repeated, the word appears in serial form at the serial output.

A basic 4-bit shift register is shown in Figure 6–51a. The shift command is accomplished by the synchronous clock connected to all the flip-flops. The output of each flip-flop is simply connected to the input of the next. In this way, a shift register of any length can easily be constructed. A shift-right, shift-left register is made by connecting a two-wide AND–OR gate to each flip-flop input. The input of one AND gate is connected to the flip-flop output at the left, and the input of the other is connected to the output at the right. A control signal to the AND gates determines whether the input data for each flip-flop come from the left or the right.

Parallel loading of the shift register is accomplished in either of two ways. One is through the Pr and Clr connections to the flip-flops. This is called an asynchronous load, or direct set, since it occurs immediately and always takes precedence over a shift operation. The other technique is to use AND–OR gates to determine whether the data at each flip-flop are from the previous flip-flop or the parallel input. If the AND–OR gates are set to accept the latter, the parallel word is loaded on the next clock edge. This form of parallel input is called a synchronous load.

Several types and lengths of shift registers are available in IC form. Many ingenious applications in counting, signal generation, and data manipulation have been devised for the versatile shift register. One early application was as the circulating memory register. Parallel data outputs are often omitted from the shift register designed for circulating applications. This means that only five or six external connections to the register are required (input, output, clock, and power) no matter how many flip-flops are used. For this reason, serial input–output (I/O) shift registers for circulating storage applications were one of the first circuits

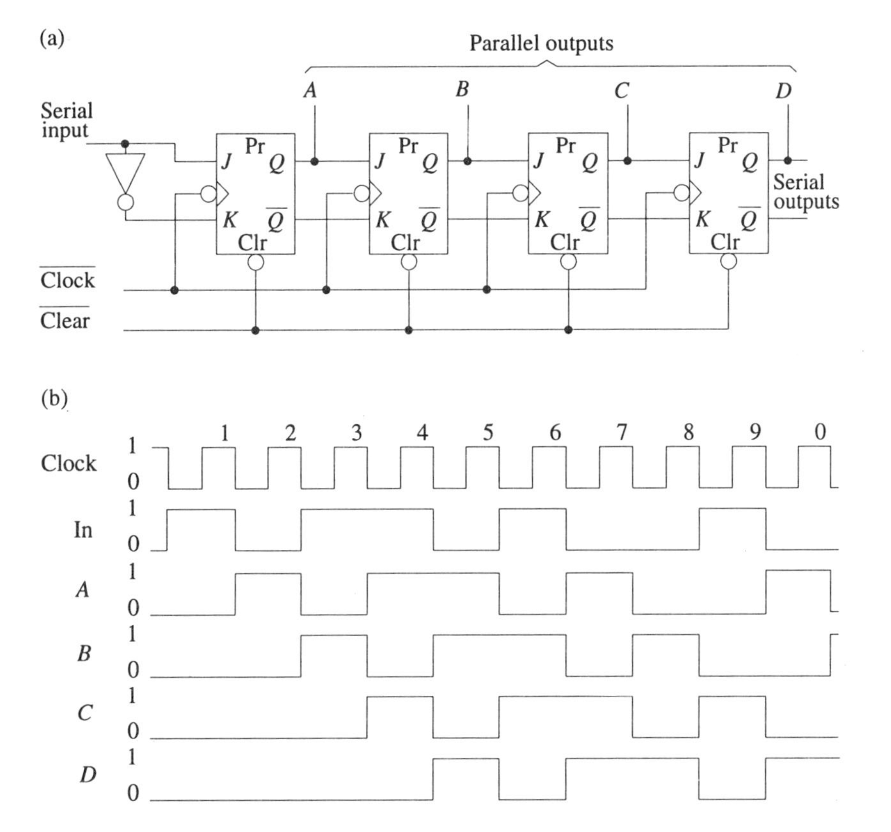

Figure 6–51. Four-bit shift register (a) and waveforms(b). The level at the J input of each flip-flop is transferred into that flip-flop at the $1 \rightarrow 0$ transition of the clock. After four clock pulses, the level originally at the serial input is at the D output.

developed for large-scale integration. Large-scale integration shift registers more than 1000 bits long are made by several manufacturers and have been a form of data storage. Magnetic-bubble memories are actually circulating serial memories that are accessed in much the same way.

6–9 Connecting Input–Output (I/O) Devices to the CPU Bus

I/O devices are connected to the data bus of the central processing unit (CPU) by input registers and output registers. An **input register** is a latch that is loaded with the input data and can put that data on the data bus when a CPU I/O read instruction is executed. An **output register** is loaded by a CPU I/O write instruction; it supplies the loaded data to the output as needed. The interaction of the I/O registers and the CPU bus is illustrated in Figure 6–52. The I/O instruction specifies the particular I/O register through the address lines and controls the data direction (read or write) and timing through the control lines. All microprocessors can accomplish the transfer of data to or from an I/O register on a **memory read** or **memory write** instruction. In such cases, each I/O register occupies an assigned address in memory address space. The address bus is decoded to provide a **device select signal** for each of the I/O registers.

One of the many possible decoding schemes is the **fully decoded I/O address system** shown in Figure 6–53. Three or four of the least significant address bits are often left undecoded, because many specialized I/O interface chips require several of the low address bits to select the register or operation being enabled. All techniques that put the I/O registers in memory space are called **memory-mapped**

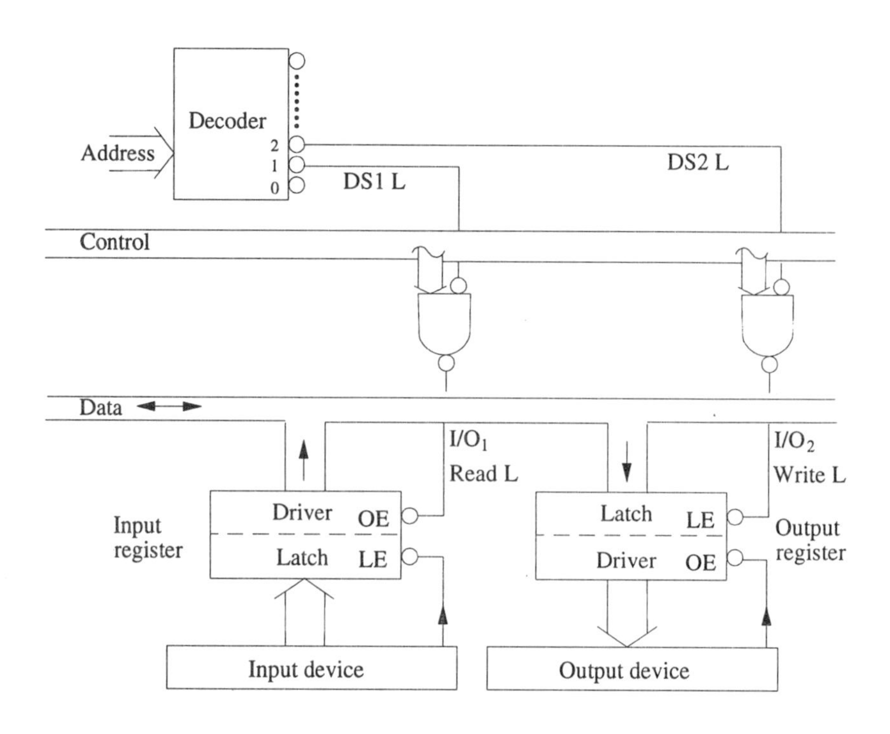

Figure 6–52. Input and output registers. Data from an input device are latched into the input register by that device. The CPU transfers these data to a CPU register by an I/O read instruction directed to the address of that particular register. The decoded address and the read command are combined to enable the drivers to put the data on the bus. Data are supplied to an output device by enabling the latch of an output register with an I/O write instruction to the addressed register. The latched data remain available to the output device. DS1 L, device select 1, LO–TRUE; DS2 L, device select 2, LO–TRUE; L, LO–TRUE; OE, output enable; LE, latch enable.

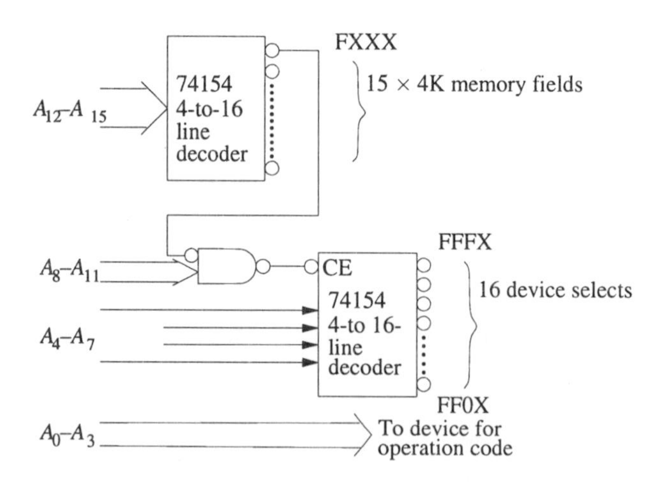

Figure 6–53. Fully coded address system for memory and I/O. The top 4- to 16-line decoder assigns I/O addresses to the top 4K section of memory space and provides select lines for 15 memory fields of 4K each. The I/O decoder is enabled for all addresses FF00 through FFFF (the top one-fourth K of memory space). Address bits A_4 through A_7 provide 16 device select lines. These can be combined with A_0 through A_3 at the device to enable up to 16 different operations for each device. CE, chip enable.

I/O. Some microprocessors have special I/O instructions for data transfers between I/O registers and the CPU accumulator. The ability to distinguish between a memory read or memory write and an I/O transfer allows I/O addressing that is independent of the memory address space.

6–10 Measuring Analog Data by Digitization and Counting

In previous sections of this chapter we focused on the logic gates, flip-flops, counter registers, and various applications that are basic ingredients in "thinking digital." It is therefore appropriate to conclude this chapter with an application that ties together a major concept presented in this chapter and a concept for analog systems described in Chapter 5. The application is the measurement of analog data by digitization and counting.

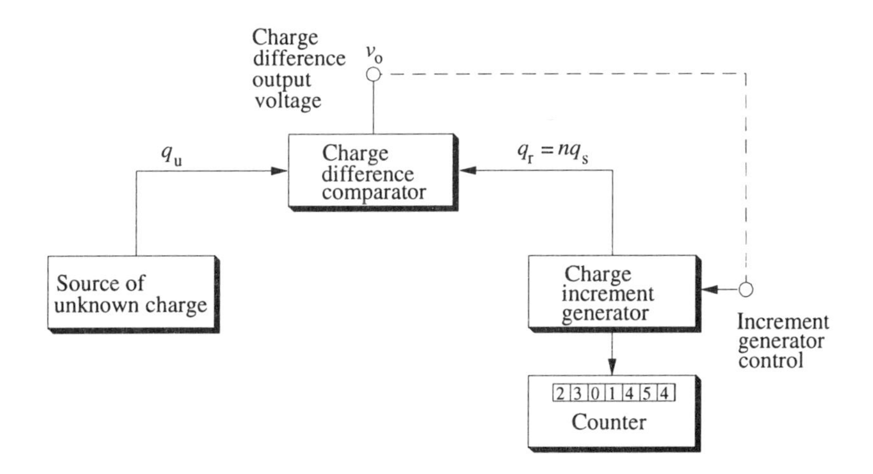

Figure 6–54. Measuring an analog quantity by digitization and counting. A source of unknown charge q_u is measured by performing an automated null comparison with an equal reference charge q_r. The reference charge is produced by generating a number n of very small charge increments q_s until nq_s equals the unknown charge. Thus n is directly proportional to the unknown charge.

In the real world we are constantly in contact with physical and chemical quantities that are typically converted into electrical data by suitable transducers for the purposes of measurement and/or control. To take advantage of the elegant microcomputer instrumentation systems, it is essential to convert the electrical analog data into digital data for processing and manipulation. Analog-to-digital conversion was introduced in Chapter 1 and will be treated in detail in Chapter 9. At this point, however, the following example of analog-to-digital conversion neatly combines the two universal concepts of null comparison (Chapter 5) and counting for measurement and/or control.

Figure 6–54 is the block diagram for a charge-to-count converter. Electrical charge makes up the analog data, which are converted into a proportional number of digital pulses. These pulses are then counted and displayed by a counting system such as those studied earlier in this chapter.

A basic concept of the charge-to-count converter is the null comparison technique described in the previous chapter. An unknown charge q_u is compared and balanced with a reference charge q_r until the comparator indicates that $q_r = q_u$.

The reference charge q_r is produced by generating a large number n of very small reference charge packets q_s so that $q_r = nq_s$. Thus $q_u = nq_s$, and the charge increments are counted by the digital counter. The quantity of charge in the small packets q_s can be adjusted and selected by the charge increment generator so that the counter can display the unknown charge directly. For example, assume that each small packet q_s contains 1.0000×10^{-10} C. Then, if we accumulate 12 520 counts in balancing the unknown charge q_u, it follows that $q_u = 12\,520 \times 10^{-10} = 1.2520$ μC.

When the circuits used to implement this charge-to-count technique are discussed in Chapter 9, it will also be shown that the same basic analog-to-digital technique can be used for current and voltage measurements.

Related Topics in Other Media

Video, Segment 6

- Use of digital logic for an interlock system
- Use of logic gates for specific functions
- Frequency divider function of flip-flops
- How a computer modem works

Laboratory Electronics Kit and Manual

- Section F
  - ¤ Connecting logic gates and indicator circuits
  - ¤ Testing logic functions of AND, OR, NAND, EXCLUSIVE-OR, and EQUALITY gates
  - ¤ Wiring an IC indicator circuit
- Section G
  - ¤ Wiring and testing flip-flop circuits
  - ¤ Connecting a shift register and studying its operation

Chapter 7

Generating, Switching, and Processing Analog Signals

Analog signals are processed with various types of electronic circuits to produce signals with the correct characteristics for performing the required functions in measurement and control instrumentation. Chapter 5 presented several basic operational amplifier (op amp) circuits that produce useful output signals related to input signals. These circuits included inverting amplifiers, current-to-voltage converters, comparators, adders, integrators, and buffers. This chapter describes several circuits that process analog signals in other specialized ways.

Programmable gain amplifiers and operational power supplies are discussed. These utilize analog switches to produce **analog signals** that are controlled by digital signals and thus by a microcomputer program. Digitally controlled analog switches are used in the **analog signal selector** (multiplexer) and **analog memory** (sample-and-hold) circuits, which are also described.

Amplifiers with specialized characteristics are introduced. These include the **difference amplifiers** and **instrumentation amplifiers**, which make it possible to selectively amplify small signals buried in large unwanted signals. An amplifier that is especially useful with certain medical and biological transducers is the **charge-coupled amplifier**. It requires a minimum of charge from the transducer source. The **bridge amplifier** is especially applicable with transducers, such as strain gauges, that change resistance as a result of dimensional changes. This type of amplifier provides a reliable output signal even though the change is only a very small fraction of the total resistance. It also isolates changes in resistance attributable to strain from those caused by temperature. **Alternating current (ac) amplifiers** isolate ac signals from direct current (dc) signals. Their frequency response limitations are considered.

Op amp filters provide selective isolation of desired signals from unwanted signals. They are also important for **tuned amplifiers** and **oscillators**.

Although the op amp has near-ideal characteristics for many circuits, there are some practical limitations. The nonideality of op amp circuits is considered in the final section of this chapter.

This chapter starts with a presentation of op amp **signal generators** for producing repetitive pulses or waveforms, such as square or sine waves. The generation of repetitive analog signals is essential for various measurement and control instruments as well as for test signals used in troubleshooting.

7–1 Generating Waveforms

A signal source for which the output voltage varies in a specific way as a function of time is often called an **oscillator**, **signal generator**, or **function generator.** These sources convert dc voltages into ac or pulsating dc voltages of various useful waveforms. One of the frequently encountered waveforms in instruments such as oscilloscopes is the sweep or linear-ramp (sawtooth) signal. Other basic waveforms include the square wave, triangular wave, and sine wave. Each of these has

2861-2/94/0191$14.00/1 Published 1994 American Chemical Society

many applications in measurement and control systems. In this section we will see how op amps are applied to produce the desired basic waveforms.

Sweep Generator

The voltage integrator was shown in Chapter 5, Section 5–6, to be applicable for producing linear voltage ramps (sweep voltages), as illustrated in Figure 5–27b. By connecting the integrator to an op amp comparator, electronic switch, and switch reset pulse circuit, as illustrated in Figure 7–1, we have a very useful sweep generator. It has many instrument applications for sweep times in a range from microseconds to minutes.

Producing a repetitive, periodic sweep signal requires an automatic periodic reset for the integrator. One technique for resetting is illustrated in Figure 7–1, which shows a generalized comparator-operated switch. As the sweep output v_o increases toward the comparison voltage v_r, the comparator output is positive. When v_r is reached, the comparator output becomes negative, triggering the pulse circuit (MS) to generate a reset pulse that resets the switch driver. The reset pulse is set to a duration that allows the integrating capacitor to be discharged completely. A pulse circuit of this type, called a monostable multivibrator, is described later in this section. In the level-control method of automatic resetting, the values of V_{in}, R_{in}, and C affect both the sweep rate and the frequency. They do not affect the amplitude, because that is controlled by the comparator. A change in the level control changes the amplitude and frequency but not the slope.

For example, if $V_{in} = -5$ V, $R_{in} = 100$ kΩ, and $C = 0.1$ μF, the voltage sweep rate ($V/t = i_{in}/C = V_{in}/R_{in}C$) is equal to 500 V/s. If the comparator reference voltage is +5 V, the integrator is reset every 0.01 s (5/500). This provides a frequency of 100 Hz and a voltage amplitude of 5 V. By changing the reference voltage to +1 V, the frequency increases to 500 Hz with an amplitude of 1 V.

Another method of automatically resetting the integrator is to use the output of an oscillator to close the reset switch momentarily at regular intervals. With this

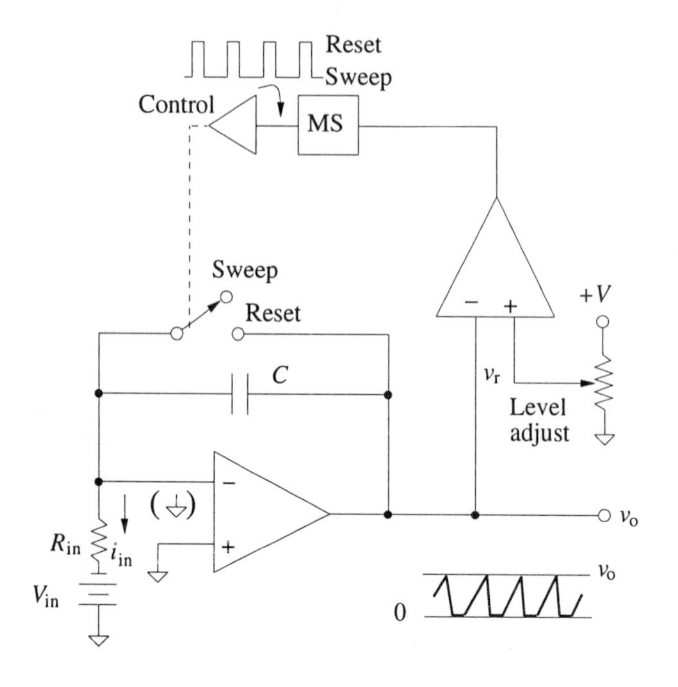

Figure 7–1. Sweep generator with automatic level switch reset. This op amp integrator produces a linear sweep until the output voltage v_o exceeds the comparator reference level V_r, at which time the reset closes momentarily and shorts the integrating capacitor. The cycle then repeats. MS, pulse circuit.

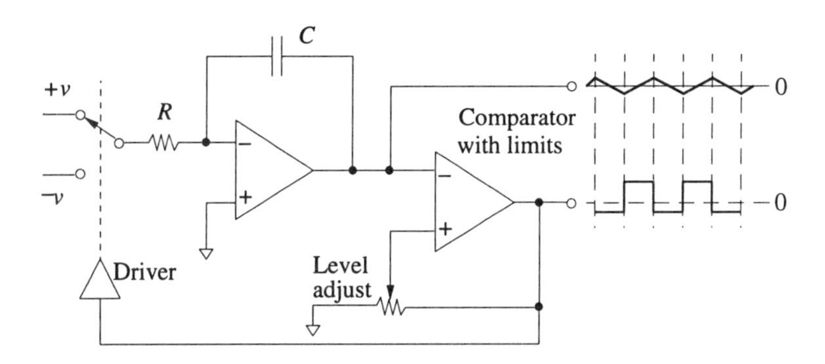

Figure 7–2. Level-controlled square- and triangular-wave generator. The input switch is used to reverse the polarity of the integrating current when the comparator threshold is reached. A fraction of the comparator output is used to obtain the threshold level.

method the values of R and C affect the sweep rate and maximum amplitude but not the frequency, because that is controlled by the oscillator.

Triangular- and Square-Wave Generator

A variation on the sweep generator of Figure 7–1 is used to generate both square and triangular waveforms. In this circuit the switch is used to reverse the polarity of the integrating current, as shown in Figure 7–2. The integrator output has a negative slope when the integrating current is switched to the positive source. The comparator output is a negative voltage in this state. Because a fraction of the comparator output is used for the comparator reference source, the reference is also a negative voltage. When the integrator output crosses the negative reference level, the comparator output becomes positive, the switch driver connects the integrator to the negative source, and the integrator output has a positive slope. The new positive comparator reference level that is established will reverse the levels again when it is reached. This generator can operate in many control modes. The amplitude can be changed at a constant slope, or the frequency can be varied at a constant amplitude. A choice of unequal values for $+v$ and $-v$ results in different positive- and negative-going slopes. A generator with voltage-controlled frequency is obtained by varying $+v$ and $-v$. This general circuit is the basis for the popular laboratory **function generator.** A sine-wave output is sometimes obtained by shaping the triangular waveform, as described next.

Obtaining Sine-Wave Output by Shaping Triangular Waveforms

In the function generator described above, the basic waveform is the triangular wave, and the sine wave must somehow be derived from it. This is usually accomplished by a diode wave-shaping circuit. The triangular waveform and the desired sine wave are shown in Figure 7–3a. In order to obtain the sine wave from the triangular wave, an attenuator is needed. This piece of equipment attenuates the larger input voltages much more than the smaller ones.

A nonlinear voltage divider to serve this purpose can be made with diodes and resistors, as shown in Figure 7–3c. The single diode in Figure 7–3b is connected as a shunt clipping switch with a resistor in series. Therefore, for $v_{in} < v_b$, the output signal is the unattenuated input signal. When v_{in} is greater than v_b, the diode will conduct a current proportional to $v_{in} - v_b$. This current causes a voltage (IR) drop in R_s proportional to $v_{in} - v_b$. Thus the slope of the output signal changes at v_b.

The point of the slope change is set by the **breakpoint** v_b adjustment. The amount of slope change depends on the value of the slope adjust resistor. Additional diodes and resistors can be added to provide additional breakpoints and slope changes until the desired waveform is approximated by a series of line segments. The line segment approximation by a multiple-break circuit (with four positive and four negative breakpoints) is shown in Figure 7–3c.

The sine-wave output produced by diode wave shaping in a good function generator is generally of high quality. The line segments are not observable with an oscilloscope. The function generator approach is a good way to produce low-frequency (below 20 Hz) sine waves, since they are difficult to produce with any quality by the regenerative sine-wave oscillator circuits described later in this section. Using a counter/read-only memory (ROM)/digital-to-analog converter (DAC) is another good method, as described in Chapter 2.

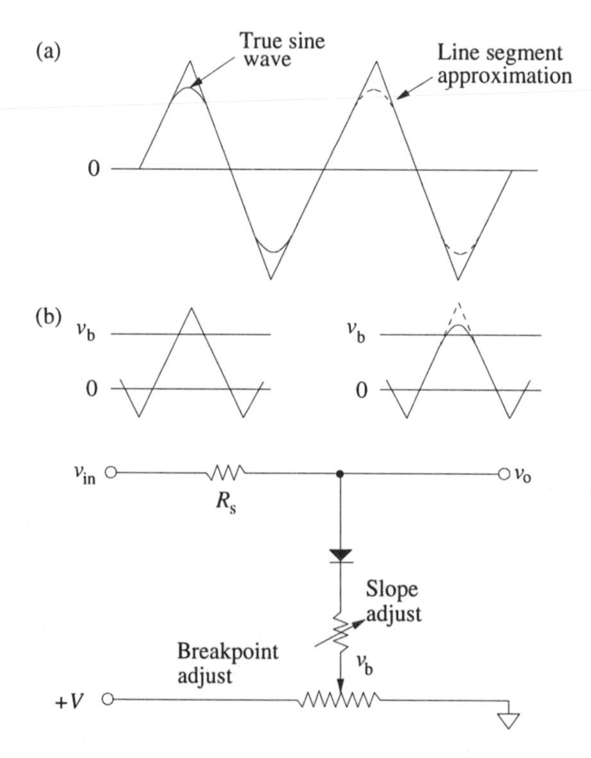

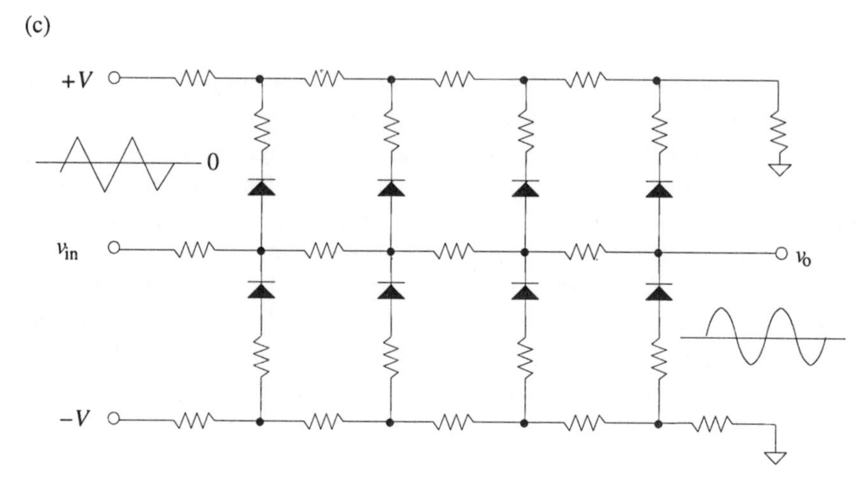

Figure 7–3. Synthesizing a sine wave from a triangular wave by line segment approximation (a), single-break diode wave shaping (b), and diode multiple-break circuit (c).

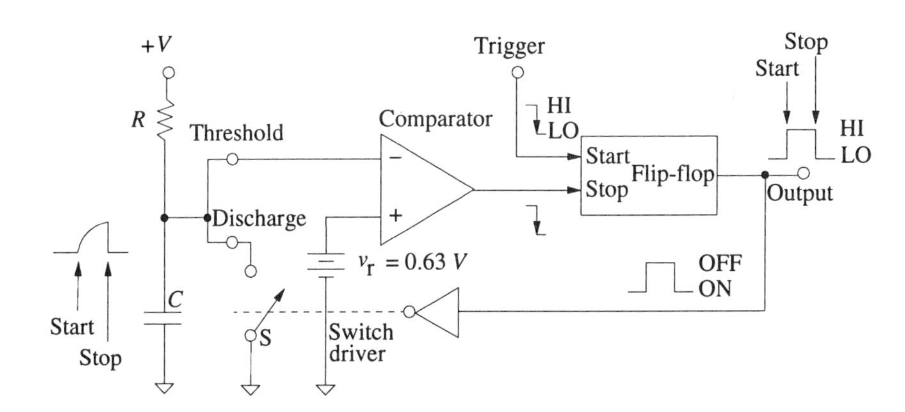

Figure 7–4. RC timer and monostable multivibrator. A HI–LO trigger signal causes the flip-flop output to go HI, opening switch S. The capacitor charges until the comparator threshold is reached. The HI–LO comparator output causes the flip-flop output to go LO, closing switch S. The HI–LO transition at the output appears 1.1 *RC* seconds after the trigger signal is applied.

Pulse Generator (Monostable Multivibrator)

A pulse is a waveform in which the voltage or current changes quickly from its normal steady value and then returns to the normal value after a period of time called the **pulse width**. Such a circuit has two states, only one of which is stable, and is thus called a **monostable multivibrator** or **single-shot pulse generator**. Pulse generators have many applications in instrumentation. Since they can produce pulse widths of a desired time period (from microseconds to minutes and even to hours or days by cascading), they can be used to delay the start of an operation or measurement until a preceding one has finished and voltage or current levels have settled to a steady value. Another application is as a sequencing element when a series of monostable multivibrators produces sequential time delay pulses for starting the next operation after the preceding one has finished.

The use of an RC (resistor–capacitor) timing circuit to produce a pulse of desired duration is illustrated in Figure 7–4. The step behavior of RC circuits is more thoroughly discussed in Supplement 2. Before the start of a pulse of timed interval, the capacitor remains discharged by the closed condition of switch S. When the timed interval is to begin, the logic level at the trigger input changes from HI to LO. This causes a change from LO to HI at the flip-flop output, which in turn causes the switch driver to open switch S. The capacitor is now free to charge from the voltage source supplying +*V* through the resistance *R*. The capacitor voltage is monitored by the comparator. The comparator output changes state when the capacitor voltage slightly exceeds the comparator reference voltage, which has been set equal to 0.63 of the supply voltage *V*. The HI–LO transition at the comparator output stops the timing by causing the flip-flop output to go LO, closing switch S. The net result of this cycle is to produce an output pulse of a duration determined by the values of *R* and *C*. With a supply voltage of *V*, the capacitor charges to 0.63 V in 1 time constant (*see* Figure S2–17 of Supplement 2), and the duration of the pulse equals 1 *RC* second.

For example, if $R = 1\ \text{k}\Omega$ and $C = 0.01\ \mu\text{F}$, the output pulse duration is $1\,(10^3 \times 10^{-8}) = 10\ \mu\text{s}$. To produce a pulse width of about 10 s, much larger values of *R* and *C*, such as 10 MΩ and 1 μF, would be selected. A 1% reproducibility of pulse duration is attainable, but it is generally limited by the stability of *R* and *C*.

The 555 IC Connected as a Pulse Generator

The popular and versatile 555 integrated circuit (IC) RC timer can be connected as a convenient monostable (one-shot) multivibrator, as illustrated for the dual in-line

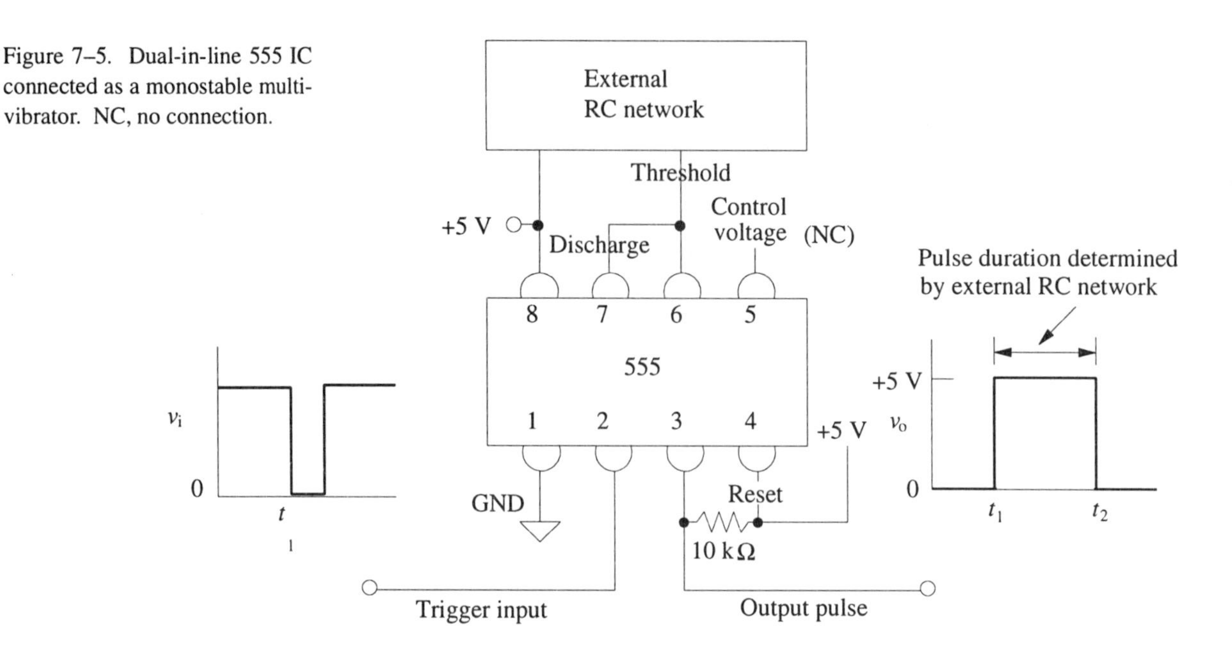

Figure 7–5. Dual-in-line 555 IC connected as a monostable multivibrator. NC, no connection.

IC package in Figure 7–5. The output voltage is LO until a HI-to-LO trigger pulse is applied, and then it switches to HI and remains HI for a duration determined by the values of *R* and *C* which are externally connected.

The operation of the 555 is similar to that described for the RC timer and pulse generator in Figure 7–4. The similarity is apparent in Figure 7–6, which shows the internal circuit of the 555 within the dashed lines. Note that the basic RC charging and comparison circuits are similar to those in Figure 7–4. Switch S is normally in the discharge (closed) position prior to the timing cycle. To begin timing, a HI–LO transition is applied externally to the trigger input. The lower comparator in Figure 7–6 senses the trigger signal and compares it to one-third of the supply voltage set by the three 5-kΩ divider resistors. When the trigger input falls below the lower comparator threshold, the comparator output changes state and sets the flip-flop to its HI output state. This causes switch S to open, and the timing capacitor of capacitance *C* now charges through the resistor of resistance *R* toward V_1 until v_C slightly exceeds two-thirds of the supply voltage as sensed by the upper comparator. When this charging level has been reached, the upper comparator output changes states, which resets the flip-flop and closes switch S to short the capacitor. The flip-flop output is thus HI for the time interval during which the capacitor was charging. This time interval is 1.1 *RC* seconds in duration. As we have noted, the product *RC* is variable over a very wide range.

The power supply voltage for the 555 can be any voltage between +5 and +18 V. The output terminal at pin 3 can be connected to either a grounded load or a supply load. Terminal 4 is a reset that allows the 555 to be disabled; it overrides signals on the trigger input. This terminal should be connected to the supply voltage when not used. A ground on pin 4 forces the output (pin 3) low. The control voltage terminal, pin 5, is connected only for changing the threshold and trigger voltage

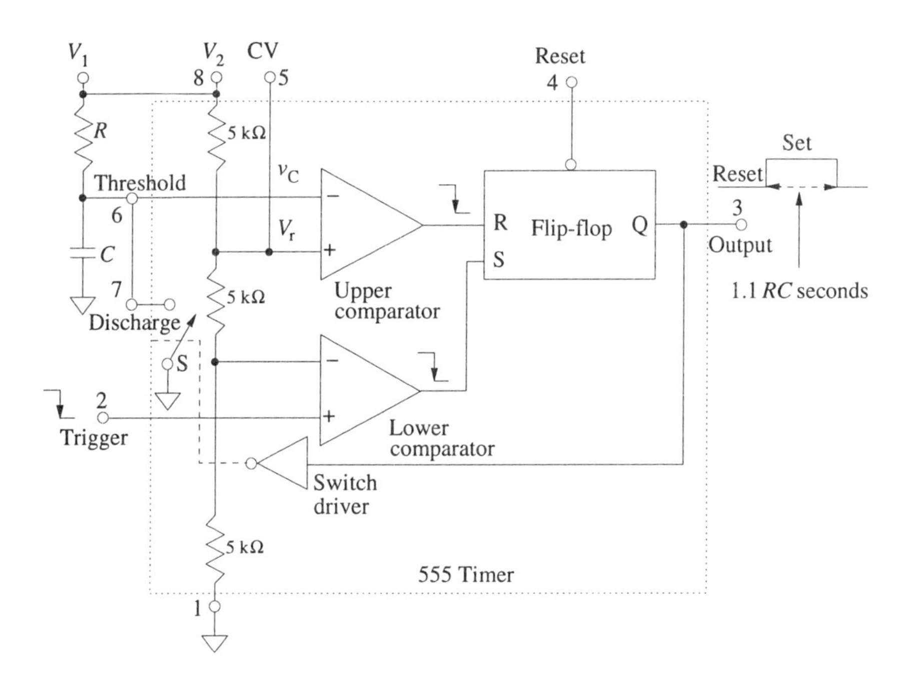

Figure 7–6. The 555 RC timer connected as a monostable multivibrator. Timing begins when a HI–LO transition occurs at the trigger input. This sets the flip-flop and causes switch S to open so that the capacitor can charge. Timing capacitance *C* charges through timing resistance *R* toward V_1 after switch S opens. When the capacitor voltage exceeds the upper comparator reference level ($0.67V_2$), the flip-flop is reset and switch S shorts capacitance *C*. The flip-flop output is HI only during the capacitor charging time or for 1.1 *RC* seconds. Thus, a single pulse of 1.1 *RC* seconds in duration is produced. R and S, flip-flop inputs; Q, flip-flop output.

levels. An external voltage applied to pin 5 can be used to modulate the output at pin 3, since it will change both the trigger and the threshold voltages.

Astable Multivibrator

By a simple change in connections, the 555 timer can produce repetitive pulses, that is, a free-running square-wave generator. Such a repetitive pulse generator is called an **astable multivibrator**, or an **oscillator**. The connections needed are shown in Figure 7–7. In this case the trigger input is not supplied externally but comes from the capacitor voltage. In other words, the timer triggers itself repetitively. To understand how this works, let us assume that the timer is triggered and the capacitor is charging through R_1 and R_2 toward V_1 from the supply voltage. Again, when v_c reaches two-thirds of V_1, the comparator changes state and the switch closes. Capacitor with capacitance *C* now begins to discharge through R_2. The flip-flop set input becomes actuated at one-third of the supply voltage, so the capacitor discharges to this point, at which time the flip-flop is set and the charging process begins again. The waveform for v_c shows this repetitive charging and discharging clearly. The flip-flop output is now a repetitive pulse train. Note that the charging time constant is $(R_1 + R_2) \times C$, while the discharge time constant is $R_2 \times C$. Hence, the flip-flop is HI (capacitor charging) for a slightly longer time than it is LO (capacitor discharging). The result is a slightly asymmetrical waveform for the flip-flop output. Note that the frequency or repetition rate of the pulse train is determined by the values of *R* and *C*.

Astable multivibrators can achieve oscillation frequencies in the range of 0.1 Hz to 1 MHz. The connection of a 555 timer with component values for a

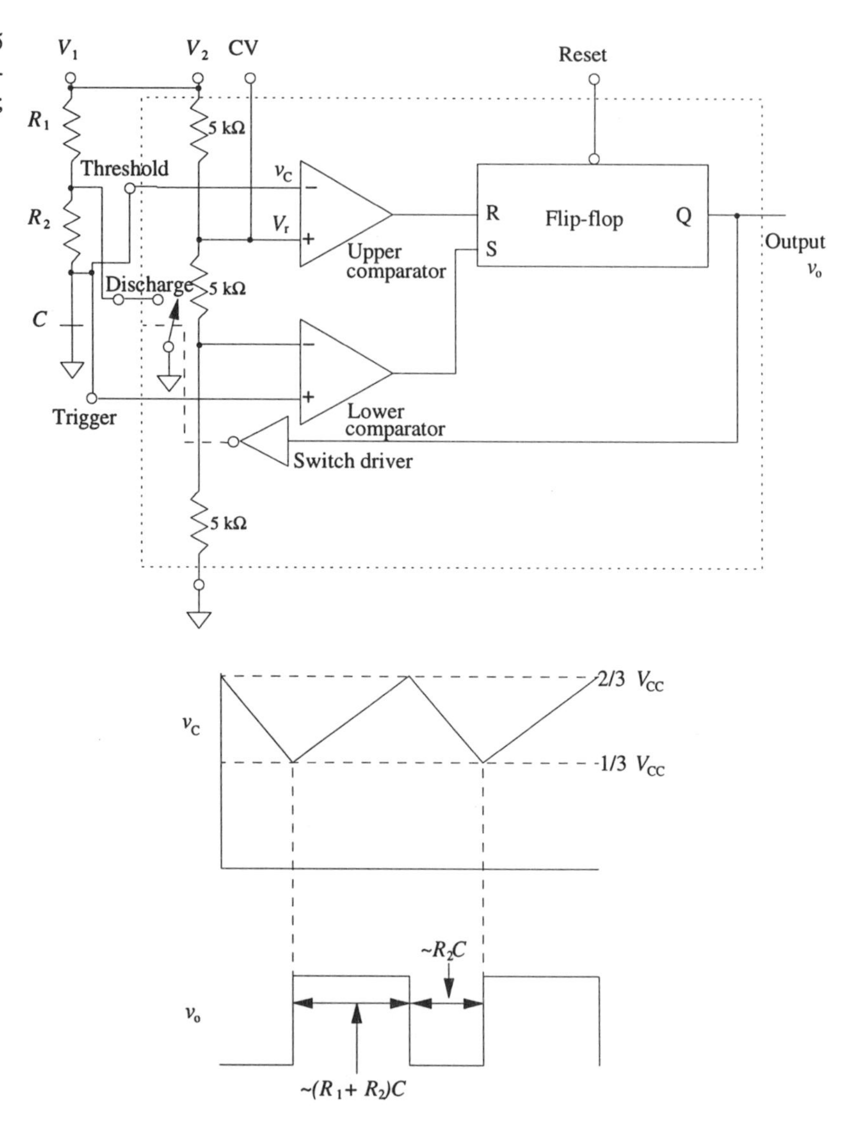

Figure 7–7. Connection of 555 RC timer as an astable multivibrator. R and S, flip-flop inputs; Q, flip-flop output.

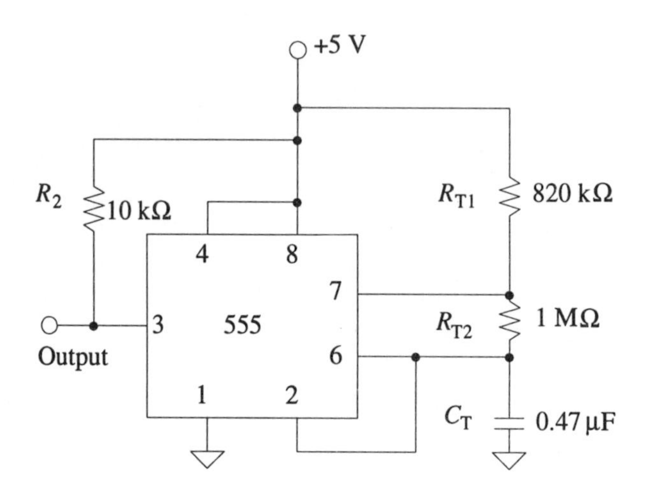

Figure 7–8. Connection of 555 timer as a low-frequency astable multivibrator.

low-frequency oscillation is shown in Figure 7–8. The frequency f equals 1.44 $[(R_{T1} + R_{T2}) C_T]$.

Regenerative Sine-Wave Oscillators

The waveform generators described in the previous parts of this section, such as the astable multivibrator and the triangular- and square-wave generators, are often referred to as **relaxation oscillators**. That is, they are based on alternately charging and discharging a capacitor, and the reversal from charge to discharge or vice versa occurs at specific charge levels. Another type of oscillator is based on regenerative (self-sustaining) feedback in an amplifier. This type is basically a sine-wave generator.

An amplifier has feedback when its output signal has an effect on the signal at its input. If the output signal causes the input signal to change further in the same sense, there is **positive feedback**. The input signal is thus augmented or regenerated by the output signal. An amplifier with positive feedback has higher gain than an amplifier with none. It is also possible to feed back the output signal so as to oppose or decrease the effective signal change at the input. This **negative feedback** has the effect of decreasing the overall gain. Op amps can be used with either positive or negative feedback. Positive feedback occurs when the feedback loop is connected to the noninverting input. Negative feedback occurs when the feedback loop is connected to the inverting input.

An op amp can be used with both positive and negative feedback to provide a classical **Wein bridge** sine-wave oscillator, as illustrated in Figure 7–9. The single output frequency of the sine wave is stable (0.1% or better) and can be changed readily by switching in different resistors and capacitors in the reactive arms of the Wein bridge. If $R_1 = R_2 = R$ and $C_1 = C_2 = C$, the frequency $f = 1/(2\pi RC)$. The frequency-selective network provides positive feedback to the noninverting input, and the resistive divider provides negative feedback to the inverting input. When the positive feedback is greater than the negative feedback, oscillations are sustained. At the frequency equal to $1/(2\pi RC)$, the positive feedback is maximum because there is zero phase difference between the series and parallel reactive arms.

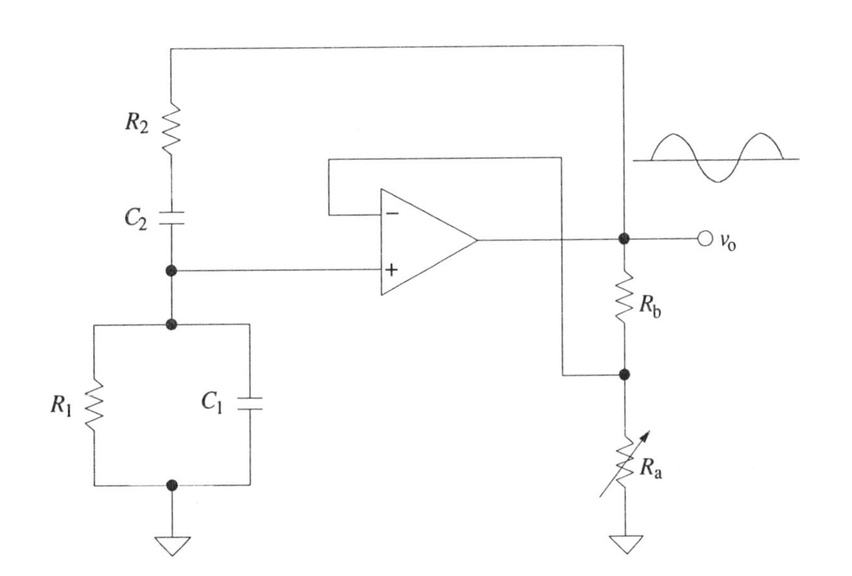

Figure 7–9. Wein bridge op amp sine-wave oscillator. The output frequency is determined by the values of C_1, C_2, R_1, R_2 and $f = 1/(2\pi RC)$ when $C_1 = C_2 = C$ and $R_1 = R_2 = R$. R_a and R_b, resistances in a resistive negative feedback divider.

If the frequency increases, the phase angle across the series *RC* arm increases and that across the parallel *RC* arm decreases.

To produce oscillation, the negative feedback is decreased to achieve the desired amount of net positive feedback. For pure sine waves, some means of automatically adjusting the negative feedback ratio to maintain constant output amplitude must be added. Several methods, such as a thermistor, lamp bulb, or back-to-back Zener diodes as part of R_a in the resistive negative feedback divider, have been used.

Other types of op amp regenerative oscillators are designed around various active filter circuits. These filters are described in Section 7–3.

7–2 Amplifying Analog Signals with Operational Amplifier Circuits

The versatile op amp has become the universal *gain block* for analog circuits of all kinds, as illustrated by the many examples in Chapter 5. In this section the applications of op amps for amplifying analog signals are expanded beyond the basic circuits introduced earlier.

Several op amp circuits require programming. For example, the gain is programmed to provide an output in the desired range, or multiple analog input signals are programmed (multiplexed) to share one op amp circuit successively. Other examples include programming power supplies or sampling an analog signal as a function of time. All of these examples require programmable analog switches with suitable characteristics. When they were introduced in previous sections, the analog switch characteristics were assumed to be acceptable, but they are considered more carefully in this section because they are critical in the programmable op amp applications that are introduced here.

After the various programmable op amp circuits have been discussed, the difference and instrumentation amplifier and various other specialized op amp circuits for recovery of low-level signals from relatively large amounts of noise are described. Op amp circuits for discriminating against dc signals and amplifying only ac signals are also introduced.

Programmable Gain Amplifier

Both the inverting amplifier and the follower amplifier with gain can be programmed to provide a wide range of gain factors by switching the feedback resistance values. High-speed IC analog switches enable the gain to be controlled by digital signals and thus by a program in the microcomputer. Several characteristics of the IC analog switches are discussed first, and then an example of a programmable voltage follower with gain is presented.

Analog switches. Both solid-state switching elements and the associated switch drivers are integrated in a single package in modern IC analog switches. Symbolism for such switches and switch drivers is shown in Figure 7–10. Because the actuating signal and the switch contacts have no terminals in common, the IC analog switch is an example of a four-terminal device. However, the switched circuit is not completely isolated from the actuating circuit as in the mechanical relay. Most IC analog switch packages contain two or more sets of switches and

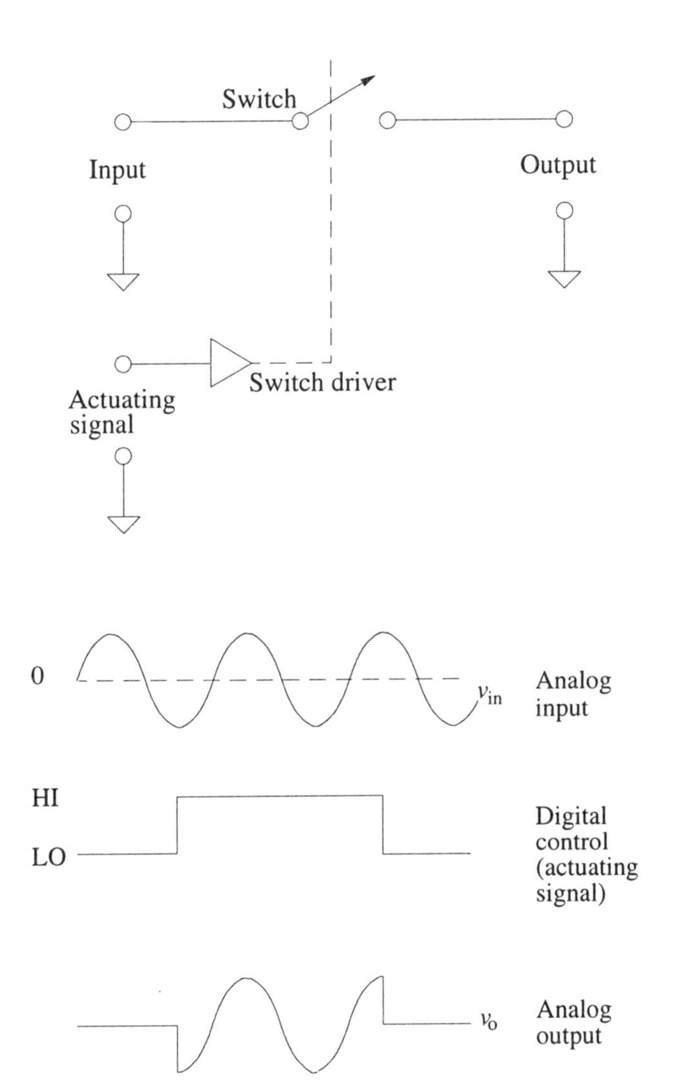

Figure 7–10. Normally open single-pole single-throw (SPST) analog switch. The switch driver converts logic-level signals to the appropriate voltage or current levels to actuate the solid-state switch.

Figure 7–11. Analog voltage switch waveform. The analog signal v_{in} appears at the output when the drive signal is at the HI logic level and the switch is closed.

drivers in the same unit. They are available with a variety of "contact" arrangements [single pole single throw (SPST), single pole double throw (SPDT), etc.] and a variety of switching characteristics.

IC analog switches are of two types: voltage switches and current switches. A **voltage switch** is used to transmit an analog voltage from input to output when the switch is closed. The switch state is determined by a logic-level actuating signal. Figure 7–11 illustrates typical waveforms for the analog voltage switch of Figure 7–10.

Analog **current switches** are made to operate with one of the switch contacts connected to the system common or to the virtual common of an op amp. Figure 7–12 shows the use of a typical current switch to switch a current source input to an op amp current follower. The small voltage drop across the switch (limited by the diode) ensures fast switching speeds and allows the switch to be opened and closed with relatively small drive voltages, such as the HI–LO logic levels of the transistor–transistor logic (TTL) family. Analog current switches can often switch states several times faster than comparable voltage switches. The input signal is shunted to ground through the diode when the switch is open. Therefore, voltage sources that should not be shorted should be connected to a current switch through a resistor.

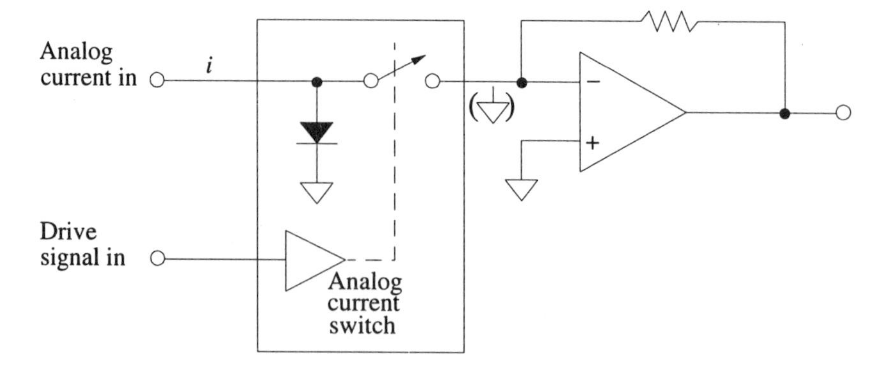

Figure 7–12. Analog current switch used to switch current signal *i* to an op amp current follower. The input diode limits the voltage drop across the switch to 0.6 V in the OFF state by shunting the input signal to ground.

Figure 7–13 shows an op amp integrator with three switched input sources and a shorting switch for the integrating capacitor. Switch S_1 controls the application of voltage v_1 and should be a voltage switch. Switches S_2 and S_3 can be current switches, since in each case one contact is connected to the op amp summing point (virtual common) and the other contact is not directly connected to a voltage output. Switch S_4 should be a voltage switch, since it is connected to the op amp output.

IC analog switches are characterized by several important parameters. First, the ON resistance R_{ON} should be very low for accurate voltage or current switching. Typical ON resistance values vary from less than 1 Ω to several hundred ohms. The OFF resistance of the switch should be very high for excellent isolation of input and output in the OFF state. Typical OFF resistance values are in the 10^9- to 10^{12}-Ω range. Another important parameter is the change in R_{ON} with applied analog voltage, sometimes called **ON resistance modulation**. For good switches, R_{ON} changes by only a few percent over the full range of allowed analog signal levels. When several switches and drivers are contained in a single IC, an important measure of the isolation of one switch from another is the switch **cross talk**. Analog signals applied to the input of one switch can appear at the output of another switch unless the level of cross talk is very low. Because the drive signal is not totally isolated from the analog signal, the drive signal can sometimes **feed through** to the analog output. Finally, the switch capacitance should be very low, because capacitance can limit the switching speed. IC analog switches have switching speeds in the nanosecond to microsecond range.

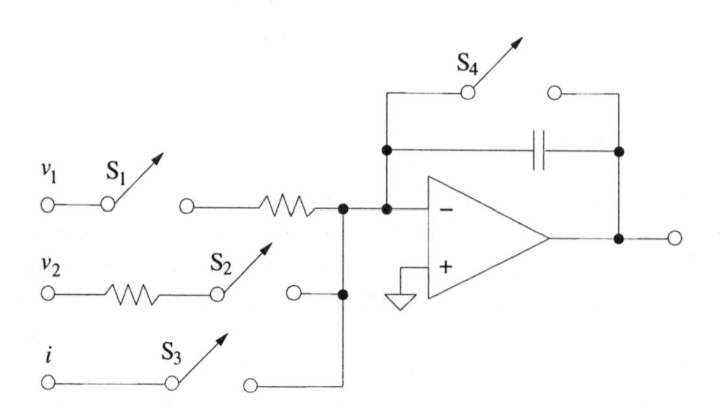

Figure 7–13. Op amp integrator with multiple switched sources. Switches S_1 and S_4 are voltage switches; switches S_2 and S_3 can be current switches.

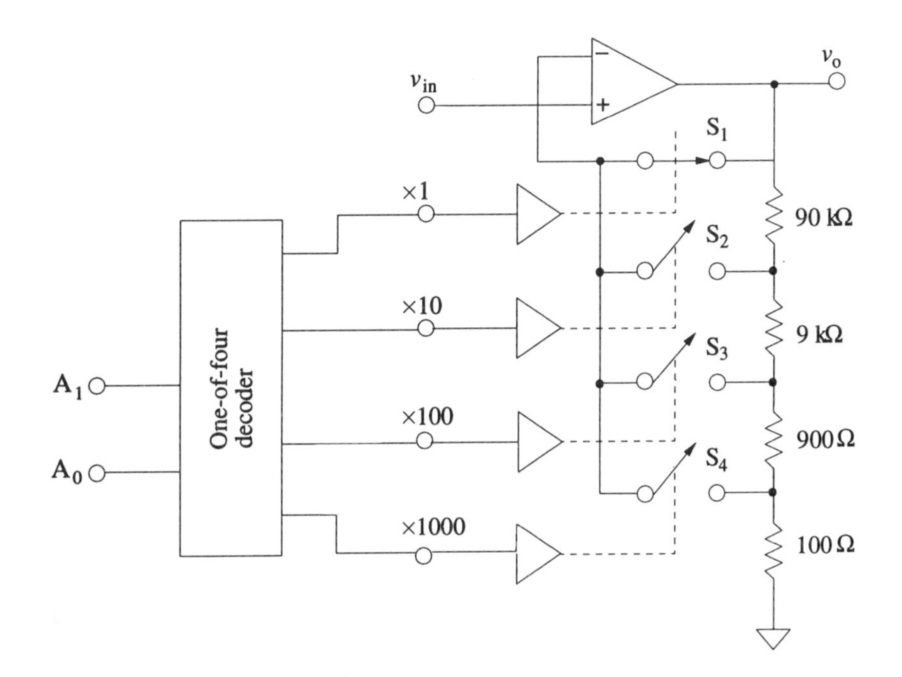

Figure 7–14. Programmable voltage follower. The gain is controlled by closing a switch that determines the feedback fraction. The switch to be closed is selected by digital signals applied to the address lines A_1 and A_0.

Programmable voltage follower. An amplifier of the type frequently used in a computer data acquisition system for automatic control of amplitudes of signals input to the analog-to-digital converter is shown in Figure 7–14. Overranging of the analog-to-digital converter could be sensed, and the overrange condition could be used to lower the gain until the highest gain without overranging was selected automatically.

With switch S_1 closed, the op amp is a voltage follower with unity gain. With S_2 closed and the other switches open, the gain is 10. The gains are 100 with S_3 closed and 1000 with S_4 closed. A one-of-four decoder, described in the previous chapter, is used to provide the switch driver signals from a 2-bit digital signal.

Programmable Analog Signal Selector (Multiplexer)

It is frequently desirable to connect analog signal sources sequentially to the input of op amp circuits that are interfaced to a computer. An example of a four-input analog signal selector, frequently called an analog multiplexer, is shown connected to an inverting amplifier in Figure 7–15. This multiplexer allows one of four analog signal inputs to be connected to the amplifier. In some applications it is desirable to cycle through the four analog inputs in turn. For example, the output of the follower with gain might be connected to a computer data acquisition system, and we might like to share this device among four different analog signals. In that case the address lines could be operated from a 2-bit counter that would cycle through the four address states and then repeat the process. The entire multiplexer switching unit with as many as eight switches and the appropriate decoder is available in a single IC package. Switching can be accomplished with such units on the microsecond or submicrosecond time scale.

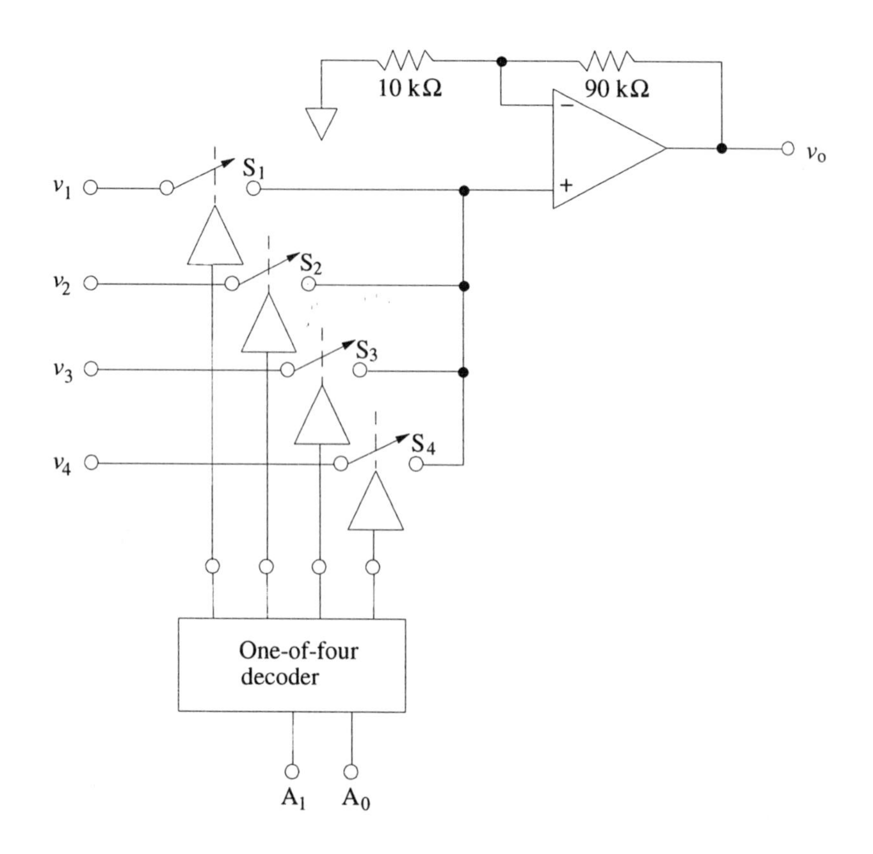

Figure 7–15. Programmable analog signal selector (multiplexer). Digital signals applied to the address lines A_1 and A_0 determine which decoder output is HI and thus which switch is closed.

Programmable Op Amp Power Supplies

A power supply in which the output voltage or current varies in response to a remote command is known as a **programmable power supply.** Some commercial power supplies, known as **operational power supplies,** are based on the inverting amplifier configuration. An example of a programmable voltage-regulated op amp power supply is shown in Figure 7–16. It can be **programmed** to automatically switch in various values of a reference voltage v_r or feedback resistance R_f so as to provide the desired sequence of power supply voltages. Many op amp power

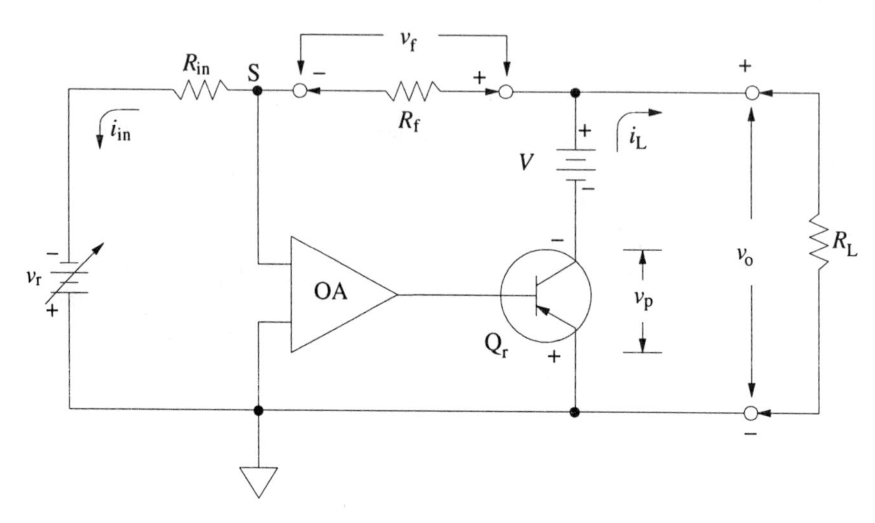

Figure 7–16. Programmable voltage-regulated op amp power supply. The output voltage v_o is given by $v_o = -v_r(R_f/R_{in})$. The combination of op amp (OA), pass transistor (Q_r), and raw supply V can be considered a high-power op amp and used for adding, subtracting, integrating, and other operations. For a bipolar output, an npn transistor and oppositely connected raw supply must be added to the circuit described above so that the output polarity depends on the direction of the input current. S, summing point.

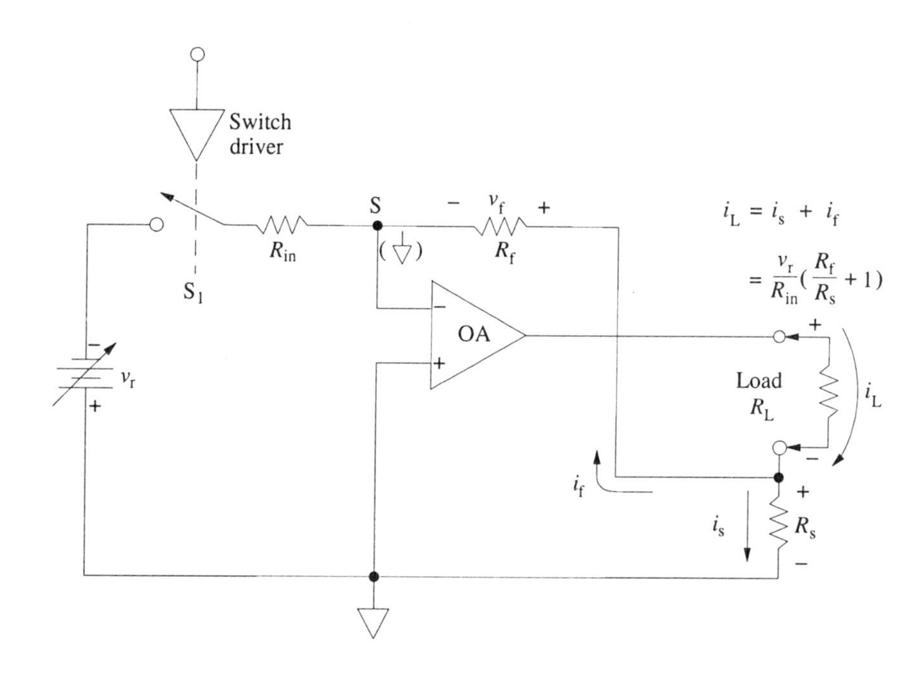

Figure 7–17. Programmable current-regulated power supply. The load current i_L is regulated at a constant value determined by the precise values for v_r, R_{in}, R_f, and R_s. The load current is readily programmed by switching any of these parameters to give the desired output current. The analog switch S_1 can be used to turn the load current ON and OFF at very high speeds.

supplies provide user access to the summing junction so that typical op amp operations can determine the voltage output.

Operational power supplies can also be readily configured so that the controlled quantity is the load current rather than the output voltage. Such a **current-regulated power supply** is highly useful in many control applications, and an example is shown in Figure 7–17. It can be seen in this figure and from the equation for the load current that i_L is regulated at a constant value determined by the preset values for the input reference voltage v_r, the input resistance R_{in}, the feedback resistance R_f, and the current-sensing resistance R_s. Any of these parameters can be automatically switched to program the load current to desired values. Also, the analog switch S_1 can be used to turn the load current ON and OFF at very high speeds. Thus, the output current can be programmed in a precise ON–OFF sequence as determined by logic-level pulses to the switch driver.

Note that the voltage drop across the current-sensing resistor must equal the voltage drop v_f across the feedback resistance R_f so as to maintain the sum of the voltages between common and the summing point (virtual common) equal to zero. Also, since $i_L = i_s + i_f$, it follows that, as shown in Figure 7–17,

$$i_L = \frac{v_r}{R_{in}}\left(\frac{R_f}{R_s} + 1\right) \qquad (7\text{–}1)$$

Sample-and-Hold Amplifier

An analog memory circuit, more commonly called a sample-and-hold circuit, is shown in Figure 7–18a. It allows an input signal v_{in} to charge a capacitor with capacitance C until it is desired to hold that voltage for measurement, conversion, or display purposes. When the switch is in the sample mode, the input signal, which is buffered by voltage follower op amp 1, charges a capacitance C through a

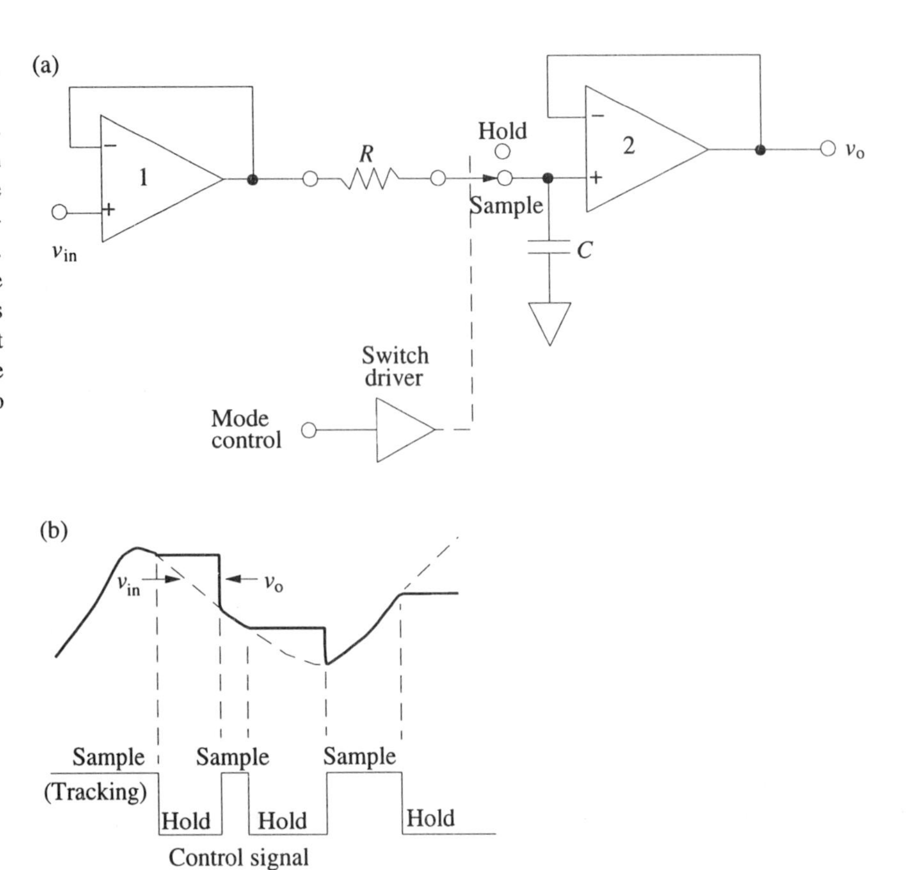

Figure 7–18. Sample-and-hold circuit (a) and waveforms of input, output, and switch control signals (b). When the mode switch is in the sample position, the voltage on the capacitor and thus the output voltage track the input signal. When the mode switch is in the hold position, the capacitor stores the value of the input signal that was present at the instant the switch changed from sample to hold.

resistance *R*. The *RC* time constant and the amplifier response time are made very short so that the capacitor voltage faithfully tracks the input signal during the sampling time. When this voltage is to be stored, a mode control signal causes the switch to change to the hold position. This isolates the input signal and leaves the capacitor voltage at the value of the input signal at the instant of switching. Voltage follower op amp 2 buffers the capacitor voltage and provides an output voltage v_o equal to the capacitor voltage. The extremely high input resistance of the follower prevents significant discharge of the capacitor. Thus, the output voltage v_o is equal to the input voltage v_{in} at the instant of the transition from sample to hold. The output voltage v_o can then be measured, converted, or otherwise treated by a slower device.

The waveforms in Figure 7–18b show the sample-and-hold action. During the sampling (tracking) time, v_o follows v_{in}. When the switch is thrown to the hold position, $v_o = v_{in}$ at this instant and remains constant until the switch is returned to the sample position. Since switching times can be in the nanosecond range, the time at which the input voltage had a specific value is known to within nanoseconds. Such a circuit is often used in conjunction with analog-to-digital converters that require the analog voltage to be constant during the finite time (typically 5 to 50 μs) required for conversion of the analog signal to digital data.

Peak Detector

The peak detector is somewhat similar to the sample-and-hold circuit except that a diode in the circuit enables the peak voltage of an input signal to be held on the storage capacitor and at the output of the voltage follower. The storage of peak information for subsequent operations is useful for measurement and control applications.

The peak detector in Figure 7–19a requires only one op amp. The input signal v_u is connected through diode D_1 to the input of an op amp voltage follower, which has a storage capacitor with capacitance C at its input. The input diode acts like a switch for positive signals. It has a very low resistance when the input voltage exceeds 0.6 V and there is no charge on the storage capacitor. The capacitor voltage (and input to the op amp follower) follows the input voltage until its peak is reached. As the input starts to decrease, the diode becomes reverse-biased and its resistance becomes very high. Thus, the peak voltage v_p is held on the capacitor. However, because of the 0.6-V drop across D_1, it is necessary to compensate by placing a similar diode D_2 in the feedback of the op amp. The voltage v_r can be adjusted so that the voltage drops across the two diodes will be equal and will provide accurate compensation. Thus, the output voltage v_o equals the peak voltage of the input. After the output peak voltage has been measured or utilized for the desired application, the capacitor is discharged by closing the switch across it. An extension of this circuit facilitates the automated logging of the peak data. The peak value could be sensed by comparing the voltage at the output with that at the input

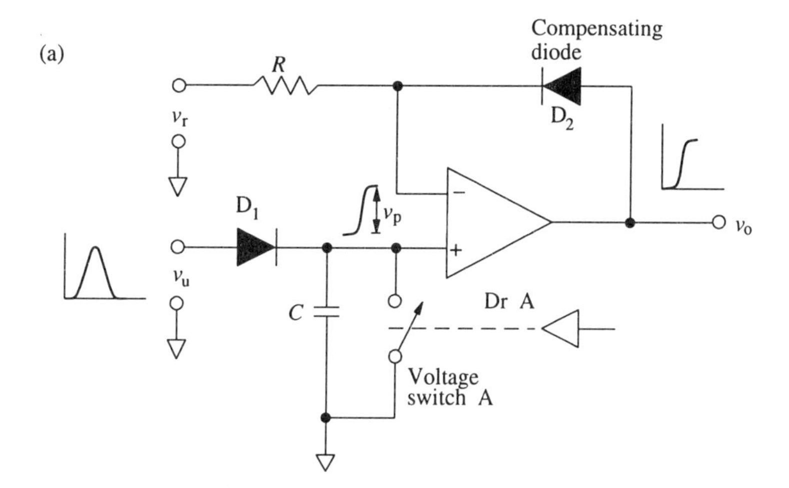

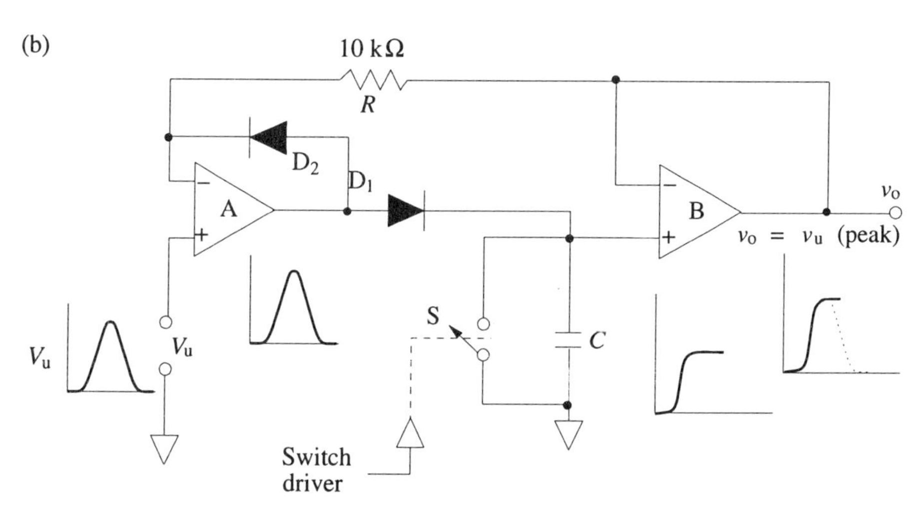

Figure 7–19. Peak detector using one (a) or two (b) op amp voltage followers. Dr, driver; S, switch.

and then closing the switch after a time delay (use of monostable multivibrator) to allow utilization of the peak data.

The peak detector in Figure 7–19b requires two voltage followers, but the circuit provides exact compensation for the diode voltage drop. The output voltage of op amp B produces a current through the feedback resistance *R*. The current fed back to the inverting input of op amp A and through D_2 causes the 0.6-V drop across D_2. Thus, the output voltage of op amp A equals the input voltage plus the voltage drop across diode D_2. This compensates for the voltage drop across D_1 so that the voltage on the capacitor follows the true value of the input voltage up to the peak value and then holds the true peak value as the input voltage decreases.

Difference Amplifier

The difference amplifier, sometimes referred to as a differential amplifier, has advantages over the ordinary inverting or noninverting amplifier. It can be used when it is necessary to measure small signals that are buried in large unwanted signals. As the name implies, the difference amplifier produces an output voltage that is proportional to the difference between two input voltages. That is, the desired measurement is the difference between two transducer outputs, the off-balance signal of a Wheatstone bridge, or the voltage difference between any two points, neither of which is at the common voltage. It is also the type of amplifier most often used to eliminate noise that occurs between the common of the signal source and the common of the amplifier, as shown in Figure 7–20.

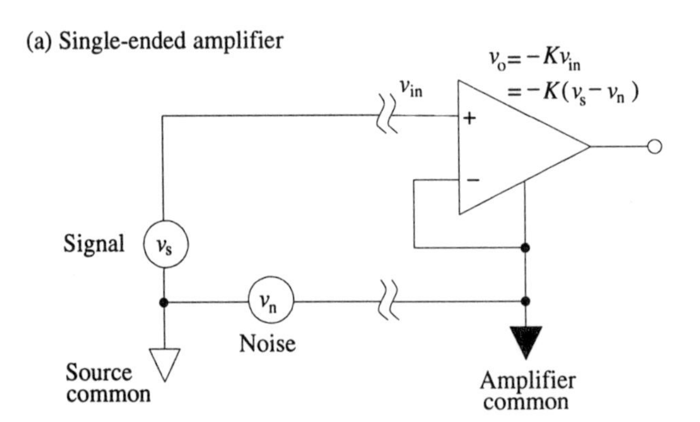

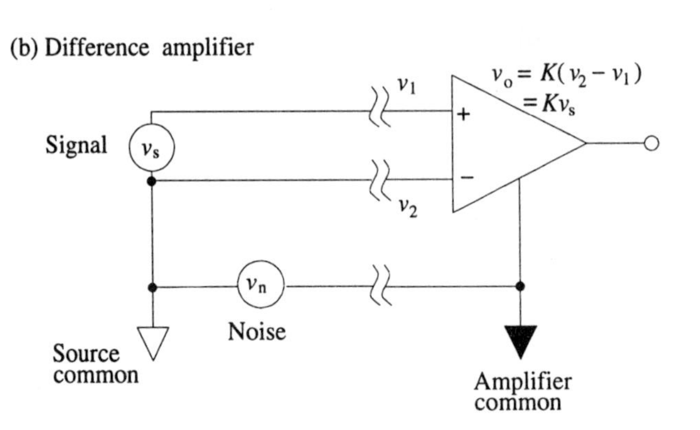

Figure 7–20. Effect of noise in amplifier input connections. (a) The single-ended amplifier amplifies signal v_s and noise v_n, which is the voltage difference between the source and amplifier common. (b) The difference amplifier can be used to exclude this noise voltage from the signal to be amplified.

The relationship $v_o = K(v_2 - v_1)$ for the ideal difference amplifier indicates that the gain is the same for signal v_2 as for signal v_1 and that this is true for all values of v_o. If the amplifier is perfectly balanced for the two inputs, an identical voltage change applied to both inputs cancels itself and has no effect on v_o. Thus, the output responds only to the *differences* in the input voltages. A measure of the degree of balance in a difference amplifier is the ratio between the response of the amplifier to a signal applied between the difference inputs and its response to a signal applied between both inputs and common. This ratio is the **common-mode rejection ratio** (CMRR), since the portion of the input signal that is applied identically to both inputs is called the **common-mode signal**. For example, if $v_o = 1000(v_2 - v_1)$ for a difference signal and $v_o = 0.1v_2$ when $v_2 = v_1$ for a common signal, then CMRR = 1000/0.1 or 10 000, which is 80 dB.

Basic Difference Amplifier

The circuit shown in Figure 7–21 is that of a basic difference amplifier. It combines the inverting amplifier and the follower with gain. A fraction of the input voltage v_2 is applied to the noninverting op amp input, and the same fraction of the difference between v_1 and v_o is applied to the inverting op amp input. The difference between v_- and v_+, the inverting and noninverting op amp input voltages, is v_o/A (A is the open-loop gain of the op amp), and it is negligibly small with respect to the input voltages.

$$v_o = K(v_2 - v_1) \tag{7–2}$$

As equation 7–2 shows, the output voltage is equal to the difference voltage multiplied by K. With the selection of a high-CMRR op amp and carefully matched resistance ratios, the amplifier of Figure 7–21 can be very useful. The common-mode response of a difference amplifier can be checked by connecting the output of a signal generator to both inputs in parallel and observing the amplifier output. If the amplifier has a balance control, it should then be adjusted for a minimum output.

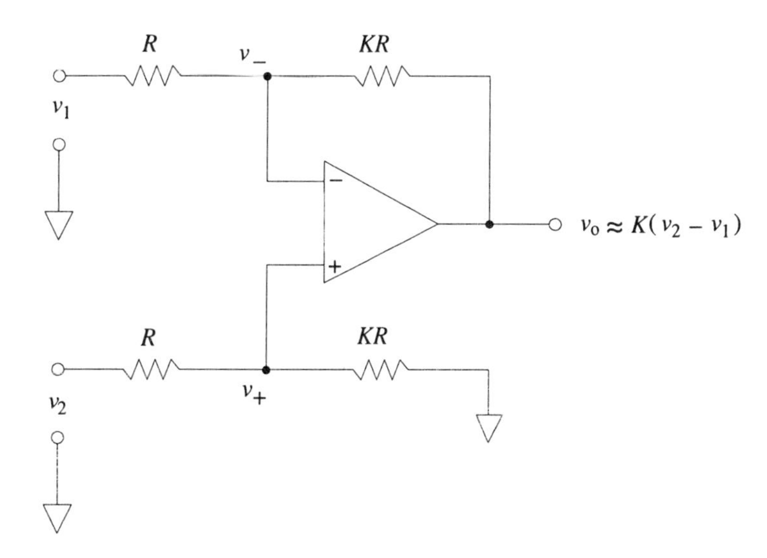

Figure 7–21. Basic difference amplifier. The resistance ratios must be carefully matched for good CMRR. A trimmer adjustment on any of the resistors would allow precise balance adjustment.

Instrumentation Amplifier

A complete instrumentation amplifier combines the advantages of the difference input with the high input resistance of the voltage follower. This combination is readily achieved by simply connecting a voltage follower amplifier before each input of the difference amplifier of Figure 7–21. The use of follower-with-gain circuits to achieve higher gain is not recommended, however, because the follower amplifier gains would have to be well matched in order to achieve a high CMRR. A very clever circuit that cross-couples the two follower-with-gain circuits so that they track each other is shown in Figure 7–22 with the difference amplifier. The gain and cross-coupling are provided by the three resistors between the two follower outputs. Follower amplifiers 1 and 2 keep the feedback points equal to v_1 and v_2 respectively. This results in a current through the three resistors of $i = a(v_1 - v_2)/R_1$ (a is a factor that varies the effective resistance and changes the follower input gain). The follower output voltages are then equal to the sum of the feedback voltage and the *IR* drop through the feedback resistor. Thus, $v_{o1} = v_1 + a(v_1 - v_2)$ and $v_{o2} = v_2 - a(v_1 - v_2)$. These expressions show that each follower amplifies its input signal by 1 and the difference signal by $+a$ or $-a$. The difference gain A_d of the cross-coupled followers is

$$A_d = \frac{v_{o1} - v_{o2}}{v_1 - v_2} = 1 + 2a$$

and the common-mode gain A_{cm} is

$$A_{cm} = \frac{\dfrac{(v_{o1} + v_{o2})}{2}}{\dfrac{(v_1 + v_2)}{2}} = 1$$

These gain equations do not depend on precision matching of R_1. Because the common-mode gain is 1, the CMRR of this stage is equal to the difference gain, which is generally between 10 and 1000. The gain of this stage can be adjusted by changing the single resistance R_1/a. In general, a resistor with $a = 4.5$ for a difference gain of 10 is wired into the circuit, and other resistors are switched in parallel with this resistor when higher gains are desired. The CMRR of the cross-coupled input amplifiers is then multiplied by the CMRR of the difference amplifier to produce an instrumentation amplifier with excellent CMRR, high input impedance, and stable, easily adjustable gain. For instance, if the gain of the input stage is set at 100 and the gain and the CMRR of the difference stage are 10 and 10^4, respectively, the resulting amplifier would have a gain of 1000 and a CMRR of 10^6

Some of the applications of differential instrumentation amplifiers include bridge, strain gauge, or any low-level voltage measurement for which a high CMRR is essential. These amplifiers are frequently used to amplify the microvolt and millivolt outputs from biological transducers. They are available in convenient IC form from several manufacturers.

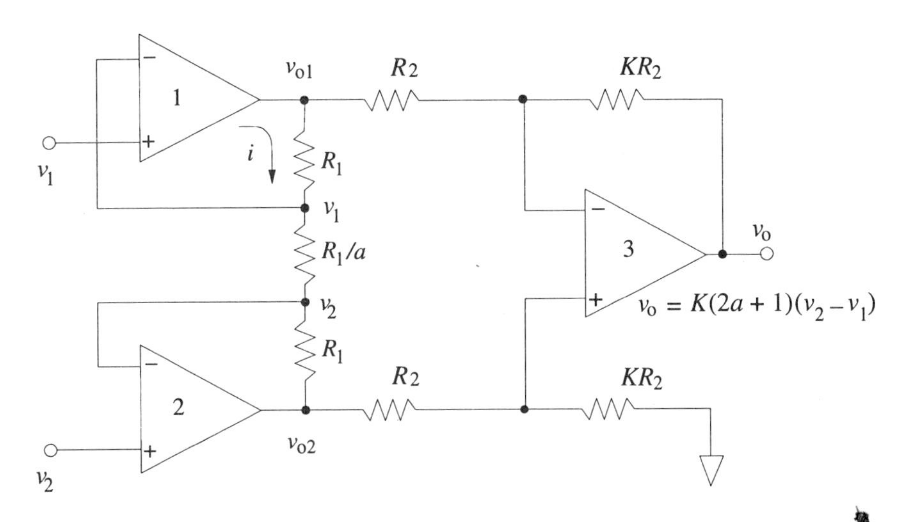

Figure 7–22. Instrumentation amplifier using cross-coupled differential follower input. The gain can be adjusted with the single resistor R_1/a. The difference gain of the input stage is $1 + 2a$.

Absolute Value Op Amp Circuit

As its name suggests, the absolute value op amp circuit produces a positive output voltage equal in magnitude to the input voltage regardless of the sign of the input voltage. A simple implementation of this function combines precision inverting and noninverting limiter circuits.

Precision inverting limiter. The ideal limiter circuit provides a linear transfer function between the input and the output over a limited range. In its precision form, the limiter function is also very useful in nonlinear analog signal conditioning. The heart of the limiter circuit is the pn junction diode, which by itself is not a suitable component for precision operations on low-level signals because of the large (~0.6-V) forward bias needed for conduction and the variability of the required bias both among devices and with temperature. However, the diode function (unidirectional conduction) can be included in an op amp circuit in such a way that the resulting function is very precise.

A precise limiter circuit that performs well even at signal levels in the millivolt range or less is shown in Figure 7–23. The circuit is very similar to the summing amplifier except that it includes two feedback paths, one for each direction of the feedback current. If the sum of the currents from V_R and v_{in} is positive, D_1 conducts the feedback current required to keep point S at virtual common. From the forward bias across D_1, v_A is –0.6 V. The output voltage v_o is exactly zero in this case, since the output is connected to common through R_L and to virtual common through R_f and since diode D_2 is reverse-biased. The voltage required to make D_1 conduct does not then appear at the output. When the sum of currents through R_R and R_{in} is negative, the only feedback path is through diode D_2 and R_f. The amplifier output v_A must be sufficiently positive to provide the required forward bias for D_2 plus the *IR* drop across R_f as shown by the dashed line in Figure 7–23. Since point S is still maintained at virtual common, v_o is simply $-i_f R_f$ where i_f is equal to the total input current $(v_{in}/R_{in}) + (V_R/R_R)$. Thus,

$$v_o = -R_s\left(\frac{v_{in}}{R_{in}} + \frac{V_R}{R_R}\right) \text{ for } v_{in} < \frac{-V_R R_{in}}{R_R} \qquad (7\text{–}3)$$

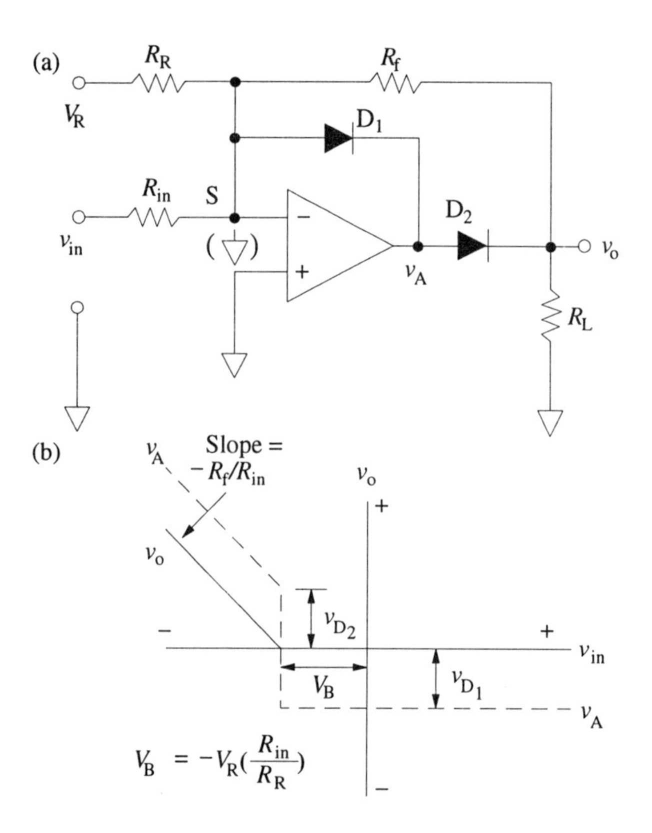

Figure 7–23. Precision inverting limiter. (a) Circuit; (b) function curve. This circuit eliminates the forward-bias errors of the diodes by including them in the feedback control loop. The op amp controls v_o in an inverting-amplifier mode when the net input current from v_{in} + V_R is negative. For net positive currents, diode D_2 is reverse-biased and $v_o = 0$. Note that the op amp supplies the extra voltage at v_A to forward bias the diodes in each case. The breakpoint voltage V_B is set by adjusting V_R or R_R. S, summing point.

It is important to include an actual load resistance R_L such that significant current (~1 mA) is drawn through D_2 when v_o is positive. This ensures good control of v_o. It should also be pointed out that when v_o is negative, the op amp is not in control of v_o. Therefore, v_o should be connected only to an amplifier input or through a resistive load to common. Both the slope and the breakpoint of this circuit are easily adjusted. When V_R is zero or unconnected and when $R_{in} = R_f$, the precision limiter acts as an ideal diode that provides unity transfer for one polarity of v_{in} and zero transfer for the other. Reversing both diodes in the circuit makes the upper bound of v_o zero and changes the condition of validity of equation 7–3 to $v_{in} > -V_R R_{in} / R_R$.

Precision noninverting limiter. A similar limiter circuit based on the noninverting voltage follower is shown in Figure 7–24. When v_{in} is positive, the feedback path through D_2 and the 10-kΩ isolating resistor enables the op amp to establish a voltage essentially equal to v_{in} at the inverting input (within the error v_o / A). The voltage difference between v_o and the inverting input is the *IR* drop across the feedback resistor due to the op amp input current. Since this difference is negligibly small, $v_o = v_{in}$ for all positive values of v_{in}. For negative values of v_{in}, the feedback path is D_1, and D_2 is reverse-biased. This would at first appear to disconnect the circuit from R_L, as in the case of the inverting limiter of Figure 7–23. However, in this case, the op amp inverting input is not held at virtual common; it follows $-v_{in}$, and the voltage v_{in} is divided between R_L and the 10-Ω feedback resistor. Thus, this circuit should be used in conjunction with some other circuit that controls v_o at 0 V should it tend to become negative. The 10-kΩ feedback resistor in this circuit limits the feedback current that must be supplied by the lower bound control circuit. The actual resistance value is not critical.

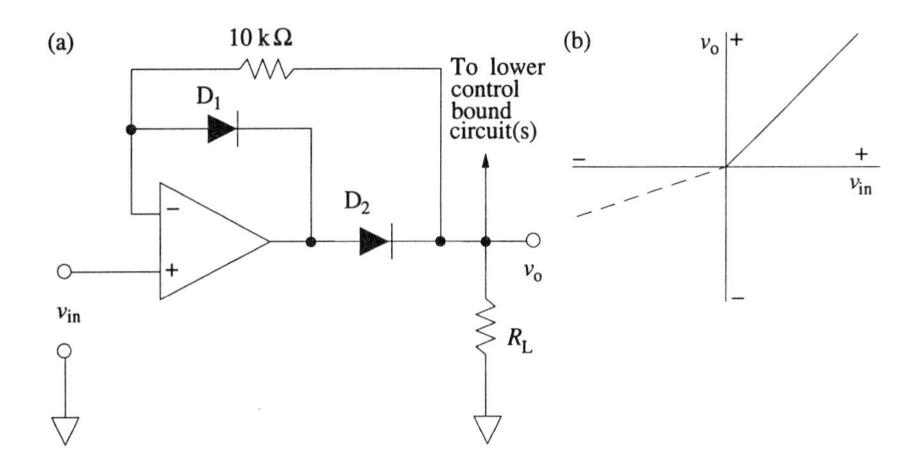

Figure 7–24. Precision noninverting limiter. (a) Circuit; (b) function curve. This circuit is based on the voltage follower. The output voltage v_o follows v_{in} precisely for all positive values. For negative values, v_o follows v_{in} loosely, with a gain of less than 1 (dashed line) unless it is controlled by another circuit.

These basic limiter circuits have been used in a variety of ingenious combinations to produce many useful nonlinear functions, one of which is the absolute value circuit.

Absolute value circuit. The precision limiters have been described and illustrated in Figures 7–23 and 7–24, and they are combined in Figure 7–25 to provide an absolute value circuit. The upper, noninverting limiter circuit establishes the voltage v_{in} at the output for all positive values of v_{in}. Similarly, the lower, inverting limiter establishes a voltage of $-v_{in}$ at the output for all negative values of v_{in}. The two resistors in the lower circuit should be carefully matched to ensure unity gain for both polarities of v_{in}. If all the diodes are reversed, the same circuit produces a negative v_o for either polarity of v_{in}.

The absolute value circuit is actually a precision full-wave rectifier circuit for use with analog signals. When the measurement information is encoded as the

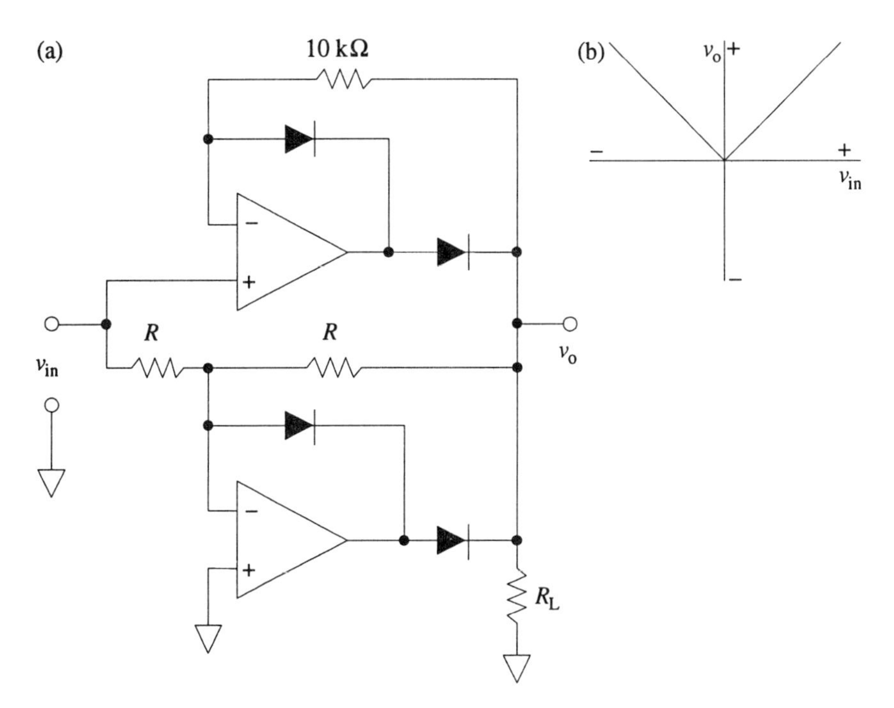

Figure 7–25. Precision absolute value circuit. (a) Circuit; (b) function curve. The inverting and noninverting limiter circuits are combined in this circuit. It is useful where signals must be unipolar (as at the input of a voltage-to-frequency converter) or when precision full-wave rectification of a small signal is required.

magnitude of voltage variations of both polarities, the absolute value circuit converts the signal variations to a proportional unipolar signal suitable for a dc voltage measurement system.

The absolute value circuit described above is a simple two-line-segment function, as its input–output function curve shows. Precision multiple-line-segment functions can be obtained by combining separate limiter circuits for each segment desired. This is most easily done by summing the outputs of inverting limiter circuits at the inputs of an op amp summing amplifier. Each limiter circuit provides an independently adjustable breakpoint and slope change. The initial signal itself can also be summed with this combination. In this way a great variety of functions can be implemented by precision line segment approximation.

Charge-Coupled Amplifier

The integrator circuit is clearly very useful for the measurement of charge or, more precisely, for the conversion of data from the charge domain to the voltage domain. Thus, it is particularly useful with transducers having an output in (or easily converted to) the charge domain. Because the charge on a capacitor is a function of the voltage across the capacitor and the capacitance value, both voltage and capacitance variations are easily converted into charge variations with a variety of transducers.

A common charge-generating and -measuring circuit is shown in Figure 7–26. Since one terminal of the capacitor with capacitance C_{in} is attached to v_{in} and the other terminal is held at virtual common by the op amp, the charge on the input capacitor is $q = C_{in}v_{in}$. If v_{in} varies by an amount Δv_{in}, the charge on the capacitor with capacitance C_{in} varies by

$$q = C_{in}\Delta v_{in} \tag{7–4}$$

This same change in charge will appear across the feedback capacitance C_f, so that

$$v_o = \frac{q}{C_f} = \frac{-\Delta v_{in}C_{in}}{C_f} \tag{7–5}$$

Thus, the input voltage variations appear at the output multiplied by the factor C_{in}/C_f. This **charge-coupled amplifier** is often used with biological transducers that produce voltage variations but have a limited charge supply capability. It is necessary to short out C_f occasionally or to bypass it with a large resistor to keep the output voltage from drifting off as a result of the long-term integration of low-level leakage currents.

The capacitance of a capacitor is directly proportional to the common area between its conductors and the dielectric constant of the insulator between the conductors and inversely proportional to the distance between the conductors. Because the capacitor is so sensitive to dimensional changes, it is often used as a transducer to measure parameters that cause physical motion. The motion can cause a variation in common area or distance between its conductors or a change in dielectric constant. Thus, there are capacitive transducers for pressure, velocity,

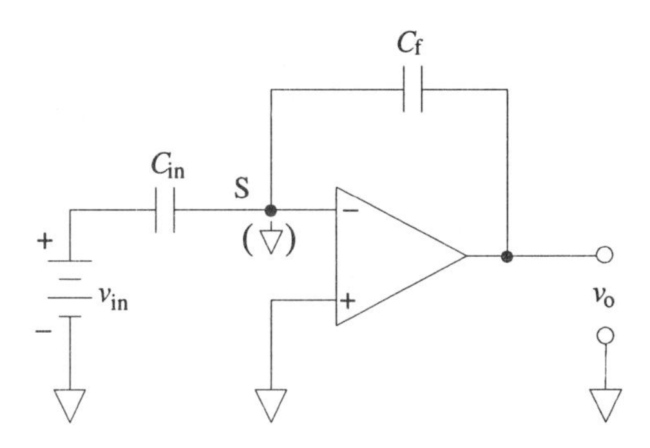

Figure 7–26. Charge-coupled amplifier. A change in the charge across C_{in} appears as a proportional change in the voltage v_o. This circuit is analogous to the inverting amplifier, but the use of capacitors for input and feedback elements makes it sensitive to charge rather than to voltage.

very small displacements, dielectric constants of materials, and other quantities to be measured or controlled.

If the capacitive transducer is used for either C_{in} or C_f in the circuit of Figure 7–26, the output voltage changes as the capacitance changes. Equation 7–4 can be rewritten to show that a change in charge Δq results from a change in capacitance ΔC_{in} with a constant input voltage v_{in}, so that $\Delta q = v_{in}\Delta C_{in}$. Thus, we have a linear **capacitance-to-voltage converter**, where

$$v_o = \frac{\Delta q}{C_f} = \frac{-\Delta C_{in} v_{in}}{C_f} \qquad (7\text{–}6)$$

Bridge Amplifier

Transducers that change resistance as a result of dimensional changes are used extensively in mechanical and biological measurements. These **strain gauges** are described in Chapter 8. The resistance change with strain is a very small fraction of the total resistance, so highly sensitive resistance-measuring techniques are required. Also, the technique must separate the changes in resistance attributable to temperature from those attributable to strain. These functions can be accomplished by the **bridge amplifier**, which can be applied to many other sensitive resistive transducers.

A basic Wheatstone bridge amplifier is shown in Figure 7–27. In the null operation of the bridge, resistances R_A, R_B, and R are standards used to determine the unknown resistance R_u. Resistance R is made variable and is adjusted until the

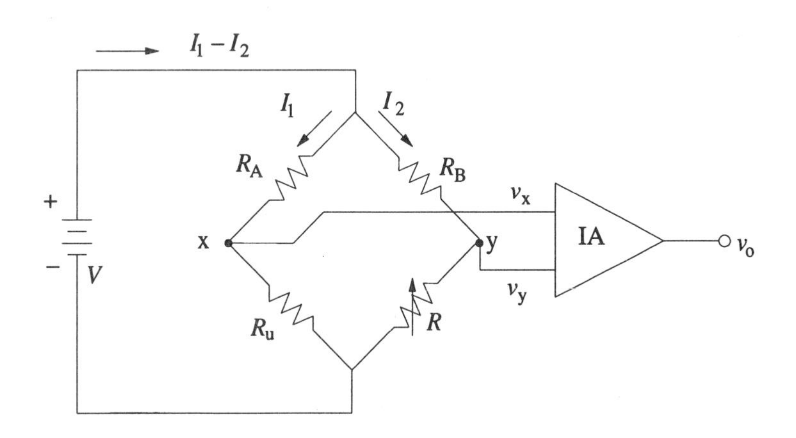

Figure 7–27. Wheatstone bridge amplifier. The value of R is adjusted to make the voltages at x and y equal, so that at balance, $R_u = R(R_A/R_B)$. Also, a pair of matched resistive transducers of resistances R and R_u (such as strain gauges) can develop a small resistance increment ΔR_u that produces a proportional off-balance voltage v_o. This voltage is readily amplified by the instrumentation amplifier (IA) to provide a readily measured signal.

null detector indicates that the bridge is balanced. At balance there is no current through the null detector; therefore, the current through R_A and R_u is I_1, and the current through R_B and R is I_2. Also, at balance the voltage between terminals x and y is zero. From the equality of IR drops on each side of the bridge, $I_1R_u = I_2R$ and $I_1R_A = I_2R_B$. These equations can be solved for R_u to give $R_u = R(R_A/R_B)$.

The unknown resistance R_u is calculated from the values of R, R_A, and R_B. Usually, the ratio R_A/R_B is a convenient factor such as 0.01, 0.1, 1, 10, or 100 and is referred to as the **multiplier** .

The major sources of error in a Wheatstone bridge are inaccuracies in the three standard resistances R, R_A, and R_B, but these can be made with errors as low as 0.001%. Other factors that could limit the accuracy are null-point accuracy, thermal voltages, and changes in resistance values due to heating by too-high currents.

The magnitude of a small off-balance signal from the bridge and amplified output can be used as a measure of the change in the resistance R_u. In this application, when matched devices are used for R and R_u or for R_A and R_u, the off-balance output obtained is related to the *difference* in the resistances of the two devices. A differential resistance measurement with matched transducers tends to cancel the effects of resistance changes caused by environmental effects other than the quantity sought. For example, with strain gauges, it is desirable to compensate for the temperature sensitivity of the gauge. If R_u and R are matched strain gauges subjected to the same temperature but different strains, the off-balance output voltage of the bridge and the amplified output will be due only to the difference in strains. That is, if R_A is much greater than R_u,

$$\Delta v_o = \frac{V\,\Delta R_u}{R_A} \quad (7\text{–}7)$$

ac Amplifiers

Use of op amps. The same op amp is applicable for both dc and ac signals. When it is important to amplify only the ac signal and reject the dc component, an input capacitor is used. This capacitor is illustrated in Figure 7–28 for an inverting amplifier. The capacitor serves to block dc from a transducer or another circuit and thus prevents the output from going to an unwanted dc level that can cause the op amp to go into saturation. Also, the capacitor helps decrease the amount of low-frequency noise being fed to the op amp input.

Figure 7–28. Inverting ac amplifier.

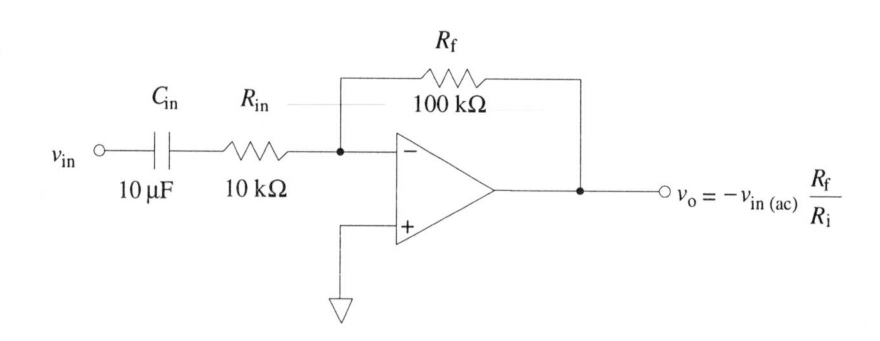

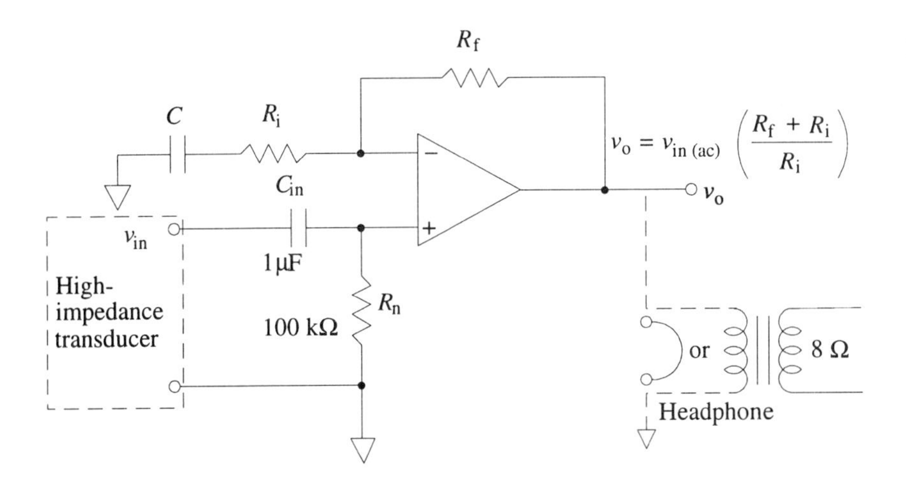

Figure 7–29. Noninverting (follower-with-gain) ac amplifier.

The input impedance of an inverting amplifier is R_{in}, so connecting a high-impedance transducer such as a high-impedance microphone is a problem. The problem can be overcome by connecting the transducer to a noninverting follower-with-gain ac amplifier, as shown in Figure 7–29. The input impedance is equal to approximately R_n. An op amp noninverting amplifier with a gain of about 100 could be used to drive a set of headphones. If a volume control is desired, the resistor of resistance R_n could be replaced with a 100-kΩ potentiometer connected as a variable voltage divider so that only a fraction of the input voltage is fed to the noninverting input. The impedance of the headphones should be at least 150 Ω. If a low-impedance headphone or speaker is to be used, it will be necessary to use an impedance-matching transformer at the output, as shown by the dashed lines.

ac Amplifier bandwidth. The combined high, low, and midrange frequency dependencies of an amplifier can be summarized by the log–log plot of Figure 7–30 (*see also* Supplement 2, Section S2–4). Here, the magnitude of the relative gain A/A_v in decibels is plotted against the log of frequency. In the midrange of frequency, where $A = A_v$, the relative gain is 0 dB. At $f = f_1$ and $f = f_2$, the relative gain is –3 dB. Beyond the 3-dB points, the gain rolls off with frequency, with an

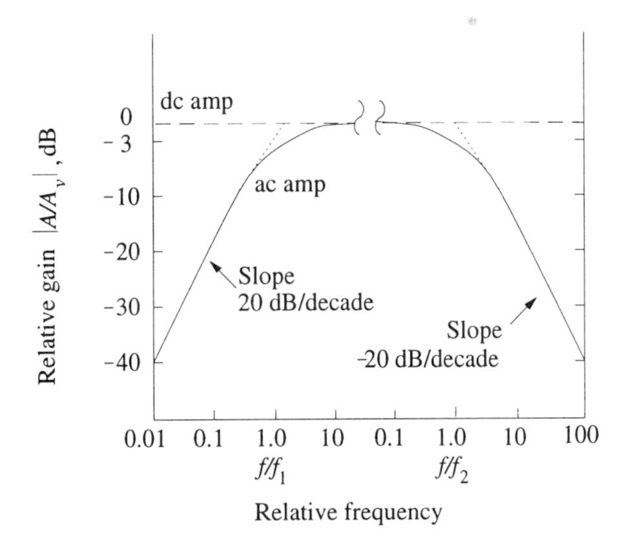

Figure 7–30. Amplifier frequency response.

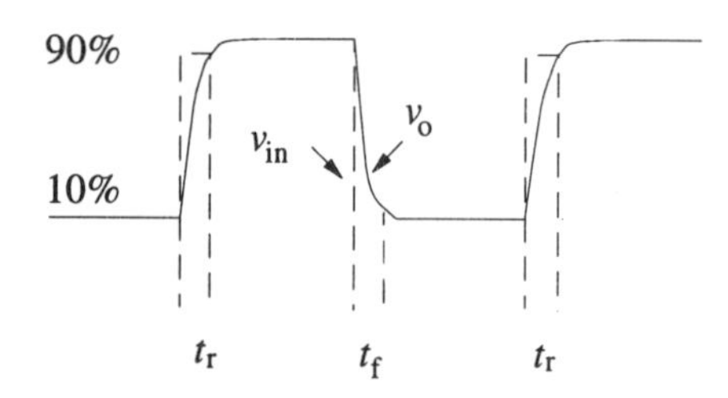

Figure 7–31. Distortion of square wave due to high-frequency response limit.

asymptotic slope of 6 dB/octave or 20 dB/decade. Such rolloff characteristics are merely those of the low- and high-pass networks. The **bandwidth** of an ac amplifier is usually taken to be $f_2 - f_1$, and although there is appreciable gain outside this frequency range, it should also be remembered that the attenuation at f_1 and f_2 is already 30%. Note that there is no low-frequency rolloff with dc amplifiers.

It seems reasonable to assume that if the frequency components of a signal fall within the bandwidth of the amplifier, the signal will be amplified without distortion. This assumption can be tested by the square wave. It has been demonstrated that the discontinuities of the square wave are made up of the highest-frequency components. Since the Fourier expansion of the square wave is a series that extends indefinitely to the high harmonics, a square waveform with an instantaneous transition from one voltage level to another contains frequency components to infinite frequency. A practical amplifier with a finite high-frequency limit distorts the signal, as shown in Figure 7–31. The ability of an amplifier to respond to instantaneous signal changes is measured in terms of the **rise time** t_r, the time required to go from 10 to 90% of the applied change. It is a useful rule of thumb that the rise time of an amplifier is approximately related to f_2 by

$$t_r = \frac{1}{3f_2}$$

An even more dramatic example of square-wave distortion is provided by the low-frequency limitations of the amplifier. The lowest-frequency term in the Fourier series expansion for a 1000-Hz square wave is the 1000-Hz fundamental. A high-pass filter with an f_1 of 1000 Hz has an RC time constant of $1/(2\pi f_1) = 160$ μs. However, a half-cycle is 500 μs long, and the amplitude of a square wave subjected to such a filter would drop to a small fraction of the initial step, as shown in Figure 7–32. Even when $f_1 = 30$ Hz and $RC = 5$ ms, there is a definite slope, or droop, to the square wave. The droop in the square wave can be detected on a scope

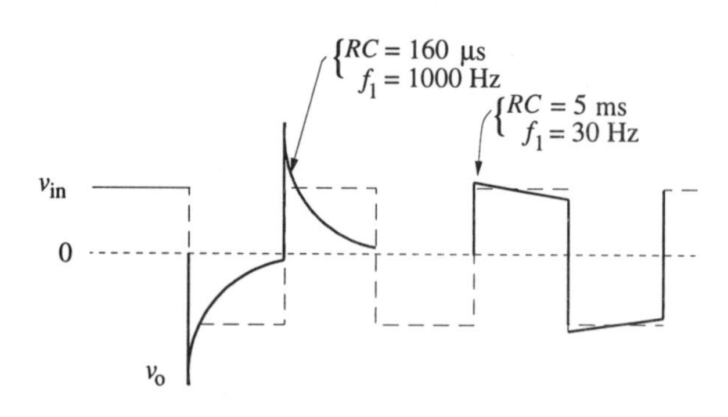

Figure 7–32. Response of a high-pass circuit to a square-wave signal.

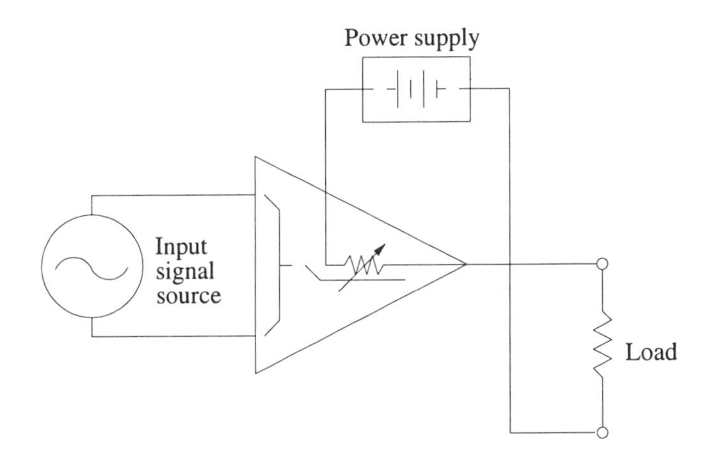

Figure 7–33. An amplifier as a signal-responsive control system. The input signal is used to control the delivery of power from the power supply to the load. If the voltage or the current in the load is greater than that of the input signal, amplification has occurred.

when the ac position of the input switch is used. When quantitative information is contained in the wave shape, careful attention must be paid to distortion caused by bandwidth limitations.

Amplifier: Signal-Responsive Control Systems

An **amplifier** is used when a load or readout device requires more current, voltage, or power than is available directly from the signal source. Signal amplification is accomplished by using the input signal to control the delivery of power from the amplifier's power supply to its output terminals, as shown in Figure 7–33. Amplification is thus seen to be not so much an operation *on* the input signal as the control of current, voltage, or power in response *to* the input signal. The power control elements used in amplifiers are generally solid-state devices such as bipolar transistors or field effect transistors. The amplification is a result of the small voltage or current required by the control element to control larger voltages or currents from the power supply. The controlled output voltage (or current) ideally follows some known and desired function related to the input signal.

The many amplifiers used in practice are aimed at meeting certain design requirements. Some possible design requirements are high-frequency response, which is the ability to provide the necessary control of power supply voltage and current even when the input signal varies at rates as high as 100 MHz or more; the ability to respond to a floating signal source, as with the differential amplifier; the ability to provide an unusually large output voltage or current, as with power or driver amplifiers; and the ability to operate with very low power, as for portable equipment.

No amplifier is ideal or universal, although the versatility of the modern op amp is very great. One must choose among the available options to meet the basic application requirements and then determine whether the deviations from ideality are within acceptable limits. The primary considerations with respect to the nonideality of op amps for several applications are illustrated in Section 7–4.

7–3 Limiting Bandwidth with Active Filters

Supplement 2 reviews how a simple passive RC or RL circuit is a frequency-dependent voltage divider. This feature can be used to favor the transmission or rejection of certain frequencies compared with signals of other frequencies. This filtering action is an essential part of many techniques that help in separation of unwanted signals (noise) from the information analog signals.

Filters generally fall into four categories: **low-pass** filters, which reject high-frequency signals; **high-pass** filters, which reject low-frequency signals; **bandpass** filters, which reject signals having frequencies either higher or lower than the desired band of frequencies; and **notch** (band elimination) filters, which pass all frequencies except a narrow selected band.

In a computer interface the analog circuits would include filters in some form. The signals from various transducers or other electronic circuits invariably contain unwanted information (noise) that must be rejected. It is also necessary to filter out the noise from external sources such as from power lines, electric motors, and communication transmitters.

First-Order Low-Pass Filters

Both passive and active low-pass filters are shown in Figure 7–34. The filter transfer functions are given in the figure. From the equation for the passive filter (Figure 7–34a, below), it can be seen that if the frequency is low, the output voltage v_o is essentially equal to the input voltage (unity gain), but when the frequency f_2 is equal to $1/(2\pi RC)$, the filter attenuates the amplitude by the factor 0.707 (the 3-dB points referred to as the upper cutoff frequency). At higher frequencies the filter attenuation increases to a limiting rolloff slope of –20 dB/decade.

The active first-order low-pass filters of Figures 7–34b (below) and 7–34c (on the next page) utilize the noninverting follower and inverting amplifier configurations, respectively. Both of these circuits have frequency response characteristics similar to those of passive low-pass filters, but both provide a gain of K, as shown by the transfer functions. Also, the active circuits make it possible to connect to a

Figure 7–34. Passive and active RC low-pass filters. (a) The passive circuits are frequency-dependent voltage dividers. (b) A buffer amplifier is added for the noninverting active filter. j, Complex operator that can be treated as though it had the value $\sqrt{-1}$ so that $j^2 = -1$; ω, angular velocity ($2\pi f$); K, gain.

(a)

R

C

v_{in}

$$v_o = v_{in}\left(\frac{1}{1 + j\omega RC}\right)$$

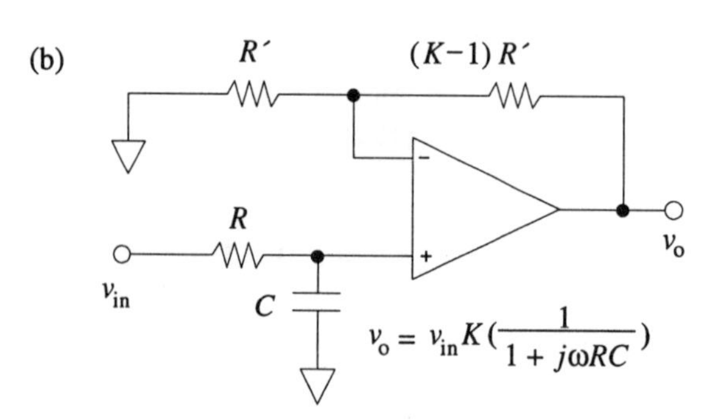

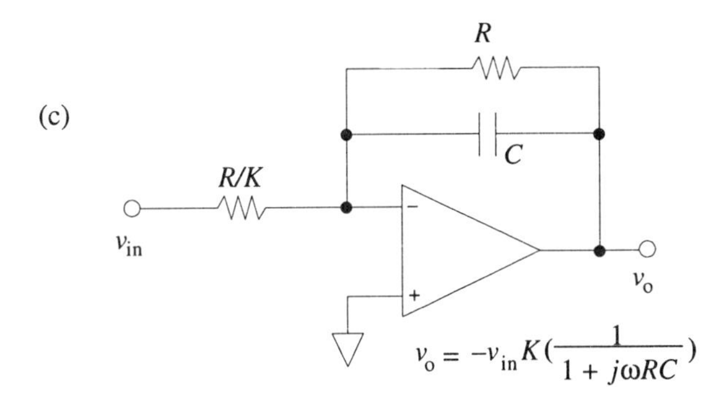

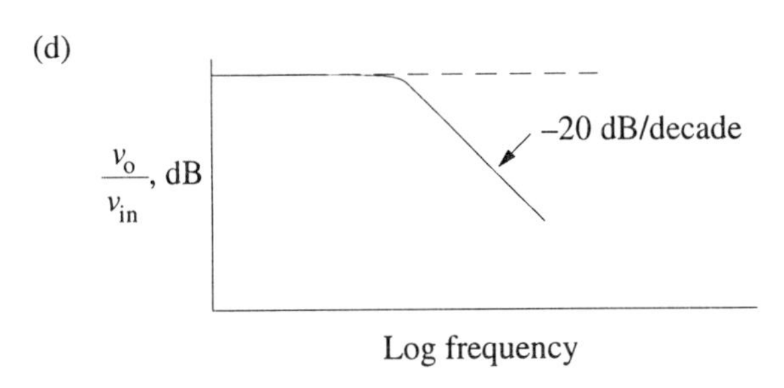

Figure 7–34—*Continued.* (c) The inverting active filter has a frequency-dependent feedback impedance. (d) Attenuation of output higher frequencies.

source and load without the problem of loading the filter, as is the case with the passive filter.

Second- and Higher-Order Active Low-Pass Filters

Filters with sharper rolloff characteristics and a better compromise between noise bandwidth and response time can be achieved with multiple time-constant active RC circuits or with RLC (resistor–inductor–capacitor) circuits. Since inductors are bulky and expensive, the multiple active RC circuits are normally preferred.

A second-order filter can be obtained by cascading two passive RC networks. However, filters from cascaded RC networks are not very practical, because the stages interact, and it is much more suitable to use active op amp second-order filters. A simple multiple-feedback low-pass filter is shown in Figure 7–35a (below). Note that the circuit is a combination of an input low-pass filter and the inverting feedback filter of Figure 7–34c. It has a rolloff of –40 dB/decade at high frequencies compared with –20 dB/decade for the first-order filter. The noise

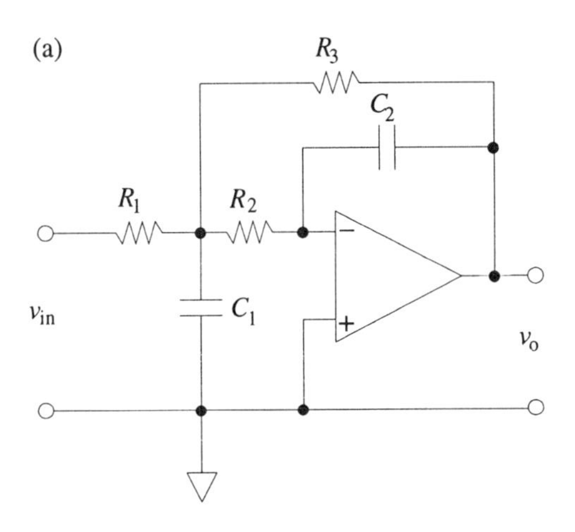

Figure 7–35. Second-order active low-pass circuits. (a) Multiple-feedback circuit that is a combination of an input low-pass filter with inverting feedback filter.

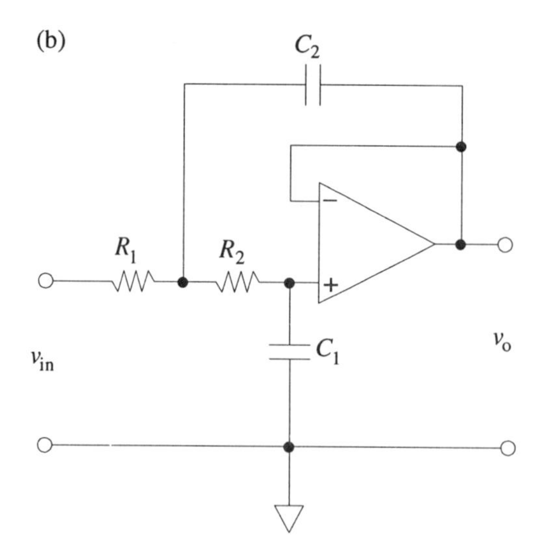

Figure 7–35—*Continued.* (b) Low-pass filter using the noninverting configuration.

bandwidth is one-half the value for the first-order filter, and the response time is also improved.

A second type of second-order filter can be obtained by using the noninverting configurations of the op amp as shown in Figure 7–35b.

To obtain even higher order filters, second-order and/or third-order filters are cascaded together. Even-order filters need only cascaded second-order filters, whereas odd-order filters are achieved with second-order sections combined with one third-order section. A sixth-order filter, for example, is achieved by three cascaded second-order active filters, whereas a fifth-order filter is achieved with one second-order section and one third-order section. Filter orders up to 10 are readily achieved with op amp active filters.

High-Pass Active Filters

High-pass active filters are also easily implemented with op amp circuits. Both multiple-feedback and noninverting configurations are shown in Figure 7–36 along with a passive high-pass filter. The transfer functions for the active filters are also given. From these equations it can be seen that when the frequency is low, the output is essentially zero, but when the frequency is high, the output is essentially equal to the gain K of the amplifier. Thus, the response is just opposite that of the low-pass filter, as illustrated by the Bode diagram in Figure 7–37. The Bode diagrams for the high-pass active filters are mirror images of those for low-pass filters. This is also similar to the voltage divider concept of the RC filter; if the applied voltage is not across the capacitor, it is across the resistor.

Second-order high-pass active filters are shown in Figure 7–38. An input high-pass network is combined with a high-pass feedback network to give a second-order response.

Narrow-Bandpass Filters

For many electronic measurements it is necessary to have circuits that respond to or reject a narrow band of frequencies or, as much as possible, a single frequency. With these circuits, tuned amplifiers and notch or band rejection amplifiers can be

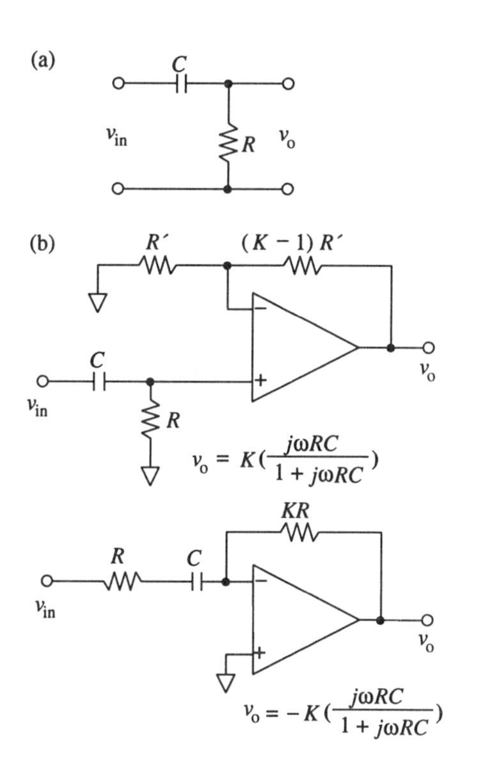

Figure 7–36. Passive (a) and active (b) high-pass filters. The noninverting circuit combines the passive RC circuit with a follower-with-gain buffer amplifier. The inverting circuit uses a frequency-dependent input impedance to produce a frequency-dependent gain.

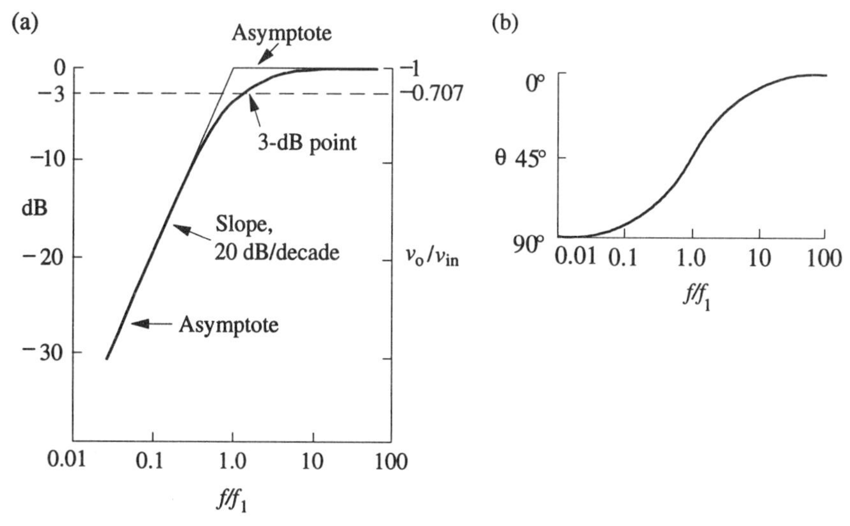

Figure 7–37. Bode (a) and phase shift (b) plots for a first-order high-pass filter. The attenuation in decibels or the phase shift in degrees is plotted against the logarithm of the frequency relative to the cutoff frequency.

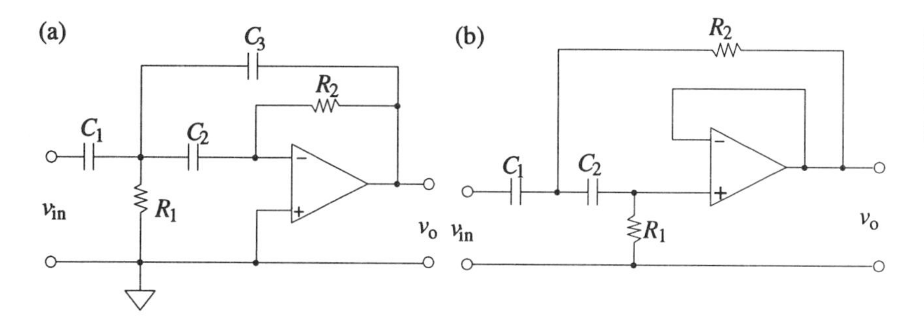

Figure 7–38. Second-order active high-pass filters. (a) Multiple-feedback configuration; (b) noninverting configuration.

realized by combining both low- and high-pass networks with amplifiers to form active filters. Simple first-order tuned systems based primarily on the twin-T

Figure 7–39. (a) Twin-T network; (b) frequency response of twin-T network.

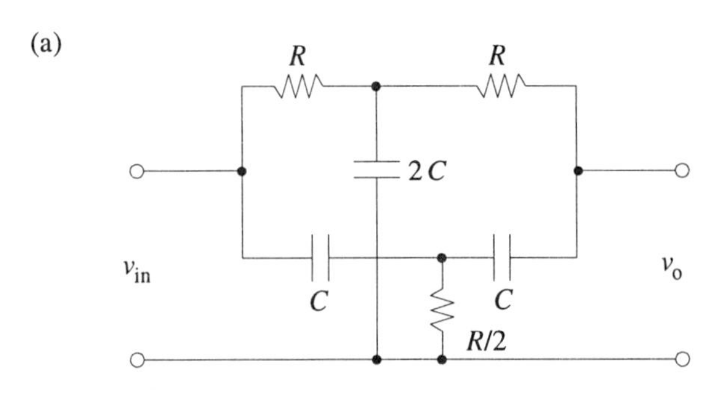

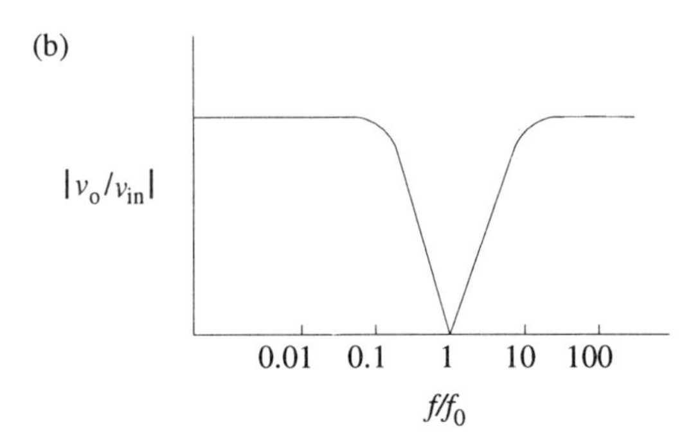

network will be emphasized, because practical circuits are readily implemented. Higher-order narrow-band filters can also be used.

Twin-T Filter

A practical frequency-selective circuit is the twin-T network shown in Figure 7–39a. Note from its response function (Figure 7–39b) that it is a rejection filter. The circuit is a combination of a low-pass filter and a high-pass filter arranged in the twin-T manner that gives the network its name. The rejection frequency f_o equals $1/(2\pi RC)$. For an input of this frequency, the impedance is very large and may be as high as 2000 MΩ if the components are well matched.

Filters Applied to Tuned Amplifiers and Oscillators

Fixed-frequency amplifiers are very effective in rejecting noise frequencies outside the narrow selected band, as illustrated with the twin-T amplifier. It is desirable in certain applications to be able to tune or vary the resonant frequency of a tuned amplifier. A simple network for tuning a narrow-band amplifier is described below. In other applications it is necessary to reject a specific noise frequency, such as 50- or 60-Hz power line noise, and to amplify lower and higher frequencies. This type of narrow-band rejection filter is also illustrated.

Fixed-frequency tuned amplifier. A tuned amplifier can be constructed by inserting a twin-T filter in the negative feedback loop of an op amp, as shown in Figure 7–40. The voltage gain of an op amp is $v_o/v_{in} = -Z_f/Z_{in}$. The feedback impedance Z_f for the amplifier shown in Figure 7–40 is determined by the parallel combination of R_f' and the impedance of the twin-T network. If Z_f was determined

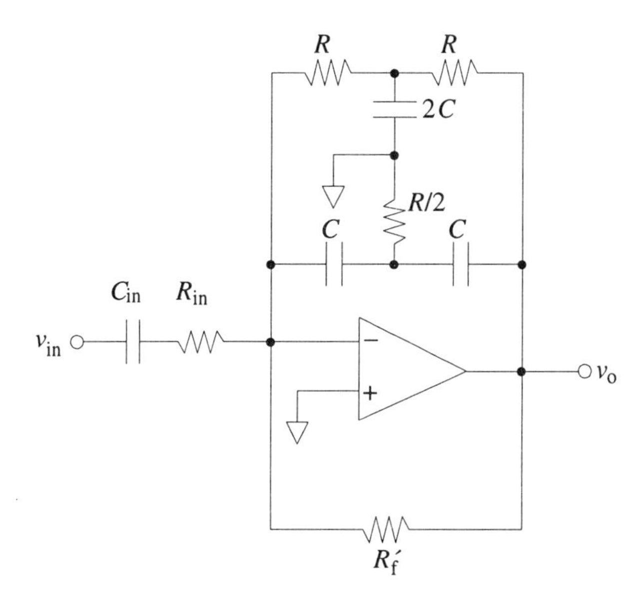

Figure 7–40. Tuned amplifier using a twin-T network.

solely by the twin-T network, the gain of the amplifier would be too large at the frequency output f_0 of the network, and the circuit would be very unstable and subject to oscillation. This problem is averted by addition of a conventional negative resistive feedback loop R_f' to control the gain at the tuned frequency to approximately R_f'/R_{in}. For frequencies other than the resonant frequency, the impedance of the twin-T network is low, and thus the gain of the amplifier is also low. The tuned amplifier is used in various types of lock-in amplifiers.

Variable-frequency tuned amplifiers. It is often useful to be able to tune or vary the resonant frequency of a tuned amplifier without redesigning the complete network. A simple network in which tuning can be accomplished by adjustment of a single component is shown in Figure 7–41a. The rejection frequency of this circuit is shown in the figure, where a is the fractional setting of the potentiometer $R/2$. The tuning range is not large. With a basic frequency $1/(2\pi RC)$ of about 800 Hz, the filter can be tuned up to about 4000 Hz. To tune over a much larger range, it is necessary to use a filter in which the values of the components are varied. Such a filter is shown in Figure 7–41b. It is a tunable twin-T filter. The two 10-kΩ potentiometers must be ganged together. The base frequency is the same as for the twin-T filter $[1/(2\pi RC)]$. With an f_0 of about 2000 Hz, this can be tuned from about 200 Hz to 20 kHz by using the ganged 10-kΩ potentiometers.

Notch amplifiers. A notch or narrow-band rejection amplifier can be implemented by inserting the twin-T network in the input circuitry of an op amp, as shown in Figure 7–42. A normal input resistance R_{in} is required in series with the twin-T network to maintain reasonable gains for those frequencies outside the rejection band. Any of the tuned and tunable networks mentioned in the last section could be used in place of the twin T. A band rejection filter could also be implemented by buffering the twin T with followers, as shown in Figure 7–43. This avoids disturbance of the characteristics of the twin-T filter by loading. Notch filters are very useful for rejecting specific interference noise frequencies that may occur in a measurement system.

Figure 7–41. Tunable twin-T networks. (a) Simple network; (b) tunable twin-T filter.

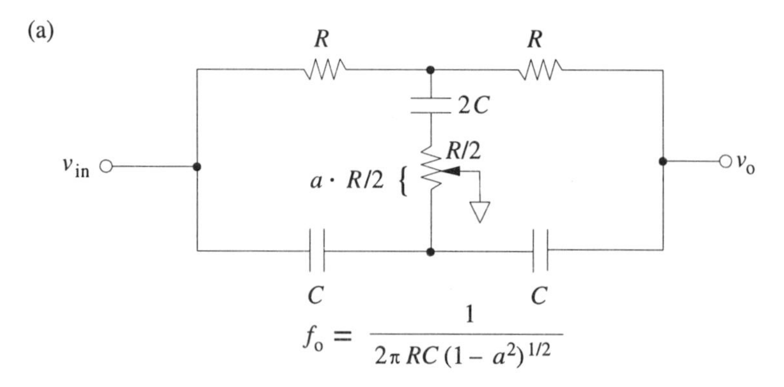

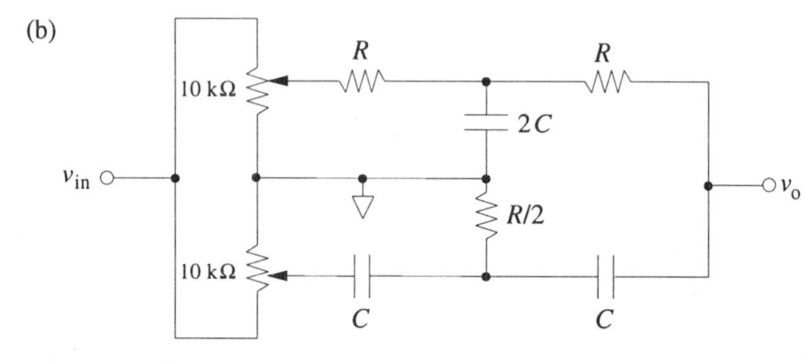

Figure 7–42. Notch amplifier using a twin-T network.

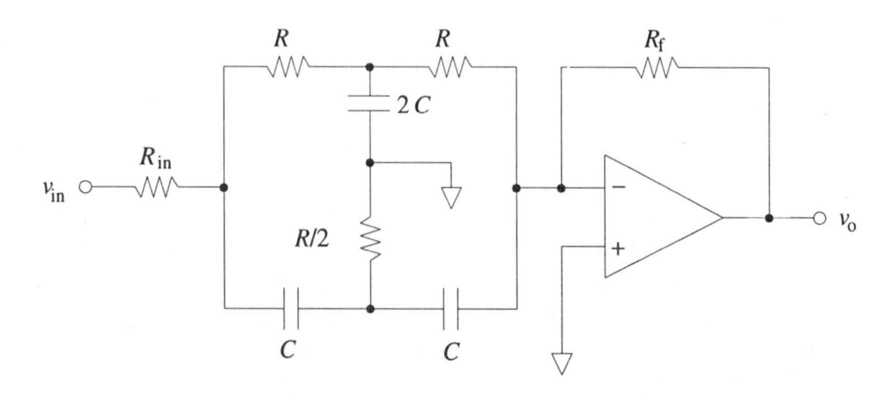

Figure 7–43. Buffered twin-T rejection filter.

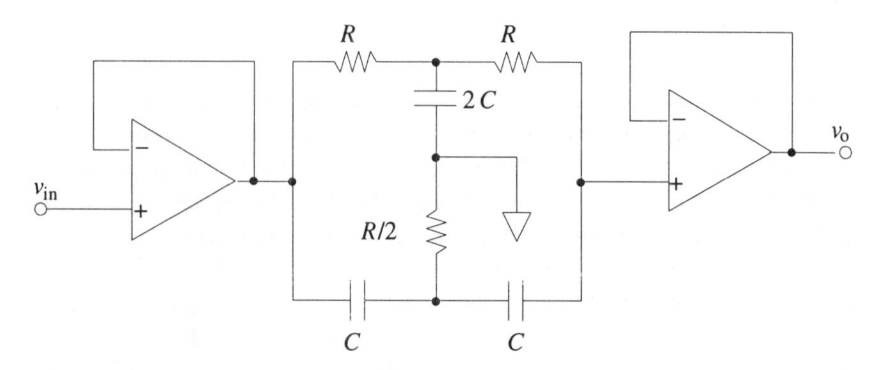

State variable filter/tuned amplifier. An especially versatile and easy to use op amp active filter is the **state variable filter.** Several manufacturers of analog circuits offer these filters as compact IC units for specific applications, including bandpass and notch filters and tuned amplifiers.

One type of state variable, second-order filter provides three outputs simultaneously: high pass, low pass, and bandpass. The cutoff and resonant frequency can be changed by varying the two external resistors of resistance R or the two capacitors of capacitance C or both. The cutoff frequency is $\omega_c = 1/RC$. Two second-order filters can be cascaded to provide a fourth-order rolloff of 80 dB per decade.

Variable tuning of the state variable filter can be obtained by varying the two external Rs with ganged potentiometers. Also, if digital control for tuning is desired, the two Rs can be replaced by multiplying DACs of the type described in Section 7–5.

Twin-T Oscillator

A twin-T network in the negative feedback loop of an op amp is used to make a tuned amplifier. Because the gain of the op amp becomes very large at the resonant frequency of the twin T, a normal resistive feedback is necessary to prevent oscillation. Without this feedback resistor, oscillation would occur, because the gain of the amplifier is very large for the rejection frequency, and the small fraction of the rejection frequency that is passed by the twin T is actually positively fed back because of the 180° phase shift of the amplifier and the 180° phase shift that the twin-T network gives to the rejection frequency. A more controlled oscillation can be provided by the circuit modifications, which are shown in Figure 7–44. The biased diode limits the output so that the amplifier does not saturate during oscillation.

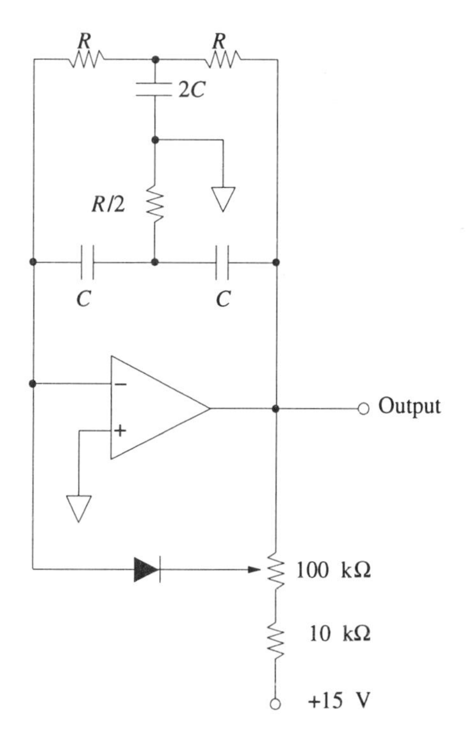

Figure 7–44. Twin-T oscillator.

7–4 Examining Nonideality of Operational Amplifier Circuits

The op amp is in many ways an "ideal" gain block. However, there are practical **specifications** of commercial op amps that determine the reliability and nonideality of the circuits in which they are used. Although most of these have been described earlier in this book, they are summarized here as a useful reference. The specifications can be more fully appreciated now, after the many types of op amp circuits have been studied. They are illustrated here with a few types of op amp circuits.

Summary of Op Amp Specifications

The major op amp specifications include the following.

1. *Open-loop gain.* The open-loop gain is the gain without feedback expressed in decibels (dB). Since gain = 20 log (v_o/v_{in}), an open-loop gain of 100 dB means that A = 100 000.
2. *Frequency response.* The frequency response characteristic reflects the variation in the open-loop gain with frequency. It is normally given as the 3-dB point or the **unity gain bandwidth**.

The relationship between the increase in frequency response of an amplifier with feedback compared with that of one with an open loop is shown in Figure 7–45 for both ac and dc amplifiers. For the ac amplifier (part a) the midfrequency gain is A, and the amplification rolls off at 20 dB/decade. If a fraction of the output B is fed back in such a way that a closed-loop gain A_f results, it is apparent that the bandwidth has been extended. The same improvement in bandwidth can also be

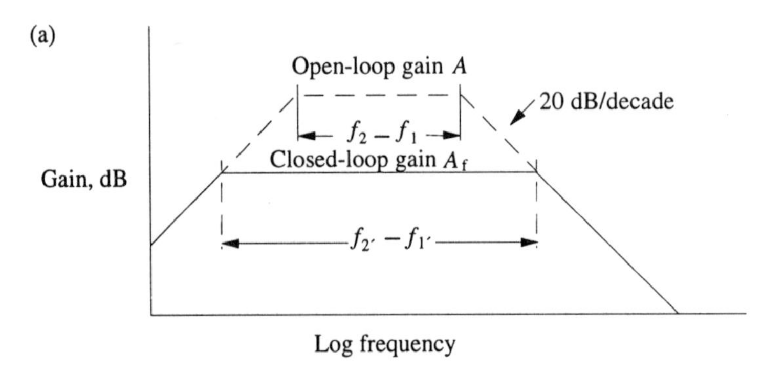

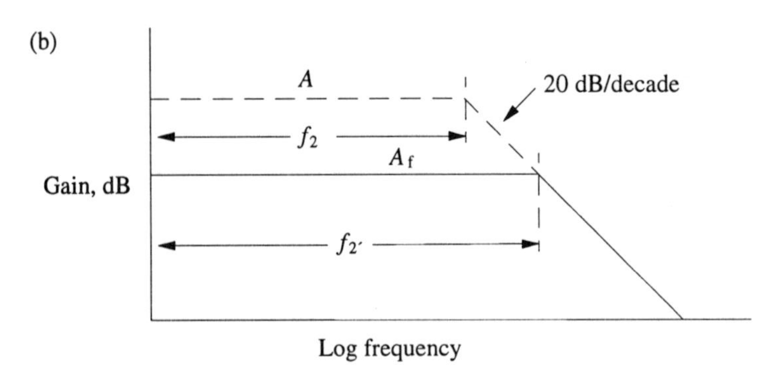

Figure 7–45. Frequency responses of feedback ac (a) and dc (b) amplifiers. Note the increased bandwidth with negative feedback.

seen in Figure 7–45b, where the bandwidth has been extended from f_2 to f_2'. The product $A_f f_2'$ is a constant and is often called the **gain–bandwidth product**.

3. *Slew rate.* The slew rate of an op amp indicates how fast the output voltage can change. Within the op amp there are capacitances inherent in the semiconductor devices and leads used in the construction of the op amp. The rate of change of voltage at each point in a circuit is limited by the available current to charge the capacitance at that point. The consideration is the same as for the response of the RC circuit to a step function, as discussed in Supplement 2. The maximum rate of change of voltage $(dV/dt)_{max}$ equals I_{max}/C. Therefore, for example, if a capacitance of 10 pF is to change voltage by 10 V in 1 ns, then the available current at that point must be $I = (10/10^{-9})10^{-11} = 100$ mA.
4. *Settling time.* In today's world of high-speed data acquisition and processing, many measurements are made in a microsecond or less. Therefore, the time required for an electronic feedback control circuit to perform its function with a specified accuracy is important and is referred to as **settling time**. For op amp feedback circuits the settling time is the time that elapses between the application of an ideal instantaneous step input and the instant at which the closed-loop amplifier output has entered and remained within a specified error band, usually symmetrical about the final value. The settling time includes any propagation delay in the circuit plus the time required to slew to the final value, recover from any overshooting of the final value associated with slewing, and finally settle within a certain specified accuracy for the final value.

Note that the settling time is associated with a closed-loop op amp system, that is, an op amp circuit with feedback, and it cannot be predicted from open-loop specifications for the op amp, such as slew rate or small-signal bandwidth.

5. *Input offset voltage.* Because the input differential amplifier is not perfectly balanced, there is a small, relatively constant but temperature-dependent offset voltage between the input terminals. The offset voltage causes an output voltage when both inputs are at 0 V. An external-balance potentiometer may be required to nullify the offset voltage and reduce offset voltage errors. The offset voltage changes with temperature, supply voltage, and time.
6. *Input bias current.* Even when the input voltage is zero, there is an input current in each input terminal that results from currents in the op amp input stages and any leakage currents within the amplifier. Typical bias currents for good op amps are in the range 0.1 to 100 pA at 25 °C. The bias current is quite sensitive to temperature. The difference between the two input bias currents is called the **input offset current.**
7. *Input resistance.* Modern op amps have extremely high input impedances, typically in the range 10^{12} to 10^{15} Ω. An amplifier with 10^{14}-Ω input resistance can keep the input current in the subpicoampere range for full common-mode voltage swings of 10 V.
8. *Input noise.* The input noise of the amplifier is the factor that limits signal resolution. It varies as a function of source impedance and frequency. Graphs of typical changes of input noise with source resistance are usually available from the manufacturer.

Transfer Functions and Errors

The transfer functions shown in Figure 7–46a and 7–46b are those of amplifiers with zero **offset voltage**, that is, those in which v_o is zero when v_{in} is zero. Figure 7–46c shows the transfer functions for several positive and negative values of offset. Most amplifiers have some offset, either unintentional (caused by circuit characteristics or component instability) or intentional, to compensate for an offset inherent in the voltage source. A "zero" or "balance" adjustment is often provided to reduce offset to within the offset stability of the amplifier.

The voltage offset characteristic of an op amp is given in terms of the input signal amplitude required to produce a zero output voltage. The offset voltage contributes directly to the error in the null point maintained by the op amp through the feedback circuit. Many op amps have less than 1-mV offset with no balance control required. With a balance adjustment, the offset can be reduced to the level of the offset instability, often about 10 μV and in some cases as low as nanovolts. Other contributions to input errors are the **input resistance** and the **input bias current**. The input resistance of a modern op amp is typically 10^{12} to 10^{15} Ω. The current through the input resistance should be almost infinitesimal. However, all transistors and other control elements have some current leakage into the control input. This leakage results in a current at the op amp input terminals even when

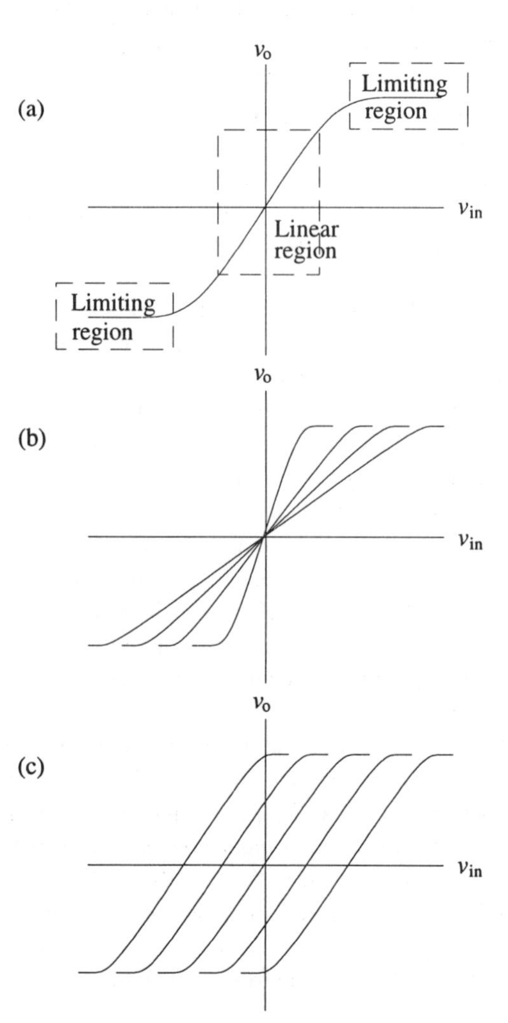

Figure 7–46. Amplifier transfer functions showing linear and limiting regions (a) for various gains (b) and various offsets (c). In the linear region, the change in v_o for a given change in v_{in} (the gain) is constant. A change in gain affects the slope of the input/output curve. The offset is equal to the value of v_{in} that produces a v_o of zero.

the input voltage is zero. The input bias current is of the order of nanoamperes for inexpensive op amps and as low as 10^{-15} A for special low-current types. This current creates an *IR* drop in the resistors connected to the op amp input, and this voltage appears at the amplifier input. A different *IR* drop in the positive and negative input circuits results in an additional source of input offset voltage. If the resistances to the positive and negative op amp inputs are made equal, the *IR* drops approximately balance, and the important characteristic is then the **difference** in input currents, called the **input offset current.** In any application in which the op amp input voltages are not at very nearly the common (or at least constant) voltage, the common-mode response of the amplifier can contribute a significant error. The evaluation of the **errors due to offset, bias current, and CMRR** is illustrated below for a few of the op amp applications discussed in this chapter.

Voltage follower amplifiers. Ideally, the voltage follower output voltage v_o exactly equals the input voltage v_u. As indicated in Figure 7–47a, the error, or difference $v_o - v_u$, is equal to v_s. The maximum value for the error is the sum of (1) the input difference required to produce v_o volts of output, (2) the common-mode error, and (3) the offset voltage. Assuming a maximum output voltage of 10 V, an amplification of 10^6, a CMRR of 10^5, and an offset voltage v_{off} of 100 μV, the maximum difference between v_o and v_u would be $10/10^6 + 10/10^5 + 10^{-4}$ V. Since the common-mode voltage is equal to v_u, the CMRR is a critical characteristic for the voltage follower application. Furthermore, the CMRR rating of some op amps decreases as the common-mode voltage increases. The figure used for v_{off} should include the voltage error caused by the input bias current.

The follower-with-gain circuit is shown in Figure 7–47b. The amplifier gain $1/b$ is $(R_1 + R_2)/R_2$. Again, the difference between bv_o and v_u, or the **input error**, is equal to the sum of the amplification error v_o/A, the common-mode error bv_o/CMRR, and the offset voltage v_{off}. The **output error** is equal to the input error times the amplifier gain $1/b$, since the total input signal is amplified by this factor. For the same amplifier used in the follower error illustration and with a $1/b$ gain of 100, the maximum input error is $10/10^6 + 10^{-1}/10^5 + 10^{-4} = 1.1 \times 10^{-4}$ V, and the maximum output error is 1.1×10^{-2} V.

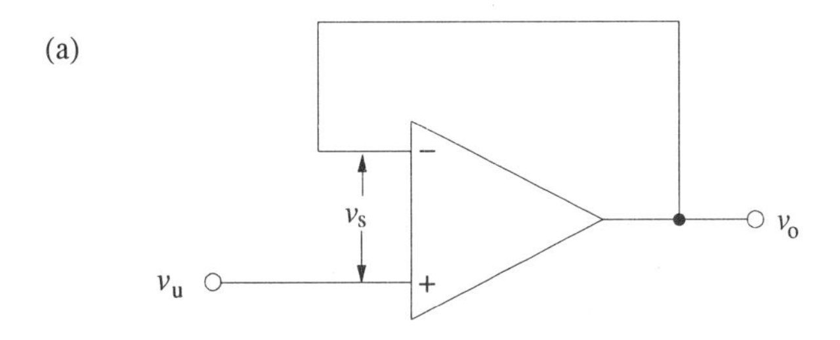

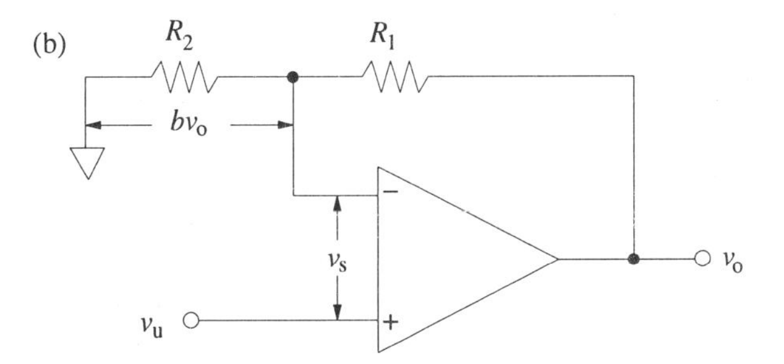

Figure 7–47. (a) Op amp voltage follower; (b) follower with gain. Errors in v_o result from the magnitude of v_s required by the gain, the common-mode rejection error (CMRR), and the input offset error.

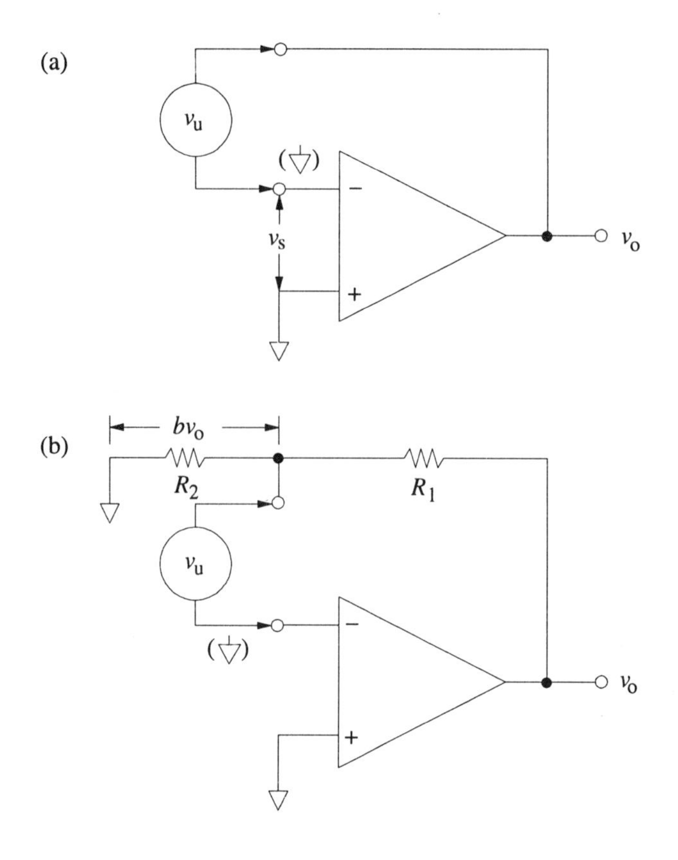

Figure 7–48. Floating signal source comparison measurement. (a) Voltage follower; (b) follower with gain. This configuration eliminates common-mode error and decreases the loading of v_u.

The CMRR error is peculiar to the floating null detector configuration for voltage null comparison circuits. It is possible to use the op amp in a floating unknown source configuration, as shown in Figure 7–48a. Note that v_{off} now contributes the major source of error. Since the common-mode input voltage does not change, there can be no error due to a finite CMRR. Because both op amp inputs are essentially at the circuit common, there is negligible voltage applied to the amplifier input resistance, and the loading of the signal source is greatly reduced. The floating signal source configuration is superior to the more common voltage follower circuit when the signal source does not require a direct connection to the op amp circuit common. The same considerations apply to the floating signal source follower with gain circuit shown in Figure 7–48b.

Current follower amplifiers. The range of currents that can be converted to voltages by the op amp current follower is limited on the high end by the output current limit of the op amp, which must supply both i_f and the current to the output load. On the low end it is limited by the op amp input bias current i_b, by v_s, or by the maximum practical value for R_f. The best way to reduce i_b is by the proper choice of op amp, but a steady value of i_b can be compensated for by applying a constant source of current opposite to i_b to point S. The value of v_s compared with that of $i_f R_f$ is sufficient to cause appreciable error only when a low-gain amplifier is used or when the output voltage v_o is very low.

The maximum practical value of R_f is limited by the state of the art in producing and using stable, accurate, high-value resistors. A circuit that has the effect of a high R_f but is made of lower-value resistors is shown in Figure 7–49. The output voltage is divided by R_2 and R_3, and the resulting fraction of v_o is applied to R_1 to

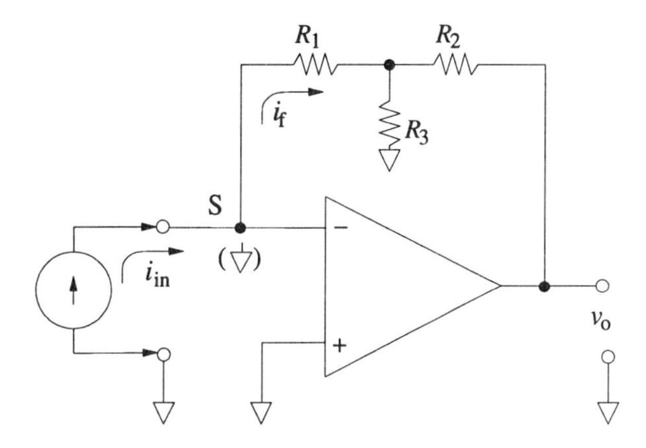

Figure 7–49. Circuit for high equivalent R_f. Three moderate resistances respond as a feedback resistance of much higher value. R_3 must be smaller than R_1 or R_2.

provide i_f. The resistance R_1 is a load on the R_2–R_1 divider. If point S is assumed to be at common potential, R_1 and R_3 are effectively in parallel, and the equivalent resistance $R_{f(eq)}$ between the output and point S is derived from

$$R_{f(eq)} = \frac{R_1R_2 + R_2R_3 + R_1R_3}{R_3} \tag{7–8}$$

To produce a larger equivalent resistance, R_3 must be smaller than either R_1 or R_2. For instance, for $R_1 = R_2 = 10^6\ \Omega$ and $R_3 = 10^3\ \Omega$, $R_{f(eq)} = 10^9\ \Omega$. This circuit should be used with the caution that the maximum value of v_s due to op amp gain, offset, and drift limitations must remain much smaller than the voltage at the junction of the three resistors. In the example given, this voltage is only $v_o/1000$.

Error considerations in other current follower-based op amp circuits are similar because all input circuits are sources of current for the basic current follower circuit. For the inverter and voltage-summing amplifiers, there is an error in i_u if v_s is not negligible compared with v_u. In the integrator, the bias current is also integrated and is often the limiting factor in the length of time over which an integration is accurate.

Comparator Considerations

The ideal comparator would have infinite gain and would change output states instantaneously. In reality, comparators have limitations due to their finite open-loop gains, their response times, and their input characteristics (bias currents, offset voltages, and CMRR). The finite open-loop gain means that for a small range of input voltages the comparator is in the linear region between its output limits. This can cause a small uncertainty in the output transition time, which for slowly changing signals depends on the rate at which the input signal traverses the linear region. For example, a comparator with logic-level outputs of +5 and 0 V and an open-loop gain of 2500 has a threshold window of 5 V/2500 = 2 mV. If the input signal changes at a rate of 10 V/s through the window region, the transition time is uncertain by 2 mV/10 V s^{-1} = 200 μs. For faster signals the transition time becomes limited by the basic response time and **slew rate** of the amplifier.

In addition to the uncertainty caused by response time and input signal rate, any noise present in the comparator, the reference voltage, or the signal will cause an additional time uncertainty, or jitter, in the comparator output. If the signal noise

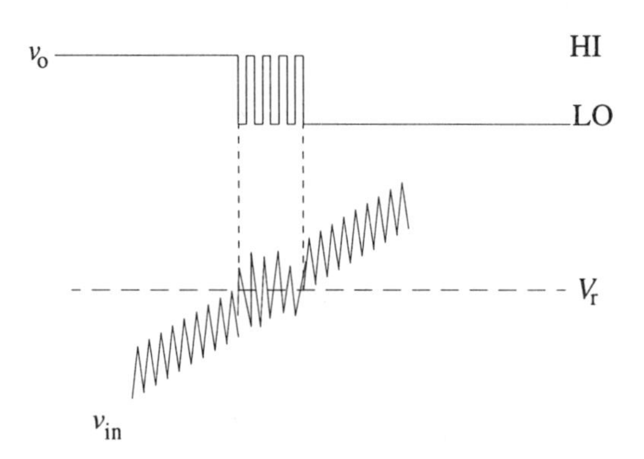

Figure 7–50. Multiple triggering of comparator by noise. This effect can be minimized by reducing high-frequency noise, increasing the rate of signal change through the threshold, or using a Schmitt trigger.

is large enough to cause the threshold region to be crossed several times, the comparator output flips from one state to the other until the difference between the signal voltage and the reference voltage is larger than the noise, as shown in Figure 7–50. This flipping causes unwanted logic-level pulses, which could cause serious errors in the digital circuits. The Schmitt trigger described below can eliminate this effect and provide faster output transition times.

The Schmitt trigger. **Hysteresis** is used to make the threshold level for a LO–HI transition different from that for a HI–LO transition in the Schmitt trigger. The Schmitt trigger and its transfer characteristics are shown in Figure 7–51. The positive feedback loop makes the threshold voltage dependent on the comparator output state. If the positive feedback loop was absent, the comparator threshold level would be the reference voltage V_r. In the presence of positive feedback, the threshold for the HI–LO output transition is increased to $V_r + \Delta V/2$ and that for the

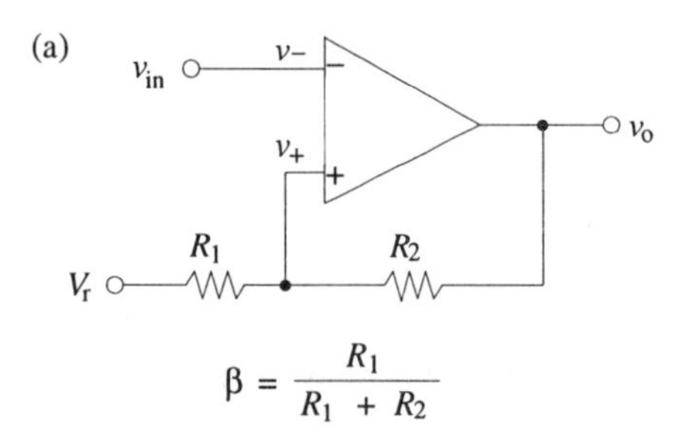

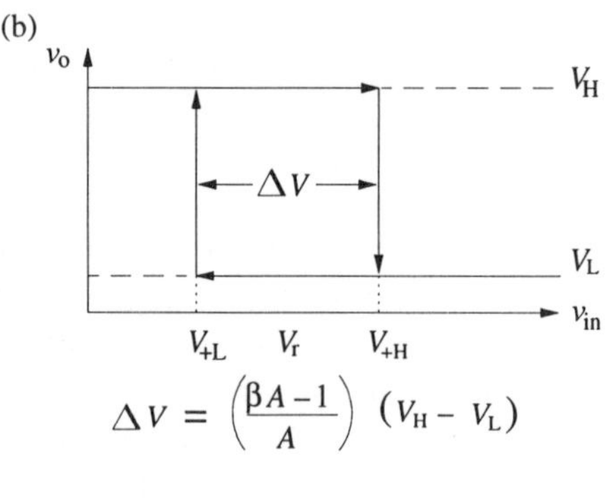

Figure 7–51. Schmitt trigger. (a) The Schmitt trigger circuit uses positive feedback to provide hysteresis to the threshold level. (b) The amount of hysteresis V depends on the feedback fraction $\beta = R_1/(R_1 + R_2)$, the open-loop gain A of the comparator, and the difference between the HI and LO logic-level voltages $V_H - V_L$.

LO–HI output transition is decreased to $V_r - \Delta V/2$. The amount of hysteresis or **hysteresis lag** ΔV is given by

$$\Delta V = \left(\frac{\beta A - 1}{A}\right)(V_H - V_L) \qquad (7\text{–}9)$$

where β is the feedback fraction and A is the open-loop comparator gain.

With a Schmitt trigger the speed of the transition is determined by the response time of the amplifier itself and not by the time required for the input signal to pass through the threshold region. For this reason the Schmitt trigger finds extensive use where input signal changes are slow and it is necessary to have very fast output transitions, as for many digital applications.

7–5 Digital-to-Analog Conversion

A digital-to-analog converter (DAC), as the name suggests, is a circuit that converts a digital signal into an analog electrical quantity directly related to the digitally encoded number. Because the input quantity is a series of bits, each with its own weight, the basis of all the conversion techniques is conversion of each bit in a number to the number of units of current or voltage corresponding to its weight and then summation of the units from each bit in an analog-summing circuit.

Digital-to-Analog Converters

Conceptually, the simplest form of the DAC is the weighted-resistor DAC shown in Figure 7–52. Resistance values for a binary-coded converter are shown. The magnitude of the current generated by each bit is directly proportional to the reference voltage V_R and inversely proportional to the resistance. If R is the resistance necessary to generate the full-scale current I and if the most significant bit (MSB) is to generate a current of $I/2$, its resistance should be $2R$. The resistance for bit 2 is $4R$ for a current of $I/4$, and in general, for the nth bit, the resistance is 2^nR. The output voltage of the current follower can be written

$$v_o = -IR\left(\frac{a_1}{2} + \frac{a_2}{4} + \frac{a_3}{8} + \frac{a_4}{16} + \ldots + \frac{a_n}{2^n}\right) \qquad (7\text{–}10)$$

where $I = V_R/R$ and a_1 is the logic level of bit 1 (0 or 1), a_2 is the level of bit 2, etc. A V_R of 10 V and an I of 2 mA are common. For these values, R is 5 kΩ. The resistance in the MSB generator is 10 kΩ, and that in the least significant bit (LSB) generator is $2n \times 5$ kΩ. For an 8-bit converter, the LSB resistance is 1.28 MΩ, and for a 12-bit converter, it is 20.48 MΩ. The large range of resistance values required seriously limits the usefulness of this simple circuit. Not only are resistance tolerances hard to maintain over this range, especially with temperature fluctuations, but also the analog switches must have very low ON resistances and very high OFF resistances.

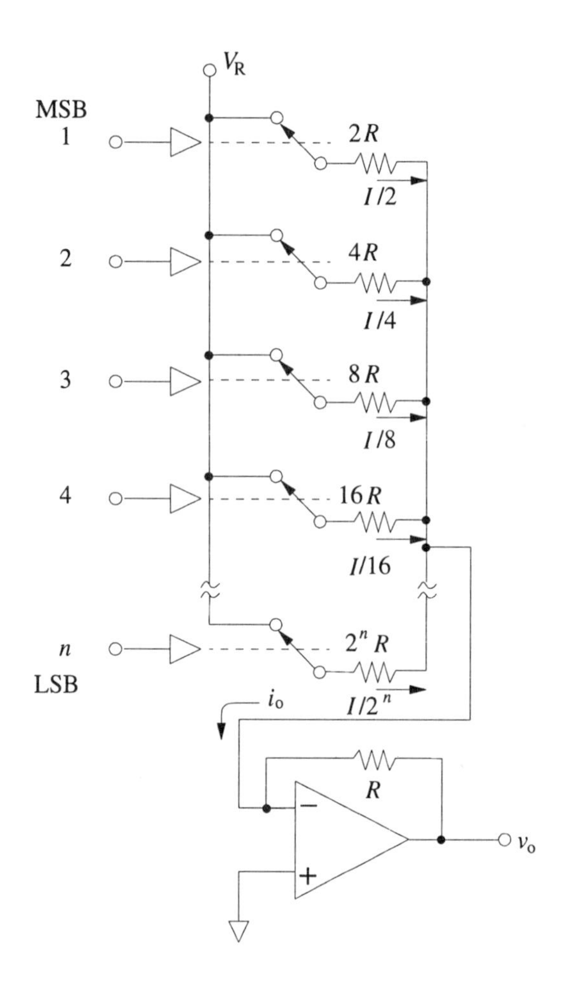

Figure 7–52. Weighted-resistor binary DAC. Precision resistors in series with the reference voltage V_R produce the current generators, which are controlled by analog switches driven in response to the logic levels at the digital inputs. Each current generator produces an appropriate fraction of the full-scale current *I*, and the currents are summed by the op amp current follower.

Ladder circuits. Another technique for greatly reducing the required range of resistance values is the **ladder network** of resistors shown in Figure 7–53. The current-mode switches used connect the individual bit currents (i_1, i_2, i_3, etc.) either to common (if the bit is 0) or to the summing point of the op amp circuit (if the bit is 1). Because the amplifier summing point is a virtual common, the value of each bit's current is independent of the bit switch position. Thus, the current supplied by V_R is independent of the value of the digital input. The current from V_R is split at N_1, with i_1 going to the MSB bit switch and the remainder going to the network composed of the resistors above N_1 in the diagram. It is a characteristic of this network that the resistance to common (or virtual common) of the resistors above each node is $2R$. Thus, the current coming into node N_1 from V_R splits equally, and the current into node N_2 is equal to i_1. This current also splits equally at N_2, with half going to i_2 and half going to the remainder of the network. The value of i_2 is thus half the value of i_1, as desired. Similarly, i_3 is half of i_2, i_4 is half of i_3, and so on. The equation for v_o for the ladder DAC is the same as equation 7–10. The ladder network is very popular for binary DACs, particularly the IC forms, because it requires only two resistance values.

Summing and output amplifiers. The DAC output amplifier serves several purposes: summing the currents, converting the current to a voltage signal, and offsetting unipolar DACs for signed digital data codes. As the switch speeds in DACs have increased, the output op amp response has become the limiting factor

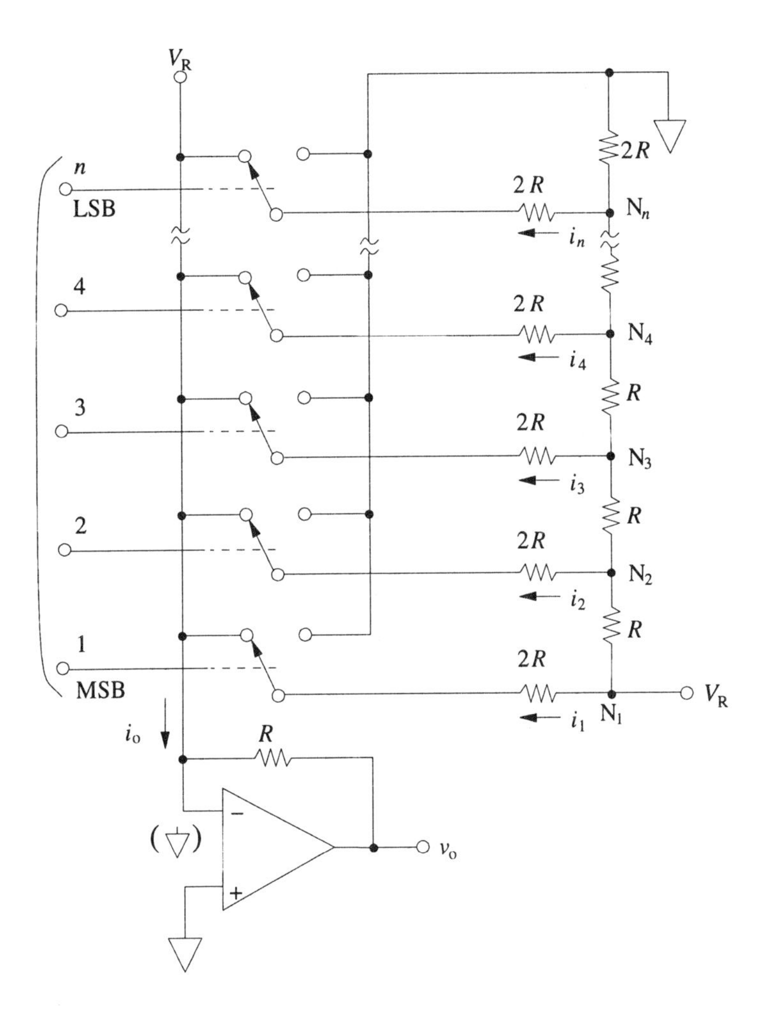

Figure 7–53. Ladder network DAC with current switching. The switches direct the generated currents to the op amp summing point or to common. All three contacts of each switch are at the common voltage. Since there is no voltage change, the switch drive signal can be small and the switching can be fast.

in DAC response speed. For this reason, many DACs now have connections that allow the substitution of a faster external amplifier if desired. To obtain the fastest response, i_o is connected directly to a load resistor. The load resistor is chosen to obtain a full-scale voltage of 0.1 V or less. For $I = 2$ mA, a load resistor of 50 Ω is ideal. If the DAC is to drive a high-speed monitor display, the i_o output is connected to the monitor input with a terminated 50-Ω coaxial cable, which serves as both connection and load.

Multiplying DACs. A multiplying DAC is one for which the reference voltage is supplied externally and the output current is accurately proportional to the value of V_R applied. From equation 7–10, the DAC output voltage is equal to the product of V_R and the input digital number. The multiplying DAC is thus a multiplier with one digital and one analog input and an analog output. It is a very convenient device for the digital control of analog signals, controlled-gain amplifiers, and so on. A full four-quadrant multiplying DAC accepts a signed digital data code and a value of V_R of either polarity and produces an output signal of appropriate polarity.

An IC DAC. The block diagram, pin connections, and external circuitry for a 16-bit IC DAC (Burr-Brown DAC 707) are shown in Figure 7–54. Digital data are input through 16 parallel digital inputs to the input latch by the simultaneous

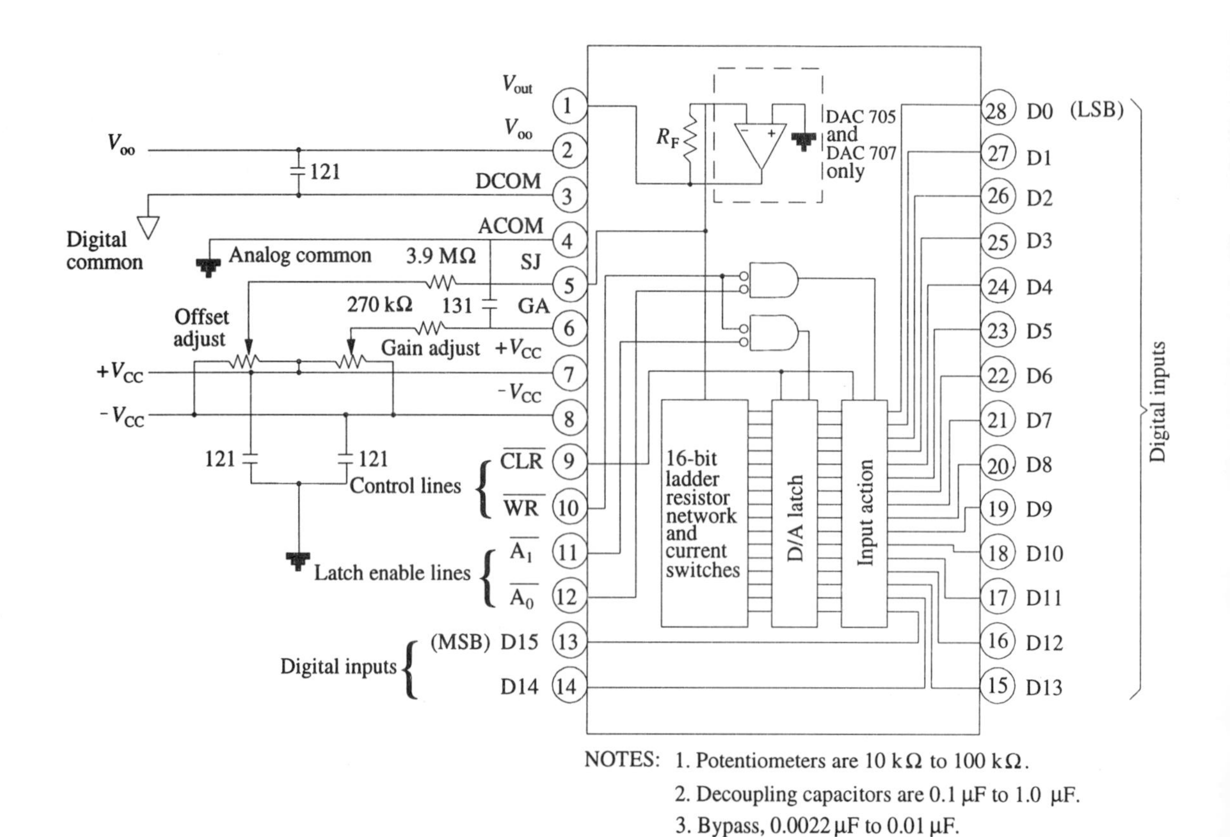

Figure 7–54. IC DAC and associated circuitry (adapted from Burr-Brown technical literature, model DAC 707). The digital data are transferred in two steps from the data inputs to the ladder network switch drivers. The output is bipolar, having a range of ±10 V. The resolution is 16 bits, but monotonicity is guaranteed for only 14 bits over the entire range of operating temperatures. DCOM, digital common; ACOM, analog common; SJ, summing junction; GA, gain adjust; CLR, clear; WR, write control; D/A, data latch.

application of a LO level to the A_0 input latch enable and write control (WR) lines. The DAC output is not yet affected, because the DAC switches are controlled by the data in the data latch (D/A). When the new output value is desired, the data in the input latch are transferred to the D/A latch by the application of LO signals to A_1 and WR. This double-buffered input arrangement is very convenient when the desired output update and CPU (central processing unit) bus write times are not the same. An example would be when two DACs are used to supply the X and Y inputs to an X–Y plotter. The next X and Y data could be read sequentially into the input latches of the two DACs, and then their outputs could be caused to update simultaneously, allowing a direct rather than a stepped transition to the next X–Y data point to be plotted.

The output current from the resistor network and switches is connected to the summing point of an on-chip op amp in the current follower configuration that provides a bipolar output with a ±10-V range. The +15-V and −15-V powers for the resistor network reference voltage and the op amp are connected to pins 7 and 8. The +5-V power for the gates and latches is connected to pin 1. Note that the commons for the analog and digital power supplies are also separate (pins 2 and 3). Where analog and digital circuits coexist intimately, as in analog-to-digital

Two's Complement	Decimal Equivalent
01111111	+127
.	.
.	.
.	.
00000101	+5
00000100	+4
00000011	+3
00000010	+2
00000001	+1
00000000	0
11111111	–1
11111110	–2
11111101	–3
11111100	–4
11111011	–5
.	.
.	.
.	.
10000000	–128

Figure 7–55. Two's complement representation of 8-bit binary numbers with sign. The MSB is 0 for all positive numbers and 1 for negative values. Positive numbers follow the normal binary notation. Increasing negative numbers follow the pattern of a binary down-counter. To change the sign of a number, complement each bit and add 1 to the result.

(ADC) converters and DACs, the noise induced in the analog signals from the changing of digital signal levels is reduced by interconnecting the commons of the analog and digital power supplies at only one point in the system. The only external circuit components required are the trimmer potentiometers for gain and offset adjustment. The former adjusts the full-scale range, and the latter adjusts the output voltage obtained when the digital input value is zero.

Signed Binary Numbers

Since the analog output of the DAC in Figure 7–54 is bipolar, a **signed number code** must be used for the digital input. The most commonly used binary signed number code is the **two's complement code** shown in Figure 7–55. As in all current signed codes, the sign is given by the MSB (0 for + and 1 for –). Note that the two's complement code for a value of zero is all zeroes and that the positive and negative codes correspond to the states of an up–down counter, where the positive values are obtained by counting up and the negative values are obtained by counting down from an initial state of zero. To change the sign of a number in the two's complement code, the number is first complemented (change each 0 to 1 and vice versa) and then incremented (add 1). Trial of this process with a few of the number codes in Figure 7–55 will confirm its validity.

Related Topics in Other Media

Video, Segment 7

- RC timers as monostable and astable multivibrators
- Function generator
- Computer-controlled waveform generator
- Bandwidth limitations of amplifiers
- Synchronous detection for signal recovery

Laboratory Electronics Kit and Manual

- Section H
  - ¤ Timing with RC circuits
  - ¤ Wiring a sample-and-hold circuit
  - ¤ Connecting a 555 time delay circuit
  - ¤ Constructing an astable oscillator
- Section I
  - ¤ Connecting a decimal UP/DOWN counter
  - ¤ Making counting and frequency measurements
- Section J–1
  - ¤ Assembling and studying a digital-to-analog converter

Chapter 8

Sensing and Controlling in Automation and Robotics

Automation is now well entrenched in most of our industrial manufacturing facilities, and highly automated systems are becoming commonplace in research and development laboratories. Why are automated systems so useful, and what are their characteristics? The replacement of human or animal labor with machines, called **mechanization**, occurred very rapidly during the industrial revolution, but the burden of control was still left to humans. Machines allow us to move objects that human muscle power alone cannot move and to perform tasks that humans find difficult or impossible to perform. **Automation**, in which the burden of control is also given to the machine, is the next logical step in technological development after mechanization. Automatic control overcomes the slow reaction time of humans and our tendencies toward boredom and distraction. Truly automated systems can perform remarkable tasks and carry out extremely complex processes on a time scale that is difficult or impossible for humans to achieve. Such systems are at the heart of modern high technology.

8–1 Controlling Physical and Chemical Quantities

The *Encyclopedia Britannica* defines automation as *the performance of automatic operations directed by programmed commands with automatic measurement of action, feedback and decision-making.* It is thus not surprising to find computers as the central elements of modern automated systems, because computers can provide highly flexible programmed commands. Automatic control, which implies measurement and feedback, is an integral part of an automated system. Control systems control either the order of events or the magnitude of some variable. As shown in Figure 8–1, a truly automated system involves both a programmable event controller (sequencer) and a programmable variable controller and has decision-making capabilities.

Figure 8–2a presents the concept of an automated system in block diagram form. The event controller determines the order in which the events are carried out

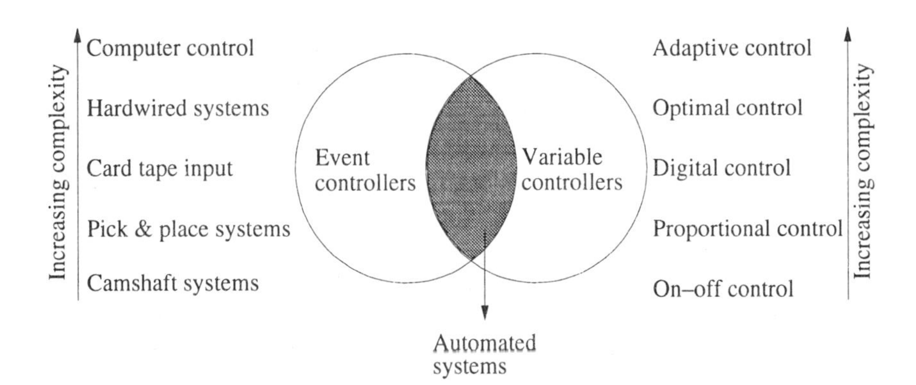

Figure 8–1. Automated systems incorporate an event controller and a variable controller.

2861-2/94/0241$15.60/1 Published 1994 American Chemical Society

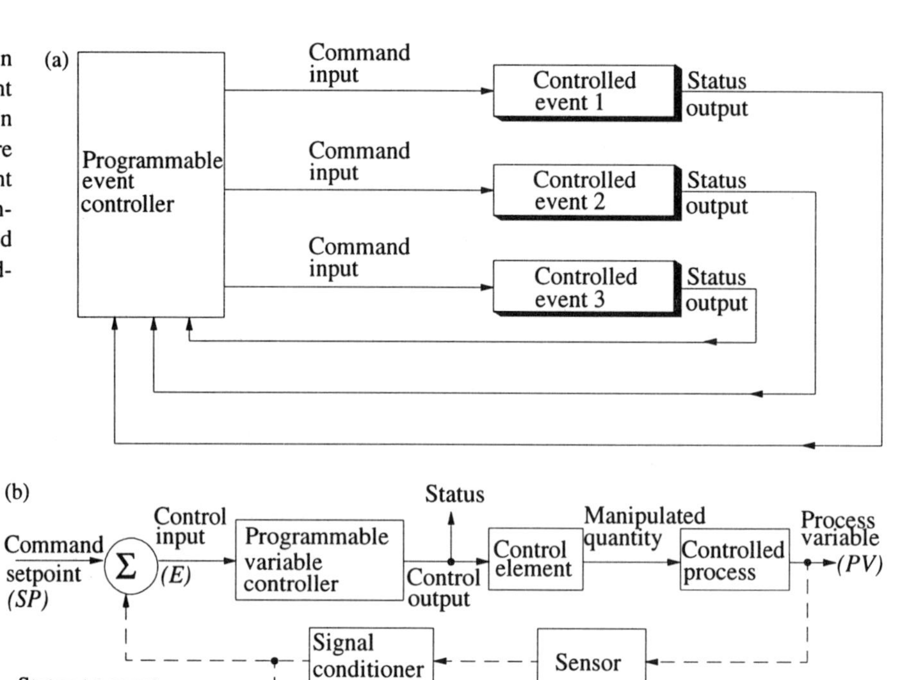

Figure 8–2. Block diagram of an automated system. (a) The event controller determines the order in which the controlled events are executed. (b) The controlled event is diagrammed. In open-loop control, the feedback shown in dashed lines is missing, while in closed-loop control, it is present.

and instructs the entire system as to how each step is to occur. In today's systems the programmable event controller is a computer. As shown in Figure 8–2a, the controller provides commands to the controlled events and receives data from the events indicating their status. Each event is also under the control of an automatic controller, as shown in Fig. 8–2b. In this case the controller is a programmable variable controller.

The various control functions can be either open loop (not shown) or closed loop as illustrated. In **open-loop control**, the setting of the controller results in the desired change in the process variable. The variable controller itself provides the status information to the event controller. For example, in the control of a cooling fan in a piece of equipment, the control element is a switch that is actuated by the controller. The manipulated quantity is the power to the fan, and the process variable is the movement of air through the equipment or perhaps the temperature of some part of the equipment. It is assumed that when the controller provides the control output signal, the switch closes and the fan is powered, providing the appropriate airflow through the equipment. In **closed-loop control** (Figure 8–2b), the process variable *PV* is monitored by a sensor. The value of *PV* is then compared with the desired (setpoint) value *SP* to produce an error signal *E*. The controller responds to the error signal by changing the manipulated quantity to bring *PV* closer to *SP*. The sensor output, after conditioning, is the status input to the event controller.

Measurement of the process variable is an extremely important part of an automatic control system. A measurement system differs from a control system in that it usually contains only the parts in dashed lines in Figure 8–2b and is missing the controller section. Instead, after signal conditioning, the sensor information is displayed or stored and used as the measurement result. In null comparison measurements (*see* Chapter 5), the distinction between measurement and control systems becomes somewhat fuzzy. Here, a variable reference standard is controlled and adjusted until its output is equal to the setpoint value determined by the unknown. An automated null comparison system, such as a servo recorder, is actually a control system in which the output of the variable reference standard at null is the measurement result. A null comparison measurement system could in fact be used as the measurement part of a larger control system which controlled some process variable. In this case, two feedback loops would be operating within the same system, one for measurement and one for control.

Sensing and controlling are integral parts of automated systems. Sensors that convert chemical and physical information from the real world (e.g., concentration, temperature, and pressure) into electrical signals (e.g., voltage, current, and charge) are known as **input transducers**. Once converted to electrical signals, the sensor information can be dealt with by the computer or instrument system. On the other hand, devices that manipulate and control real-world quantities (e.g., position, light intensity, volume delivered) based on electrical signals supplied by the controller are called **output transducers** or, sometimes, output devices or activators.

As an essential link in all electronic measurement and control functions, the transducer and its characteristics can easily be the limiting factor in the overall system performance. Hence, we begin by describing some of the ways in which transducers are characterized. A knowledge of some of the terminology allows one to read and interpret transducer specifications. In subsequent sections the specific input and output transducers used in measurement and control systems are described. These include input transducers for measuring temperature, pressure, flow rate, velocity, light intensity, and pH and output transducers such as electric motors, mechanical positioning devices, and light-emitting diode (LED) displays. Automatic controllers are then described, and the principles of control systems are developed. The chapter ends with a consideration of laboratory robots, which are devices that combine sensing, manipulating, and positioning elements with a computer controller. The robots discussed are beginning to profoundly affect the way many laboratories operate.

8–2 Transducer Principles

The overall quality (e.g., accuracy and precision) of a given measurement or the ability to control a specific variable can depend directly on the input and output transducers used and their response characteristics. For example, the accuracy of a temperature measurement may depend on how well a thermocouple is calibrated and what its linearity is. Likewise, there are many electronic systems in which the output transducer determines the overall system quality. In stereo systems, it is often the loudspeaker characteristics that determine the overall reproduction quality. It is thus important to understand the way in which transducers respond and the functional relationship between the input and output quantities. This section examines first the features that are commonly used to describe transducers and how these features are determined. Characteristics that apply specifically to input or output transducers are also presented along with the classification scheme used here for input transducers.

Transducer Characteristics

Transducers can be characterized in many ways. The input/output relationship (transfer function) is very important, as are the sensitivity, the accuracy and precision attainable, and the response speed. Any one of these factors can become limiting in a given application.

Transfer function. One important specification of a transducer is the relationship between the input and the output quantities. If the input quantity is designated I and the output quantity is designated O, we can write $O = f(I)$. The functional relationship between the input and the output quantity is known as the **transfer function**. This function was introduced in Chapter 2, Section 2–2. It can be a theoretical relationship or an empirical relationship established experimentally. Often the transfer function is presented graphically, as illustrated in Figure 8–3. The transfer function is used with both input and output transducers. For an input transducer, it expresses the dependence of the output electrical quantity on the input quantity (temperature, pressure, light intensity, etc.). For an output transducer, on the other hand, the transfer function expresses the dependence of the output quantity (position, light intensity, sound level, etc.) on the electrical input quantity.

It is desirable, although by no means necessary, for the transfer function to exhibit **linearity**. A linear transfer function, particularly one that goes through the origin, makes it easy to interpret results and to calibrate the system. There is not as much need for the transducer to show linearity if it is used in a system containing a microcomputer. With such systems, a nonlinear transfer function can be stored and then used to modify the input or output response. Transducers usually exhibit linearity over a limited range of input values. The total range, expressed in powers of 10, for which the transfer function is linear is called the **linear dynamic range**. The **dynamic range** refers to the total range of input quantity values over which the transducer is responsive.

Sensitivity. The slope of the transfer function is called the transducer **sensitivity** S ($S = dO/dI$). It tells us what change in the output we can expect for a given change in the input. A very sensitive transducer gives us a large change in the output quantity for a small change in the input quantity. The sensitivity of a transducer is often a function of variables other than the input quantity. For example, the sensitivity of a radiation transducer, such as a phototube, depends on wavelength. It may also depend on bias voltage, temperature, and other quantities. Hence, transducer sensitivities are usually measured with several additional variables held constant.

Sensitivity is often confused with **minimum detectability** or **detection limit**. The detection limit is the smallest input value that produces an output value that can be detected with a certain level of certainty.

The constancy of sensitivity over time is known as the **stability** of the transducer. Stability can be expressed as short term (hours) or long term (days, or weeks). Often the stability is highly dependent on maintaining the constancy of

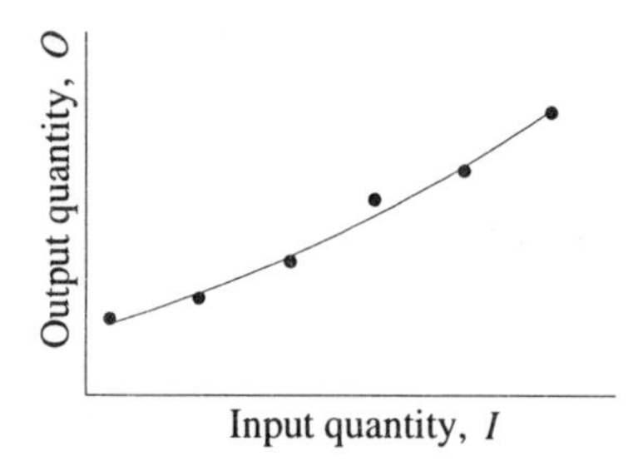

Figure 8–3. Graphical representation of the transfer function of a transducer. The points represent measured output values for the indicated input quantities.

the variables upon which the sensitivity depends. A long-term change in S with time is called **degradation**. Some transducers exhibit **hysteresis**, in that the responsivity at a particular value of the input quantity changes if the input value is suddenly increased and then brought back to its original value.

The transducer **responsivity** R is the magnitude of the output quantity for a given input quantity evaluated at a particular value of the input quantity ($R = O/I$). If a transducer is linear and its transfer function passes through the origin, the sensitivity is equal to the responsivity.

Accuracy and precision. The **accuracy** of a measurement refers to the agreement between the measured value and the true or accepted value. When a transducer is used in a measurement system, the accuracy may be limited by the transducer or by other components. Usually, when the accuracy of the transducer itself is discussed, reference is being made to how well the device follows the accepted transfer function or its theoretical behavior. **Precision**, on the other hand, refers to the agreement of replicate values. For a transducer the precision is a measure of how close the output values are for the same value of the input under identical conditions.

Response speed. Transducers vary widely in their abilities to respond to rapid changes in their input quantities. The response time is usually given quantitatively as the **time constant** $\tau = 1/(2\pi f_c)$, where f_c is the cutoff frequency at which the responsivity R has fallen to 0.707 of its maximum value (3-dB point). The **rise time** of the transducer is the time required for the output to rise from 10 to 90% of its final value when the input quantity is increased in a step function manner.

Input Transducers

Input transducers are often called sensors. They can be classified in several ways. One classification scheme is to group transducers according to the chemical or physical information being converted to an electrical signal. In this scheme, light input transducers would be grouped together, as would temperature transducers. Another classification scheme is to group transducers that produce the same type of electrical output. Thus, we could have voltage transducers, current transducers, etc. Unfortunately, neither of these popular classification methods gives much insight into the principles by which a transducer functions. Hence, in this book we will group transducers that operate by similar physical principles.

Some devices are **energy conversion transducers**; that is, they convert other forms of energy directly into electrical energy. Because these transducers generate electrical energy, they require no external power source. An energy conversion input transducer can produce an output voltage or current related to the phenomenon of interest. In some cases, linearity of the transfer function can be optimized by operating in an open-circuit mode (voltage), while in other cases it may be necessary to use a short-circuit mode (current). On the other hand, many of the same devices (e.g., solar cells) can be used to provide power, and the power output can be maximized by a different combination. Among the energy conversion transducers are photovoltaic cells, thermocouples, velocity transducers, Hall effect transducers, and electrochemical cells.

Another class of transducers is based on changes in a physical property of a device with some external parameter. These devices are often **resistive, capaci-**

tive, or **inductive transducers**, which produce a change in property as a result of a change in the parameter of interest. Such devices require an external source in order to measure the change in the physical property. They are often called **inactive transducers**. The strain gauge and the thermistor are but two examples of resistive transducers. Capacitive and inductive transducers are used in the measurement of position, pressure, and velocity.

Limiting-current transducers constitute the final class of transducers. These transducers require an external voltage source in order to operate. They are nonohmic devices that are operated in the limiting-current region. With these devices, charge carriers are generated at a rate determined by the phenomenon of interest. Biased electrodes collect the charge carriers. If the bias voltage is high enough that all generated carriers are collected, the current is limited by the rate of arrival of the charge carriers and independent of the applied voltage. The flame ionization detector, the oxygen electrode, and the photomultiplier tube (PMT) are examples of important limiting-current transducers.

Output Transducers

The output transducer or output device is the control element in Figure 8–2b. It converts the electrical control information into the desired manipulation. With input transducers, only a few electrical quantities can be produced at the transducer output (current, voltage, charge, etc.). By contrast, the manipulated quantities in the chemical and physical domains at the outputs of output transducers are nearly limitless. Hence, we will consider only a few of the most important output devices. Direct current (dc) and alternating current (ac) motors are widely used as output transducers. Motors are used to manipulate medium to heavy payloads. Pneumatic and hydraulic devices are used to provide force over a greater distance than is convenient to provide with electromagnetic devices alone. Such devices are almost always controlled by electromagnetic devices, however. For example, pneumatic devices use air pressure that is controlled by a solenoid valve to move, position, or in general translate an object. Pneumatic power is used for relatively light loads, while hydraulic power is used for heavy loads.

Some output transducers produce light or heat at their outputs. LEDs, seven-segment displays, and Peltier effect devices are examples of such output transducers.

Most output devices require some type of driver to switch and control the appropriate electrical power for their operation. Drivers represent important elements for interfacing output devices to computers.

8–3 Energy Conversion Transducers

Energy conversion devices are often called **active transducers** because they require no external power source. They can thus be completely self-contained. Also, the precision and stability of the outputs of energy conversion transducers are not dependent on the stability of an external power supply.

Photovoltaic Cells

The **photovoltaic cell** or **barrier-layer cell** is a transducer that converts radiant energy (light) into electrical energy. When input photons are incident on the

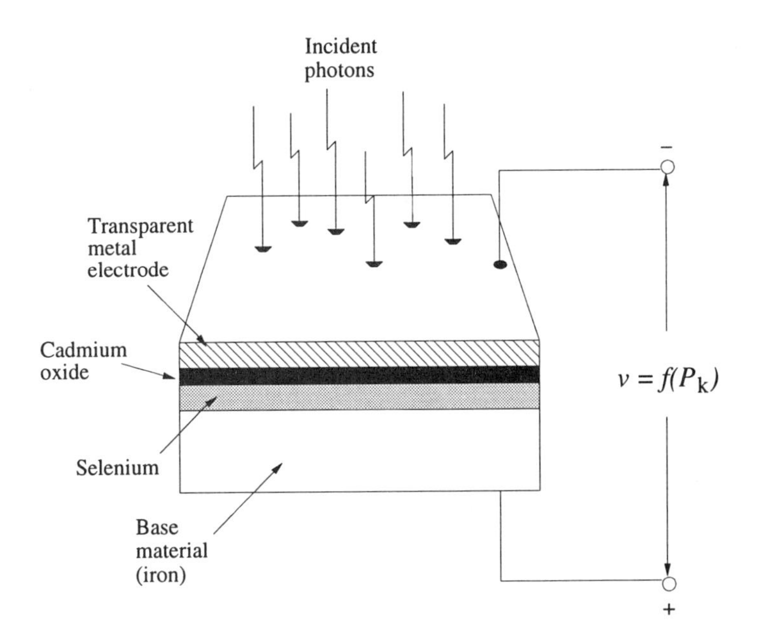

Figure 8–4. Photovoltaic cell. The cadmium oxide insulation layer is a barrier layer across which a potential difference develops when photons are incident on the device.

junction of certain dissimilar materials, the energy in the radiation can displace charge carriers and produce a voltage across the junction. A typical photovoltaic cell is illustrated in Figure 8–4. The potential difference v is a function of the incident photon flux P_k. In many cases the open-circuit voltage as a function of light intensity rapidly saturates, and the short-circuit current shows better linearity.

Selenium cells are sensitive to radiation in the spectral range 300 to 700 nm, with maximum sensitivity at about 560 nm (Figure 8–5). It is therefore easy to combine the selenium cell with a filter in such a way that the spectral response of the combination is similar to that of a human eye. Selenium cells are especially useful in camera exposure meters and simple colorimeters.

Another prominent type of photovoltaic cell is made with a silicon semiconductor. Because it can provide relatively large amounts of current, the silicon cell can be used as a power source or solar battery, as described in Chapter 4, Section 4–9.

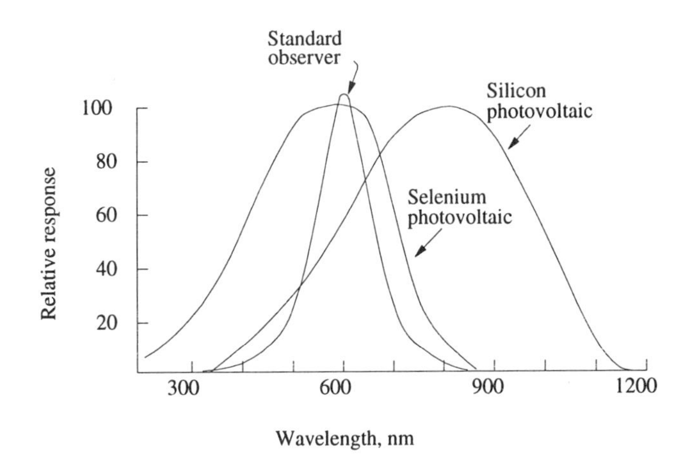

Figure 8–5. Spectral response curves for selenium and silicon photovoltaic cells and for a human eye.

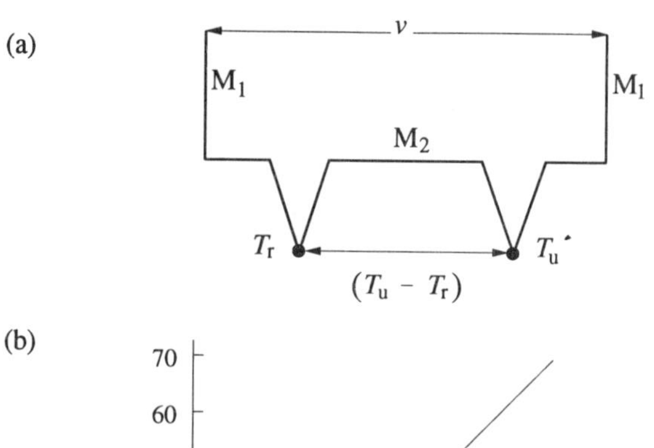

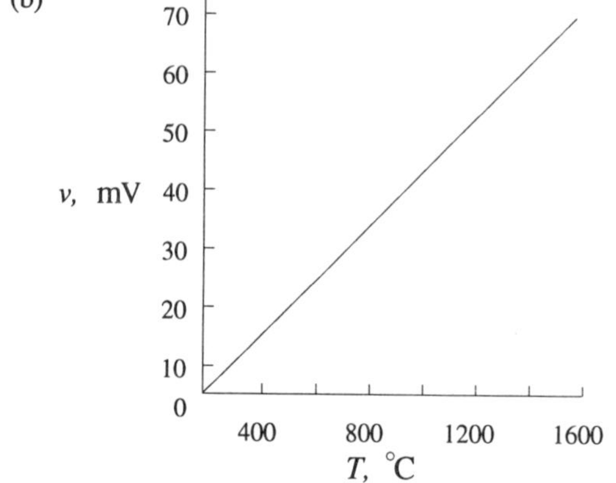

Figure 8–6. (a) Schematic representation of a thermocouple; (b) graph of the transfer function. A metal or metal alloy M_1 (e.g., chromel) is joined to a dissimilar metal or alloy M_2 (e.g., alumel) to form the junctions. To measure the temperature T_u, the second junction must be at a known (reference) temperature T_r.

Thermocouple

The **thermocouple** is a heat energy to electrical energy converter that is widely applicable in temperature measurements. When two dissimilar metals, M_1 and M_2, are joined together as shown in Figure 8–6a, the voltage v developed between the open ends is a function of the temperature difference between the junctions. For some metal pairs, this thermoelectric effect provides a reproducible relationship between the voltage v and the temperature difference between the two junctions $(T_u - T_r)$. The transfer function for a chromel–alumel thermocouple, given graphically in Figure 8–6b, can be expressed by the equation

$$v = AT_u + \frac{1}{2}BT_u^{\ 2} + \frac{1}{3}CT_u^{\ 3}$$

when T_r is 0 °C. Because the coefficients B and C are small in most cases, the transfer function for the thermocouple can be approximated by the simple linear equation $v = A(T_u - T_r)$. The coefficient is approximately 4×10^{-5} V/ °C for the chromel–alumel thermocouple. Other combinations of materials have considerably different coefficients, as shown in Figure 8–7. Combinations of thermocouples, called thermopiles, can be used as power sources.

The output voltages of thermocouples are relatively small (in the microvolt to millivolt range), and they are often either amplified before the voltage measurement step or measured with highly sensitive null comparison methods.

Electromagnetic Transducers

If a conductor is moved in a magnetic field, the voltage induced in the conductor is proportional to the rate at which the conductor traverses the magnetic field lines.

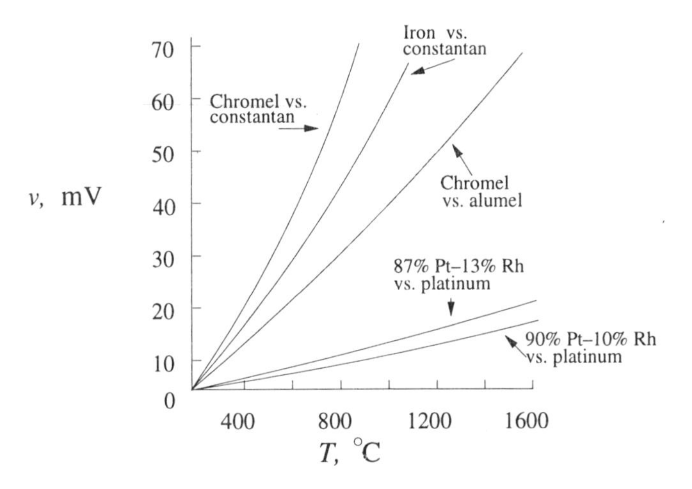

Figure 8–7. Graph of the transfer functions for several common thermocouples. The voltage for a given temperature difference depends on the composition and physical treatment of the materials used to make the thermocouple.

This is the basis of the electromagnetic voltage generator as well as of the linear- and angular-velocity transducers.

Linear-velocity transducers. A simple form of an electromagnetic linear-velocity transducer is shown in Figure 8–8. The induced voltage is proportional to the velocity at which the magnetic core moves in or out of the coil. The object whose velocity is to be measured is attached to the permanent magnet. Alternatively, the magnet may be held stationary and the coil attached to the object.

Angular-velocity transducers. Devices that are designed to determine the rotation rates of mechanical systems by measuring angular velocity are called **tachometers**. The most common types are the ac induction tachometer and the dc tachometer.

The ac tachometer is made of two sets of coils separated by a rotating cylinder of nonferrous metal. The sets of coils are arranged at right angles to each other so that there is no inductive coupling between them. When the rotating cylinder is at rest, an ac excitation current applied to one set of coils induces no voltage in the other set. The rotation of the metallic cylinder in the magnetic field produced by the excitation coils produces local current loops in the cylinder that induce a voltage in the output coils. The output voltage is of the same frequency as the excitation signal, and its amplitude is proportional to the rate of rotation of the cylinder. The direction of rotation determines whether the output signal is in phase with or 180° out of phase with the excitation signal.

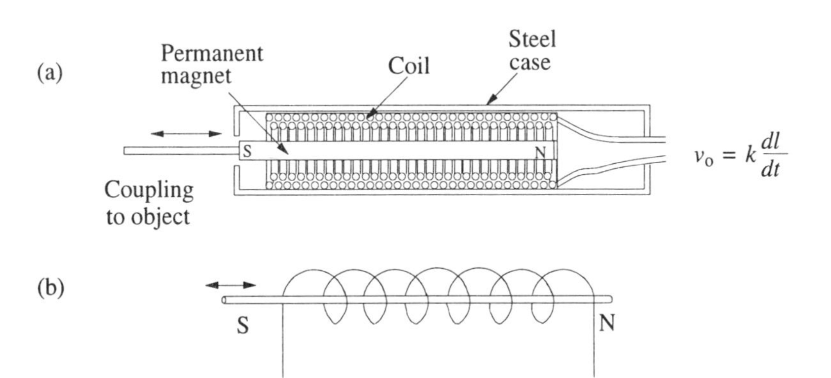

Figure 8–8. Electromagnetic velocity transducer; pictorial (a) and schematic (b) representations. The motion of the magnetic field relative to the coil induces a voltage in the coil proportional to the rate of motion.

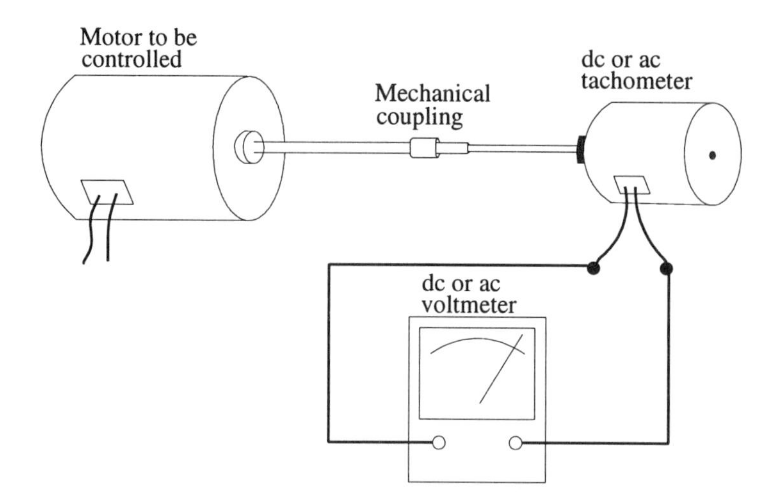

Figure 8–9. Application of a tachometer generator to measure motor speed. The power to the motor can be adjusted to maintain a constant speed regardless of load.

The dc tachometer is a small dc generator. A coil or set of coils is rotated in a steady magnetic field to produce a voltage in the coils as the magnetic field lines are traversed. Wiping contacts (commutators) are used to connect each coil to the output as it moves through the maximum field density. The output voltage has an ac component of some multiple of the rotation frequency, but the average dc level is directly proportional to the rotation velocity. The output polarity depends on the direction of rotation.

An application of a tachometer is illustrated in Figure 8–9. The tachometer output voltage can be compared with a reference voltage equal to the tachometer output at the desired speed. The voltage difference is an error signal that can be used to actuate a controller to restore the speed to the desired value. Motor speed control is further discussed in Section 8–6.

Hall Effect Transducers

An important type of interaction known as the **Hall effect** occurs when a current-carrying conductor is introduced into a magnetic field. When a metal strip is fixed in position with its plane perpendicular to that of a magnetic field and a control current i_c is applied in one direction through the strip (Figure 8–10), then a potential difference called the Hall voltage v_H is developed across the strip at right angles to the current direction and the magnetic field.

The Hall coefficient k depends on the material and the temperature. It is very large for n-type germanium and for indium arsenide or antimonide, and it is quite small for most metals other than silicon, bismuth, and tellurium. Therefore, most devices are made with n-type semiconductors. The Hall voltage is typically in the range of millivolts per kilogauss at the rated control current. Although the Hall voltage equation indicates that v_H can be increased by increasing the control current i_c or by decreasing the thickness d, either change would cause the strip to become hotter and change its characteristics. The internal resistances of typical Hall devices (generators) vary from a few ohms to several hundred ohms.

A change in magnetic field strength B causes a proportional change in the Hall voltage v_H; therefore, Hall effect devices can be used in several configurations to transduce the strength or the position of a magnet into a related electrical voltage. For example, Hall effect sensors are used to detect when a key has been depressed on a computer or instrument keyboard. A magnet attached to the key causes a

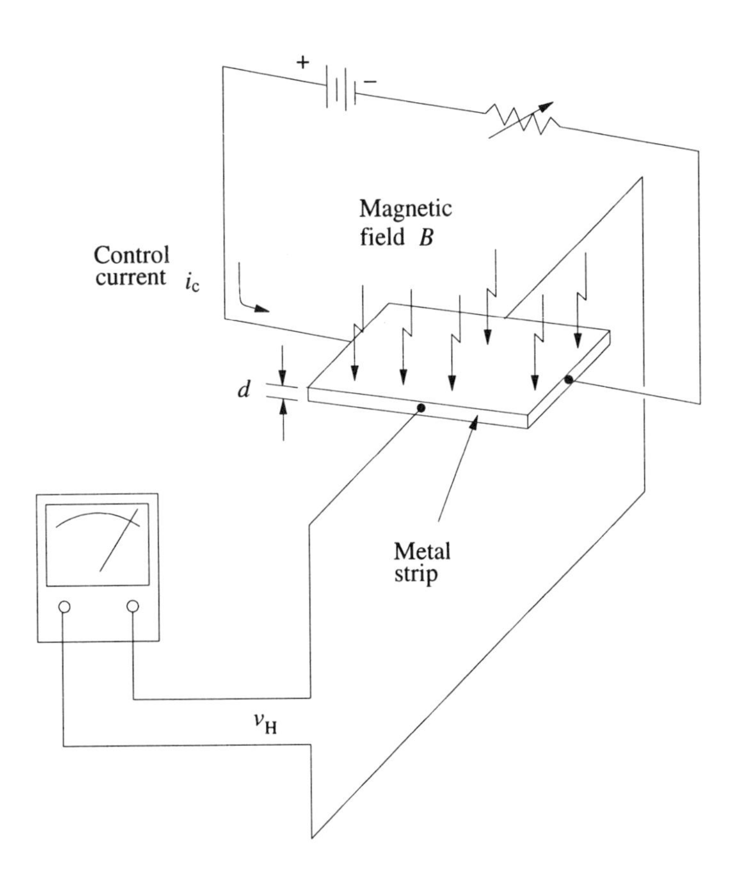

Figure 8–10. Hall effect transducer. The Hall voltage is $v_H = ki_cB/d$, where B is the magnetic field strength in gauss, d is the thickness of the strip in centimeters, and k is the Hall coefficient.

change in the Hall sensor voltage when the key is depressed. Keyboards made with Hall sensors usually have a longer lifetime than mechanical contact keyboards.

Ion-Selective Electrodes

Electrochemical cells convert chemical energy into electrical energy. They are the basis for concentration transducers and chemical power sources (*see* Chapter 4, Section 4–2). Ion-selective electrodes are electrochemical cells that respond to certain chemical constituents in a solution. For many years it has been possible to insert a pair of electrodes into a solution and obtain an output voltage related to the solution pH (acidity). Because of the success of pH measurements, other ion-selective electrodes have been introduced in recent years. These transducers respond to specific cations and anions, as shown in Table 8–1. As can be seen from the table, ion-selective electrodes use different types of membranes to achieve their selectivity.

The potential of an ion-selective electrode v_{ise} is measured with respect to a reference electrode, as illustrated in Figure 8–11 for a pH electrode. The potential is related to the activity of the ionic species to which the electrode responds a_{ion} by the Nernst equation. The cell potential can be written as

$$v_{cell} = V' + \frac{RT}{nF} \ln a_{ion} \tag{8–1}$$

Table 8–1. Common Ion-Selective Electrodes

Cation Electrode		Anion Electrode	
Ion	**Membrane Type**	**Ion**	**Membrane Type**
Cadmium	Solid state	Bromide	Solid state
Calcium	Liquid ion exchange	Chloride	Solid state
Copper	Solid state	Cyanide	Solid state
Hydrogen	Glass	Fluoride	Solid state
Lead	Solid state	Nitrate	Solid state
Mercury	Solid state	Perchlorate	Liquid ion exchange
Sodium	Glass	Sulfide	Solid state
Univalent	Glass	Thiocyanate	Solid state
Divalent	Solid state, liquid ion exchange		

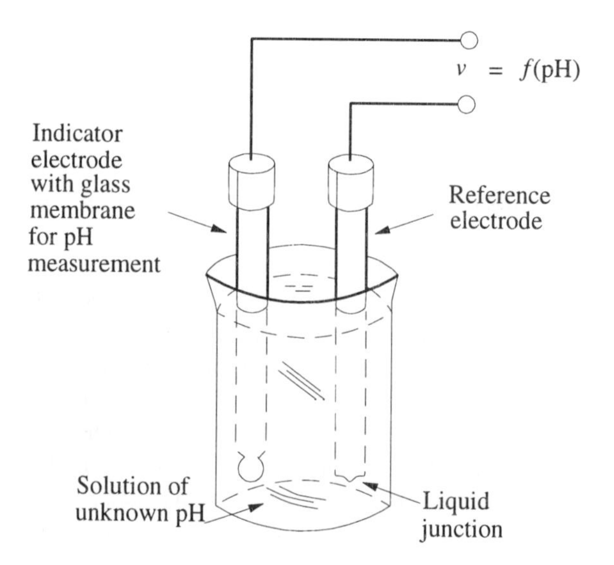

Figure 8–11. pH–reference electrode pair. Electrodes for pH measurement are available as individual indicator and reference electrodes (shown) or as a single combination electrode (not shown).

where V' is a constant that includes the reference electrode potential, R is the gas constant, T is the temperature, n is the ionic charge, and F is Faraday's constant. Because V' is constant, the measured voltage varies with the logarithm of the ionic activity of a specific ion.

More-detailed views of a glass membrane electrode and a liquid ion exchange membrane are shown in Figure 8–12. The glass pH-sensitive electrode has a high resistance, typically 50 to 500 MΩ. Although other ion-selective electrodes are not as selective as the glass pH electrode, they have become widely used in chemical analysis.

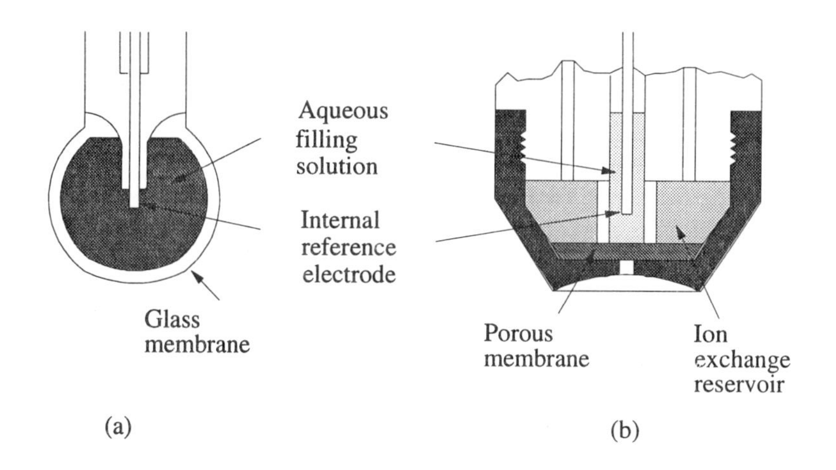

Figure 8–12. Ion-selective membrane electrodes. (a) Exchange reactions of hydrogen ions between solution and glass surface are responsible for the glass pH electrode response. (b) The ion exchange membrane electrode has a thin layer of an ion exchanger that provides a membrane potential that varies as a function of divalent cation activity.

Piezoelectric Transducers

When certain crystalline substances are mechanically deformed (by twisting, bending, or shearing forces), charge carriers are displaced and separated along certain crystalline axes. If electrodes are placed on specific faces of the crystal, then the mechanical deforming force in a suitable direction will be transduced into a linearly related output voltage across the electrodes. This characteristic of crystalline materials is called the **piezoelectric effect.** The phenomenon is reversible in that the application of a potential difference across the electrodes causes a distortion of the crystal. The two effects are responsible for the use of piezoelectric devices in many applications.

Piezo crystals are relatively stiff, and the tolerable distortions are small. For example, the deformation of crystals used in phonograph pickups is about 10 μm/g of crystal. Small crystals of this type develop outputs of a fraction of 1 V, whereas some very large crystals under high forces can develop outputs of several hundred volts.

Because of the leakage resistance of piezo crystals, the output voltage is not maintained when a constant force is applied. The device acts like a capacitor with a resistive load that slowly discharges after the charging source has been removed. Consequently, piezo crystals can be used only when the mechanical forces change periodically, usually at frequencies of a few hertz to a few megahertz.

In the biomedical field, piezoelectric transducers have been used to convert heart sounds, muscle pull, respiration, and arterial and venous pulses to related output voltages. The use of piezo crystals in many types of phonograph pickups, crystal microphones, and industrial pressure transducers is widespread.

Faraday Cage

A transducer that can be used to monitor electron or ion beams and convert the number (and charge) of incident particles to a proportional output voltage is shown in Figure 8–13. It is known as a Faraday cage or cup and is made of a well-insulated metallic cup. If the incoming beam consists of electrons and the capacitance between the cage and common is C, the output voltage is

$$v = \frac{q}{C} = \frac{n_e Q_e}{C} = k n_e \quad (8\text{–}2)$$

Figure 8–13. Faraday cage. The particle beam enters through a small hole. Inside the cage the beam of ions or electrons can be reflected and will produce secondary electrons. The resulting separation of charge gives rise to the output voltage.

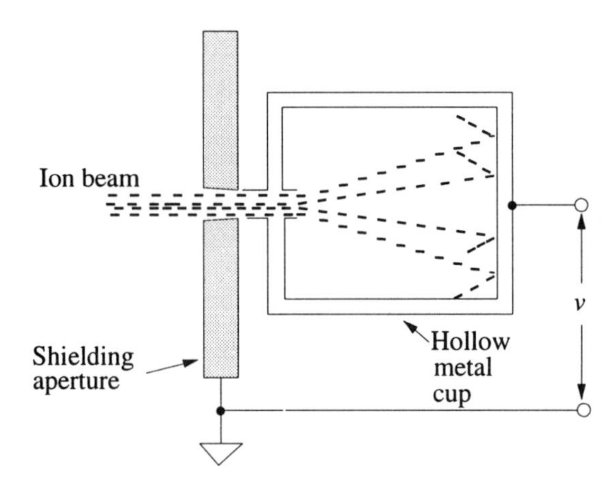

where n_e is the number of incoming electrons over the collection period, and $k = Q_e$ (charge of electron)/C(capacitance).

If the incoming beam is composed of positive ions, then

$$v = \frac{q}{C} = \frac{n_1 Q_1 + n_2 Q_2 + \ldots + n_n Q_n}{C} \tag{8–3}$$

where n_1 and Q_1 are the number and charge per ion of ion 1, and n_n and Q_n are the same for other ionic species in the incoming beam. As few as 10^3 charged particles can be measured with a Faraday cage.

Charge Excess Transducer

In studies of atmospheric electricity, the excess charge in air is a significant quantity. This charge can be determined by a charge collection device known as the **charge excess transducer**, as shown in Figure 8–14.

The net difference in the positive and negative charges that are drawn into the transducer produces an output voltage proportional to the difference so that

$$v = \frac{q}{C} = \frac{n_d M t Q_e}{C} \tag{8–4}$$

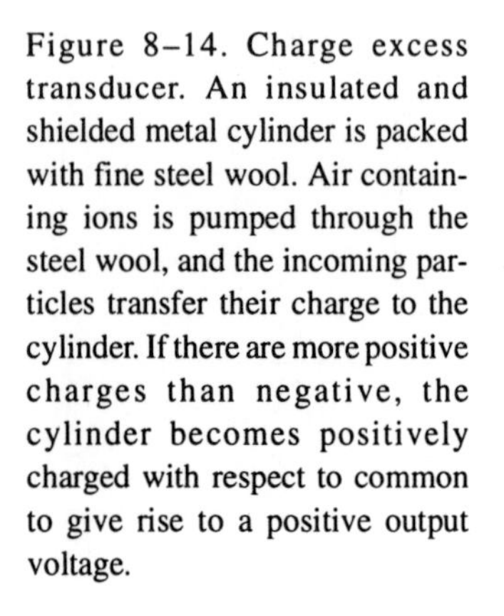

Figure 8–14. Charge excess transducer. An insulated and shielded metal cylinder is packed with fine steel wool. Air containing ions is pumped through the steel wool, and the incoming particles transfer their charge to the cylinder. If there are more positive charges than negative, the cylinder becomes positively charged with respect to common to give rise to a positive output voltage.

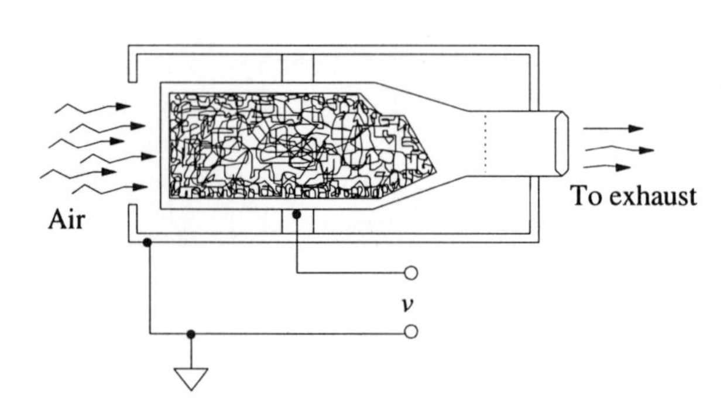

where n_d is the excess of positively charged particles ($n_+ - n_-$) per cubic meter of air, M is the gas volume flow velocity in cubic meters per second, and t is the collection time.

8–4 Resistive, Capacitive, and Inductive Transducers

Resistive transducers are based on changes in the resistance of a device as a function of a physical parameter. Likewise, capacitive and inductive transducers are based on changes of capacitance (or capacitive reactance) or inductance (or inductive reactance) as some physical quantity changes. The resistance of a device is given by $R = \rho l/a$, where ρ is the resistivity and l and a are the length and cross section, respectively. Changes in resistance can be effected by changing the dimensions l and a or by changing the concentration of charge carriers, which changes the resistivity ρ. The geometry of conductors in a capacitor determines its capacitance according to $C = \varepsilon_0 K_d a/d$, where a is the area between the conductors, d is the distance between the conductors, K_d is the dielectric constant of the insulator, and ε_0 is the permittivity of free space. Changes in capacitance can be affected by changes in geometry (a or d) or dielectric constant. Inductive transducers are based on the relationship $L = N(d\phi/di)$, where L is the inductance, N is the number of turns cut by the magnetic flux, and $d\phi/di$ is the rate of change of flux with current. A physical movement of a ferrous core causes changes in N and thus L, which can be used to sense position. Likewise, changes in inductance can be caused by a change in the chemical composition of a portion of the flux path or a stress on ferromagnetic material in the path.

In order to appreciate better how resistive transducers operate, we will first review what determines the conductance of materials. A description of the most important resistive transducers follows. The Wheatstone bridge, which allows resistance measurements to be made by the null comparison technique, is introduced. This section ends with a discussion of capacitive and inductive transducers.

Conductance of Materials

The outermost and most easily removed electrons of an atom are frequently called the valence electrons. The interaction of valence electrons is primarily responsible for the formation of crystals and molecules. When similar atoms combine to form a crystalline solid, the discrete energy levels of the valence electrons broaden into bands of allowed energy levels. The electrons may have any energy within the band limits. One such band is the **valence band** or **bonding band**. In this band, atoms share electrons in covalent bonds. If there is an odd number of electrons in each atom and there are two available energy states per atom in each band, one of the energy bands will be only half filled. The unoccupied energy states in the energy band allow the electrons in that band to be accelerated by an electric field. Thus electrons in a partially filled band are conductive electrons, and a partially filled band is a **conduction band**.

In materials with an even number of electrons in the lowest energy state, the valence bands are filled, and therefore the electrons in them are not free to move. Any allowed energy band higher than the highest valence band will be empty. This is illustrated in Figure 8–15. If some of the valence electrons can be promoted into the next-higher band, they will be free to move in this partially filled band. The next-higher band is the conduction band in this case. The magnitude of the energy

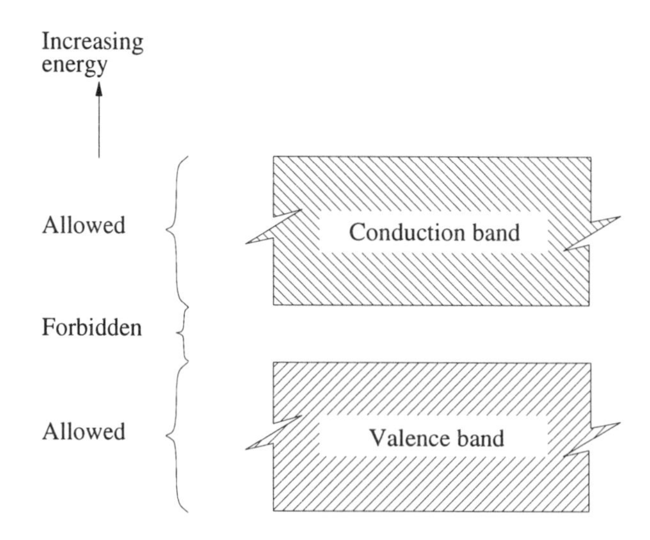

Figure 8–15. Electron energy bands in a solid. If an energy band is only partially filled, the electrons in it are mobile charge carriers.

gap between the filled valence band and the conduction band determines the conductance type of a given material. If the energy gap is zero, the material is a metallic conductor; if the gap is small, it is a semiconductor; and if the gap is large, it is an insulator.

Semiconductors

A material with an even number of valence electrons and a forbidden energy gap between the valence and conduction bands cannot conduct electricity at 0 K. At absolute zero temperature, the valence band is filled and the conduction band is empty. A source of energy is required to promote valence electrons to the conduction band. This energy might be provided by quanta of radiation (photons) as in photoconductors or by thermal excitation at higher temperatures as in thermistors. An energy at least equal to the forbidden gap is needed for excitation to conduction. The result of promoting an occasional valence electron to the conduction band is shown in Figure 8–16. For each electron in the conduction band, there is a hole, or a lack of one electron, in the valence band. It must be remembered that although the conduction band electrons are free to move throughout the material, a valence electron can move only to an adjacent unoccupied site, a hole, thus creating a hole where it was. The area of a missing valence electron is a localized but movable region of positive charge.

The number of electrons per unit volume of material that are in the conduction band at any given temperature depends on the magnitude of the energy gap. The number of conducting electrons per volume is greater for materials with lower forbidden gap energies E_g and increases with increasing temperature in any

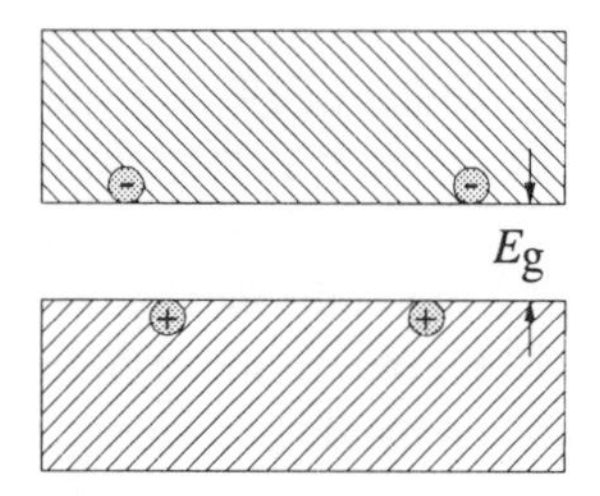

Figure 8–16. Band structure of a semiconductor or insulator with valence electrons excited to the conduction band. Both the valence and conduction bands are now partially filled, and conduction can occur.

material. For some semiconductors the conductivity increases approximately exponentially with increasing temperature. This thermal behavior is the basis of the temperature-dependent resistor (thermistor).

A semiconductor that conducts by equal numbers of holes and electrons created by the excitation of valence electrons is called an **intrinsic semiconductor**. All pure semiconductors and insulators are intrinsic semiconductors. Whether a material is a semiconductor or an insulator depends on the magnitude of the energy gap E_g and on the operating temperatures or energy of excitation. At low enough exciting energies, all such materials are insulators, and at high enough exciting energies, all such materials are conductors.

Thermistor

The **thermistor** is made from an intrinsic semiconductor. As we have seen, the concentration of charge carriers in intrinsic semiconductors increases rapidly with increasing temperature. Thus the resistance of the thermistor decreases quickly as the temperature increases. A typical resistance-versus-temperature curve is shown in Figure 8–17 together with some typical configurations of thermistor probes. The relatively large resistance change per degree makes the thermistor a useful device for temperature measurement in control devices requiring high accuracy or resolution. Temperature changes as small as 5×10^{-4} °C can be detected. Thermistors are used in the range from –100 to +300 °C. Any current in a device with resistance does work that appears as heat. Care must therefore be taken that the amount of heat produced by the current in the thermistor does not affect the temperature of the measured system.

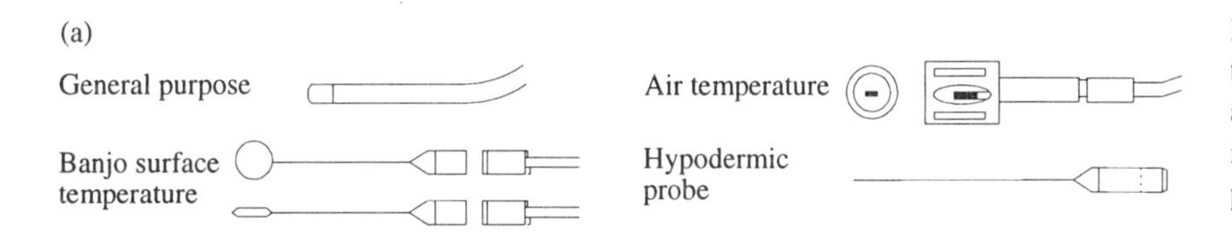

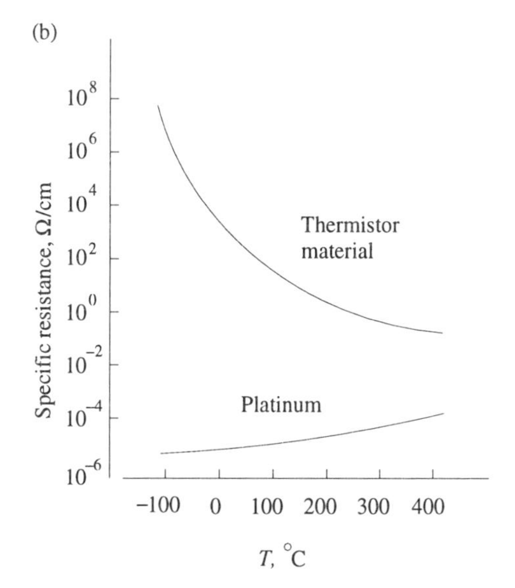

Figure 8–17. (a) Practical thermistor probes; (b) resistance–temperature response of a typical thermistor material compared with that of platinum.

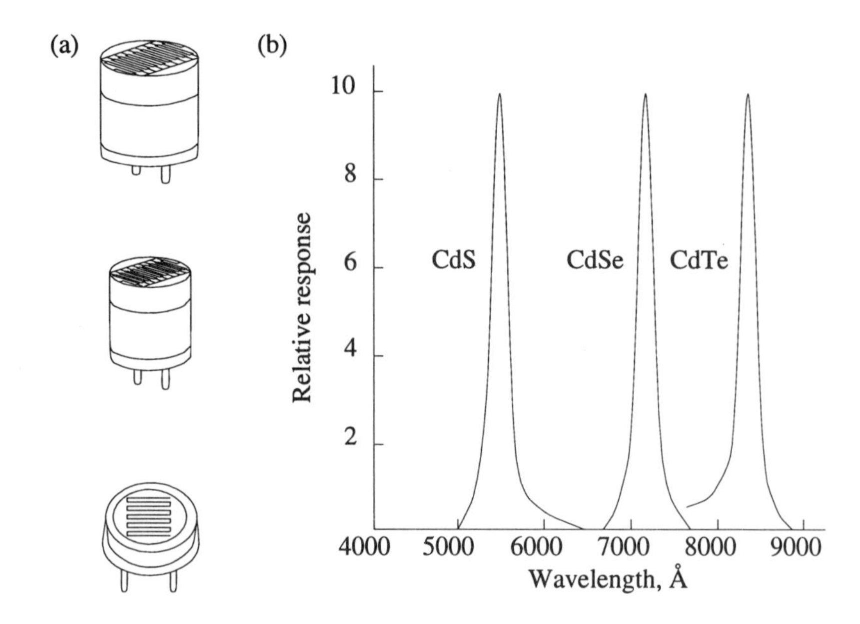

Figure 8–18. (a) Practical photoconductive cells; (b) photoconductivity of CdS, CdSe, and CdTe versus wavelength.

Thermistor resistances can be measured in several ways. Differential temperature measurements are most often made with the Wheatstone bridge discussed later in this section.

Photoconductive Cell

When light falls on a semiconductor, the photon energy may be sufficient to move valence electrons into the conduction band. The resulting increased density of charge carriers due to light makes the material more conductive. Thus, the resistivity decreases as the light intensity increases. The semiconducting sulfide, selenide, and telluride salts of cadmium are most often used for this purpose. Some practical **photoconductive cells** and the spectral response curves of such devices are shown in Figure 8–18. For each material to photoconduct, the incident light must have a short enough wavelength (high enough energy) for the photons to promote electrons from the valence to the conduction state. As the energy of the light increases beyond the minimum required for promotion of valence electrons to the conduction band, the absorption of the light by the material increases rapidly. This increased absorption results in a marked decrease in the fraction of the light that penetrates the active region of the device.

Strain Gauge

If a resistor made of a fine wire is distorted, its resistance changes as a result of dimensional changes. Its resistivity may also change. The result is a resistor for which the resistance is related to the strain. Such a device is called a **strain gauge**. Strain gauges have extensive applications in mechanical and biological measurements. Often the fine wire is bonded to a flexible insulating substrate in such a way that the wire dimensions change as the substrate is flexed. Some strain gauges are made by depositing a pattern of thin metal film on the substrate material. Several practical strain gauges are shown in Figure 8–19. Because the resistance change with strain is a very small fraction of the total resistance, highly sensitive resis-

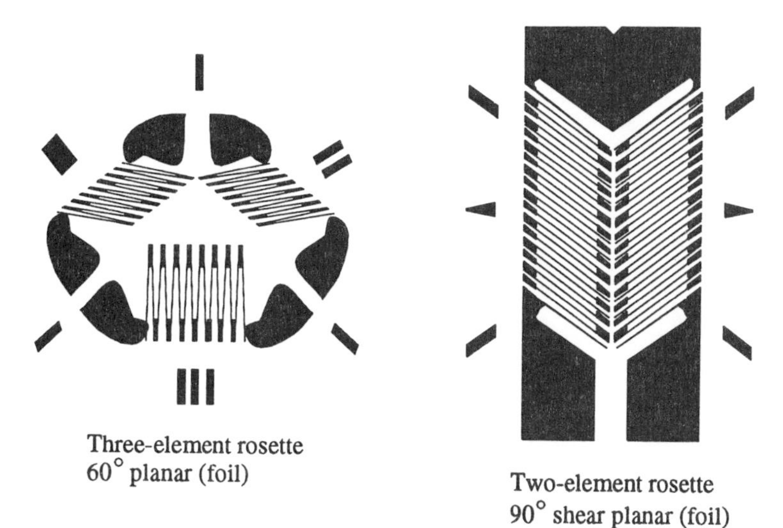

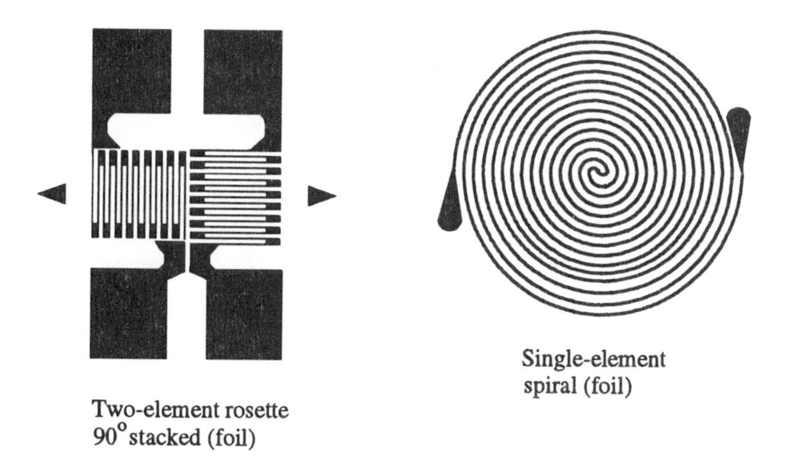

Figure 8–19. Practical strain gauges. Various patterns are used to enhance sensitivity for particular directions of strain or to allow the effects of less desired strains to be balanced out.

tance-measuring techniques are required. Another difficulty is separating the changes in resistance due to temperature from those due to strain. The Wheatstone bridge described below can be used with dual strain gauges kept at the same temperature to minimize the effect of changing temperature.

Other Resistive Transducers

Several other resistive transducers are in common use. The platinum resistance thermometer is an extremely accurate, sensitive, and stable transducer with a resistance that increases with temperature (*see* Figure 8–17) over the range of –270 to 1100 °C. Linear- and angular-displacement transducers can be based on the change in some characteristic of an adjustable circuit element such as a potentiometer. The device whose displacement is being measured is connected to the movable contact (wiper) of a potentiometer, and the effective length of the resistance element (proportional to resistance) is changed in proportion to the displacement.

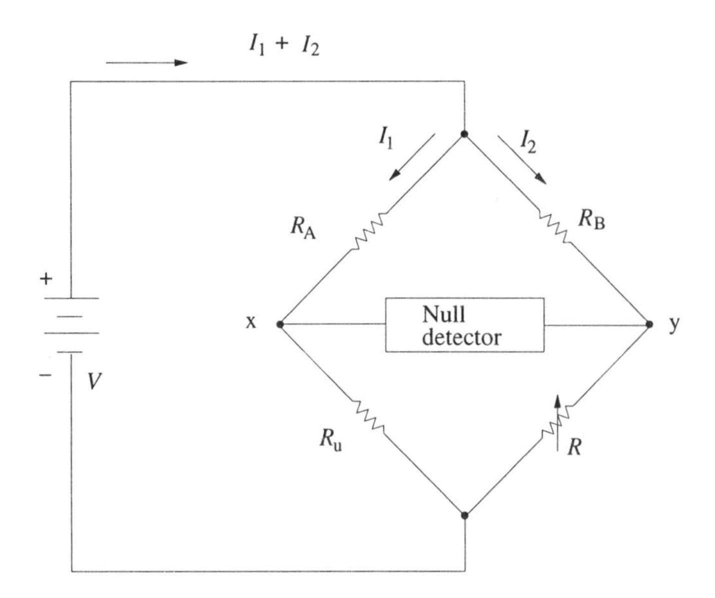

Figure 8–20. Wheatstone bridge. The value of R is adjusted to make the voltage at x and y equal, as shown by the null detector. At balance, $R_u = R(R_A/R_B)$.

Wheatstone Bridge

The most direct method of comparing unknown resistances is with a Wheatstone bridge (Figure 8–20). In the null operation of the bridge, resistances R_A, R_B, and R are standards used to determine the unknown resistance R_u. Resistance R is made variable and adjusted until the null detector indicates that the bridge is balanced. At balance there is no current through the null detector; therefore, the current through R_A and R_u is I_1 and the current through R_B and R is I_2. Also, at balance the voltage between terminals x and y is zero. From the equality of IR drops on each side of the bridge, $I_1R_u = I_2R$ and $I_1R_A = I_2R_B$. These equations can be solved for R_u to give $R_u = R(R_A/R_B)$.

The unknown resistance R_u is calculated from the values of R, R_A, and R_B. Usually, the ratio R_A/R_B is a convenient factor such as 0.01, 0.1, 1, 10, or 100 and is referred to as the *multiplier*.

The major sources of error in a Wheatstone bridge are inaccuracies in the three standard resistances R, R_A, and R_B, but these resistances can be made with errors as low as 0.001%. Other factors that could limit the accuracy are null point accuracy, thermal voltages, and changes in resistance values due to heating by passing too much current.

The magnitude of the off-balance indication of the null detector is sometimes used as a measure of the change in the resistance R_u. In this application, R_u is usually due to a resistive transducer. When matched devices are used for R and R_u or R_A and R_u, the off-balance output obtained is related to the *difference* in the resistance of the two devices. A differential resistance measurement with matched transducers tends to cancel the effects of resistance changes caused by environmental effects other than the quantity sought. For example, with strain gauges it is desirable to compensate for the temperature sensitivity of the gauge. If R_u and R are the resistances of matched strain gauges subjected to the same temperature but different strains, the off-balance output voltage of the bridge will be due only to the difference in strains.

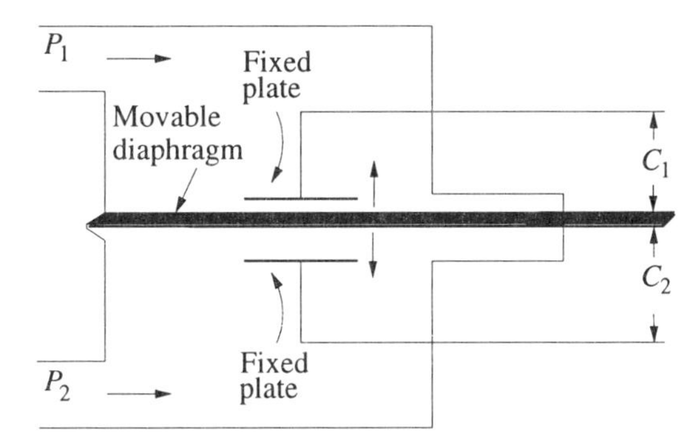

Figure 8–21. Capacitive pressure transducer. A metal diaphragm or thin metal film is stretched between two plates such that two capacitors result, with the diaphragm as a common element. The difference in the capacitance of the two capacitors $C_1 - C_2$ is related to the pressure difference across the diaphragm $P_1 - P_2$.

Capacitive Transducers

Capacitive transducers are often used where the parameter to be measured can cause a physical motion. This motion can cause a variation in the distance between the plates, a change in the common area of the plates, or a change in the dielectric constant of the insulator. Thus, there are capacitive transducers for displacement, velocity, pressure, fluid level, and so on.

A capacitive pressure transducer is shown in Figure 8–21. This transducer is sensitive to the pressure difference across a diaphragm. The difference in capacitance could be measured with a bridge circuit or with an oscillator circuit. Capacitive transducers can also be used to measure the dielectric constant of materials.

Inductive Transducers

Inductance can be varied in proportion to chemical, thermal, or mechanical displacement or velocity information. Two types of position-sensitive transducers are considered here: the Wheatstone bridge transducer and the linear variable differential transformer (LVDT).

Wheatstone bridge transducer. An inductive Wheatstone bridge is shown in Figure 8–22. The bridge contains two resistances R_1 and R_2 in the fixed arms and two inductors with inductances L_1 and L_2 in the variable arms. An ac excitation voltage is applied. Typically, a root mean square (rms) voltage of between 5 and 10 V is used at a frequency of 400 Hz to 5 kHz.

Inductances L_1 and L_2 represent variable inductors. These devices contain a common ferrous core that can be moved in and out of the coils. This changes the number of effective turns N and, from $L = N(d\phi/di)$, the inductance. With the core positioned equally inside both coils, the bridge is balanced. Movement of the core, as shown in Figure 8–22b, increases L_2 and decreases L_1, which unbalances the bridge. The core is normally linked mechanically to the device whose position is being sensed.

The LVDT. The LVDT is quite similar to the inductive Wheatstone bridge except that a primary winding is also present, as shown in Figure 8–23a. The ac excitation

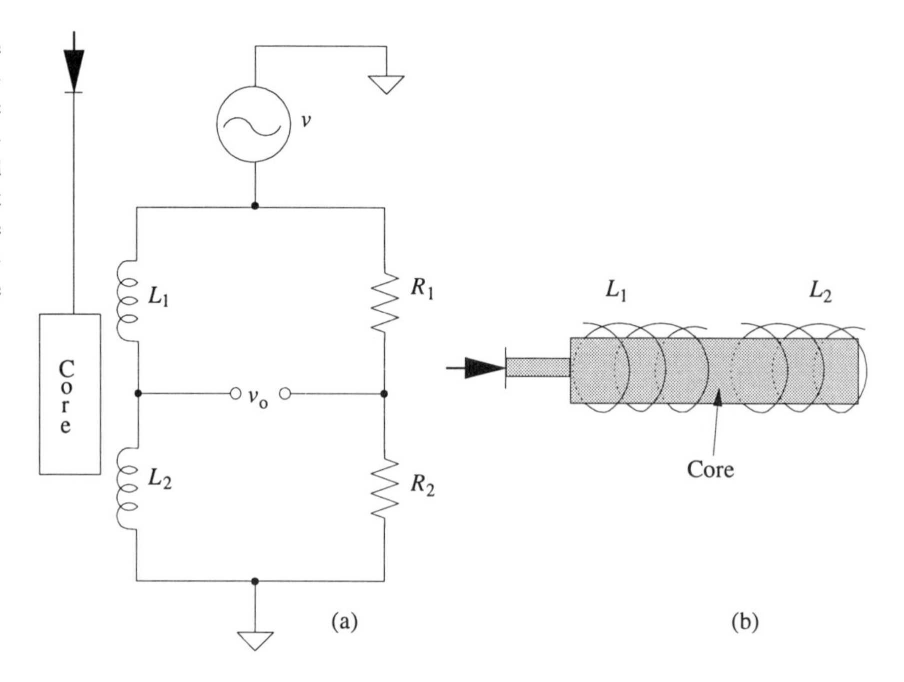

Figure 8–22. (a) Wheatstone bridge inductive transducer circuit; (b) pictorial of the magnetic core. Two inductors with inductances L_1 and L_2 are arranged coaxially with a common core. At rest the core is equally inside the two coils. When the core is displaced to the right, the inductive reactance L_2 is increased while L_1 is decreased.

voltage is typically in the range of 1 to 10 V rms at a frequency of 40 Hz to 20 kHz. The output voltage is directly proportional to the core displacement, as shown in the transfer function in Figure 8–23b.

The LVDT produces a higher output voltage for a given change in position than does the Wheatstone bridge inductive transducer. Commercial models are available that produce rms output voltages in the range of 50 to 300 mV/mm of displacement. The outputs are linear to within ±0.5%. However, when used with a computer, these ac voltages must be converted to dc before acquisition.

8–5 Limiting-Current Transducers

The resistive transducers described in the preceding section are all examples of ohmic devices. That is, they obey Ohm's law. Another class of transducers does

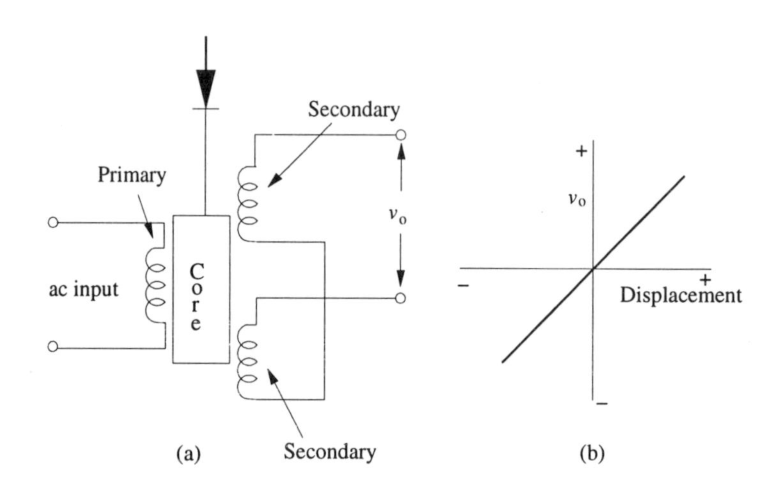

Figure 8–23. LVDT. (a) Schematic; (b) output transfer function. Under zero-displacement conditions, the core is placed symmetrically inside the two secondary windings, which are arranged in series and opposing each other so that the output voltage is zero. If the core is moved, the coupling between the primary and the two secondary windings is no longer equal, and a difference output proportional to the magnitude of the displacement is obtained.

not obey Ohm's law. In these devices the current is not proportional to the voltage but is instead independent of the applied voltage over a relatively wide range of values. These devices are operated in the limiting-current region. In **limiting-current transducers**, charge carriers arrive at electrode terminals at a rate determined by the quantity or phenomenon being measured. The important limiting-current transducers discussed in this section include the photodiode, the flame ionization detector, the oxygen electrode, the vacuum phototube, and the photomultiplier tube (PMT).

Photodiode

The current through a reverse-biased pn semiconductor junction arises from the energetic creation of electron–hole pairs in the depletion region at the junction. Like thermal generation of electron–hole pairs, electromagnetic radiation of the proper energy (wavelength) can be absorbed by the material in the depletion region, with the creation of electron–hole pairs as illustrated in Figure 8–24. If the rate of charge carrier production by the electromagnetic radiation greatly exceeds the rate of thermal generation, the reverse-biased current is directly proportional to the photon flux (light intensity). Under limiting-current conditions, the photocurrent I_p is given by $-I_p = b(\lambda)Q_eP_k$, where P_k is the photon arrival rate (flux), $b(\lambda)$ is the quantum efficiency of the material (number of emitted photoelectrons per incident photon), and Q_e is the charge on the electron.

The **photodiode** is an extremely fast responding transducer. Transit times of charge carriers across the junction are often in the subnanosecond range. The photodiode is therefore often used to detect extremely rapid, high-intensity events such as pulsed-laser outputs. Because photodiode output currents are usually in the microampere range, photodiode–amplifier combinations are manufactured in a single unit to preserve the fast response. Phototransistors, which have internal gain, are also useful, particularly as light-activated switches. In recent years, integrated circuit packages containing arrays of individual photodiodes have become available. They typically contain 256, 512, 1024, or even 2048 elements arranged linearly. Although diode arrays have no internal gain, they show dynamic ranges of two to four decades. Diode arrays are also available with front-end microchannel plate image intensifiers, which gives them gain and allows lower light levels to be detected. Diode arrays are used as multiwavelength spectroscopic detectors, optical

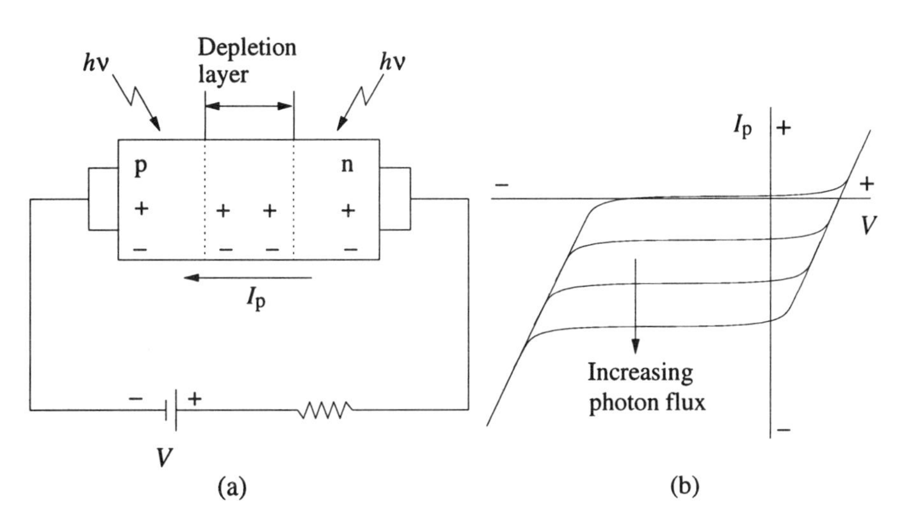

Figure 8–24. Photodiode. (a) Pictorial representation showing electron–hole pair production; (b) current–voltage characteristics for different photon fluxes. The reverse bias applied in the limiting region is sufficient to collect all the charge carriers generated before they can recombine in the depletion region. The current is then equal to the rate at which the charge carriers are produced by the light flux.

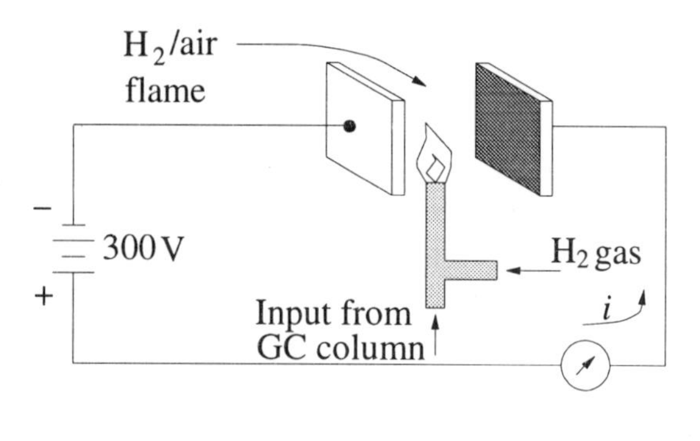

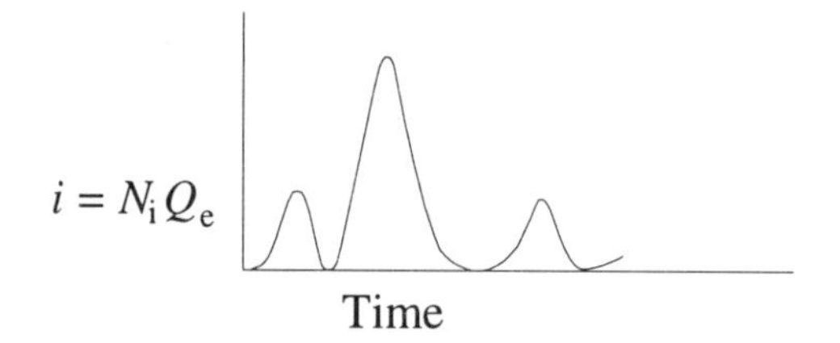

Figure 8–25. Flame ionization detector for gas chromatography (GC). The current i is equal to the number of ions produced by flame ionization that reach the electrode plates per second N_i times the unit charge per charge carrier Q_e. Typical currents measured with the flame ionization detector are from about 10^{-6} to 10^{-11} A. Often the electrodes are a wire loop collector and the burner base instead of two metal plates as shown.

scanners, optical positioners, and the detection elements of some solid-state television cameras. They are being supplanted in many applications by two-dimensional devices such as charge-coupled devices.

Flame Ionization Detector

If two metal plates are placed opposite each other and just above a small hydrogen–air flame, as shown in Figure 8–25, and if molecules entering the flame are readily ionizable by the flame energy, there will be a flow of charge carriers (both electrons and positive ions) between the electrodes. The rate of flow of electrons in the external circuit depends on the rate of formation of charge carriers by the flame, which in turn is determined by the concentration of ionizable molecules in the flame. The **flame ionization detector** is widely used in gas chromatography. The various constituents in a complex sample are separated as they pass through the chromatographic column, and as each exits from the end of the column, it passes into the flame ionization detector. The peak heights or peak areas of the current–time waveform are a quantitative measure of the concentration of each separated species.

The Oxygen Electrode

The determination of oxygen concentrations in lakes, rivers, oceans, and many laboratory solutions and biological systems is often based on an electrode system that is a limiting-current transducer. The electrode and experimental systems were previously shown in Chapter 3, Figure 3–1a. If the voltage from the source is varied between 0 and 1.0 V, the current–voltage (i–v) curves obtained (Chapter 3, Figure 3–1b) have the voltage-independent plateau characteristic of limiting-current devices. The height of the current plateau is a function of the oxygen concentration, as illustrated by the i–v curves. Therefore, the current-producing process must be related to the oxygen concentration.

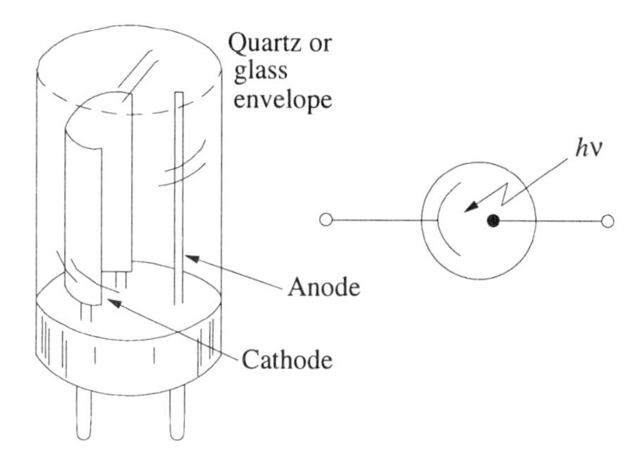

Figure 8–26. Vacuum phototube in pictorial and schematic representation. Light striking the cathode causes the emission of photoelectrons, which are collected at the anode.

When the platinum microelectrode is made about 0.6 to 0.8 V negative with respect to the reference electrode, the oxygen molecules (O_2) that reach the platinum surface immediately accept electrons and become reduced. The current in the external circuit is then proportional to the rate of arrival of oxygen molecules at the platinum surface. This rate is limited by the rate of diffusion of O_2 molecules to the electrode surface and is directly proportional to the oxygen concentration (Chapter 3, Figure 3–1c).

Vacuum Phototube

The vacuum phototube contains a photosensitive cathode and an anode in an evacuated quartz or glass envelope. A typical **phototube** and its schematic symbol are shown in Figure 8–26. Input radiation incident on the photocathode ejects photoelectrons with an efficiency $b(\lambda)$ that depends on the energy of the photon ($h\nu$) and the type of cathode surface. If the anode is held at a sufficiently positive potential with respect to the cathode (Figure 8–27a), all the emitted photoelectrons are collected at the anode. The resulting photocurrent i in the external load is shown in Figure 8–27b. When the phototube is operated in the limiting-current region, the output current is directly proportional to the light intensity (photons incident on the photocathode per second). The proportionality constant is nearly unaffected by the applied voltage.

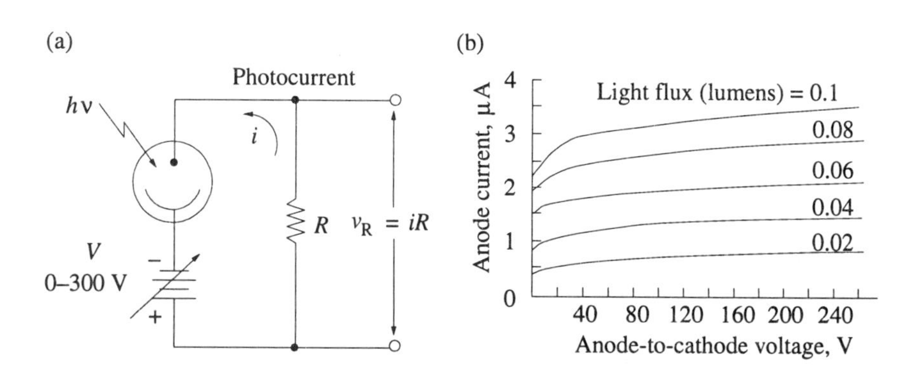

Figure 8–27. (a) Phototube circuit; (b) current–voltage curves. When V is sufficient to collect all photoelectrons, i is proportional to the light flux and independent of V. Often an operational amplifier i–v converter is used to measure i and avoid a perturbation error from resistance R.

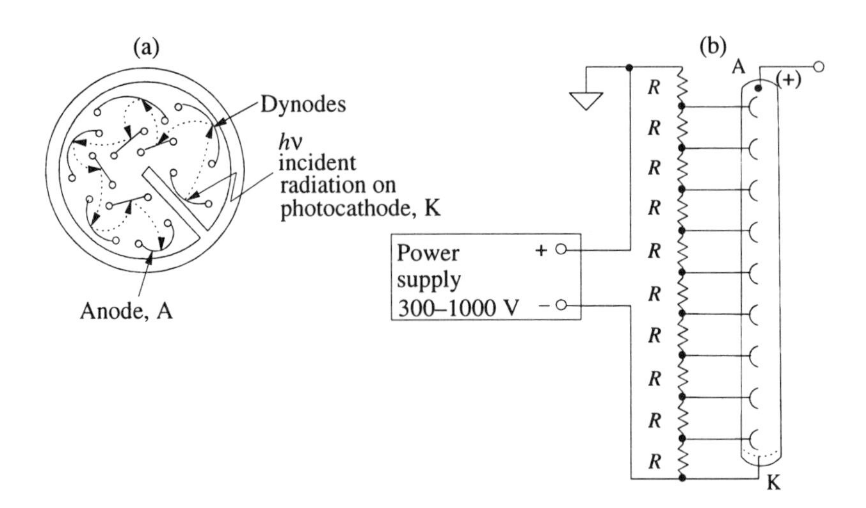

Figure 8–28. Photomultiplier transducer. (a) Cross-sectional view of side-illuminated PMT; (b) schematic showing connection to power supply. The voltage divider provides the required successively higher voltage at each dynode. K, cathode; A, anode.

Photomultiplier Tube (PMT)

The photometric transducer commonly used in many types of spectrometers is the **photomultiplier tube** (PMT). It is also used in specialized rapid-scan, sub-microsecond time-resolved, temperature jump, and chromatogram-scanning spectrometers; multichannel direct-reading spectrometers; densitometers; and many other instruments.

The electrode arrangement of a typical PMT is shown in Figure 8–28a. Like the vacuum phototube, the PMT contains a photosensitive cathode and a collection anode. However, the cathode and the anode in the PMT are separated by several electrodes, called **dynodes**, that provide electron multiplication or gain.

Radiation incident on the photocathode gives rise to photoelectrons with an efficiency $b(\lambda)$. Most of the photoelectrons emitted by the photocathode are accelerated toward the first dynode by its more positive voltage. Each successive dynode is maintained at a more positive voltage by the voltage divider shown in Figure 8–28b. The fraction of photoelectrons collected by the first dynode is the collection efficiency b_c, typically more than 75%. The energy imparted to the surface of the first dynode by each accelerated photoelectron causes several (e.g., four or five) secondary electrons to be ejected from the dynode. The secondary electrons from the first dynode are attracted to the second dynode, where each electron ejects more secondary electrons, and so on from dynode to dynode until the electrons are collected at the anode. Thus, each photoelectron collected by the first dynode is greatly multiplied within the PMT by the process of secondary-electron emission to form a packet, or pulse, of a relatively large number of electrons at the anode. This pulse is typically only a few nanoseconds long.

The number of anode pulses N_a over certain boundary conditions is

$$N_a = b_c b(\lambda) N_k$$

where N_k is the number of photons that strike the photocathode within the boundary conditions. Examples of boundary conditions might be the time of a spark discharge, or a shutter opening, or simply a unit time interval. If the anode pulses are sufficiently infrequent to avoid overlapping each other, N_a can be determined by digital counting techniques. The coulombic content of the average anode pulse is GQ_e, where G is the effective PMT gain (electrons per anode pulse). The gain depends greatly on the power supply voltage but is typically 10^5 to 10^7. For a gain of 10^6, the average pulse contains 1.6×10^{-13} C, which gives an average current of 32 μA over a 5-ns pulse. The counting of anode pulses is the measurement technique known as **photon counting** (*see also* Chapter 9, Section 9–4). When used for photon counting, the output of the PMT is in the digital domain for counts per event or in the frequency domain for counts per unit time interval.

The PMT is also used as an analog transducer with outputs in the current or charge domains. If the desired measurement is the total number of photons N_k that strike the photocathode during a particular event, the output is the charge delivered to the anode during that event Q_a. Thus,

$$Q_a = N_a G Q_e = N_k b_c b(\lambda) G Q_e$$

The desired measurement may instead by the light intensity or photon flux at the photocathode (N_k/t). Now the desired output is the anode current i_a:

$$i_a = \frac{Q_a}{t} = \frac{N_k}{t} b_c b(\lambda) G Q_e$$

For example, if the photon flux at the photocathode is 10^7 photons per s, and if b_c, $b(\lambda)$, and G are 0.9, 0.1, and 10^6, respectively, then the anode current is $i_a = (10^7)(0.9)(0.1)(10^6)(1.6 \times 10^{-18}) = 1.44$ μA.

The PMT owes its great popularity to its versatility, great sensitivity, and extremely wide (10^6 or greater) dynamic range.

Related to the PMT for light measurement is the electron multiplier for charged-particle detection. Here, charged particles strike a cathode, where they give rise to secondary electrons. These electrons are accelerated in an electric field to give rise to an amplified packet of electrons at the anode. Such multipliers are used as ion detectors in mass spectrometry and other areas.

8–6 Output Transducers

Output transducers are widely used in measurement and control systems. In control systems the output transducer or output device is usually the final control element for the process variable *PV*. It receives the signals generated by the controller (*see* Figure 8–2b) and converts these into the manipulated quantity (position, light intensity, heat, etc.). In measurement systems the output device is often some type of display or recording unit. It receives the conditioned signals from the input transducer (sensor) and produces a readout that a human can understand or interpret.

Output devices can also profoundly affect the quality and performance of a measurement or control system. In designing a computer-controlled system, it is very important to know the characteristics of the output device. These include the data domain and magnitude of the input signal required by the transducer and the data domain and magnitude of the output response. In this section some of the most common output transducers are considered. These include ac, dc, and stepper motors; hydraulic and pneumatic actuators; light-emitting devices; and thermal-output elements. Output drivers that are used in interfacing computers to output devices are also considered here.

Electric Motors

An electric motor converts electrical energy into mechanical energy. Electric motors can be dc motors, stepper motors, or ac motors. The dc and ac motors used with instruments and laboratory robots are often servomotors. These are especially constructed for use in control system applications. Motors inherently produce rotary motion. They are thus directly useful for turning an object (e.g., a mirror or a platform) and for rotating objects continuously (e.g., a light chopper). Motors can be combined with a screw drive mechanism to produce linear motion. Precision screw drives, called leadscrews, are used for precise linear positioning with motors. To understand the operating principles and characteristics of motors, we will first look at dc motors. Then we will consider brushless dc motors, stepper motors, and ac motors.

Principles of dc motors. A **dc motor** contains interacting magnetic fields. One magnetic field is created by permanent magnets or field windings in the **stator**, and the second is created by a wire-wound **armature**, as shown in Figure 8–29. The armature is also called the **rotor** because it is free to rotate. The **commutator** acts a double-pole, double-throw switch. The commutator has a conducting part and an insulating part, which allow the rotor coil to develop a reversible magnetic field. Each end of the rotor winding is connected to one of the conducting sections on the commutator.

The magnetic field from the armature windings is developed from a dc power supply connected to components called **brushes**. As the rotor turns, the brushes make contact with the conductive and insulating sections of the commutator. The operation of the dc motor is illustrated in Figure 8–30.

Although Figures 8–29 and 8–30 show permanent magnets for the exciting field, electromagnets are also used. Wound field magnets can be of the series, shunt, or compound variety. Normally, if the load to a dc motor increases, the motor must develop higher torque to move the load. As shown in Figure 8–31a, the higher the

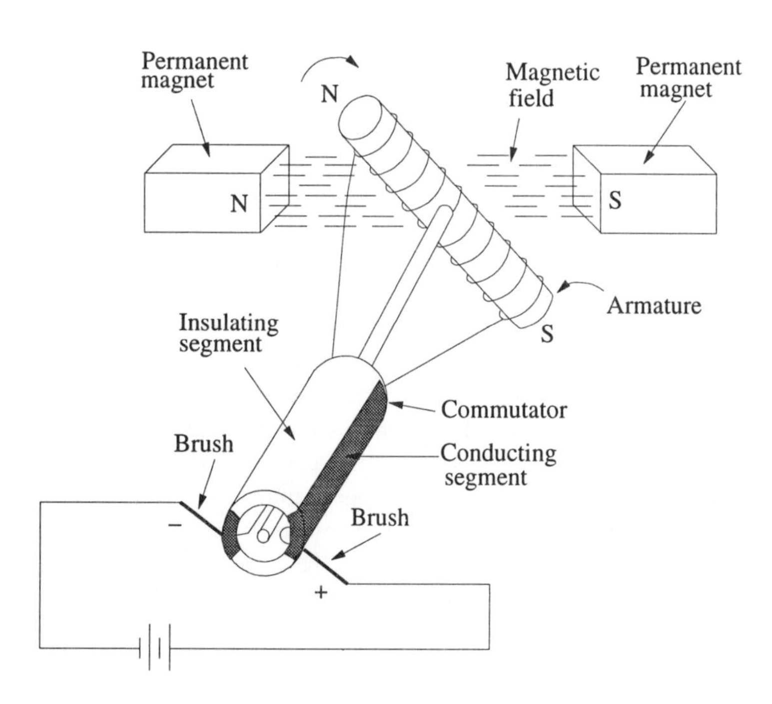

Figure 8–29. Basic components of a dc motor.

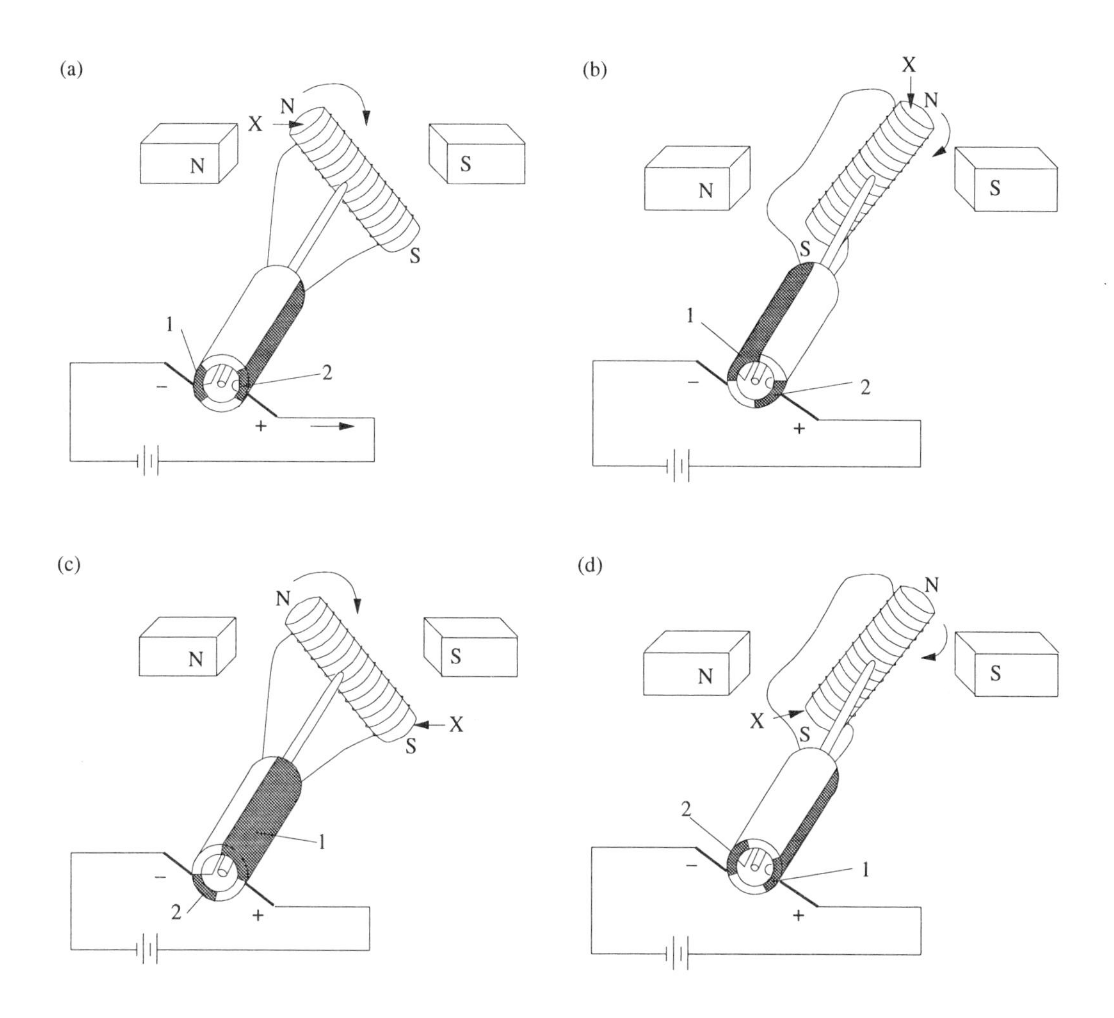

Figure 8–30. Operation of a dc motor. (a) A dc current is applied to the brushes, which contact conducting sections 1 and 2 of the commutator. Current through the armature windings allows a magnetic field of the same polarity as the permanent magnet to be developed. Armature end X becomes the north pole. These two fields repel each other and the rotor turns in a clockwise direction. (b) The armature rotates 90° until the north pole of the armature comes under the influence of the south pole of the permanent magnet. This causes a further 90° rotation. (c) Rotation of the commutator causes the polarity of the dc applied to the armature windings to change. End X of the armature, which was the north pole in part a, is now the south pole. (d) Now the magnetic field of the armature and the permanent magnetic field repel each other again and the rotor continues to turn clockwise. This action continues until the dc power supplied to the armature is removed.

torque, the slower the motor speed at constant applied voltage V_{dc}. To overcome this difficulty, the current applied to the field windings or the current applied to the armature (i.e., V_{dc}) can be increased. Armature control is more common. Figure 8–31b shows constant torque curves of motor speed versus V_{dc}. Motor speed control allows the motor to develop the needed torque and maintain its speed. Solid-state motor controllers based on silicon-controlled rectifier circuits are often used in controlling motor speed.

Most dc motors are reversible by a change either in the direction of the armature current or in the direction of the field current. A reversible motor must have additional control circuitry designed to cause direction reversal. Motor reversing should occur only when the motor is stopped.

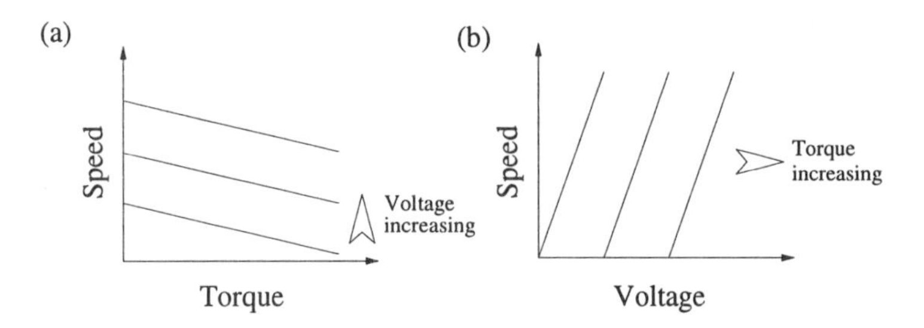

Figure 8–31. Steady-state speed and torque characteristics for a dc motor. (a) Speed versus torque at constant applied dc voltage; (b) speed versus voltage at constant torque.

Stopping a dc motor is also an important operation. The rapid stopping of motor rotation is called **braking**. Braking can be accomplished by a dynamic-braking circuit or by a technique called plugging. In dynamic braking, the dc power source is removed and a braking circuit is connected, often by some type of relay circuit. In plugging, the armature is connected with opposite polarity across the power source. After braking, the motor is reconnected for normal operation.

When a dc motor is used in a control system, it is desirable to feed back information concerning position to the controller. This is ordinarily accomplished with a shaft encoder, as illustrated in Figure 8–32. The incremental encoder produces a series of pulses at the output. These can be counted from a reference position to produce positional information or counted per unit time to produce speed information. The absolute encoder generates a parallel digital output related to shaft position (*see* Chapter 2, Figure 2–1). The incremental encoder is simpler to construct; however, more hardware and software are needed to interpret the output signals. Position information can also be encoded with a potentiometer, a variable transformer, or a positional transducer called a **resolver**. The resolver converts the angular position of the rotor into a proportional voltage.

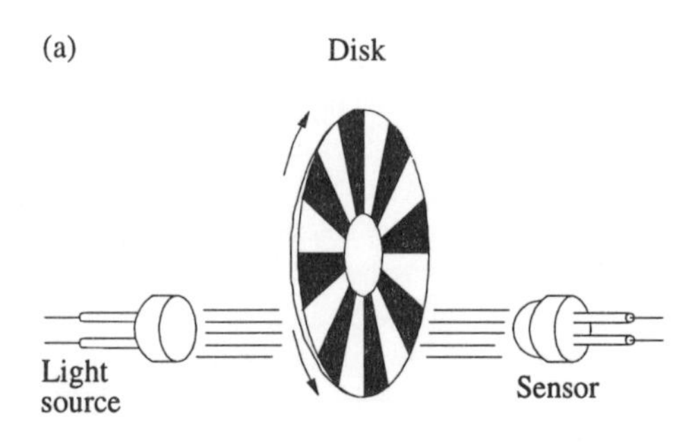

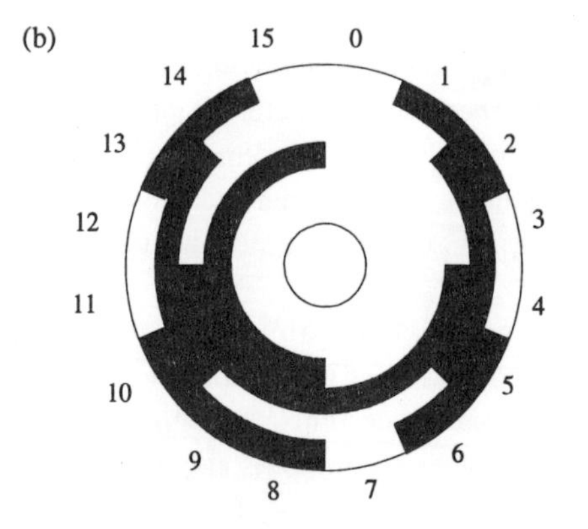

Figure 8–32. Optical shaft encoders. (a) Incremental encoder; (b) absolute encoder.

Brushless dc motors. Brushes are often the weak link in dc motors. They tend to wear out and then cause arcing in the interior of the motor. The wear can also cause brush dust, which can get into the bearings of the motor and cause damage. Instead of brushes, an electronic switch can be used to change the armature current. Two types of brushless motors are used: the split-phase permanent magnetic motor and the Hall effect motor. The latter uses a permanent magnet rotor and is based on the Hall effect (*see* Figure 8–10).

Stepper motors. Stepper motors are very popular in computer-controlled systems because they eliminate the need for feeding back positional information. Stepper motors operate by means of pulses to the motor. Each pulse causes the motor to rotate by a certain fixed amount. By keeping track of the number of pulses applied, the computer can calculate the angular movement of the rotor and from this the position of the load. Stepper motors are available with step sizes in the range 1.5° to 30° per pulse.

The operation of unipolar and bipolar stepper motors is illustrated in Figure 8–33. The unipolar stepper motor (Figure 8–33a) has a 90° step size. When windings A1 and B1 are energized (left position) they become north poles, which attract the south pole of the rotor. Application of a pulse switches B1 off and B2 on (right position), which attracts the south pole of the rotor to a new position 90°

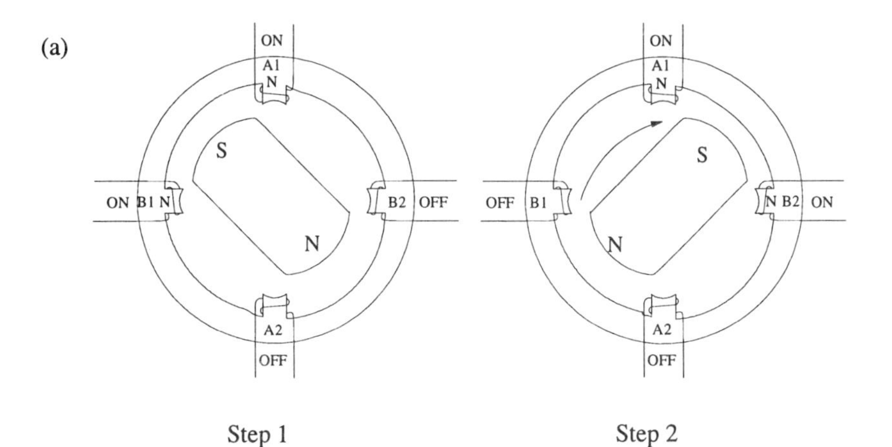

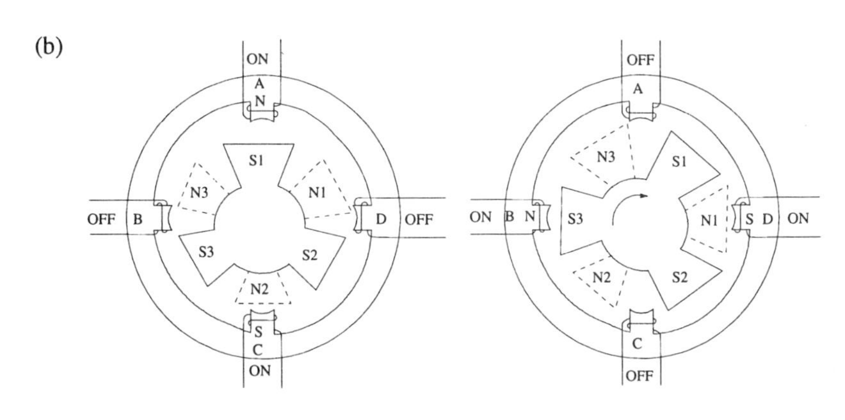

Figure 8–33. Stepper motors. (a) Unipolar; (b) bipolar.

clockwise. The four-phase bipolar stepper motor (Figure 8–33b) has a 30° step size. The rotor has three north poles and three south poles, and the windings are energized in the sequence shown. Even smaller step sizes can be obtained with more poles. A stepper motor has the ability to maintain a static holding torque when energized.

Stepper motors cannot produce the torque associated with dc and ac motors. However, they are readily interfaced to computers and are nearly ideal for positioning relatively light loads.

Alternating current (ac) motors. Alternating current servomotors are also widely used. These are based on the ac induction motor. In the induction motor, the stator establishes a rotating magnetic field that causes the rotor to turn and develop the torque for moving the load.

The most commonly used type of ac servomotor is the two-phased induction motor. The stator in the two-phased motor has two sets of windings: the reference winding and the control winding, as illustrated in Figure 8–34. The two-phase motor must have a supply that provides reference and control voltages that are 90° out of phase. A special two-phase supply can be used, or a capacitor can be placed in series with the reference winding.

As shown in Figure 8–35, when the ac current in coil A leads that in coil B by 90°, the magnetic field rotates counterclockwise (top row of circles). At time t_0, the current is maximum in A and zero in B, and the direction of the magnetic field is shown by the small arrow in the first circle. At time t_1, the currents A and B are equal, and the magnetic field has rotated 45° counterclockwise during the time $t_1 - t_0$. At time t_2, the current is maximum in B and zero in A so that the magnetic field has rotated 90° counterclockwise during the time $t_2 - t_0$. The magnetic field continues to rotate counterclockwise as long as the current in stator winding A leads that in B by 90°. As soon as the voltage is removed from coil A, the rotating magnetic field is eliminated and the rotor stops. Now, by shifting the phase of the voltage applied to winding A by 180° from the previous case, the current in winding A lags that in B by 90°. The magnetic field will now rotate clockwise, as indicated by the bottom row of circles in Figure 8–35. The magnetic field rotates at a

Figure 8–34. Two-phased ac motor. (a) Wiring of windings; (b) schematic diagram.

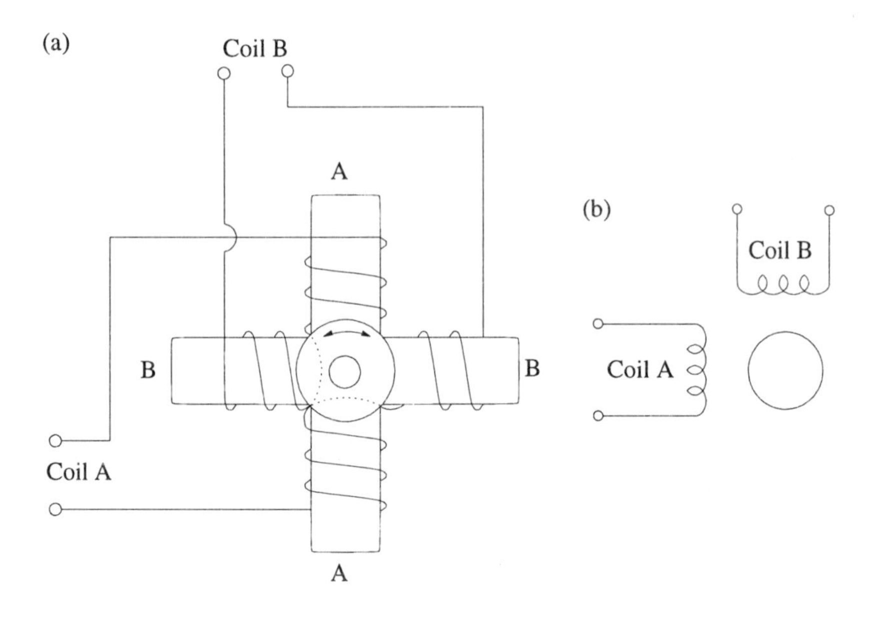

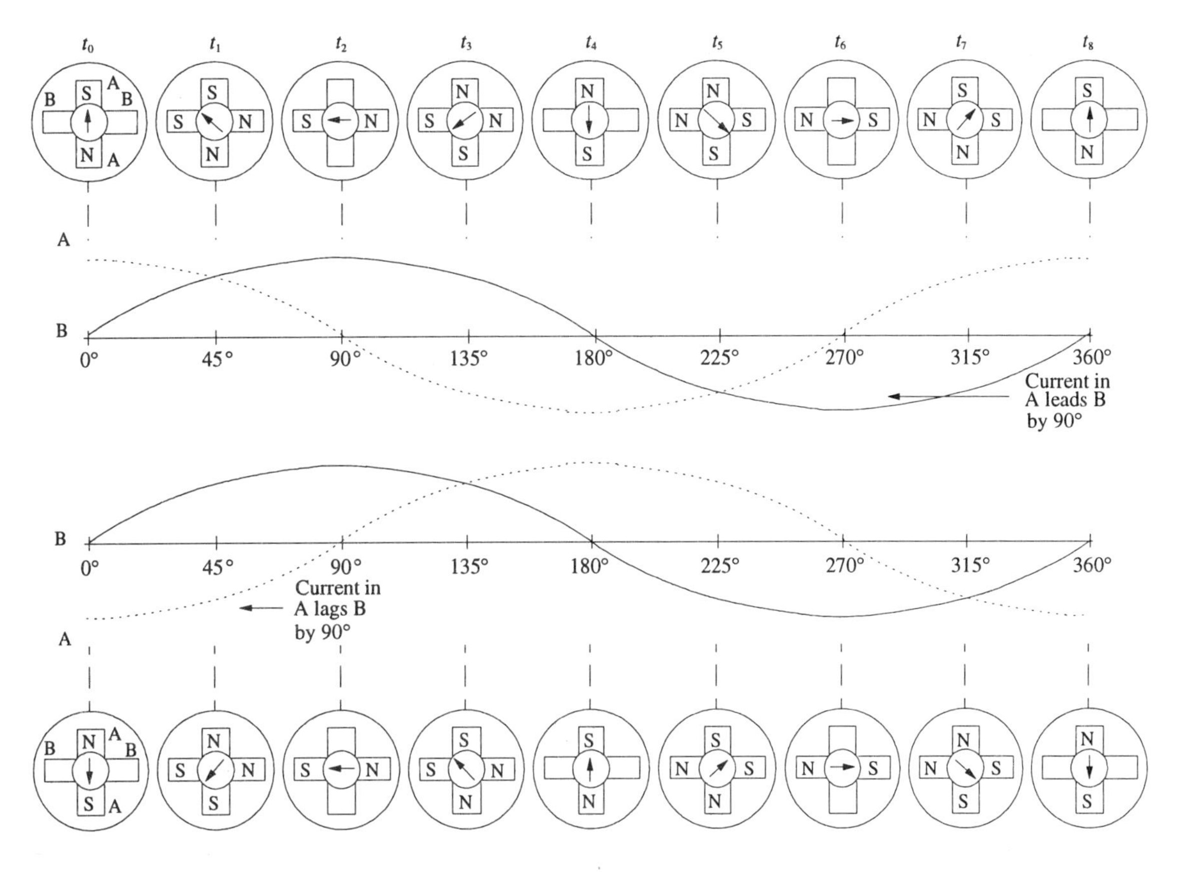

Figure 8–35. Rotating magnetic field in a two-phased induction motor.

synchronous speed given by 120 *f/p*, where *f* is the frequency of the ac power applied to the stator and *p* is the number of poles of the stator. The interaction of the resultant stator magnetic flux with the rotor conductors induces a rotor current and generates a rotor magnetic field that is in opposition to the stator magnetic field, which provides torque to the rotor.

The operating speed of an induction motor can never equal the synchronous speed of the stator field, for the rotor conductors would then be stationary with respect to the stator field and no current would be induced. Instead, the rotor travels at that lower speed needed to keep the rotor and stator fields stationary with respect to one another. This produces torque and maintains rotation. The difference between the synchronous speed and the rotor speed is called the **slip** of the rotor. To produce a higher torque, the motor speed must diminish. This is, then, one of the problems with normal ac induction motors, namely, that the speed decreases with increasing load.

The rotor can be of the squirrel cage design or wire wound. The former consists of bars of heavy-gauge aluminum or copper conductors separated by ferromagnetic material. The latter consists of windings on an iron core.

Special ac motors called **synchronous motors** can overcome the speed–torque limitations of normal induction motors. These motors are in fact made to operate at the synchronous speed. The ac windings on the stator of a synchronous motor are identical to those on an induction motor. The rotor, however, has a winding similar to the field winding of a dc motor. Direct current is usually supplied from a small dc generator called the **exciter**. With such a motor, the stator and rotor fields are stationary with respect to each other only when the rotor turns at the synchronous speed. Synchronous motors are thus made to operate at only a single speed, which is determined by the frequency of the ac voltage applied and the number of poles. A synchronous motor does, however, have constant speed regardless of load.

The ac motors discussed can provide relatively high torque, and they operate without brushes. They are ideal as control devices when ac coupling must be used. On the other hand, they are relatively inefficient and heavy and somewhat more difficult to interface with computer controllers than dc or stepper motors.

Hydraulic and Pneumatic Actuators

Hydraulic and pneumatic actuators are widely used in automated systems. Hydraulic systems are particularly useful for moving very heavy loads, whereas pneumatic systems are most useful for light loads. Such actuators are used in automated factories and assembly plants. In addition, we are seeing the increasing laboratory use of hydraulic and pneumatic systems in robots and robotic manipulators.

Hydraulic systems. Hydraulic systems convert hydraulic energy into mechanical energy. They can generate very large forces because of the high working pressures used (e.g., up to 280×10^5 N/m^2 [280 bar]). An important advantage is the high force-to-weight ratio that can be achieved. This makes hydraulic systems attractive for use in remote applications such as aircraft, spacecraft, and robots.

Hydraulic actuators are either **hydraulic cylinders** or **hydraulic motors**. Both are based on the use of hydraulic flow to provide motion to an actuator. Hydraulic flow is developed by a pump. Although the word "hydraulic" comes from the Greek word for water, oil is the most commonly used fluid in the drives considered here.

Hydraulic cylinders are linear actuators because they develop a linear force at the output of the cylinder. Several common types of cylinders are illustrated in Figure 8–36. The ram cylinder has only one fluid inlet and thus can achieve only an upward motion when fluid is flowing. Retraction may be achieved by means of the load itself, by gravity, or by a spring. Because the cylinder has to do work against the spring, spring returns are normally limited to short-stroke applications. The double-acting and double-rod cylinders have two ports so that fluid can enter and exit in two directions. This allows positive cylinder control during both extension and retraction cycles.

Hydraulic motors are rotational actuators. Pressure from the hydraulic fluid causes the motor to develop torque, which provides a continuous rotational motion. A hydraulic motor is shown in Figure 8–37. Such motors are rated by their displacement and their torque. Typically, a hydraulic motor can develop a displacement of 10 in.^{3} (ca. 164 cm^3) per revolution.

Hydraulic or electrohydraulic servo systems are widely used in position control devices. The former have mechanical feedback, while the latter have electrical feedback. Such systems allow force amplification and precise control over mechanical position.

The advantages of hydraulic systems, namely, high force and high force-to-weight ratio, are somewhat offset by their expense. There must be a reservoir for storing the hydraulic fluid, and there is always the possibility of fluid leakage.

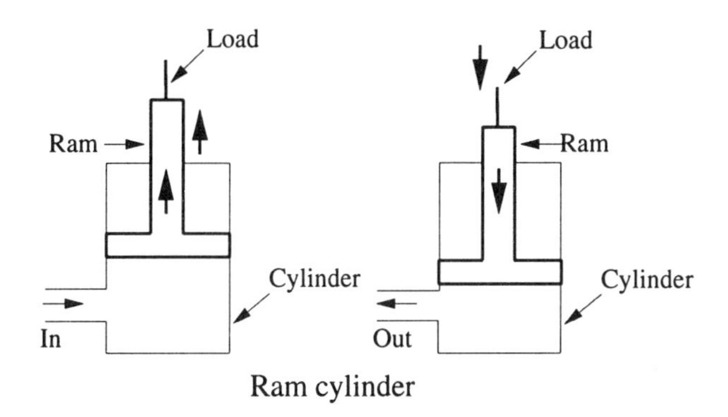

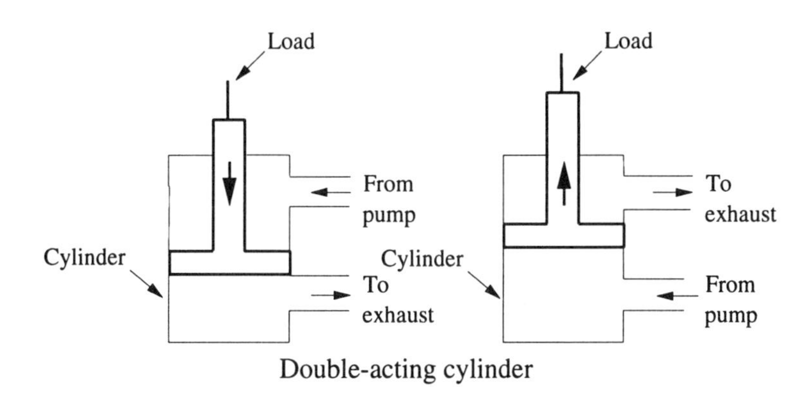

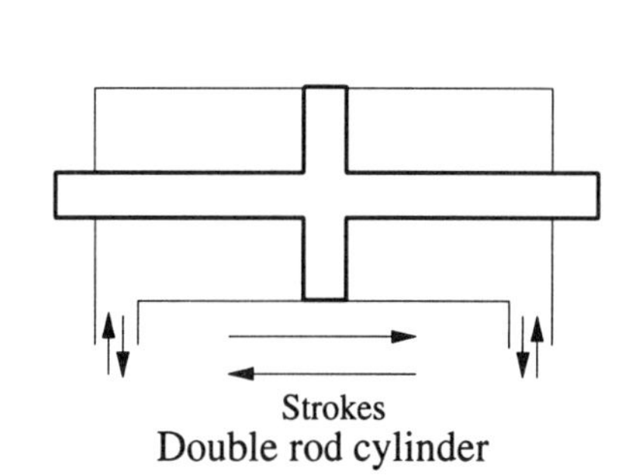

Figure 8–36. Hydraulic cylinders. Fluid flow in and out of the cylinder causes the actuator to move.

Pneumatic systems. Pneumatic drive systems are similar to hydraulic systems except that they use compressed gas, usually air, instead of oil as a medium of power transmission. Because of lower operating pressures, pneumatic actuators are used for positioning light loads.

Pneumatic cylinders often include a deceleration cushion at the end of the stroke. This cushion helps dissipate the kinetic energy of the load smoothly and prevent mechanical damage to the cylinder. The linear output of a pneumatic cylinder can be converted to rotary motion with various external linkages. Alternatively, there are commercial vane-type actuators and rack-and-pinion designs that give a limited rotary output.

Both hydraulic and pneumatic drive systems require valves to direct the fluid or gas to various points. A **check valve** can operate either as an on–off directional

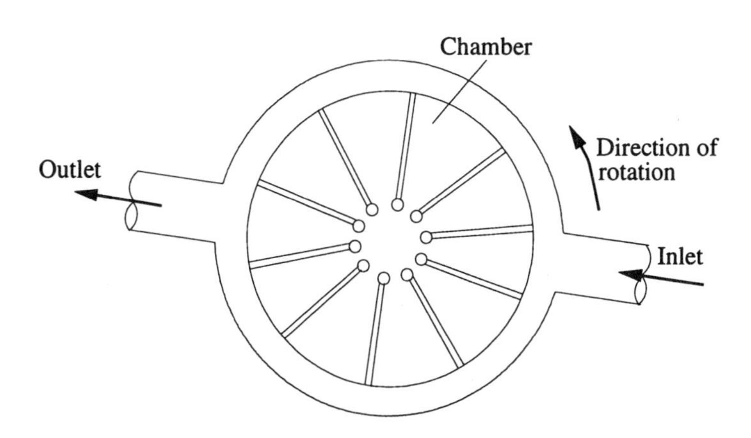

Figure 8–37. Hydraulic motor. Flow of hydraulic fluid from inlet to outlet causes the motor to develop rotational motion. Reversing the direction of fluid flow causes the motor to turn in the opposite direction.

control or as a pressure control. The check valve is a type of fluid diode in that it allows passage of fluid in only one direction when the fluid flow is sufficient to unseat a ball. Multiport valves are used to control the movement of cylinders. A three-port valve is needed to control a single-ram cylinder, whereas two three-port valves are needed to control double-acting cylinders. When continuous control is needed rather than on–off control, a servovalve is used. These valves are capable of an infinite number of positions.

Pneumatic systems are relatively inexpensive compared with hydraulic systems. Compressed air is readily accessible, and pneumatic systems are clean. System components are commercially available. They are highly reliable and also easy to maintain. On the other hand, compressed air is somewhat expensive to produce; on a power basis, it is more expensive than electricity or hydraulics. Because of the elastic nature of compressed air, it is difficult to achieve extremely high accuracy with pneumatic drives. If large forces are needed, pneumatic systems can become bulky and expensive. For these reasons, hydraulic systems are preferred for very accurate positioning of large loads.

Light-Emitting Devices

The function of light-emitting devices is opposite that of photodetectors in that they convert electrical signals into light intensity. Light-emitting devices are used as output transducers for control, isolation, or display purposes. Sources used as transducers in automated systems should be stable, relatively long-lived, and of high output brightness. Incandescent lamps, neon bulbs, electroluminescent devices, LEDs, and various types of semiconductor lasers have been used.

Incandescent lamps. Incandescent lamps are made of resistive materials that are heated electrically. The tungsten filament lamp is inexpensive and has a wide spectral output similar to blackbody radiation, as shown in Figure 8–38. Tungsten lamps are readily available in a variety of sizes and shapes. They are used as sources in light-controlled switching devices, spectroscopic instruments, visual lighting systems, and many other applications. Tungsten lamps are quite slow in responding (rise times are typically 100 ms) and susceptible to vibration damage and burnout. They are used primarily in systems that do not need high-speed intensity changes.

Neon lamps. The neon lamp is inexpensive and has a long life. It requires less power than the incandescent bulb and is much faster in responding (typically, microsecond rise times). However, a high-voltage power supply (e.g., 170 V) is needed. Most of the light output is limited to two orange spectral lines, and the maximum output intensity is fairly low. This limits the flexibility of the neon bulb in optical coupling, because a detector that can respond to the orange lines is needed.

Light-emitting diodes. LEDs are pn junction devices; when forward biased, they emit photons as a result of electron–hole recombinations. Gallium arsenide (GaAs), gallium phosphide (GaP), and silicon carbide (SiC) are commonly used in LEDs. Mixtures of these compounds are used to shift the wavelength of maximum emission to anywhere in the region from 540 to 900 nm (*see* Figure 8–38 for gallium arsenide phosphide [GaAsP]).

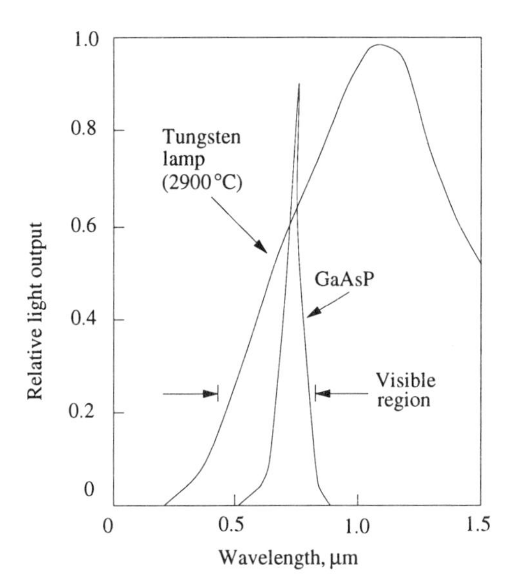

Figure 8–38. Relative outputs of light sources versus wavelength.

The LED is usually operated at low voltages (e.g., 5 V) and with currents in the range of 1 to 260 mA. Other important advantages of LEDs include their small size and fast response time (nanoseconds). They also have good dynamic range and an output intensity that is fairly linear with applied forward current. They can be operated in a pulsed mode to increase the output intensity; the LED can withstand higher peak currents in a low-duty-cycle pulsed mode.

In the seven-segment LED display, seven independent LEDs are used. Combinations can be lighted to form the numerals 0 through 9 (*see* Chapter 6, Figure 6–28). Dot matrix displays containing LEDs are used in scoreboards and bank time–temperature billboards. These 5 × 7 matrices consist of 35 sources. They can be illuminated to form all alphanumeric characters.

Electronic displays can also be made from gas discharge devices (plasma devices), electroluminescent devices, and cathodoluminescent devices. Liquid crystal displays (LCDs) do not emit light. They depend on reflected light or light sent directly through the display. When a liquid crystal material receives an electrical charge, the electrons in the material align and the material becomes transparent and allows light to pass. Uncharged areas are opaque and become visible to the eye. Backlighting allows the display to be used even in low-ambient-lighting conditions. Laptop and notebook computers use LCDs to produce a complete video display.

Diode lasers. In a diode laser, population inversion is achieved between the conduction band and the valence band of a pn-junction diode. Stimulated transitions of the electrons from the conduction band to the valence band are responsible for laser action, and stimulated emission occurs because of electron–hole recombinations. The frequency of the light emitted is directly related to the band gap energy. Hence, different combinations of semiconductors can be used to obtain different wavelengths. Because the lasing occurs between bands, diode lasers can be tuned over a small wavelength interval. Typical materials are GaAs, which lases at 840 nm, and PbSnTe, which lases in the midinfrared region.

Optical coupling. The combination of a light source and a photodetector inside an opaque enclosure can produce a totally isolated analog switch often called an

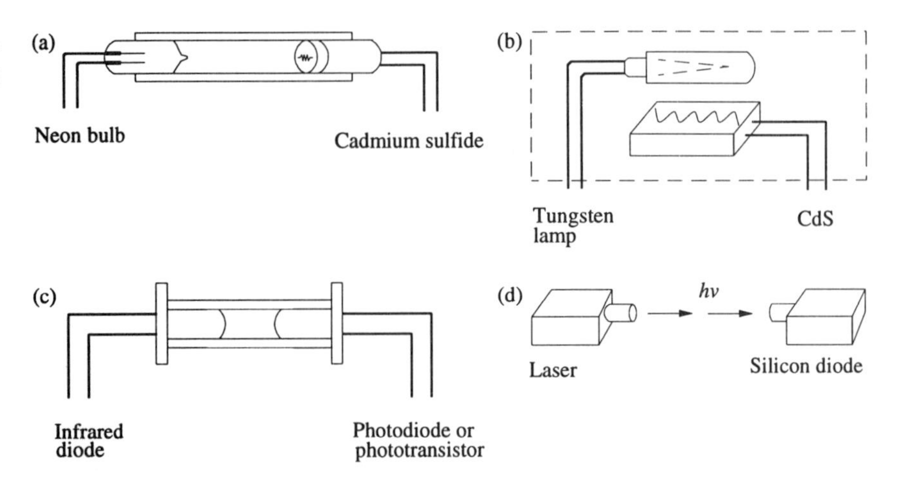

Figure 8–39. Source–detector pairs used in optically coupled switches.

optoisolator. The actuating circuit is the light source, and the photosensitive device acts as the switch. No electrical coupling between the actuating circuit and the switch is needed. Light is the coupling link. Because the light source terminals and the photodetector terminals are electrically isolated, the pair forms an almost ideal four-terminal device in which there is almost no interaction between switch and driver circuits.

Some typical optical links between matched source and detector pairs are illustrated in Figure 8–39. In recent years, completely encapsulated **optical couplers** have become available commercially. Many of these use LEDs as sources and phototransistors as detectors. In addition to providing electrical isolation of switch and actuator, optical coupling eliminates ground loops and isolates noise sources.

Other Output Transducers

Many other output devices are used in control and measurement systems. It is frequently desirable to regulate temperature in experiments or buildings or for other applications. In some cases temperature control is provided by refrigeration and/or heating units that circulate fluids to the desired location. Temperature control of a laboratory reactor, for example, can be accomplished by circulating water or some other fluid from a constant-temperature bath. The actual heating of the fluid is often accomplished by resistive (Joule) heating. Here, the controller supplies current through a resistive device that serves as a heating element. Heaters are available in a variety of sizes and shapes for different applications.

Thermoelectric heating and cooling devices are often based on the **Peltier effect**. Whenever current is forced through a pair of dissimilar junctions, one becomes warmer and the other becomes cooler. Peltier effect devices usually consist of p and n semiconductors connected by conductors. The passage of current results in heat being pumped from one junction (the cold junction) to the other (the hot junction). If heat is removed from the hot junction, the cold junction becomes cool. When operated as coolers, the hot junction becomes warm not only because of the Peltier effect but also because of Joule heating. This heat must be removed to get efficient cooling. Usually, Peltier devices are used when localized low-capacity cooling is needed.

Output Drivers

In an automated system the controller must be interfaced to the required control elements. With microcomputer control systems, this interfacing is usually accomplished with a parallel output port. Because each bit in the latch of an 8-bit parallel port is independently controllable by the word written into it, eight different devices could be turned on and off by that one port. For on–off control, the interface to the device is completed by a **digital driver** circuit that can control power to the device in response to the logic level at its input. In some cases on–off control may not be desirable, and we may want control in the analog domain.

Buffer driver. A simple digital driver is the open collector gate shown in Figure 8–40. The gate output is the collector of a transistor switch (*see* Supplement 3, Section S3–1) driven by the gate logic. The output is thus a logic-controlled switch for positive current to common. The voltage and current in the switched circuit are limited by the characteristics of the output transistor.

Many low-voltage dc devices can be controlled by the open collector buffer driver just by connecting them in the load position shown in Figure 8–40. Such loads include small incandescent lamps and LEDs (with appropriate current-limiting resistors in series). In other cases, devices may have to be driven by a different power source (e.g., an ac motor operated from the power line), and this may require isolation from the computer power supply.

Electromagnetic relays. Electromechanical relays can be used to isolate a controlled circuit from the controller. Relays are remotely controlled mechanical switches. Electromagnetic relays use a current through a coil to provide a magnetic field that moves the switch contacts as illustrated in Figure 8–41 for an **armature relay**. The minimum current that is required to move the armature is called the **pull-in current**, because at or above that current the armature "pulls in" to close the normally open contact. At somewhat lower current, the armature "drops out" and the normally open contacts open. Switching circuits are normally designed to exceed the pull-in current by several times the minimum to ensure operation of the relay. The electromagnetic relay is a four-terminal device in which the actuating

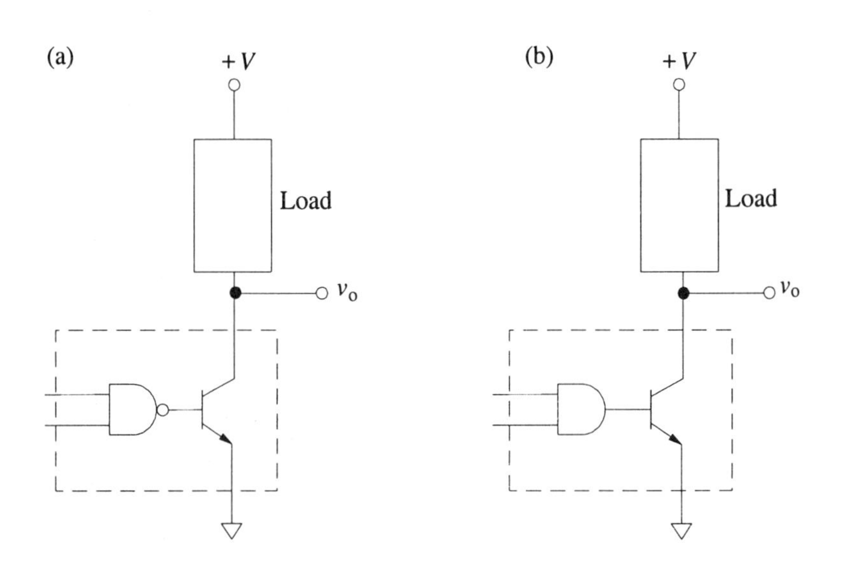

Figure 8–40. Open collector gates. An output transistor switch is turned on or off by the output of a logic circuit. When the logic output is HI, the transistor is ON and the connection from $+V$ through the load is complete; that is, the load device is turned ON. If the load is a resistor, v_o is LO when the logic output is HI and vice versa. Thus, the circuit in part a is called an AND gate, and the circuit in part b is called a NAND gate. If there is only one input to the gate circuit, the circuit in part a is a buffer driver and the circuit in part b is an inverting buffer driver.

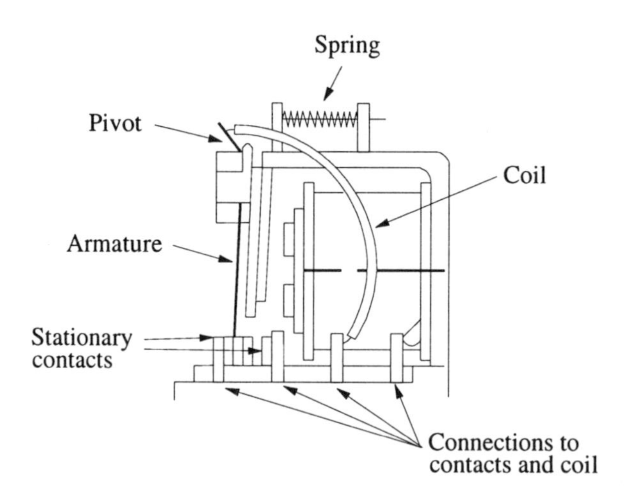

Figure 8–41. Pictorial diagram of a single-pole double-throw (SPDT) relay. If the current in the coil (and thus the magnetic force) exceeds a minimum value, the armature moves the movable contact until it touches the stationary contact on the right.

terminals are electrically isolated from the switched signal terminals. They have activation response times of a few milliseconds, very high open-circuit resistance, very low contact resistance, and, often, the ability to switch high currents and/or voltages.

The **reed relay** contains two or more metal reeds enclosed in a hermetically sealed glass capsule. The reed contacts can switch to actuate an electromagnet.

Relays for various applications differ in the number of contacts and contact arrangements. The nomenclature and symbolism for the four most common contact forms are given in Figure 8–42. Many other contact forms are available. These may be combined in a variety of multiple-form arrangements.

Most relays require several milliseconds to complete the transition from one contact state to another. When a new contact is made, they exhibit **contact bounce,** which can seriously distort the switched signal and limit switching speeds. Mercury-wetted relays are sometimes used to overcome the contact bounce limitation.

When an inductive load, such as a motor, relay, or solenoid, is to be driven from a computer, the circuit of Figure 8–43 should be used. The supply voltage V is chosen to match the requirements of the load. Normal relays (and also solenoids) require that the current in the coil remain on as long as the armature is to remain in the ON position. This uses the power during the ON time, and the device auto-

Figure 8–42. Four common forms of relay contacts with designations. The heavy arrow indicates the position and direction of the force from the coil when energized. SPST, single-pole single-throw relay; SPDT, single-pole double-throw relay; NO, normally open; NC, normally closed.

Form	Description	Symbol	Form	Description	Symbol
A	Make or SPST, NO	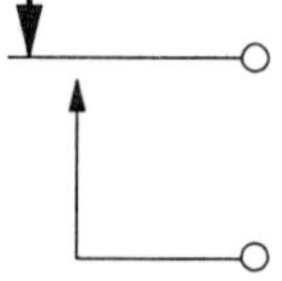	C	Break, make, SPDT (B-M), or transfer	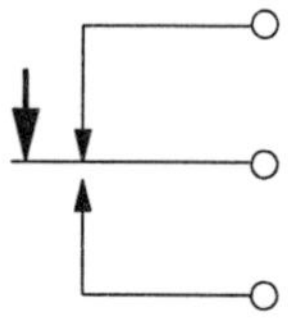
B	Break or SPST, NC	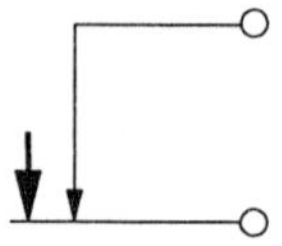	D	Make, break, make-before-break, or SPDT (M-B)	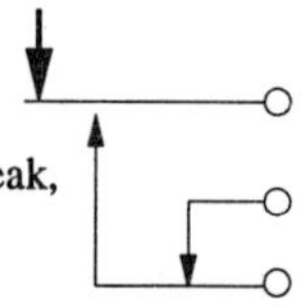

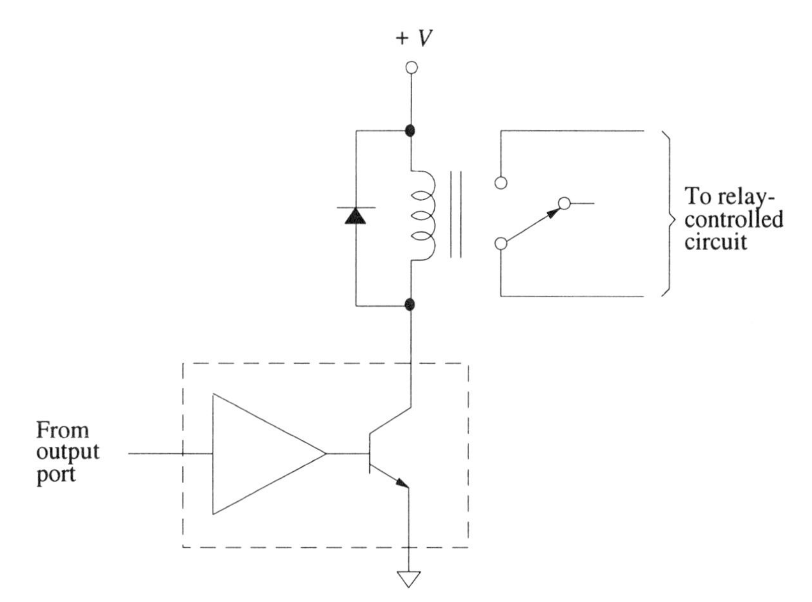

Figure 8–43. Driving an inductive load with an open collector output. The diode protects the transistor from the voltage spike produced when the current through the coil is turned off.

matically reverts to OFF when the system power is turned off. A **latching relay** (or latching solenoid) remains in its ON position once put there by a momentary current. To return it to the OFF state, a momentary current is applied to an opposing coil, or current is removed from a holding coil.

Solid-state relays. The advantage of solid-state relays is that they can be switched more rapidly than their mechanical counterparts. Also, they exhibit a longer life, have no contact bounce, are immune to vibrations, and are usually smaller. With these devices a small dc signal can switch high-power dc or ac loads. Several types of solid-state relays are available. Most involve semiconductor devices called **thyristors**. Power devices in this class include the silicon-controlled rectifier and the triac (*see* Supplement 3, Section S3–4). The triac is the most popular semiconductor device for replacing mechanical relays. This application is so common that all the needed gating and control circuitry is combined in a single encapsulated package. Many of these are transistor–transistor logic compatible. These relays are usually optically isolated so that they act as true four-terminal devices. For switching ac power, zero-voltage switches are often used. These **solid-state ac relays** are activated only when the line voltage crosses zero to minimize radio frequency noise.

Analog control. The drivers discussed above all provide simple two-level, or ON–OFF, control. Additional levels of control can be achieved by using more bits of the output port and some means of converting the digital output word into a multilevel control signal. For control in the analog domain, a digital-to-analog converter (DAC) is generally used (*see* Chapter 7, Section 7–5). A DAC is not designed to provide significant power at its output. Therefore, a DAC output that is to control a load such as a dc motor or a heater requires a power amplifier. A simple power amplifier can be made from an operational amplifier (op amp) and a booster amplifier, as shown in Figure 8–44. Incorporating the booster amplifier in the follower-with-gain feedback loop as shown maintains precise control over the booster output voltage. The booster amplifiers available from op amp manufac-

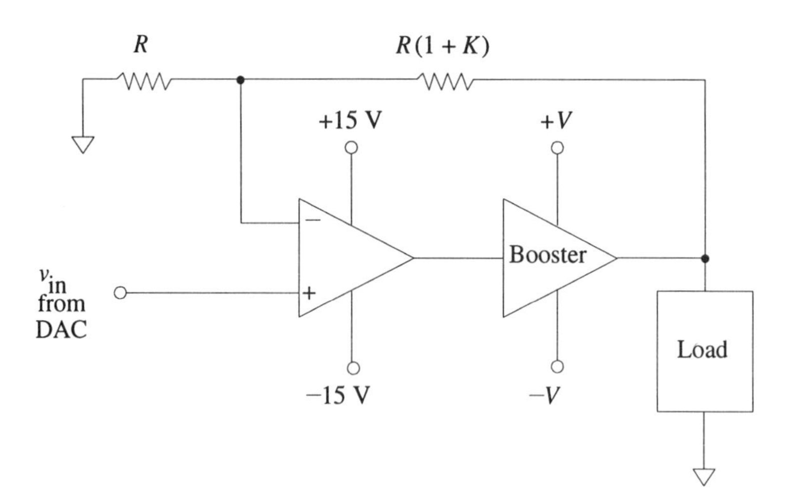

Figure 8–44. Power amplifier for driving a load from a DAC. The booster amplifier acts as a higher-current (and sometimes higher-voltage) output stage for the op amp. Because it is included in the gain-controlling feedback loop, the circuit acts as a high-output voltage follower with a gain of K.

turers can provide many watts of output power to a load. For still greater power, the DAC output can be used to control a voltage-programmable power supply (*see* Chapter 7, Section 7–2).

8–7 Automatic Control Systems

We have seen that truly automated systems require both a programmable event controller or sequencer and a programmable variable controller. In this section we will discuss how the controller can be combined with control, feedback, and interface elements to produce automated systems. Several examples of computer-controlled systems for controlling light intensity, motor speed, and other variables are explored.

Event Controllers

In automated systems, repetitive sequences of operations are often performed. Event controllers can be time based, event based, or fully programmable.

Time-based controllers. With time-based sequencers, the various operations are initiated at predetermined times. An electromechanical time-based controller can be made by operating switch contacts by cams attached to a motor shaft. One cam and one switch are required for each operation. Commercial electromechanical controllers are available with adjustable cams for varying the sequence as required.

An electronic **time-based controller** can be made with several monostable multivibrators connected in series. The output pulse of each monostable unit triggers the input of the next. Thus, pulses appear at the outputs in succession; these pulses can be used to actuate the desired operations. Because the monostable multivibrator is not a highly accurate or precise timer, this type of sequencer is limited to uses where accuracy and precision are not critical. Much more accurate time-based sequencers can be made from crystal oscillators and digital counting circuitry.

Event-based controllers. Time-based controllers are somewhat limited, particularly if operations require variable amounts of time. In such cases it is often desirable that one operation be completed before the next is initiated. This type of control is termed **event-based control**. Electronic event-based controllers can also be based on monostable multivibrators. Instead of connecting the output of one monostable multivibrator to the input of the next, a signal representing the completion of one operation can be used to trigger the next monostable unit. One could, of course, use a combination of time- and event-based control by connecting the monostable multivibrators in the appropriate manner.

Programmable event controllers. The sequencers described so far are all said to be "fall-through" sequencers in that they merely proceed from one operation to the next. A more versatile sequencer would be capable of branching and thus would be able to skip, repeat, or alter sections of the basic sequence. Such controllers are said to be **programmable event controllers.** The programmable event controller is capable of both time- and event-based control. There are both hardware-based controllers and software-based controllers.

One hardware-based programmable controller is the state counter sequencer. The heart of the state counter sequencer is a counter, a multiplexer, and one or two decoders. As shown in Figure 8–45, an eight-state sequencer with both primary and secondary functions can be made with four integrated circuits. Some additional circuitry is needed, of course, for generating delays, producing the input condition signals, and driving the various devices being controlled.

Programmable logic controllers (PLCs) for switching a number of electrical outputs on and off according to a preprogrammed sequence are available commercially. Most of these units are microprocessor based with programs stored in read-only memory (ROM). Many are based on erasable programmable read-only memories (EPROMs) and can be field programmed for added versatility. Of course, the most versatile sequencer is the full microcomputer system, where the software determines the sequence. With such systems, one merely changes the program to alter the sequence or incorporate new functions. Computer systems can also readily carry out conditional branch and loop operations. The combination of this sequenc-

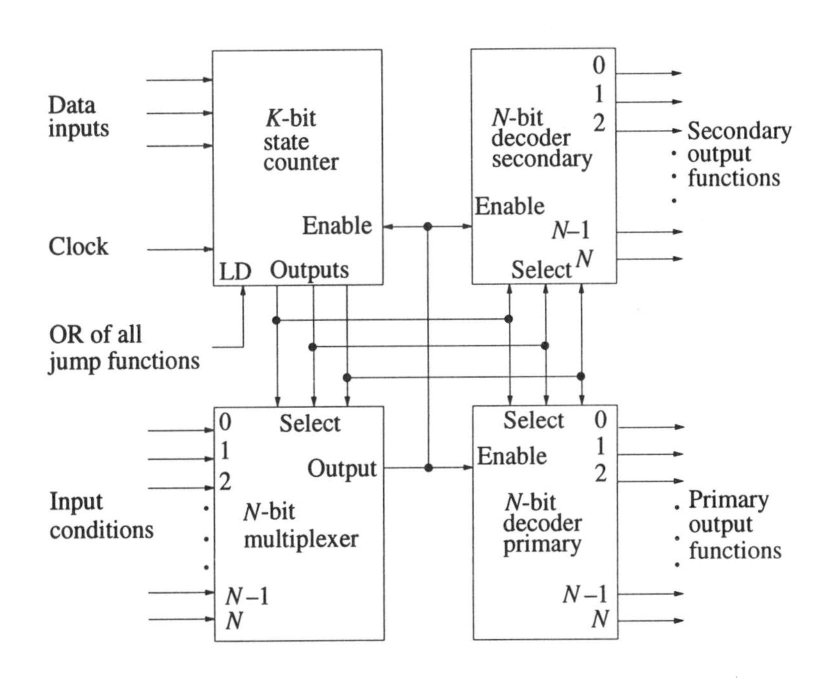

Figure 8–45. General state counter sequencer. The K-bit counter keeps track of the state of the system. The outputs of the state counter determine which input condition is supplied by the multiplexer to the decoders. The presence of a true input condition at the multiplexer output causes the appropriate primary output function to go true and the state counter to increment on the next clock pulse. A false input condition causes either a secondary output function line to go true or a wait period until the condition is true. When a jump or branch in the sequence is necessary, the load (LD) line of the state counter is activated, and a new state is entered through the data inputs.

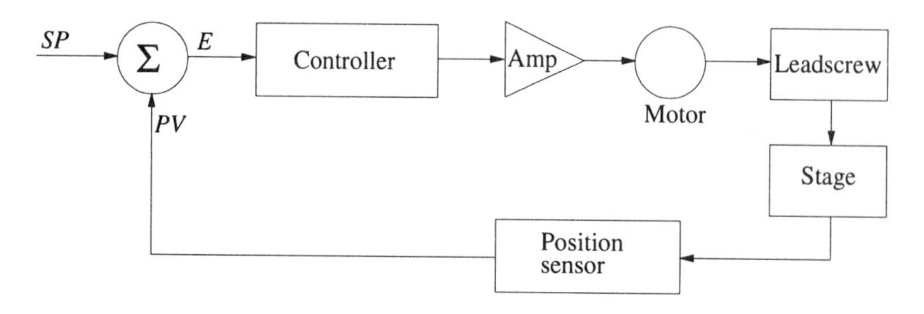

Figure 8–46. Block diagram of a servo system for controlling position. The command signal from the event controller is compared with the position feedback signal from the stage being positioned. Any difference between these signals is processed by the variable controller, amplified by the servo amplifier (Amp), and applied to the motor to adjust the position of the stage.

ing capability with the ability to control external variables and devices is a major reason that microcomputer systems are so commonly used in automated systems.

Process Variable Controllers

As mentioned earlier, control systems are either open loop or closed loop. The most accurate control systems are closed-loop systems in which the process variable being controlled is monitored and its value is fed back to the controller to produce an error signal. A home furnace is an example of a closed-loop system. The thermostat compares the room air temperature with the desired setpoint value and closes a switch if the temperature is below that value. The controller responds to the closed switch by activating the burner to heat the air. This continues until the setpoint temperature is reached. Closed-loop control systems fall under the general definition of servo systems. We will first investigate some of the components of servo control loops and then consider the dynamics of servo systems and their modes of operation. The section concludes with several examples of practical control systems.

Control system dynamics. A block diagram of a simple servo system for controlling position is shown in Figure 8–46. In simple control systems without a computer, the controller is actually a combination of the comparator and servo amplifier and not a separate element. With computer-controlled systems the computer can be outside the loop, where it is used for setpoint adjustment, or within the loop, as shown in Figure 8–46. In some systems a single computer serves as the event controller, the setpoint command controller, and the variable controller.

The dynamics of control system operation are extremely important. In the positioning example of Figure 8–46, instability could readily occur if the control system hunts for the appropriate position. Hunting occurs if the system is improperly damped. The servo system overshoots the control point in one direction and then undershoots it in the other direction, resulting in oscillation. An appropriately damped servo system comes to balance in the minimum time without oscillation.

In a control system, every element in the loop can affect the system response. Consider the temperature control system of Figure 8–47. There is **lag** in the heater, because the temperature of the heating element and the pipe cannot change instantly with changes in the heater current. Then there is a delay called **dead time** between the heating of a portion of the liquid and the transport of the liquid to the sensor. The sensor, too, has lag, because its temperature cannot instantly follow the changes in liquid temperature. The controller can be designed to respond in any of several ways to an error input E. The current applied may be fully on, proportional to E, proportional to the rate of change of E, or proportional to the accumulated error, or

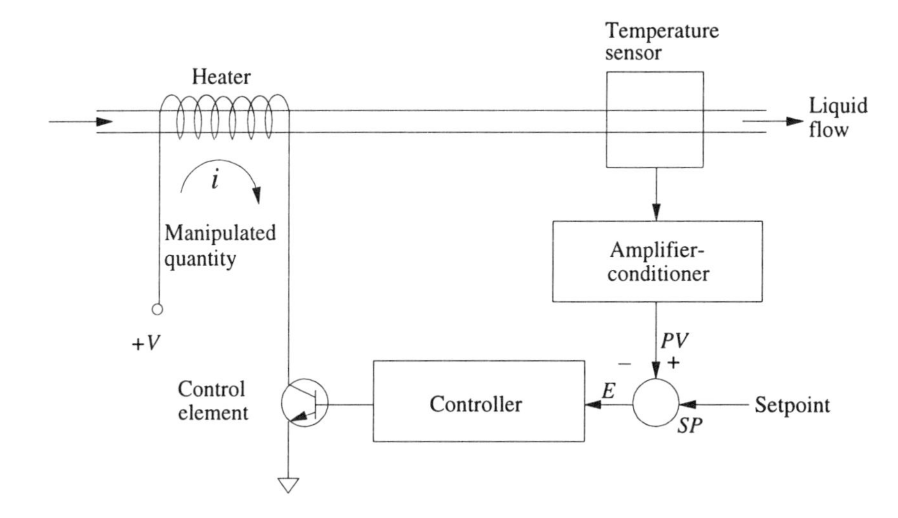

Figure 8–47. Temperature controller for flow system. The current in the heater is controlled to equal the setpoint temperature in the flowing liquid. Response delays in the heater and sensor and dead time in the flow system between the heater and sensor contribute to the dynamics of the control system. The ideal controller would compensate for undesirable dynamic characteristics of other components to produce a stable, responsive, and accurate control system.

it may respond in some other way. The controller's response to the error signal is called its **control law** or **control algorithm**. The choice of control law can affect the system dynamics. If the controller is in the loop, as shown in Figure 8–47, it can either compensate for or exaggerate the effects of the dynamic characteristics of the other elements.

The simplest control algorithm is **ON–OFF control**. In this algorithm the manipulated quantity *MQ* is simply on or off, depending on whether the process variable value *PV* (e.g., temperature) is greater or less than the setpoint value *SP*. The dynamics of ON–OFF control are illustrated in Figure 8–48. The lag and dead time in the system result in overshooting the setpoint from both directions. Cycling and overshoot are characteristic of ON–OFF control. If the controller has fast response and negligible ON–OFF gap, the amplitude and frequency of the *PV* variation are not affected by the controller but by all other lags and delays in the system.

A more-sophisticated control algorithm is **proportional control**. In this strategy the manipulated quantity is set at a value proportional to the error

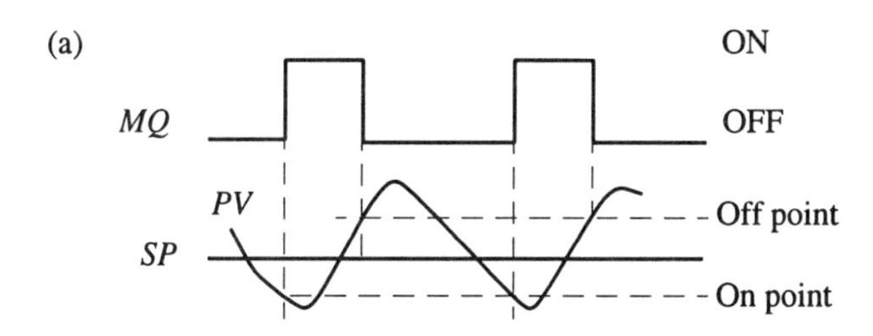

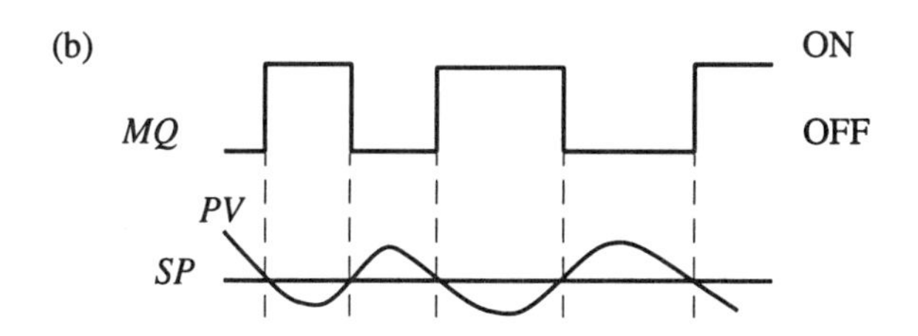

Figure 8–48. ON–OFF control dynamics. (a) The turn-on and turn-off values for process variable *PV* are separated. When *PV* drops below the ON point, the controller turns on the manipulated quantity *MQ*. The decrease in *PV* continues, however, because of lags and delays. The manipulated quantity is turned off when *PV* exceeds the OFF point, and an overshoot is again produced by the delay. (b) Eliminating the gap between ON and OFF points reduces the variation in *PV* and increases the control cycle frequency.

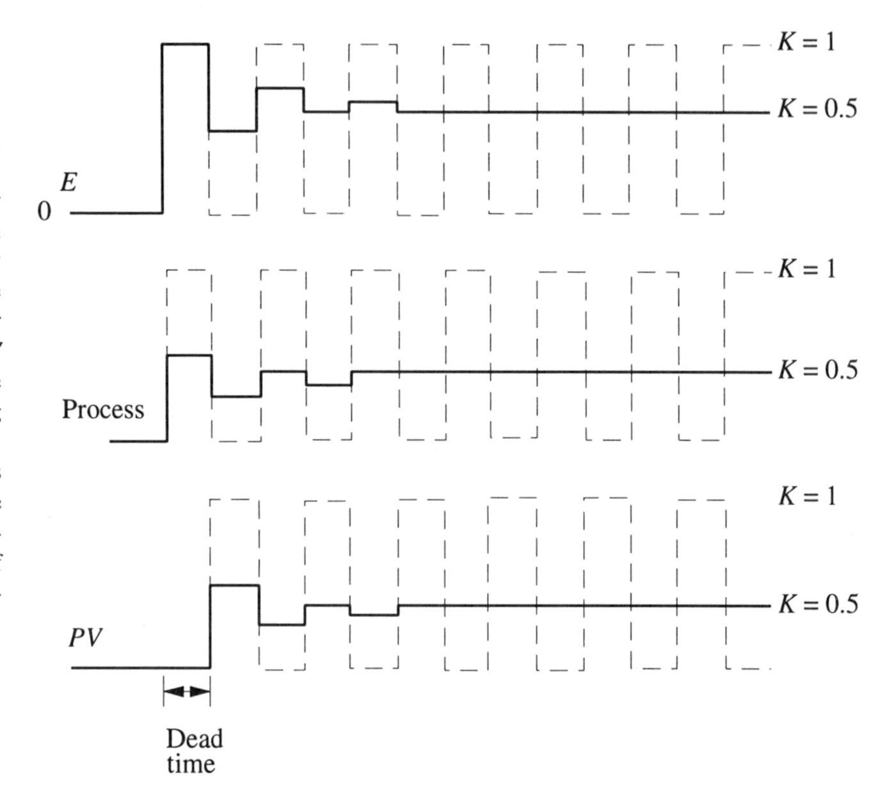

Figure 8–49. Response of proportional controller to step change in error E. Two cases are illustrated: loop gain $K = 0.5$ (solid lines) and $K = 1$ (dashed lines). The system has a finite dead time that is longer than the lags. The step change in E affects the process immediately according to the gain. This change does not appear at *PV* until after the dead time. An increase in *PV* results in a decrease in E and the process. This decreases *PV* 1 lag time later and increases E again. For $K < 1$, a steady value is reached with less than half the error corrected. For $K = 1$, a continuous oscillation results, and if $K > 1$, the amplitude of the oscillations increases.

magnitude E. Thus, as *PV* approaches *SP*, E decreases, the drive quantity decreases, and *PV* changes more slowly. This increasingly gradual approach to *SP* can eliminate the overshoot and cycling problems of ON–OFF control. The dynamic response of such a control system depends on the overall gain, dead time, and lag in the control loop. The **loop gain** K is the change in *PV* that results from a unit change in E when the loop is opened by disconnecting the *PV* signal from the comparator. (In Figure 8–47, $K < 1$ if a value of E equivalent to a 10 °C temperature error results in a temperature change of less than 10 °C.) The response dynamics of a proportional control system are shown in Figure 8–49 for $K \leq 1$. The perturbation in E could have come from a change in *SP* or from a change in the process conditions. As the dynamics show, a loop gain of less than 1 results in imperfect compensation, whereas a loop gain of 1 or more produces oscillations with a period of twice the dead time. The uncompensated fraction of the perturbation is $1/(1 + K)$, which is 2/3 for $K = 0.5$, 1/2 for $K = 1$, and approaching zero for K approaching infinity. It is clear that the dead time in the system prevents the use of a large enough value of K to provide accurate control. In some systems this difficulty is not observed because the lags in the process and detector responses are much longer than the dead time. For such systems the loop gain is very low at the frequency at which dead time oscillations would occur. Accurate control is achieved on a longer time scale by the much higher loop gain at low frequencies.

An example of a proportional controller is shown in Figure 8–50a. Two servo loops are present in this system to control the position and velocity of a machine axis such as a robot manipulator. Positional data from the axis are compared with the command data, and any difference is amplified and sent to the speed com-

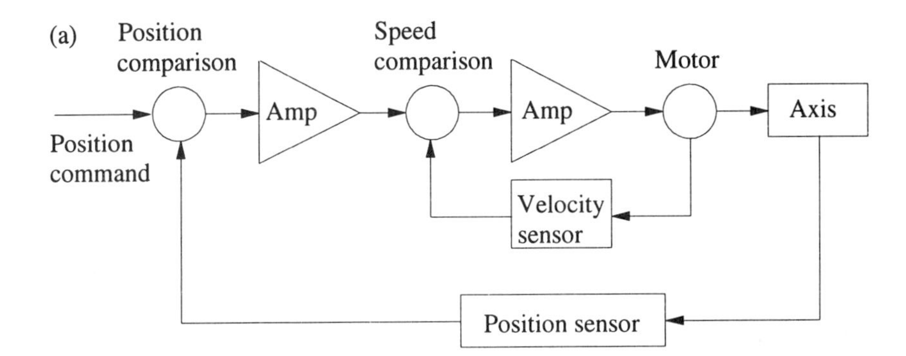

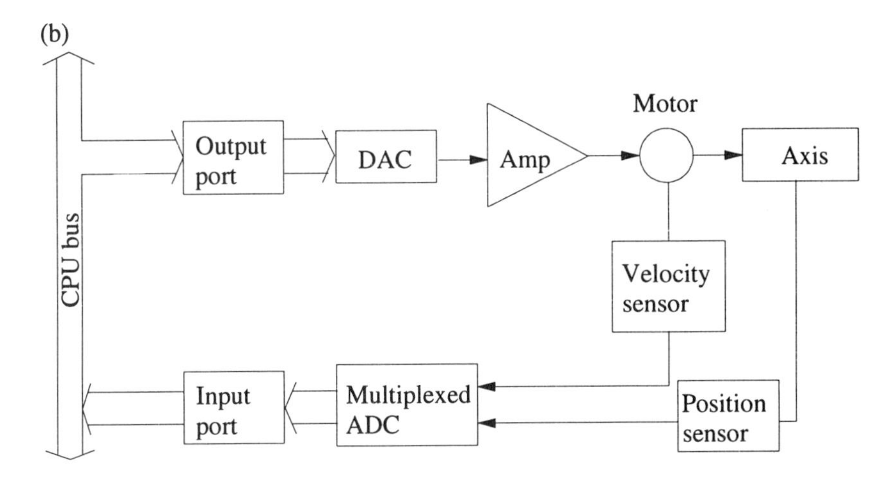

Figure 8–50. Servo systems for controlling the velocity and position of a machine axis. (a) Two servo loops. One (outer loop) controls the position, while the other (inner loop) controls the velocity at which the motor varies the position. (b) Computer implementation.

parator. The farther away from the desired position the axis is, the larger the amplified difference signal will be. This difference signal is now used as the command signal for the velocity servo loop. The latter compares velocity feedback information from the velocity sensor (e.g., a tachometer) with the command signal. As the axis gets close to the desired position, the velocity servo causes the motor to slow. When the desired position is reached, both servo systems are at balance and no movement of the axis occurs. This is, then, an example of a proportional control system. In Figure 8–50b, a computer implementation of the same control system is shown. In the computer the positional command information is stored in one register, while the digital feedback information about the actual position is stored in another register. During a movement operation the computer samples the two registers. Any difference is then sent to the velocity command register. The data in this register are compared with the digital feedback signal representing the velocity sensor output. As movement continues, the error signals become smaller, indicating that the axis is reaching its programmed position. When this position is reached, the axis drive motor stops and signals the computer that the axis is in position.

Algorithms more complex than that for proportional control can also be implemented, particularly with a computer controller. A controller can be tailored to give close to optimum response for a system by adding averaging and rate terms to the control algorithm. The controller response function is then

$$MV = K_P E + K_I \int E dt + K_D \frac{dE}{dt}$$

where MV is the manipulated variable. This control algorithm is called a **PID controller** for the three facets (*proportional, integral,* and *derivative*) of its response function. The integral or averaging term is particularly useful in oscillation-prone systems, where K_P must be kept low. A correction term proportional to the integral of all past values of E is applied to change MV. As long as an error exists, the integral term operates to reduce it. The integral term is therefore good for long-term accuracy. The derivative or rate term is designed to aid the response speed on the basis of giving the controller a stronger response to sudden changes in the error signal. The proportionality constants for the three terms must be carefully chosen in a given system to provide quick and accurate response to perturbations as well as good stability against oscillations.

One advantage of using a microprocessor for the controller is that once it is interfaced to the sensor and control element, any useful algorithm can be implemented or adjusted by simply changing the program. Even more complex algorithms that respond to combinations of sensors are practical. Or, as described earlier, a microcomputer can readily manage multiple control loops. Subsystem control can be economically implemented so as to optimize sections of a process independently. Overall control is greatly simplified, because each of the perturbing variables is separately controlled.

Practical control systems. Modern electronic devices and techniques offer a variety of solutions to practical problems in measurement and control. The systems discussed in this section illustrate the kinds of options available and their relative merits.

Two approaches to a computer-controlled temperature regulator are shown in Figures 8–51a and 8–51b. In the first approach (Figure 8–51a, below), the computer

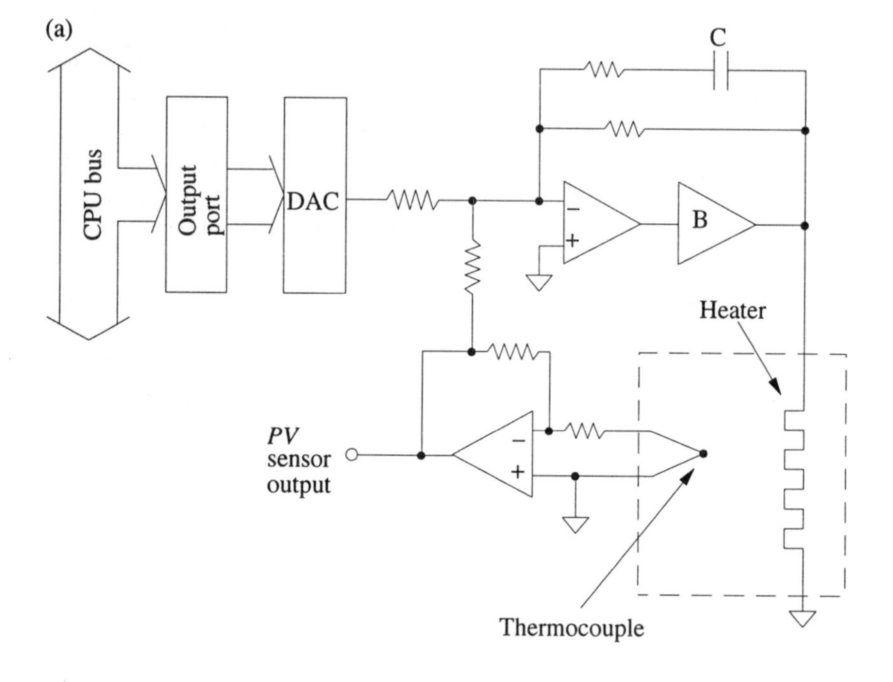

Figure 8–51. (a) Computer-controlled temperature regulators. The computer determines the setpoint for an analog temperature control system. The loop gain is adjusted by the control amplifier input and feedback resistors. The capacitor provides an integrating function to the control algorithm. B, booster amplifier.

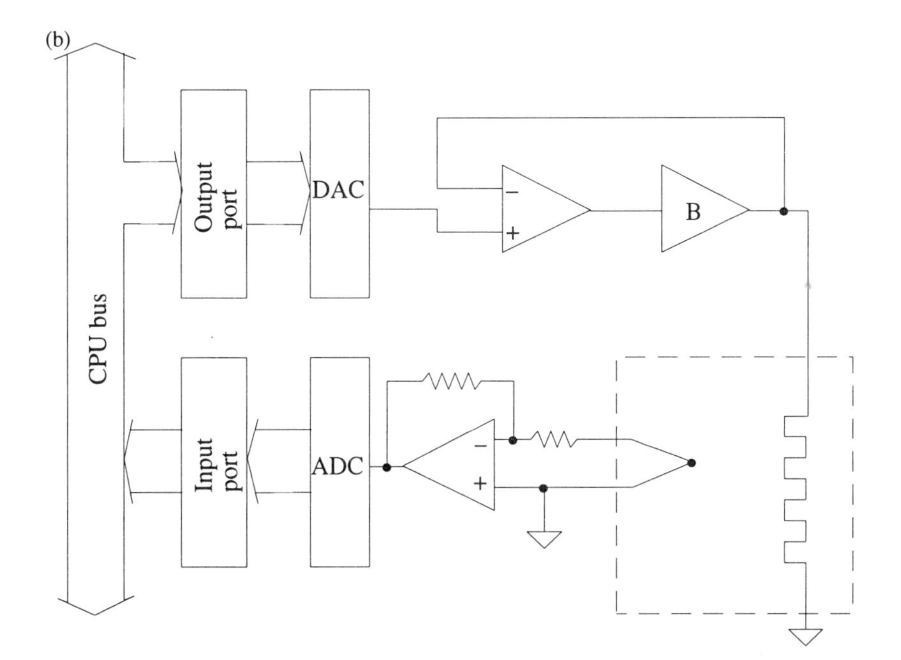

Figure 8–51—*Continued.* (b) Computer-controlled temperature regulators. The computer controls the manipulated quantity directly and reads the value of the process variable through the ADC. The computer is part of the control loop, and the control algorithm is executed through the computer program.

is not part of the control loop but provides the setpoint value. This controller assumes that the *SP* command is correctly executed by the analog controller. In the second approach (Figure 8–51b, above), the control system depends on the computer to maintain the correct value for the heater voltage, which it does by reading the value of *PV* through the interfaced analog-to-digital converter (ADC) and calculating the appropriate response. The advantages of including the computer in the control loop are that the control algorithm is readily changed without altering the system hardware, continuous information about the process dynamics is available to the computer, and the computer detection of control error is inherent in the system. The disadvantage is that the computer must execute the control loop subroutine often enough to maintain adequate control.

A motor speed controller is illustrated in Figure 8–52. The control loop signal is in the frequency domain at the sensor output and is converted to an analog signal by the multiplier phase comparator. The computer can provide the speed setpoint as shown, or it could take over the multiplier, filter, and controller functions in the control loop. In a completely computer-controlled system, the computer real-time clock would provide the frequency reference, and the only interface would be the motor drive control and a flag input from the opto-interrupter.

The control of light intensity is used in Figure 8–53 to illustrate the desirability of deriving the feedback control signal from the quantity that is actually being controlled. Controlling the voltage across a lamp certainly produces a more constant illumination than operating the lamp from an uncontrolled supply. However, this approach assumes that the power input is constant (lamp resistance is constant) and that the relationship of power input to light output is constant (i.e., there are no aging effects in the lamp). Sampling of the light intensity with a photodiode as in Figure 8–53b allows a control loop to keep that quantity constant at the setpoint value. In the spectrophotometer (Figure 8–53c), even with constant illumination, the light intensity at the sample varies with movement of the filament, changes in the monochromator wavelength setting, and solvent. Measuring the light intensity just before it illuminates the sample allows the controller to com-

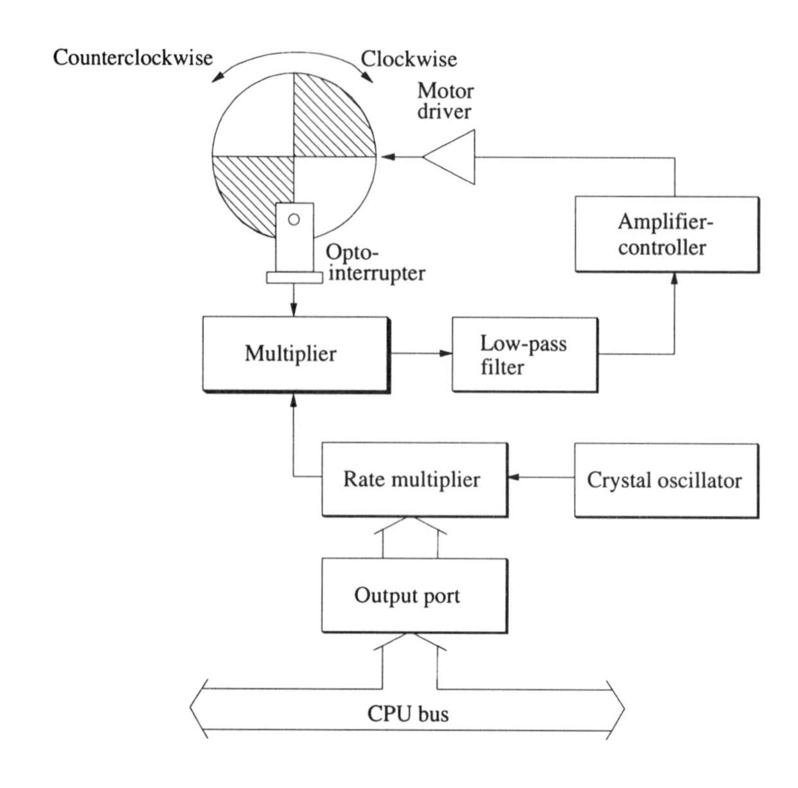

Figure 8–52. A dc motor speed controller. The output frequency from the opto-interrupter is compared with a CPU-controlled fraction of the frequency of a crystal oscillator. The result of that comparison determines the control signal to the motor driver. The control loop is of the phase lock loop type and involves both analog and frequency domain signals.

pensate for all but the solvent effect. The effect of solvent is equalized in the feedback beam by a solvent-only reference cell. The constant beam intensity greatly aids this system in accurately determining a sample absorbance A_s.

Dedicated microprocessor controllers. In the previous discussion the advantages of using the central processing unit (CPU) as the controller in feedback control systems are weighed against the burden this adds to the CPU. The development of inexpensive microprocessors and memory has made it economical to design hardware controllers using dedicated microprocessors. In other words, the main microcomputer is interfaced to a controller that incorporates another microprocessor to achieve the control function. Thus, hardware simplicity and software flexibility are achieved without burdening the main CPU with the control task. Some microprocessors have been designed specifically for peripheral control.

Large-scale integrated circuit (LSI) controllers that include microprocessors are available for control of CRT (cathode ray tube) displays, floppy-disk drives, and stepper motors. The control algorithms are often in ROM, which is part of the LSI controller. Except for the device handlers, these "smart" control chips are all that is needed to interface the main CPU bus to the controlled device.

Self-Optimizing Systems

Instruments, machines, and peripheral devices are increasingly referred to as being "smart" or having local intelligence. Usually, this means that there is a CPU controlling the device operation, but it can also mean that the instrument is adaptive; that is, it does not continue following a routine that is inappropriate for the circumstances. The development of intelligent systems is one of the most exciting areas of modern technology. Electronic **adaptive control systems** provide

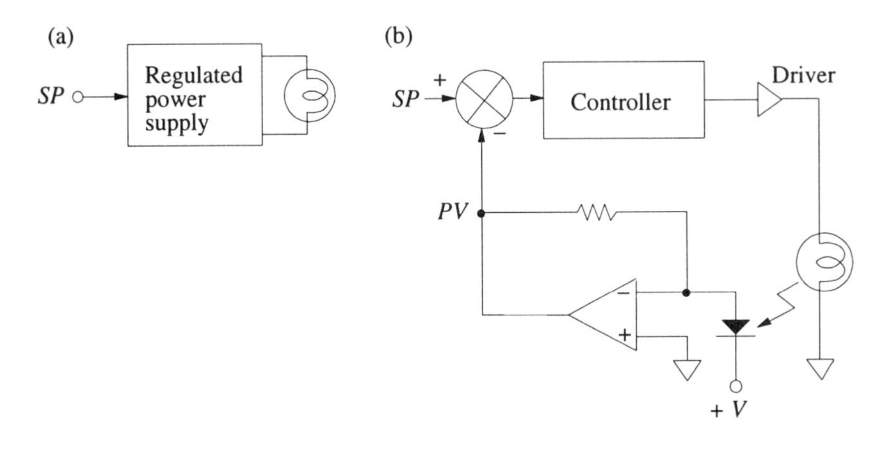

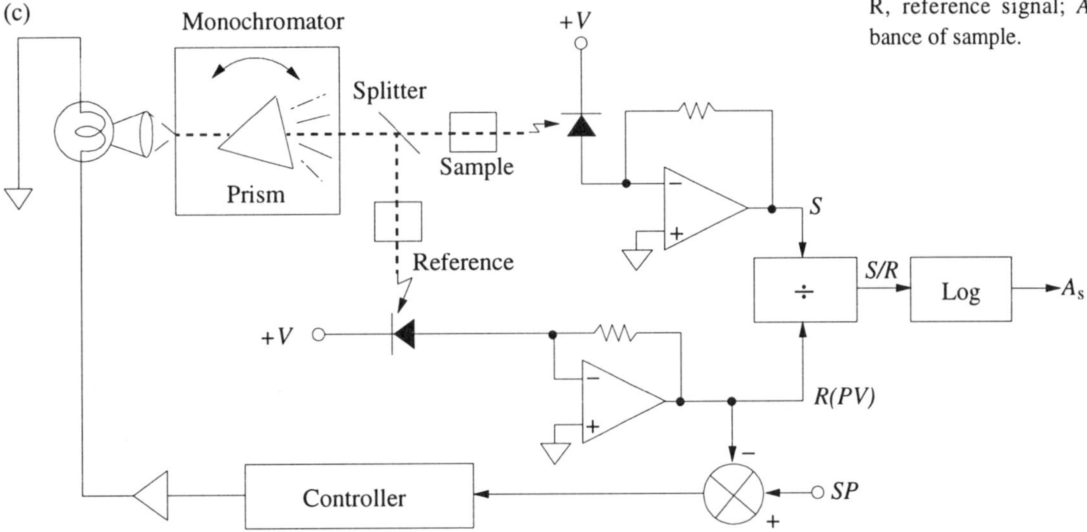

Figure 8–53. Light intensity control. (a) The regulated power supply controls the voltage (or current) to the lamp. (b) Adding a light detector controls the actual light output rather than the power input to the lamp. (c) In a spectrophotometer, many other factors affect that portion of the light used in the measurement. To keep that portion constant, the feedback quantity should be sampled as close to the point of use as possible. S, sample signal; R, reference signal; A_s, absorbance of sample.

a level of automation beyond mere control. Through the speed and precision of modern electronics, adaptive systems can greatly extend the capabilities of virtually any operation or measurement.

Autoranging. One relatively simple operation that a smart controller can accomplish is keeping a measuring instrument in range. In a voltage measurement system, for example, a programmable gain amplifier is often used between the signal input and the ADC. When the ADC indicates an overrange, the amplifier gain is reduced. A digital comparator monitors the ADC output. If the conversion falls below the comparison value, the amplifier gain is increased. Note that the resulting gain setting is part of the conversion information. In a digital meter, the autoranging circuit keeps the decimal point in the display in the correct place. In a computer system, the gain control lines must be interfaced to an input port to complete the information about the signal amplitude.

Autocalibration. A smart controller can be used to test and compensate for drift in device characteristics. For example, a principal limitation in high-resolution ADCs and DACs is the problem of keeping the drift in offset and gain to less than the value of the least significant bit (LSB). A controller can be used to test for drift

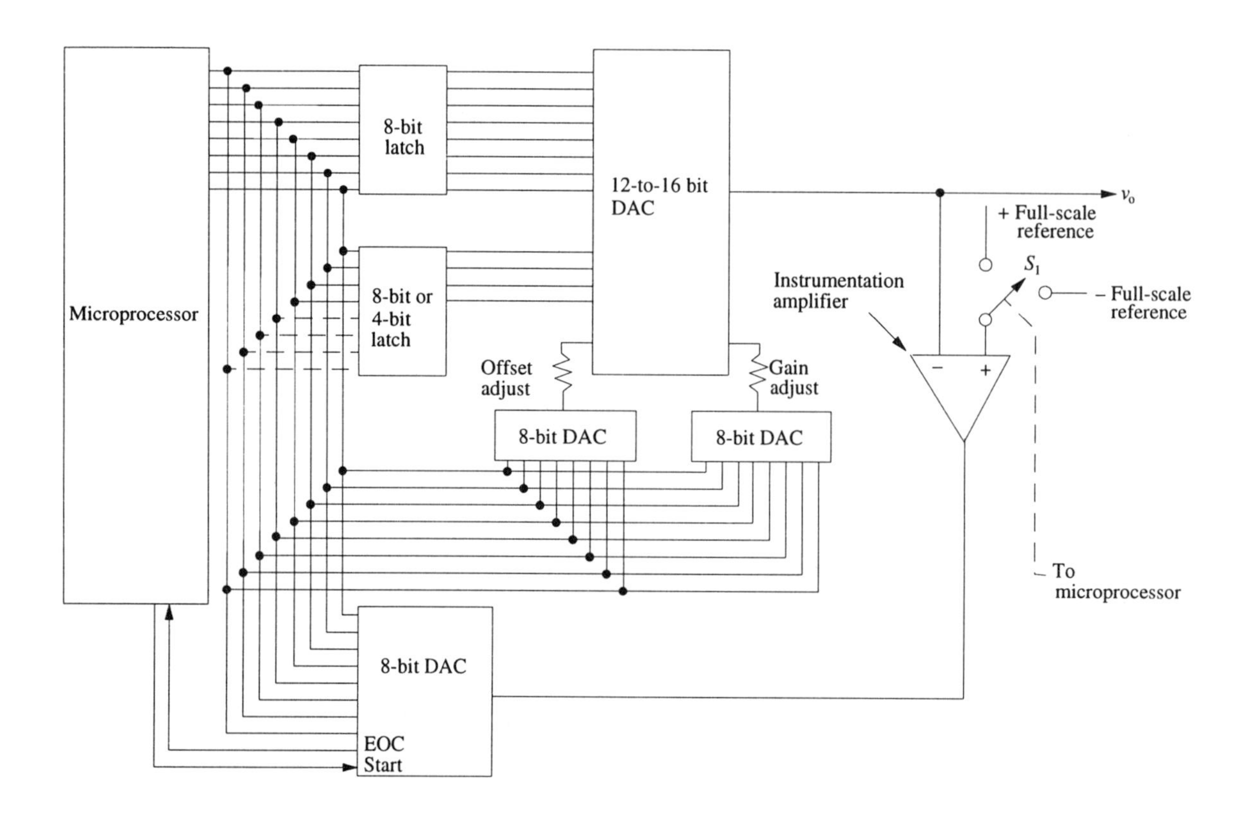

Figure 8–54. Autocalibrating DAC. The microprocessor periodically compares the DAC output extremes with positive and negative reference values. The error is read from the 8-bit ADC output and used by the computer to determine the needed change in gain or offset. These are applied by altering the values in the low-resolution gain or offset DACs. EOC, end of conversion.

in these parameters and make necessary adjustments. A DAC designed on this principle is shown in Figure 8–54. Note that low-resolution control is sufficient to calibrate a high-resolution system, because only a few of the LSBs are affected by the drift.

Autocompensation. Most traditional measurement systems attempt to minimize the change in any conditions that might cause an error in the measured value. Even so, the dominant source of error in many measurements is the variation in quantities that affect the measurement system characteristics. For example, if temperature is a factor in a strain gauge or a pH measurement, its effects could be compensated for by measuring the temperature accurately, experimentally determining the effect of temperature on the measured quantity, and applying an appropriate compensating correction to each measured result. A rough temperature control could be used to limit the degree of compensation needed. This approach has been used in kinetic methods of analysis to compensate for the influence of temperature on rate constants.

Compensation can also be used for known errors in generated signals. For example, a charge pulse generator can be made by gating a controlled current pulse for a controlled time. The charge, which is the current–time product, can readily be made accurate to within 1% for various combinations of current and time. However, it is much more difficult to obtain 0.1% accuracy of charge generation.

In one example, the error in charge was quite constant for each combination of current and time. The computer controller was then used to determine a table of errors for each current–time combination by measuring the charge delivered to a known load. The computer used this table to correct the charge error, and a better-than-10-fold improvement in accuracy resulted without any change in the hardware.

Optimization of strategy. As shown previously, the best choice of control algorithms depends greatly on the characteristics of the system being controlled. Thus, if conditions in the system change considerably, the algorithm may not only be suboptimal: it may also produce an unstable system. There are two solutions to such a problem. One solution is to make a controller with sufficiently low gain and response time that it is stable under all expected conditions (and optimum under none). The other is to analyze the system's response to changes in the driven variable and then adjust the control algorithm as conditions change in the system. This approach is widely used in the flight control systems of commercial airliners.

Adaptive control is a useful concept in measurement as well. The optimum measurement strategy often changes with signal conditions. An example is in instruments that measure the spectrum of the light emitted from a sample or a star. If the light source is weak, photon counting might be used. Waiting at each setting of the grating until a set number of counts have accumulated would give a constant standard deviation for each value. However, if the source does not emit light at some wavelengths in the spectrum, the integration time of the system is longest at the wavelengths of no interest. An adaptive strategy could be used to determine first whether there is enough light to be of interest and then whether to wait for the desired number of counts or to move to the next part of the spectrum. If the counting (integration) is to continue for a long time, the count rate may drift during this period because of slow variations in the system (background counts, amplifier gain, and so on). In such cases an adaptive measurement strategy could anticipate the need for long integration times based on low signal levels and switch to a synchronous (lock-in) detection mode to eliminate possible drifts. The implementation of adaptive strategies allows the operator to choose between speed and accuracy in a given measurement and then to have the chosen goal implemented in the most efficient way.

8–8 Laboratory Robotics

Go into many industrial plants and laboratories and you will find robots performing a variety of tasks. Robots are a relatively recent development, and they represent a high level of automation. Despite the robots' early and unique role in science fiction, the first industrial robots were not introduced until the early 1960s. Robots specifically intended to operate in the laboratory were introduced only in the mid-1970s. Today, laboratory robots perform such tasks as aliquoting, diluting, grinding, weighing, filtering, dissolving, extracting, and adjusting pH as well as handling hazardous materials or operating in hostile environments. Industrial robots, of course, also perform assembly line tasks such as welding, positioning components, loading and unloading machines, and much more.

Like the word automation, the word robot has been defined in different ways. The most widely accepted definition seems to be that adopted by the Robot Institute of America: *an industrial robot is a reprogrammable multifunctional manipulator*

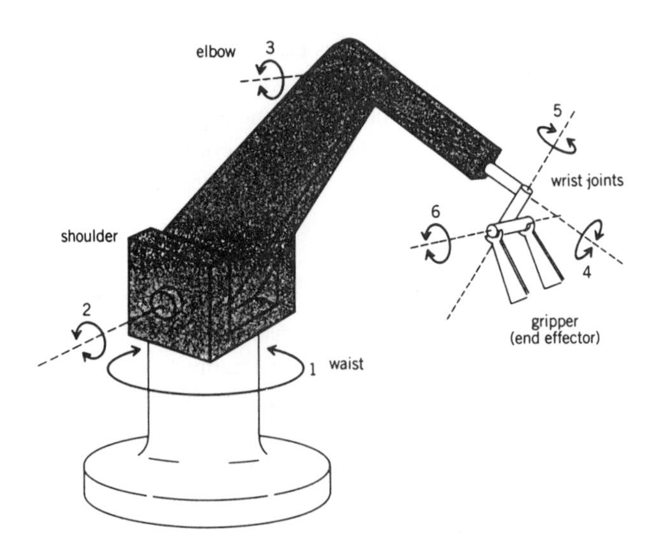

Figure 8–55. Basic components of a robot manipulator. There are 6 degrees of freedom of movement excluding the gripper. This allows the gripper to approach an object from any orientation. (Reproduced with permission from *Computer Science Source Book*; Parker, S. P., Ed. Copyright 1987 McGraw-Hill.)

designed to move materials, parts, tools, or specialized devices through variable programmed motions for the performance of a variety of tasks. The important words in this definition, the words that serve to distinguish robots from other machines, are "reprogrammable" and "manipulator." As will be shown in this section, current laboratory robots consist of manipulators, sensors, drive systems, and controllers. The term robotics, then, covers research and development activities related to the design, construction, and applications of robots.

Robot Manipulators

Almost all robots produce some form of mechanical motion for the purpose of manipulation. Manipulation is the act of grasping an object and changing its position and orientation in space. The **manipulator** is the arm, wrist, and hand of the robot, as shown in Figure 8–55. It allows the robot to bend, reach, twist, and grasp. The movement of the manipulator is provided by **axes**, also called **degrees of freedom**. An **axis** is the rotational section of the manipulator arm that rotates around the centerline of the body. A robot can have from 3 to 16 axes. The arm may move so that the gripper sweeps out a space that is Cartesian (x, y, z), cylindrical, or spherical. Joints, such as wrist joints, give additional motion.

The three-axis wrist joints shown in Figure 8–55 can orient the gripper by three independent motions (roll, pitch, and yaw) around the x, y, z location of the wrist. The robot's mechanical design can include sliding joints as well as the rotational joints shown in Figure 8–55.

At the end of the mechanical arm of the robot is the end effector, which is frequently a gripper. The standard robot gripper has two or more fingers that move together upon closure. Hands based on suction and magnetic attraction are also used. Many laboratory robots are designed so that different end effectors can be attached to the arm. In addition to the hands described, highly specialized devices such as stepper motor-driven syringes and pipettes can be used. Often the robot can change end effectors under program control when changing from one type of task to another.

Most laboratory robots work in a confined area known as the work cell or work station. Figure 8–56 illustrates a typical sample preparation work cell. The size of the work cell corresponds to the area that can be reached by the manipulator. Although only fixed-area manipulators have been discussed, there are also mobile

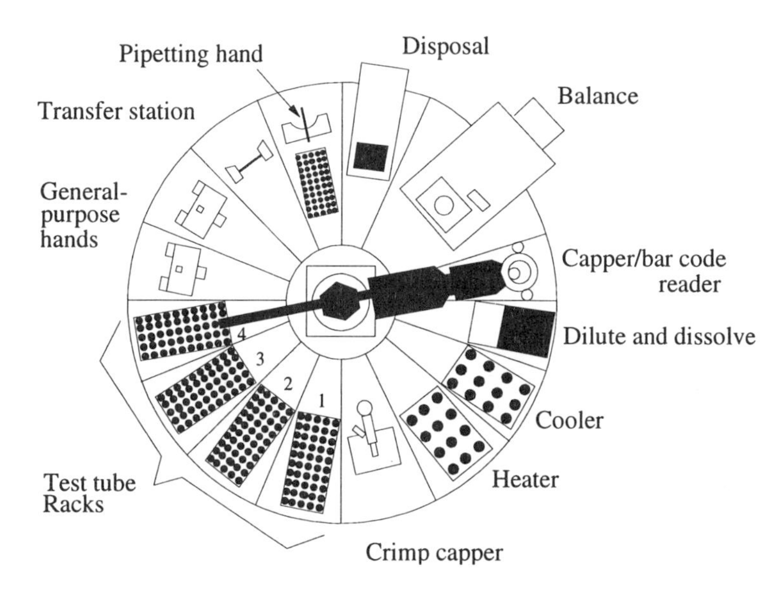

Figure 8–56. Top view of a work station for a laboratory robot involved in sample preparation. Samples in test tubes are placed in racks 1 through 4. Each sample tube has a bar code with sample preparation instructions encoded. Sample test tubes can then be transferred to a bar code reader and capped, to a dilution and dissolution stage, etc. Samples can be weighed on a balance, diluted with solvent, dissolved in acid, heated or cooled, adjusted to an appropriate pH, etc.

robots that can move from one area to another on a track or by using wheels and treads and there are walking robots.

Drive Systems and Controllers

The drive systems of robots use hydraulics, pneumatics, electric motors, or some combination of these (*see* Section 8–6). Industrial robots often use hydraulic actuation for manipulating large, heavy objects. Pneumatic actuation is often used when speed is important. For example, hands are often controlled pneumatically. The small tabletop robots used in laboratory applications usually use servomotors or stepper motors. They may also use pneumatic actuation. Regardless of the type of drive system, positional control is normally accomplished by a feedback control system (*see* Section 8–7).

In the servo systems used for robots, a transducer, such as a motor shaft encoder, feeds back positional information that is compared with the desired position. Any difference signal causes movement of the manipulator so as to eliminate the difference (*see* Figure 8–46). Analog, digital, and hybrid servo systems are used.

Although low-technology robots use hardware controllers, all modern laboratory robots are computer controlled. The computer can be inside or outside the servo control loop, depending on the application and design of the robot. With computer-controlled systems, the robot's trajectory is generated automatically by the computer. Positional data for an entire task or even for many different tasks can be stored in memory or on disk and recalled later. Programming a robot can involve a written program in a suitable control language or a lead-through approach. In the latter method, keyboard control is used to effect each desired motion of the robot. The computer stores each keystroke and later produces a program for the user.

The most advanced robot control systems use hierarchical control. In these systems there are multiple levels of control stages. Each level accepts from a higher level a statement of the goal to be achieved. The goal and its complexity depend on the position of the control stage in the hierarchy. The lowest level might be, for example, the joint position servo system. Here the goal might simply be to move to the position commanded. With hierarchical control, each level of the hierarchy acts as a servo control for the actions of the next-lower level by supplying it with commands that will reduce the difference between the current state and the state defined by its own goals. Such systems can be sensory interactive in that the robot can respond in different ways depending on the conditions reported by sensors.

Sensory Systems

For robots to perform tasks done by humans, they must have sensing capabilities. Sensors give the robot a measuring capability and allow verification that it has properly executed its commanded task. The position sensors used in feedback control systems provide a minimal sensory system. However, it is desirable to include other sensors to gather data about the external world. These can be visual, proximity, acoustic, tactile, or force sensors.

Tactile sensors. Tactile (touch) sensors can be mounted in the gripper to detect contact with objects. These sensors can be as simple as microswitches or as complex as an analog transducer to indicate degree and direction of pressure. Thin-film capacitive transducers (*see* Section 8–4) have been used to create matrices of pressure-sensitive materials. Transducer arrays can provide sensing pads for fingers to allow the robot to distinguish types and orientations of objects. Force and torque sensors allow the robot to detect resistance encountered by the gripper and to adjust its pressure so as to grasp the object without destroying it. Such sensors also allow the robot to properly mate one part with another or with a surface.

Visual sensors. A common way to sense objects at a distance is with some type of visual sensor. The simplest of these is an LED/light pipe/photodetector in the gripper to provide proximity information via reflectance or transmittance of light. Multiple source/detector units in fingers allow precise manipulations as an object is grasped.

Photodiode arrays and solid-state television cameras have also been used to give the controller visual feedback. With these systems we want the robot to be able to discriminate shapes and orientations of various objects. Vision systems use either ambient light or structured lighting. Structured-lighting systems can simplify and speed up the pattern recognition and image interpretation processes. In either case the task of emulating human vision is formidable.

The most advanced robot systems are multimodal in that they use many different sensor types. These are also hierarchically structured to generate increasingly complex levels of sensory feedback. These advanced systems are ideally suited for hierarchical control.

Intelligent Robots

Most laboratory applications of robots to date involve repetitive, nonvarying procedures. Usually, the laboratory robot is operated in an "off-line" mode with respect to laboratory instruments. In analytical chemistry, for example, most applications of robots have been aimed at sample preparation. The robot merely weighs, grinds, dissolves, and treats the samples in a repetitive manner. The robot is not part of the analytical instrument feedback loop. In one approach to incorporating robots in an intelligent system, sample preparation can be included in the overall loop. Thus, if a sample is too concentrated for an instrument to measure, the computer controller can instruct the robot to prepare a diluted sample. By bringing the robotic sample preparation system into the loop, dynamic matrix matching could be used to compensate for interferences with complex samples based on information gathered by the analytical instrument.

To be even more useful in the future, robots will have to be intelligent. One approach is to equip them with some degree of artificial intelligence. An expert system, for example, can act as a supervisor for a robotic system. Here, the user communicates with the robot through the expert system. For a chemical analysis procedure, the user gives information to the system regarding the sample type, matrix, possible interferences, and possible approaches to the analysis. The user input along with information in the expert system knowledge base is then used to choose the specific procedure to be applied. If the information available is insufficient to permit a decision, the user can be asked for additional information.

The expert system first checks to see if the same problem has been previously solved. If so, it recalls the previous solution and programs the robot to perform the analysis. If no previous solution was available, the expert system uses its problem-solving rules to propose a solution. The expert system checks the proposed solution against other solutions that have failed to produce appropriate results. If the proposed solution clears this screening, the expert system consults with the user regarding the proposal. A few preliminary experiments might be needed to test new conditions. Once the user, in consultation with the expert system, has agreed upon the procedure, the robot is automatically programmed to execute it.

Obviously, we have only begun to scratch the surface of the potential of robotic systems in the laboratory. Much research remains to be done to establish the role of the operator and the role of the robot. An exciting future awaits us.

Related Topics in Other Media

Video, Segment 8

- Feedback control systems
- Automation by computer-controlled pH system
- Automated mass spectrometer system

Laboratory Electronics Kit and Manual

- Section A
  - ¤ Using transducers for temperature and light measurement
- Section C
  - ¤ Connecting and testing voltage regulators for controlling power supply voltages
  - ¤ Control by electronic servo systems
- Section D
  - ¤ Manual servo measurement with a comparator
- Section E
  - ¤ Connecting electronic servo control systems
- Section H
  - ¤ Generating electronic waveforms by feedback

- Section J
  - ¤ Constructing an automated comparison system that illustrates analog-to-digital and digital-to-analog conversions for voltage measurements with digital readout
- Supplement 1
  - ¤ Experiments related to servo systems

Chapter 9

Getting the Most from Real-World Signals

We have seen in previous chapters how analog- and time-encoded signal values can be acquired by a computer and how the data thus acquired can be processed and displayed. In this chapter, the converters used to acquire analog signals are explored in greater depth with the goal of aiding optimum choice and implementation for each application. Once data are acquired, various powerful signal-processing techniques, formerly reserved for only the most sophisticated and specialized instrumentation, can be applied routinely. These processes can be of considerable value in getting the most information from the data we have collected.

9–1 Optimizing Data Acquisition

The optimization of analog data acquisition involves making the right choice of data conversion rate and resolution. No single type of analog-to-digital converter (ADC) covers the full range of possible rates and resolutions; each of the several types of converters available is the best choice under certain conditions. In this section, the three principal types of converters, integrating ADCs, digital servo ADCs, and flash ADCs, are examined with respect to the fundamental differences in their operating principles and preferred areas of application.

Integrating ADCs

Integrating ADCs are based on a comparison of the charge generated by the analog signal input and a digitally controlled reference charge generator. Within this class of ADC, there are two types: those that maintain a continuous charge balance and are based on the current-to-frequency converter (IFC) and those that balance an accumulated input charge and are called dual-slope converters.

The IFC. The IFC is an application of the charge balance principle in which the input signal current is balanced by the rate of reference charge pulse generation so that the charge difference detector output remains near zero. This process is illustrated in the block diagram of Figure 9–1. The reference current i_r is generated by repeated triggering of the charge increment generator. The average current is the charge per pulse q_s multiplied by the pulses per second f. Thus i_r is equal to fq_s. The currents i_u and i_r are connected to the charge difference detector simultaneously. Any difference between i_u and i_r causes a change in the charge difference output voltage v_C. The null condition is achieved when v_C is constant. At null, i_u is equal to i_r is equal to fq_s. A number proportional to the unknown current is read from the frequency meter display.

A practical IFC is shown in Figure 9–2. The charge q_s in each reference charge pulse is $t_p i_s$. The average rate of reference charge addition is $f_o t_p i_s$, which at balance is equal to i_{in}. Therefore, f_o is equal to $i_{in}/t_p i_s$. The parts of the circuit that affect

2861-2/94/0299$10.00/1 Published 1994 American Chemical Society

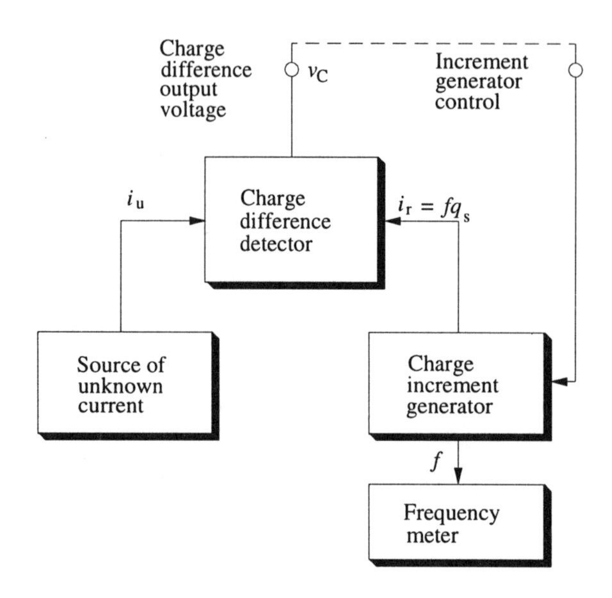

Figure 9–1. Current-to-frequency conversion. The charge difference detector is at balance when v_C is a constant value. In this state the unknown and reference currents are equal, and the reference current can be measured as the frequency f of generating charge increments q_s.

the current-to-frequency ratio and that should thus be stable are the parts that determine i_s and the pulse width t_p of the monostable multivibrator. It is interesting that the integrating capacitor and the comparator threshold are not critical. With reasonable care in component selection, linearity to a few hundredths of a percent is achievable.

The upper frequency limit for the IFC is determined by the rate at which the reference charge pulses begin to deviate significantly from their low-frequency magnitude. Full-scale frequencies f_{fs} of 10 or 100 kHz are common, but converters that operate up to 10 MHz are available. The value of f_{fs} sets some limits on t_p and i_s. The pulse width t_p must be shorter than one output cycle at f_{fs}, and i_s must be larger than the full-scale input current i_{fs}. A typical value for i_{fs} is 1 mA.

Though it is inherently an IFC, this circuit is often provided with an input resistor and called a **voltage-to-frequency converter (VFC)**. In this case, i_{in} is equal to v_{in}/R_{in}. Therefore, f_o is equal to $v_{in}/t_p i_s R_{in}$, and the output frequency is proportional to v_{in}. If the converter has a full-scale input current of 1.0 mA, a 10-kΩ resistor at R_{in} would give a full-scale input voltage of 10 V. Note that the input voltage or current cannot be bipolar and that at zero input current, the output frequency is zero.

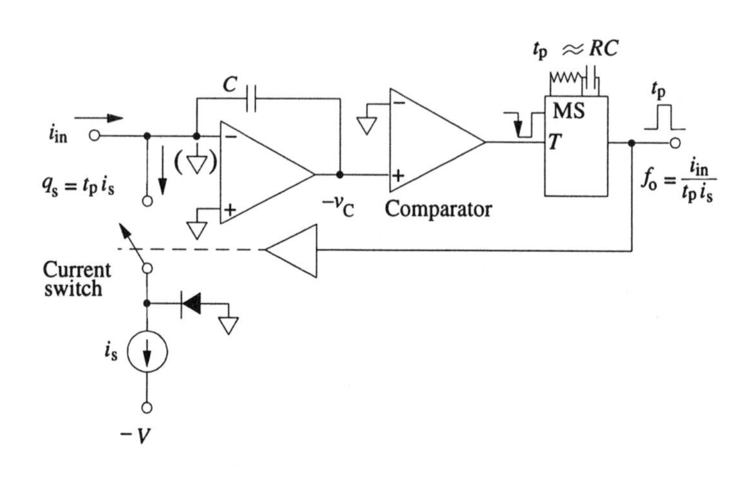

Figure 9–2. IFC. The integrator output is negative-going while i_{in} is being integrated. When the comparator threshold is crossed, the monostable multivibrator (MS) is triggered, and the current switch connecting it to the integrator is closed for a time t_p. At balance, the average rate of charge addition from the reference charge pulses is equal to i_{in}. Therefore, the frequency of charge pulse addition f_o is proportional to i_{in}. For voltage-to-frequency conversion a resistor is used in series with the input.

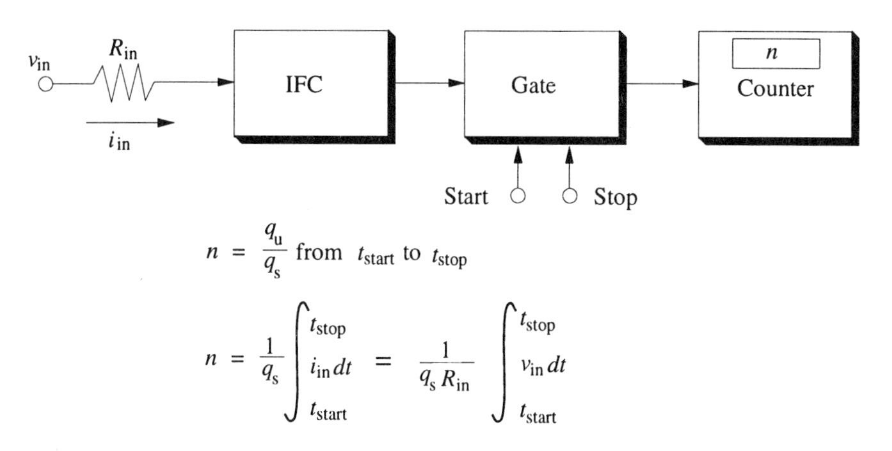

Figure 9–3. Digital measurement of charge with an IFC. The count total is equal to the number of charge increments q_s used to balance the charge input to the IFC over the time interval between the start and stop commands to the gate. The total charge is related to the integral of the input current or voltage over that same period as shown. A clock is used as a gate control for precision control of the integration period.

VFCs have many applications in modern electronic systems. They are used as voltage-controlled oscillators of exceptional linearity and dynamic range, as analog-to-frequency converters for reliable data transmission over a single connection, and as a basic function in ADCs. These applications have been made all the more attractive by the availability of the complete VFC in integrated circuit (IC) form.

Because the IFC produces a pulse for every increment of input charge, the total number of pulses produced in a given period is equal to the total charge (the integral of the current) applied to the input over that period. As shown in Figure 9–3, all that is required for digital integration is an IFC and a counter. The input current, output frequency, and count value versus time are shown in Figure 9–4 for four

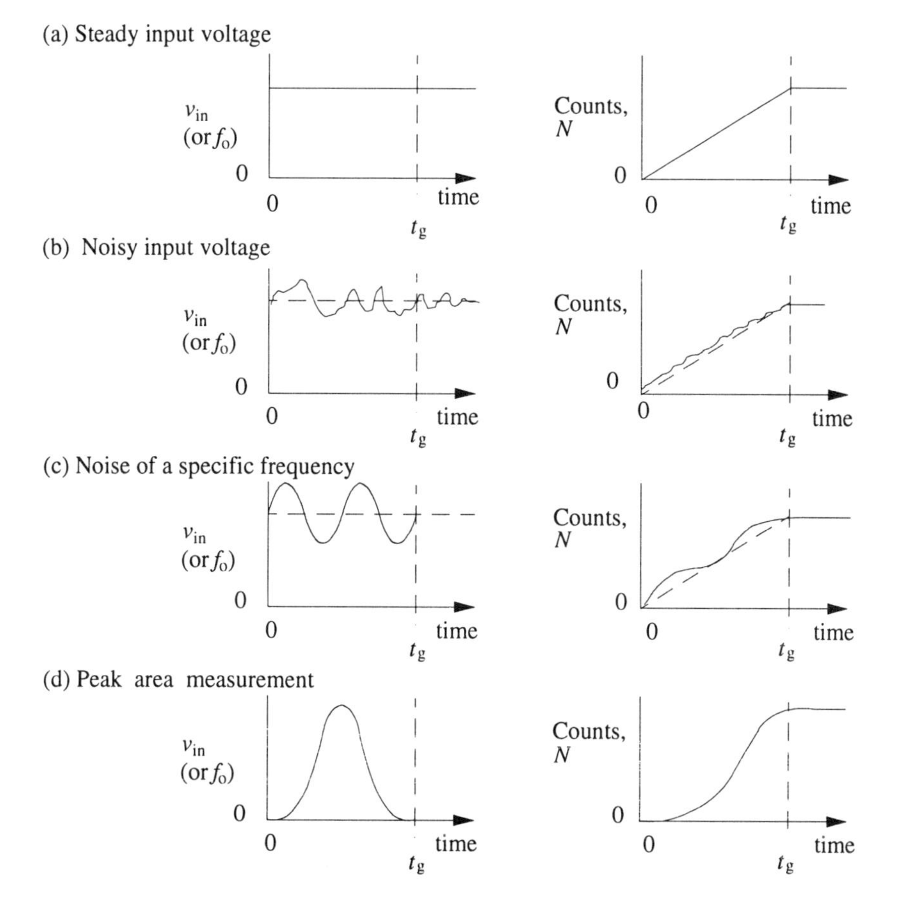

Figure 9–4. Examples of input signal integration. The output frequency of the VFC follows variations in the input voltage, but the total count at the end of the gate period t_g is the average frequency over that period. The accumulation of counts over the gate period is shown on the right for each example of input signal variation given on the left.

different input waveforms. In most applications of the VFC as an ADC, a crystal oscillator-controlled clock signal would be connected to the start and stop inputs of the counter gate in Figure 9–3. The number of VFC output pulses that occur during one clock period t_c is determined. The longer the clock period, the greater the number of counts accumulated for a given input signal level and the greater the period over which the input signal is integrated. The shorter the clock period, the more frequently the input signal is sampled. Because the clock period must be long enough to accumulate enough counts to provide reasonable measurement resolution, the VFC/frequency meter type of ADC is not suitable for signals that have to be sampled more than 50 or 60 times per s.

A particularly interesting quality of the integrating type of ADC is illustrated in waveform c of Figure 9–4. The signal in this waveform contains fixed-frequency noise such that the integrating period is an integral number of periods of the noise. Because the integral of a sine wave over exactly 1 cycle is zero, the effect of the presence of this noise is eliminated. For example, if a signal has a major noise component of 60 Hz and if t_c is 0.1 s, then the 60-Hz component completes exactly 6 cycles during the integration time and the noise from that source is rejected. In fact, all noise frequencies that are integer multiples of 10 Hz will be rejected.

Dual-slope converter. In the dual-slope technique (first introduced in Chapter 2), the input voltage is first converted to a charge by integrating a proportional current over a set period. The quantity of charge is then determined by counting the number of charge units required to discharge the integration capacitor. A typical dual-slope ADC is shown in Figure 9–5a. The complete conversion takes place in three phases: auto zero, signal integration, and reference integration. The timing of the conversion cycle is controlled by the oscillator and the counter. The duration of each phase is equal to the time required for the counter to go from zero to full scale, counting output pulses from the oscillator.

At the beginning of the conversion, in the auto zero cycle, switch S_1 connects the input of the converter to common, and S_2 closes to allow the auto zero capacitor with capacitance C_{AZ} to charge. The auto zero capacitor then charges from the analog offsets in the system until the integrator output v_C is zero and the rate of change of v_C is zero. At the end of the auto zero cycle (counter overflow), the control logic connects v_{in} to the input and opens switch S_2. The input signal is then integrated for one complete counter cycle. During the signal integration phase, the integrating capacitor is charged by a current equal to v_{in}/R. Counter overflow again indicates the end of the signal integration phase. As shown in Figure 9–5, the integrator output v_C is directly proportional to v_{in} at the end of this phase. The control logic then changes switch S_1 to the reference voltage position, and the integration capacitor is discharged toward zero with a constant current V_r/R. The number of oscillator pulses required for the integrating capacitor to discharge to zero is determined and is proportional to v_{in}. The end of the conversion occurs when the comparator indicates that v_C has reached zero. The dual-slope converter is so named because of the shape of the integrator output voltage shown in Figure 9–5b.

The waveforms in Figure 9–5b are shown for a positive v_{in} and a negative V_r. Signals of either polarity can be accommodated if V_r can be made bipolar. In many converters this is done with a single reference voltage by charging a capacitor to V_r during the auto zero cycle. The polarity of v_{in} is then sensed by the comparator during the signal integration cycle, and the capacitor charged to V_r is connected by

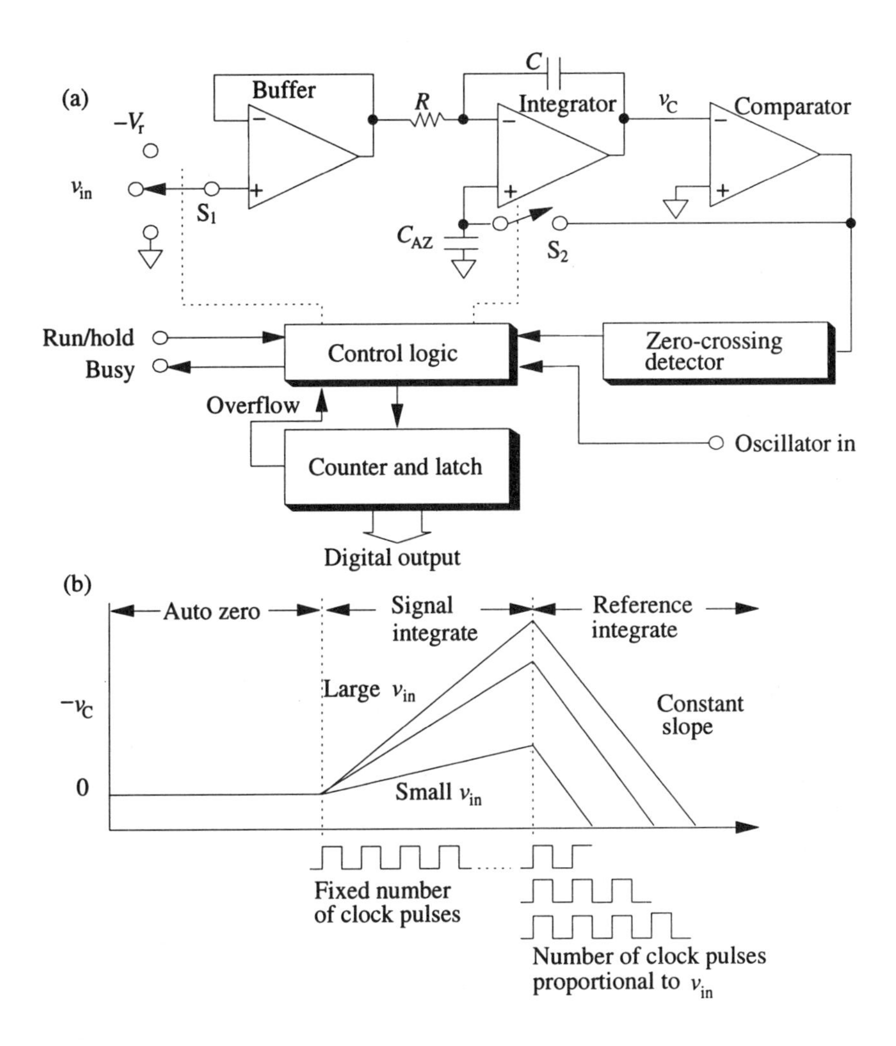

Figure 9–5. Dual-slope ADC (a) and waveforms (b). During the auto zero phase, error information (buffer and integrator offset voltages, etc.) is stored on the auto zero capacitor with capacitance C_{AZ}, and the integrator output v_C is forced to zero. During the signal integration phase, v_{in} is integrated for a fixed number of clock pulses. In the reference phase, the reference voltage $-V_r$ is connected to the integrator, and the integration capacitor discharges toward zero. The number of clock pulses required to discharge the integrating capacitor to zero is directly proportional to v_{in}.

switches with the polarity required to discharge v_C during the reference integration stage.

The relationship between the readout count n_r and the input voltage v_{in} reveals many of the advantages of the dual-slope technique. At the end of the integration cycle, the charge on the integrating capacitor (if offsets are nulled during auto zero) is $q_C = Cv_C$, which is given by

$$q_C = \frac{v_{in}}{R}\Delta t = \frac{v_{in}\, n_m}{Rf}$$

where Δt is the integration time equal to the maximum number of counts n_m divided by the oscillator frequency f. Typical values for n_m and f are 20 000 and 200.0 kHz for an integration time of 100.0 ms. During the reference cycle the charge q_C is discharged to zero in a time n_r/f, where n_r is the readout count. Thus

$$q_C = \frac{V_r\, n_r}{Rf} = \frac{v_{in}\, n_m}{Rf} \quad \text{or} \quad n_r = \frac{v_{in}}{V_r} n_m \tag{9–1}$$

For a 20 000-count cycle, n_r is equal to 20 000 v_{in}/V_r. Note from equation 9–1 that the capacitance C, the resistance R, and the oscillator frequency f do not influence the readout. The reason, of course, is that the reference and input signals are integrated by the same integrator, timed by the same oscillator, and referenced to the same comparator threshold voltage.

The dual-slope technique inherently provides excellent noise rejection because of the signal integration. Interference noise from the 50- or 60-Hz power line can be almost eliminated by choosing the integration time to be an integral number of power line cycles. The major disadvantage of the dual-slope technique is the rather long conversion time (~300 ms for a 3-cycle converter of 100 ms/cycle). Inexpensive IC forms of the dual-slope converter are the basis of many of the digital panel meters and digital multimeters produced today. They also are used on some analog interface boards for applications requiring high precision but not a high sampling rate.

Digital Servo ADCs

In the **digital servo ADC**, a digital-to-analog converter (DAC) is used to generate an analog signal for comparison with the signal to be digitized. The digital input to the DAC is changed in a direction determined by the results of the comparison until the input signal and the DAC output are equal, as shown in Figure 9–6. This type of ADC is called a digital servo because it is a feedback system in which the feedback information is digitally encoded. There are several types of digital servo ADCs based on different digital registers and different methods for adjusting their contents. The two predominant types of digital servo ADCs, the tracking and the successive-approximation converters, are described below.

Tracking ADC. The **tracking ADC** is a true digital servo in the sense that it can incrementally adjust the output in either direction in order to follow, or "track," the input quantity changes. This is achieved by using an up–down counter as the register in the block diagram of Figure 9–6. The clock is connected directly to the up–down counter input, and the comparator output controls the count direction. As the input voltage varies, the counter counts up or down to keep the DAC output

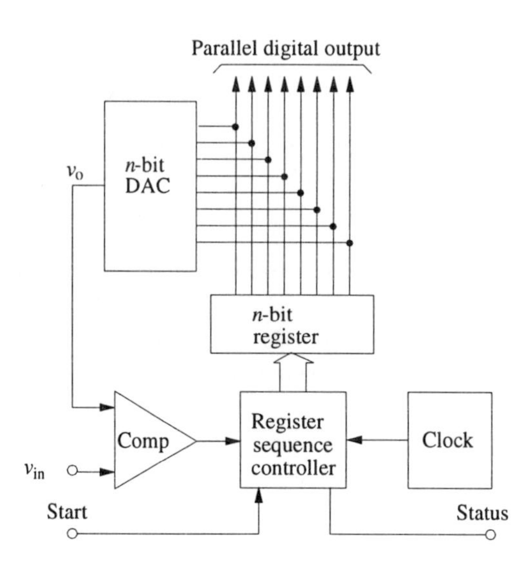

Figure 9–6. Digital servo ADC. On application of a start signal, the register sequence controller alters the contents of the register until the DAC output is within 1 LSB of the analog input voltage. The comparator (Comp) determines whether the number in the register will be increased or decreased, and the clock determines the rate at which the changes are made. The status (end-of-conversion) line provides an appropriate logic-level signal when the converter is busy and a logic-level transition when the conversion is complete.

equal to the input voltage. The tracking ADC cannot track instantaneous changes in input voltage, having to count its way to the new value. In the worst case, this could require as many as $2^n - 1$ clock cycles for a full-scale change in input voltage. However, the tracking ADC can follow small variations in the analog signal within a few clock cycles. It has the advantages of a continuous output and rapid updating at the clock rate. If a 10-bit DAC with a 10-V output range is used and a 10-MHz clock is connected to the counter, the converter can follow input voltage changes as fast as 10^5 V/s. Because the system is continuously following the input voltage, no start signal is needed and no status signal is generated. When a tracking ADC is used to convert a steady input signal, the digital output alternates, or "waffles," between the two adjacent quantizing levels that span the theoretically correct output value. The quantizing error present in all ADCs is very noticeable in the tracking ADC because the "hunting" or alternation in the least-significant bit (LSB) is observable on successive clock pulses.

Successive-approximation ADC. The major difference among types of digital servo ADCs is in the method by which the number in the register is adjusted to give a DAC output equal to the input. The most efficient way is to divide the range of possible numbers into a small number of segments (usually two, but sometimes four or more) and to identify the segment with the desired number in it. Then that segment is divided into smaller segments, and so on until the final result is determined. This procedure is illustrated in Figure 9–7. The first test in effect tests whether v_{in} is in the upper or lower half of the range. Having determined that v_{in} is in the upper half of the range, the upper half-segment is divided in half and tested by making the next most significant bit (MSB) a 1. In this test, the desired number is not in the upper segment, so this bit is determined to be 0. Next, the quarter between 1/2 and 3/4 is tested. Then the eighth between 5/8 and 3/4 is tested, and so on. The block diagram of a typical **successive-approximation converter** is shown in Figure 9–8.

The successive-approximation procedure requires only one clock cycle per bit of conversion. The conversion time t_c is constant and is given by $t_c = n/f$, where n is the number of bits in the converter and f is the clock frequency. A 10-bit converter with a 10-MHz clock can complete a conversion every microsecond. However, this requires a DAC and a comparator that can settle to within 0.5 LSB from a half-scale step input in 0.1 μs, and indeed, such high-speed converters with

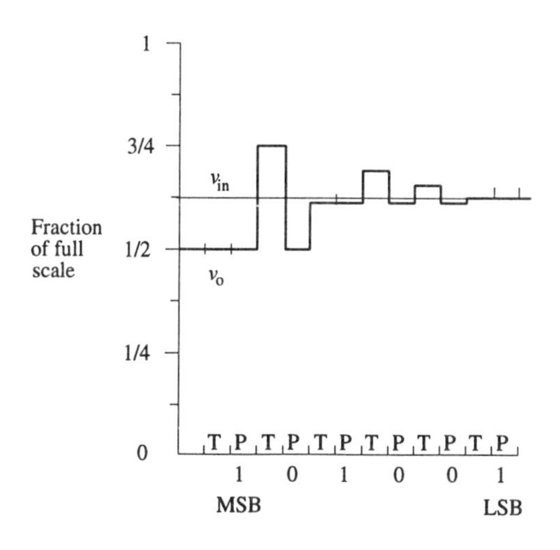

Figure 9–7. Successive-approximation search. At the start of the conversion, the register is set to half scale (1000 . . .). This gives a half-scale output for v_o during the first test period (marked T), and the comparator indicates whether v_o is too high or too low. Because the test shows the v_o to be too low, the 1 in the MSB is retained during the posting interval P. Next, the upper half of the range is divided into two parts and tested by making the next MSB 1. This test shows v_o to be too high, so the 1 is replaced by a 0 during the second posting period. The process continues until the LSB of the converter has been posted.

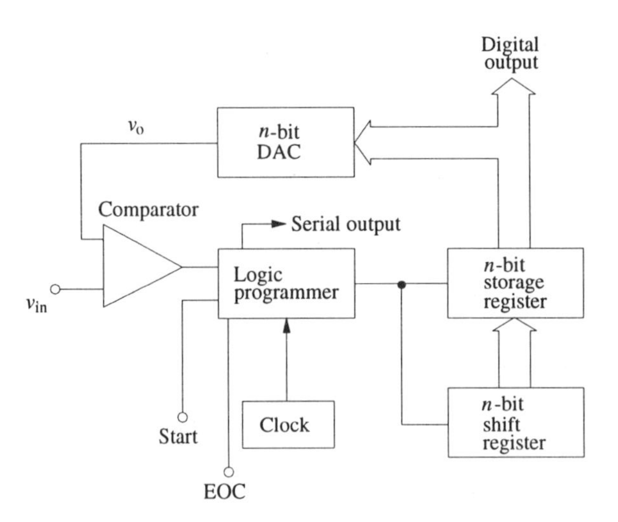

Figure 9–8. Successive-approximation ADC. A conversion cycle begins with a start pulse, which sets the output of the two registers to half scale (1000 . . .). The output of the storage register is converted by the DAC to give an analog voltage v_o. If $v_o < v_{in}$, the 1 in the MSB of the storage register is retained, and the shift register shifts its 1 to the next MSB. If $v_o > v_{in}$, the logic programmer resets the MSB of the storage register. Each remaining bit is then tested, in succession, until the LSB has been tested with the comparator output indicating whether to retain or reset the storage register test bit. An end-of-conversion (EOC) pulse signifies that the conversion is complete.

conversion times in the microsecond range are available. More common and less expensive, however, are 10- and 12-bit converters with conversion times of 4 to 30 μs, which is still very fast. One requirement of the successive-approximation converter is that the analog input voltage must remain absolutely constant during the conversion time. If it does not, errors in the intermediate bit tests can occur. Analog sample-and-hold circuits are usually used to acquire the voltage to be converted and to hold it constant during conversion.

The successive-approximation converter is of course subject to the quantizing error of ±0.5 LSB. Successive-approximation converters are available in monolithic IC packages as well as in hybrid form (analog ICs, digital ICs, and discrete components in the same package). Successive-approximation registers, which contain the logic programmer, the shift register, and the storage register, are also available in IC form for use with an external DAC and comparator. A serial output is available on some successive-approximation ADCs. The serial information is obtained from the comparator output during the posting interval (*see* Figure 9–7). Note from the waveform of Figure 9–7 that the sign of the comparator output follows the posted bit value.

Flash ADC

Conversion rates in the 1- to 400-MHz range can be achieved at the expense of resolution by the **flash ADC**, as shown in Figure 9–9. The flash, or parallel, ADC consists of a bank of $2^n - 1$ comparators that perform the quantizing and encoding logic, where n is the number of bits in the conversion. The comparators all have different thresholds set by the reference voltage and its divider. The encoder logic produces a normal binary code from the seven output lines plus the overrange indication. The high speed of the flash converter results from the simultaneous comparison of all output levels. Very rapid encoding can be achieved with high-speed logic gates. The flash converter is essentially a continuous converter and needs no strobe to begin conversion. However, an "output-stable" pulse is provided to ensure reading only when the encoded output is valid. The major disadvantage of the flash ADC is the need for 2^n comparators to achieve n bits of conversion. For an 8-bit converter, 256 comparators are needed. Despite this very

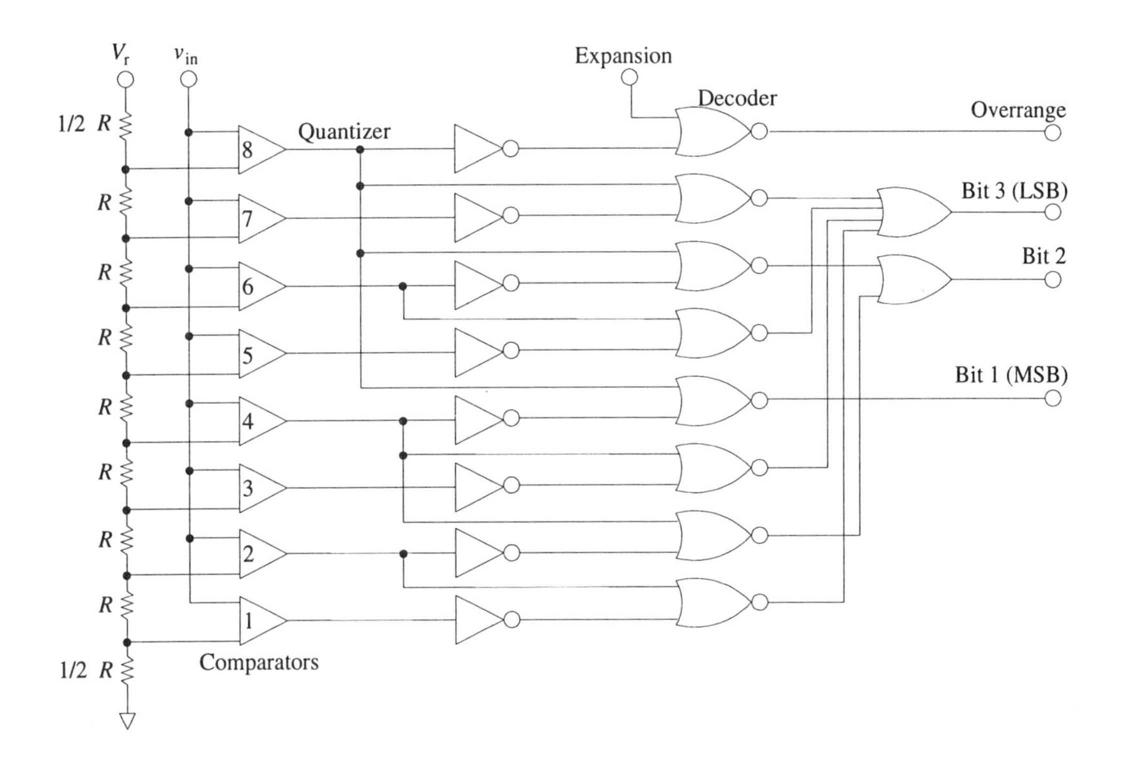

Figure 9–9. Flash ADC. Eight comparators and an encoder provide a 3-bit conversion plus overrange indication. The reference voltages for each comparator come from the divider V_r. The quantizing level Q is $V_r/8$. The threshold for comparator 1 is $Q/2$ or $V_r/16$; that for comparator 2 is $3Q/2$ or $3V_r/16$, and so on.

large number, several manufacturers produce 8-bit flash converters with bandwidths of several hundred megahertz.

Although the flash converter provides the fastest rate of electronic analog-to-digital conversion available, its complexity and power dissipation increase rapidly with the number of bits of resolution provided. Currently, 10 bits is the practical limit. The flash converter produces data at rates that exceed the maximum acquisition rates of standard computers.

The flash converter concept also can be applied in successive stages, converting first the more significant bits and then the less significant bits. A two-step converter based on this principle is the **half-flash converter** shown in Figure 9–10. Two flash ADCs, each supplying half the bits in the final result, are used. The first converter, operating directly from the analog input, produces the most significant half of the bits. This result is fed not only to the output but also to a DAC, the inverted output of which is summed with the analog input. The difference voltage,

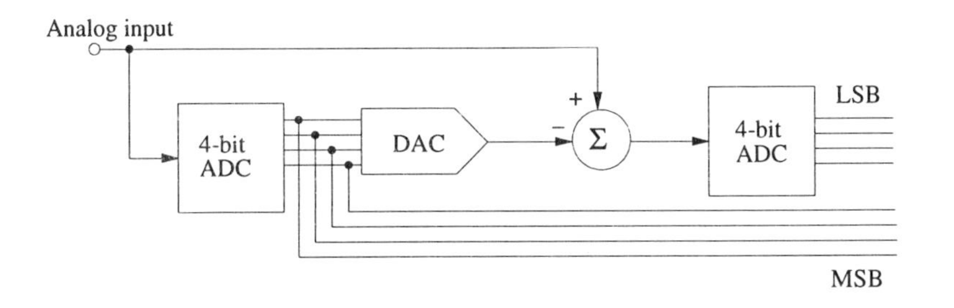

Figure 9–10. Half-flash converter. The analog input is first converted by the flash converter on the left to give the 4 MSBs and produce a DAC output equal to the converted value. The output of the summer, now equal to the difference between the analog input value and the value of the 4 MSBs, is converted by the flash converter on the right. This provides the 4 LSBs to complete the conversion. For an n-bit conversion, two $(n/2)$-bit converters are required.

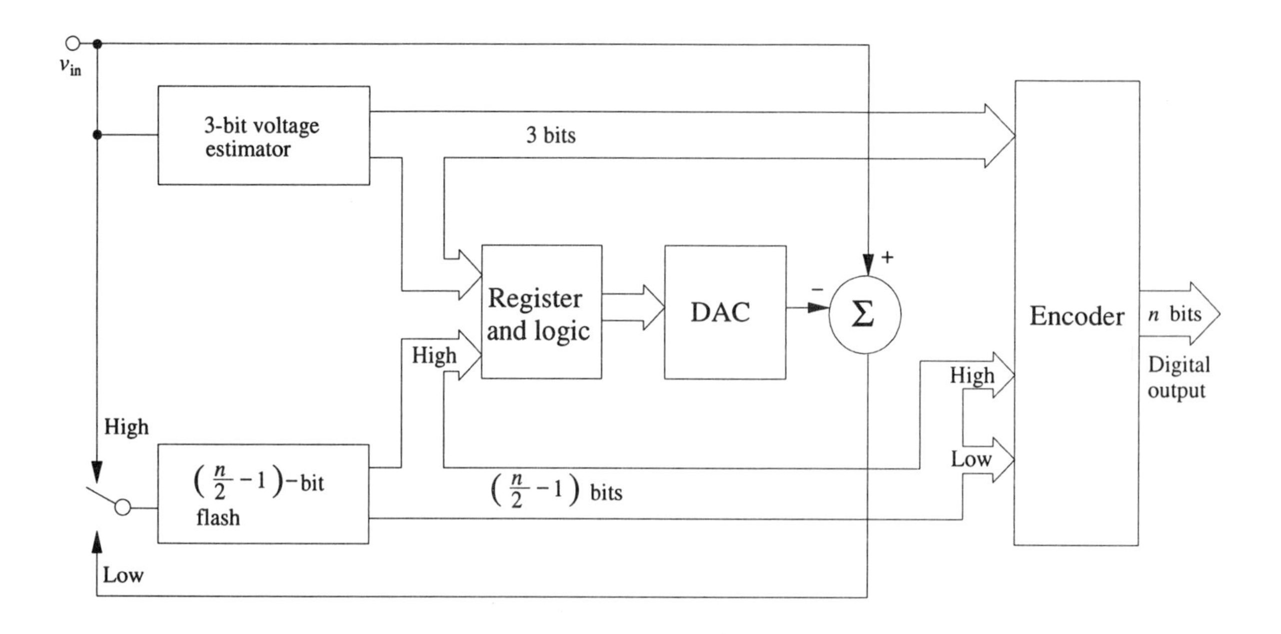

Figure 9–11. Multistep converter. In this variation of the half-flash converter, the results of the first partial flash conversion of v_{in} are held at the DAC output, and the same flash converter is used to convert the remaining voltage difference. This process is made still simpler and more comparators are saved by the 3-bit estimator, which provides the MSBs and reduces the number of bits the flash converter needs to produce.

equal to the unmatched part of the input voltage, is connected to the second flash converter for the determination of the least significant half of the digital output. The two 4-bit flash ADCs in Figure 9–10 contain only 16 comparators each, compared with the 256 comparators needed for an 8-bit flash converter. The DAC in this example has only a 4-bit input but needs to be precise to 8 bits for the LSBs to be significant. With the half-flash converter, a great saving in power and complexity is achieved at a cost of only a factor of 2 in speed of conversion.

A variation on the half-flash converter that performs the sequential flash conversions with the same flash converter unit rather than two separate ones has been developed. This type of converter is called the multistep converter and is shown in Figure 9–11. In this device the analog voltage is simultaneously connected to the flash converter, a 3-bit voltage estimator, and the analog summing circuit. The voltage estimator is a six-comparator circuit that determines the 2 MSBs of the result and estimates the third. The remaining $n - 2$ bits are converted in two steps, the more significant half with the flash converter input switched to v_{in} and the remaining bits with the flash converter connected to the summer output. A 10-bit converter based on the multistep concept has only 22 comparators (6 in the estimator and 16 in the flash) and a voltage divider with 80 taps. Ten-bit converters in the multistep family can provide a throughput of over 10^6 conversions per s with errors below 1 LSB. Thus, the half-flash and multistep converters bridge an important gap between the high resolution of successive approximations and the high speed of flash devices.

9–2 Enhancing Signal-to-Noise Ratio by Bandwidth Reduction

In modern measurement and control systems, it is increasingly necessary to measure weak electrical signals in the presence of noise. As sources and detectors are improved and weaker physical effects are used to provide information, the problem of discriminating between an information-conveying signal and extraneous, unwanted noise components becomes increasingly difficult. Fortunately, several elegant hardware- and computer-based techniques have been developed to aid measurement where the signal-to-noise ratio (S/N) is quite small. This section and the one following are devoted to exploring the principles of these bandwidth reduction (frequency discrimination) methods of S/N enhancement. Among the techniques included are low-pass filtering, hardware and software averaging (integration), and digital filtering in the time domain (smoothing) and the frequency domain. Waveform correlation techniques, which take advantage of the phase coherence of repetitive signals, are the subject of Section 9–3.

Noise Sources

The total noise in an electronic system results from two distinct types of noise: fundamental noise and excess (nonfundamental) noise. **Fundamental noise** arises from the motion of discrete charges in electrical circuits and cannot be completely eliminated. Excess noise arises from imperfect instrumentation or nonideal component behavior and can in principle be reduced to insignificant levels by careful practice and instrument design. Noise is also introduced in the process of converting an analog signal into a digital representation. This type of noise is called **quantizing noise**.

The two most important types of fundamental noise are Johnson noise and shot noise. **Johnson noise**, also called **thermal noise,** is produced by the random motion of electrons in resistive elements because of thermal agitation. Johnson noise has a white power density spectrum; that is, its power density is independent of frequency. Because Johnson noise is due to thermal motion, the magnitude of the noise voltage increases with the temperature T. Thermal noise also increases as the resistance R of the component increases. Because the noise power density is equal at all frequencies, the total noise voltage observed across a resistor depends upon the range of frequencies that the measurement system passes, that is, the system bandwidth Δf. The quantitative relationship describing the root-mean-square (rms) noise voltage $\bar{v}_{rms}$ is known as the Nyquist relation and is

$$\bar{v}_{rms} = (4kTR\Delta f)^{1/2}$$

where k is the Boltzmann constant.

Shot noise results from the random movement of discrete charges across junctions. Examples include the flow of charges across semiconductor junctions or between cathode and anode in a vacuum tube or phototube. Shot noise also has a white power density spectrum. The shot noise current increases with increasing average current and system bandwidth.

The rms shot noise $\bar{i}_{rms}$, due to an average current i observed during a time interval t, is

$$\bar{i}_{rms} = \left(\frac{Q_e i}{t}\right)^{1/2}$$

where Q_e is the charge on an electron. This shot noise equation, or Schottky equation, can also be expressed in terms of bandwidth, because the signal bandwidth equivalent to an observation time t is $\Delta f = 1/(2t)$. Thus, the rms shot noise current can also be written

$$\bar{i}_{rms} = (2Q_e i \Delta f)^{1/2}$$

Any noise above and beyond Johnson or shot noise is considered **excess noise**. In contrast to fundamental noise, excess noise is almost always frequency dependent. Interference noise from the 60-Hz power lines and noise with a $1/f$ power spectrum are excess noise. The ***1/f* noise** is often considered synonymous with drift. It can be introduced by long-term power supply fluctuations, changes in component values, temperature drifts, and other sources whose exact nature is poorly understood.

Quantizing noise is the result of the finite resolution of an analog-to-digital conversion. It is usually thought of only in terms of ADCs, but this type of noise is present in any process that converts a continuous infinite resolution signal to a finite number of digits. Thus, quantizing noise can be introduced in the manual conversion of a strip chart recorder deflection or a meter scale position to a numerical value as well as in an electronic ADC. The quantizing noise in an ADC can be visualized by the process illustrated in Figure 9–12. If the DAC output is compared with the ADC input, it is apparent that the quantization process adds

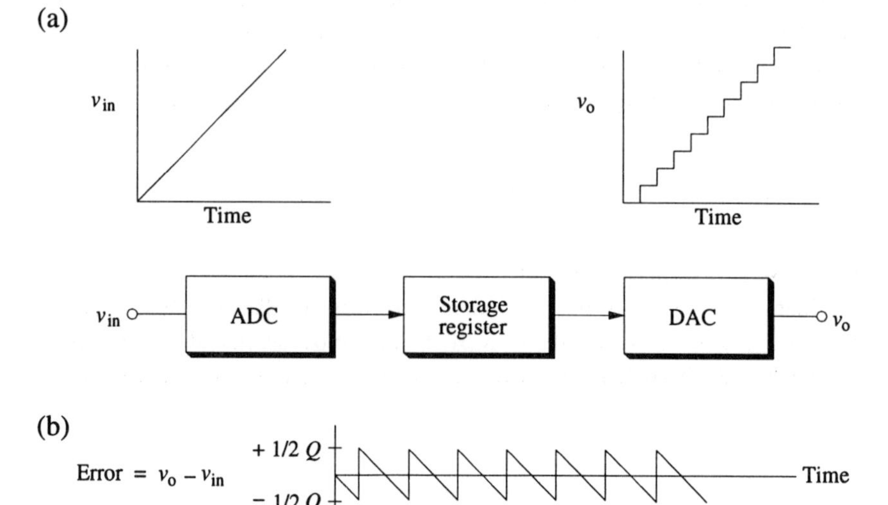

Figure 9–12. Quantizing noise. The devices connected in part a are used to reconstruct a slowly varying analog input signal by playing the ADC output back through a DAC. If the analog input is subtracted from the DAC output as shown in part b, the result is the quantizing noise. This noise has an average value of zero, a peak-to-peak value of Q (the quantizing level), and an rms value of $Q/\sqrt{12}$.

noise to the original signal. If the quantizing interval of the converter is Q, the root-mean-square (rms) value of the quantizing noise is $Q/\sqrt{12}$.

The S/N. In most applications it is the total noise present that is of interest. For dc signals, the mean-square noise can be defined as the average squared deviation of the signal from its mean value:

$$\text{mean-square noise} = \left[\frac{(S - S_1)^2 + (S - S_2)^2 + \ldots + (S - S_n)^2}{n}\right]$$

where S is the mean value of the signal; $S_1, S_2, \ldots, S_n$ are the instantaneous values; and n is the total number of values. The mean-square noise is also called the variance of the signal, and the rms noise is its standard deviation.

If noise sources are completely independent, the total mean-square noise voltage v_t^2 is the simple sum of the individual mean-square noise components (variances) $v_{n1}^2, v_{n2}^2, \ldots$, according to

$$v_t^2 = v_{n1}^2 + v_{n2}^2 + v_{n3}^2 + \ldots$$

The **signal-to-noise ratio** S/N can be expressed as

$$\text{S/N} = \frac{\text{average signal}}{\text{rms noise}}$$

The S/N is also commonly expressed as a power ratio in decibels or as a voltage or current ratio in decibels.

For dc signals the S/N is the reciprocal of the relative standard deviation of the measurement if electrical noise is the factor limiting measurement precision. For alternating current (ac) signals the relationship between S/N and precision is less straightforward. However, in many cases ac waveforms are converted to dc before they are displayed or digitized. In these situations, the S/N and the relative standard deviation are reciprocally related, as they are for dc signals.

Low-Pass Filtering

Perhaps the most common method of enhancing the S/N of a measurement is low-pass filtering. Many signals of interest have major frequency components at dc (0 Hz), with bandwidths extending only a few hertz. In these cases a simple low-pass filter can effectively limit the measurement system bandpass to that necessary to pass the signal frequencies. The characteristics of first- and higher-order active low-pass filters are discussed in detail in Chapter 7. It is simply necessary to choose the RC time constant and hence the bandwidth and phase shift characteristics such that the signal frequencies are affected as little as desired. The improvement in S/N by filtering comes at the expense of a decreased response time, which can lead to distortion of the signal. Thus, a compromise between measurement precision (S/N) and preservation of signal shape must be made.

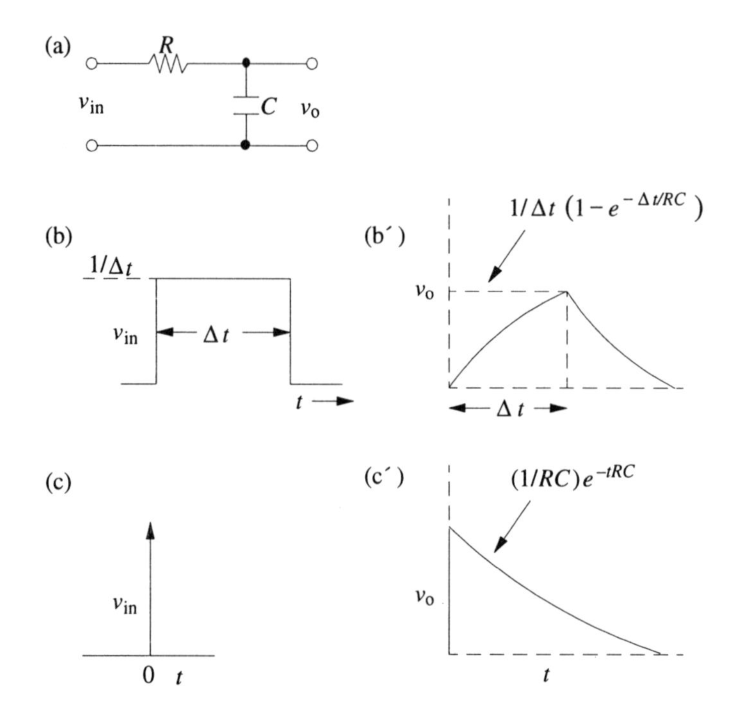

Figure 9–13. Impulse response of RC low-pass filter. A pulse of height $1/\Delta t$ and width Δt (b) is applied to the low-pass filter of part a. The output of the filter (b′) rises exponentially toward $1/\Delta t$ during the pulse and decays exponentially after the pulse. The input pulse width is made shorter and shorter (at constant area) until the pulse width goes to zero (c). The result is the impulse response of the filter (c′).

The distortion in the signal can be accurately predicted if the **impulse response function** of the filter is known. The impulse response of a simple resistor–capacitor (RC) low-pass filter is illustrated in Figure 9–13. A unit area pulse of decreasing pulse width produces, in the limit of zero pulse width, the response shown in Figure 9–13c′, which is the discharge curve for an instantly charged capacitor. The impulse response function is important because any arbitrary input signal shape can be considered the summation of a series of impulses. Thus, the total response of the filter to an arbitrary signal is the superposition of the separate responses due to each impulse. This can be seen more clearly in Figure 9–14. As the number of impulses becomes very large, the output response approaches the continuous response of Figure 9–13b′.

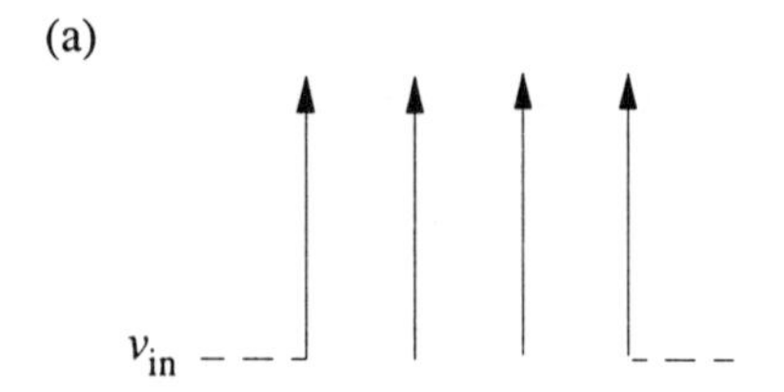

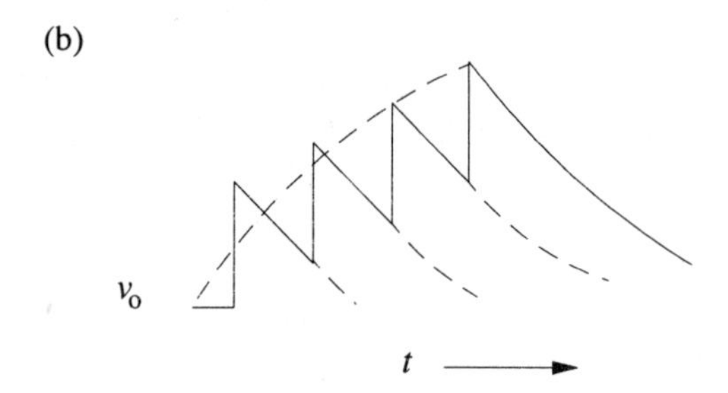

Figure 9–14. RC filter response to a rectangular pulse. The input pulse (a) is shown as a series of impulses. The filter output (b) is the superposition of the response toward each impulse.

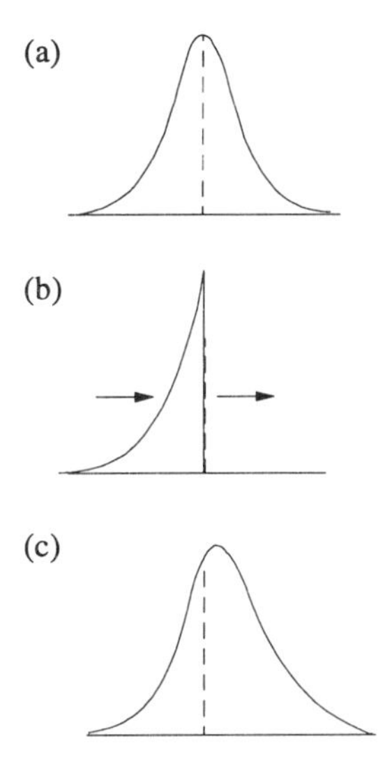

Figure 9–15. Distortion of a peak signal by an RC filter. The distortion in the filtered signal (c) can be visualized by the convolution operation. The impulse response function (b) is reversed and moved to the right. At each displacement, the raw signal (a) is multiplied by the response function, and the area of the product is taken. The resulting area versus displacement is the distorted signal (c).

The process of obtaining the superposition of a sequence of impulse responses is known as **convolution**. Convolution is a multiplication–integration operation used widely in science. It relates the input and output of a system by the impulse response function. Mathematically, convolution involves multiplying the input signal $X(\tau)$ by a reflected and displaced version of the impulse response $I(t-\tau)$ and time averaging or integrating the product as a function of the displacement t. The output $C(t)$ is obtained from the convolution integral

$$C(t) = \int_{-\infty}^{\infty} X(\tau)\, I\,(t-\tau)d\tau$$

where τ is a dummy variable of integration.

The convolution of a peak-shaped signal with the impulse response of an RC low-pass filter is illustrated in Figure 9–15. When the convolution operation is carried out mathematically or graphically, the impulse response is reflected on the time axis, and the area of the product of the signal and the impulse response is evaluated for various relative displacements. The reflection is necessary to keep the time sense correct in that the early time edge of the response function must be applied first to the early time edge of the signal. In this way, previous values contribute to the current value with exponentially decreasing weight. The distortion shown in Figure 9–15 includes altering the peak height, shifting the peak maximum, and skewing the peak with the generation of a trailing edge. These effects can be minimized by ensuring that the *RC* time constant is small relative to the time required to scan across the peak. The situation in Figure 9–15 is for a Gaussian peak with a half-width of 1 s and a filter with a 0.25-s *RC* time constant.

Integrating Analog and Digital Signals

Integration is a widely used technique for S/N enhancement. Integration differs from RC filtering in that the response of an integrator to a rectangular pulse is a linear function of time. A linear integrator has a step function impulse response, with the step lasting for the integration time. This gives the linear integrator constant weighting of all previous values in their contribution to the current value.

Linear integration of analog signals can be accomplished in several ways. Active and passive low-pass filters approach linear behavior when the time constant is much larger than the integration time. An operational amplifier integrator with timed switching of course makes an excellent linear integrator. Integrating ADCs such as the dual-slope converter and the VFC are also linear integrators. Another approach is to use digital integration in hardware with a fast adder following the sampling ADC or in software by summing successive ADC values. Time domain signals, such as pulse outputs from transducers, can be integrated by counting techniques.

Integrating ADCs. Use of an integrating ADC, using the dual-slope approach or VFC, is an excellent way to digitize analog signals if high conversion speed is unnecessary. As mentioned earlier in this chapter, a significant advantage of integrating ADCs is their ability to reject noise at certain frequencies. For the dual-slope converter, for example, the integral obtained during the signal integration period can be thought of as the sum of the integrals of the true signal and the noise. Because a sine wave has an average value of zero over one period, noise that has a period equal to the signal integration period has no effect on the output value. Thus, noise signals of these frequencies are greatly attenuated. The ability to reject input signals at certain frequencies is called **normal-mode rejection**. Normal-mode rejection differs from common-mode rejection in that a common-mode signal is present at both inputs of a differential amplifier, whereas a normal-mode signal is present at only one input.

Their excellent normal-mode rejection is one reason that dual-slope converters are frequently used in digital voltmeters. Because many discrete frequency noise sources can be traced ultimately to the power line, noise rejection at the power line frequency is highly desirable. This filtering characteristic is also evident at any multiple of the power line frequency, and normal-mode rejection ratios as large as 70 dB are practical.

Filters of differing frequency response characteristics are often compared on the basis of their equivalent bandwidths. The **equivalent bandwidth** Δf is a rectangular bandwidth of area equal to that of the power spectrum of the filter. The equivalent bandwidth of a linear integrator of integration time t is $\Delta f = 1/(2t)$. This is compared in Figure 9–16 with the equivalent bandwidth of an RC low-pass filter, which is $\Delta f = 1/(4RC)$. Thus for a linear integrator to have a bandwidth equivalent to that of an RC low-pass filter, the integration time must be twice the time constant of the filter ($t = 2RC$). For white noise and dc signals, the S/N improves with $(\Delta f)^{-1/2}$, or for an integrator, the S/N is proportional to $t^{1/2}$

Digital integration. Digital integration in hardware or software can provide excellent signal-averaging characteristics. Many computer-based systems use multiple-point averaging of successive analog-to-digital conversions for this purpose. The number of points to be averaged, the total averaging time, and other variables can be program-controlled. Many computer- or hardware-based digital averaging systems use the data acquisition scheme illustrated in Figure 9–17. Here, the ADC operates at nearly its maximum throughput rate. An anti-aliasing input filter limits the bandwidth of the input signal to that necessary to provide accurate sampling. If the ADC had a conversion time of 8 μs and the sample-and-hold

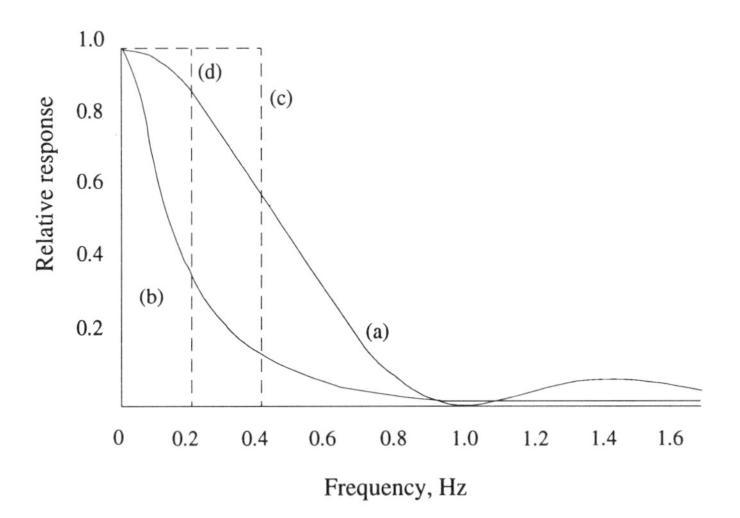

Figure 9–16. Frequency response of an integrating ADC of 1.0-s integration time (line a) compared with that of an RC low-pass filter with a 1.0-s time constant (line b). The equivalent bandwidth of the integrating ADC at line c is compared with that of the RC filter at line d. In these cases, the bandwidth extends from 0 Hz (dc) to the upper cutoff frequency.

settling time was 1 μs, for example, the monostable multivibrator might be set up for a 2-μs delay. The sampling rate of the data acquisition system would then be 100 kHz. The input filter would be chosen to limit the bandwidth to less than 50 kHz ($RC > 3.2$ μs). If no input signal changes faster than 1 ms were important, then it would be feasible to average 100 data points.

This approach provides essentially complete averaging over the 1-ms period even though the sample-and-hold aperture time may be as short as 50 ns, because the input filter averages the raw input signal, and the computer or hardware adder averages the analog-to-digital conversions. For slowly changing signals and white noise, the S/N should improve with the square root of the averaging time or with the square root of the number of points averaged. Thus, averaging 100 points should improve the S/N by a factor of 10 compared with using a single point.

Digital integration can also provide high rejection for noise signals with periods equal to the averaging time. With a 1-ms averaging time, for example, rejection nodes should occur at 1 kHz and multiples of 1 kHz.

Digital Filtering

Because filtering a signal can be considered a process of weighting the data (convolution), filtering can readily be carried out in the digital domain. With computer systems, virtually any weighting factor (filter function) desired can be achieved in software. Hardware systems can also perform the digital filtering operation. There are two general schemes by which the filtering operation can be performed: time domain weighting and frequency domain weighting.

In the time domain approach, the digital filtering operation is often referred to as a data-smoothing operation. The smoothing operation is normally carried out

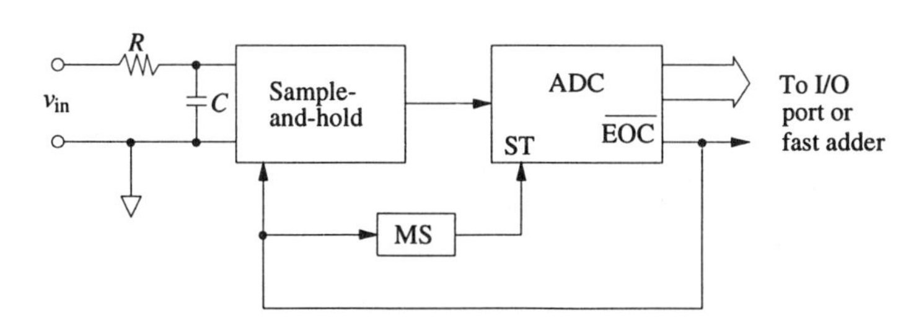

Figure 9–17. Fast data acquisition system. An input RC filter limits the bandwidth of the input signal. The end-of-conversion ($\overline{EOC}$) signal from the ADC puts the sample-and-hold circuit in the hold mode. The ADC start convert signal (ST) is delayed by a monostable multivibrator (MS) to allow the sample-and-hold circuit to settle before conversion.

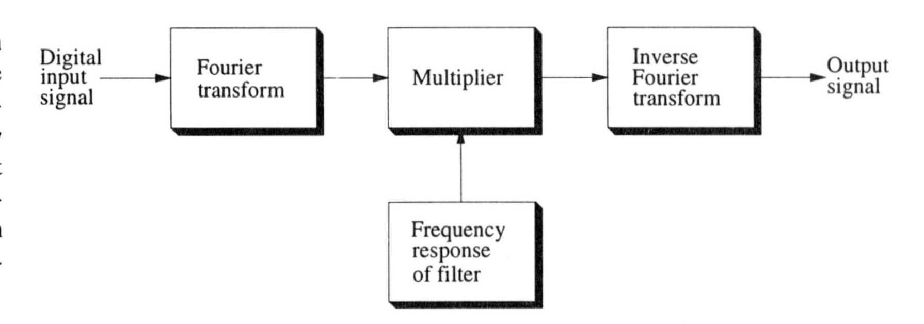

Figure 9–18. Digital filtering in the frequency domain. The Fourier transform of the input signal is computed and multiplied by the desired frequency response at all pertinent frequencies. The inverse Fourier transform is then computed to give the filtered output.

on data sampled at evenly spaced time intervals. One approach to smoothing is the n-point **moving averaging smooth** (Savitzky, A.; Golay, M. J. E. *Anal. Chem.* **1964,** *36,* 1627). Here, n points are averaged to provide a smoothed data point that represents the central value, and the averaging function is moved along the time axis. For example, in a five-point moving average smooth, points 1 through 5 would be averaged to provide a smoothed data point replacing point 3 in the raw data array. Points 2 through 6 would then be averaged, replacing point 4, etc.

In more complex smoothing operations, a polynomial can be used to approximate local sections of data, and the fitted polynomial then provides a smoothed value for the central point. Tables of weighting coefficients for a variety of smoothing functions including higher-order filtering operations are available. All of these smoothing techniques improve the S/N through bandwidth reduction. Unfortunately, because the smoothing function is only an approximation of each local section of data, the true signal can undergo distortion unless the smoothing parameters are carefully chosen.

In the frequency domain approach, the desired frequency response of the filter is applied to the Fourier transform of the signal. It was shown earlier that filtering can be considered the result of convolving the amplitude–time waveform of the signal with the impulse response function of the filter. An important Fourier transform theorem states that *convolution in the time domain is equivalent to multiplication in the frequency domain*. Hence the frequency domain digital filter is applied as illustrated in Figure 9–18.

Various fast Fourier transform algorithms are available to calculate the Fourier transform of the input digital signal. Essentially any desired frequency response curve can be set up, including many that are impossible to design with hardware. Filters with no phase shift, square cutoff filters, differentiating filters, and unique discrete frequency filters are all readily implemented. Filtering is implemented by multiplying the real output of the transformed input signal by the frequency response of the selected filter and regenerating the signal by inverse Fourier transformation. The effect on frequency response can be appreciated by reviewing the Fourier transform pairs in Figure 2–8.

A simple example that demonstrates the effectiveness of this approach is shown in Figure 9–19. Here the desired filter response is a low-pass filter with an abrupt cutoff and no phase shift, characteristics that are impossible to achieve with analog filters. Because of the abrupt rather than gradual cutoff, the noise level of the filtered signal is much lower than that of the original signal. Analogous reduction of the noise level with analog techniques would have been difficult without distortion of the signal because of the large phase shifts near the cutoff frequency. Distortions can result in digital filtering if signal information is also attenuated by the filter. However, because the frequency domain representation of the input signal can be displayed, it is often quite simple to choose the filter

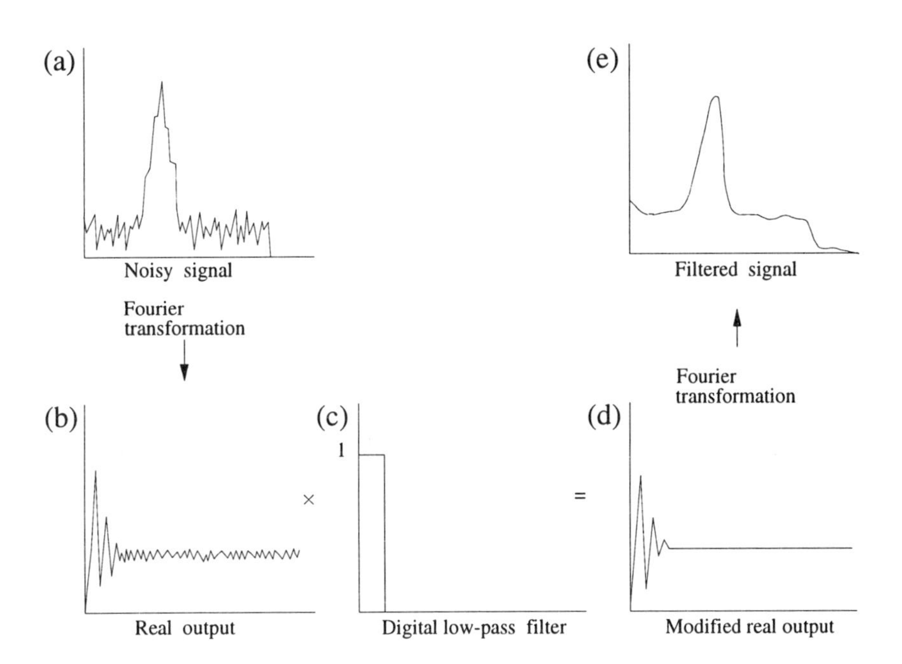

Figure 9–19. Digital filtering. The Fourier transform of the input signal (a) produces the real output shown in part b. This is multiplied by the desired filter response (c) to yield the modified real output (d). The inverse Fourier transform is then calculated to regenerate the filtered signal (e).

bandwidth to avoid distortions. It is important to note that digital filtering in the frequency domain is exactly analogous to smoothing operations in the time domain.

9–3 Enhancing Signal-to-Noise Ratio by Correlation Techniques

The preceding section emphasized S/N enhancement techniques that depend on the different frequency characteristics of signal information and noise power. A second important difference between the frequency components of signals and noise is their phase relation. The frequency components that make up a signal are in general phase related; noise frequencies, on the other hand, typically are not related in phase to the signal frequencies or, for that matter, to other noise frequencies.

These two distinguishing properties of noise and signal frequencies (relative distribution and phase relation) are illustrated in Figure 9–20 for two noisy peak-shaped signals. From the amplitude spectra, it is clear that attenuation of the higher frequencies by filtering would significantly improve the S/N. In addition, the phase spectra show that both noisy signals have in-phase components at low frequencies. The similarity of the phase spectra in the low-frequency region indicates that the phases of the frequency components responsible for the peak are essentially the same in these two noisy signals. It is also important to note that the two phase spectra are quite dissimilar in the higher-frequency region, where random noise is expected to predominate.

This phase coherence of the signal frequency components can be used to enhance the S/N *if the signal is or can be made repetitive*. For example, if the amplitude–time waveforms in Figure 9–20a could be obtained repetitively (by repetitive scanning or by repeated triggering of the signal initiation process) and if the signals obtained on each repetition were added together, then the frequency components that make up the peak would add in phase, while the noise frequencies would add randomly and tend to cancel out. The S/N enhancement techniques

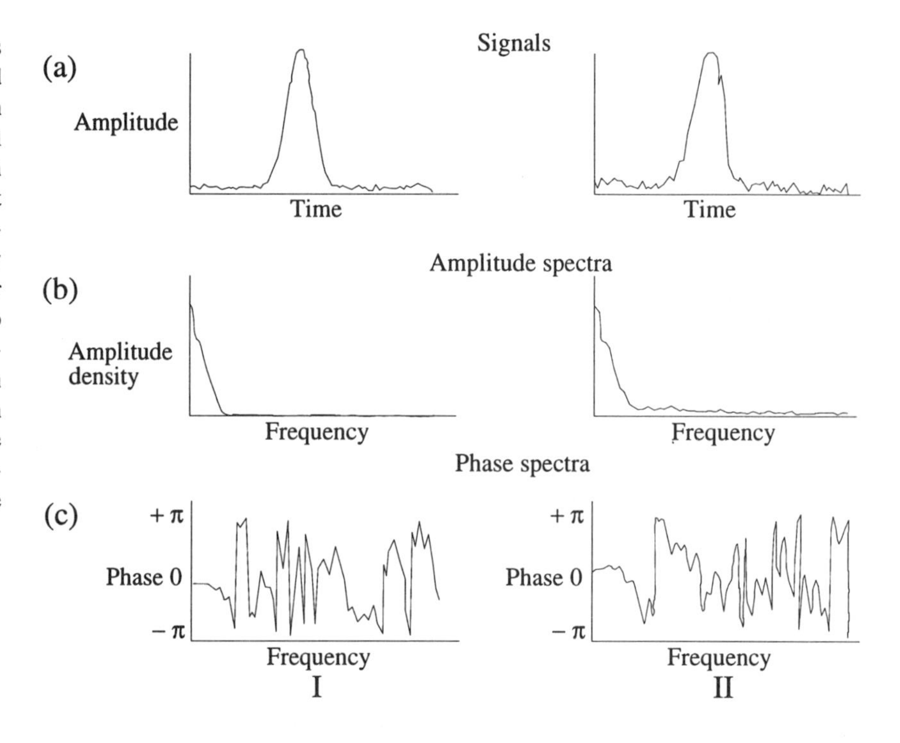

Figure 9–20. Noisy peak signals (a) and their amplitude (b) and phase (c) spectra. The signal in column II has a higher noise level than that in column I. Note from the amplitude spectra (b) that most of the signal frequency components are located in a narrow band near 0 Hz and that the higher frequencies are primarily due to noise. In part c most of the lower-frequency components of both noisy signals are seen to be in phase (0 phase), while the phase of the higher-frequency components varies randomly about the zero value.

presented in this section all depend on this basic principle to discriminate between signals and noise.

The section begins with a brief discussion of correlation, because it is the correlation of two signals that can provide the necessary phase discrimination. Then several important S/N enhancement techniques are described. These include lock-in amplification, boxcar integration, multichannel signal averaging, and waveform correlation techniques.

Correlation

A generalized block diagram of an S/N enhancement technique is shown in Figure 9–21. The bandwidth control step discriminates against noise on the basis of its frequency distribution, and the multiplication–integration step discriminates against noise on the basis of the predictable time behavior of the signal information (phase coherence). The multiplication–integration operation is best described in terms of correlation. **Correlation** involves multiplying one signal by a delayed

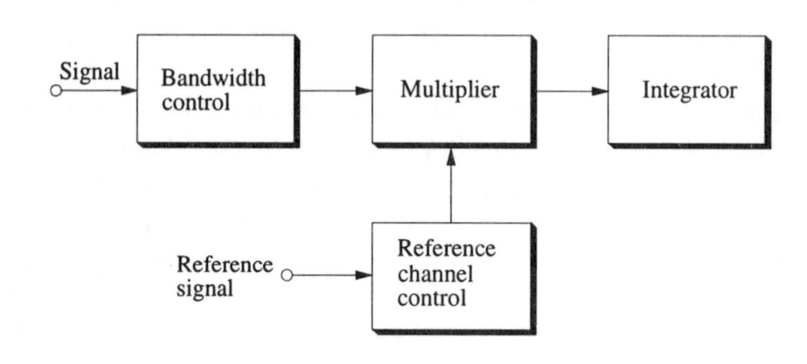

Figure 9–21. Generalized block diagram of an S/N enhancement technique. The noisy signal is amplified with a band-limited amplifier and multiplied by a reference signal. Finally, the multiplier output is time averaged or integrated.

version of a second signal and integrating or time averaging the product. Correlation with continuous functions can be described by the following integral:

$$R_{xy}(\tau) = \lim_{T \to \infty} \frac{1}{2T} \int_{-T}^{+T} x(t)\, y(t-\tau)\, dt$$

Here, $R_{xy}(\tau)$ represents the correlation function of the two signals $x(t)$ and $y(t)$ over the interval $-T$ to T, and τ is the relative displacement. Many correlations are carried out on sampled waveforms, where correlation is described by

$$R_{xy}(n\Delta t) = \sum_t x(t)\, y(t - n\Delta t)$$

Here, Δt is the sampling interval, and $n\Delta t$ is the relative displacement. Evaluating this time-averaged product over a range of relative displacements or delays generates a correlation pattern that is a function of the relative displacement.

Two general types of correlation are commonly distinguished. If the two signals are different, the process is called **cross-correlation**; if they are the same, it is **autocorrelation**. The correlation of two functions is very similar to their convolution. Correlation and convolution both involve the multiplication–integration operation. In convolution the impulse response function must be reversed from left to right before the multiplication–integration operation is carried out. When the scanning function is symmetrical, correlation and convolution give equivalent results. Correlation of two signals is equivalent to multiplication of their frequency spectra, as illustrated in Figure 9–22. The Fourier transform of a single rectangular pulse is the $(\sin x)/x$ function, and the product of the two transforms is $(\sin^2 x)/x^2$. This product of two spectra is often called a **cross spectrum**. Inverse Fourier transformation yields the triangular autocorrelation function. The effects of a particular correlation operation are often more easily discerned by thinking in terms of multiplying spectra rather than correlating waveforms.

The multiplication–integration operation (correlation) provides discrimination between phase-related signal components and randomly phased noise components. Some of the techniques described below involve only a simplified correlation operation in the sense that the correlation function is evaluated at only a single relative displacement. In general, a measurement technique is referred to

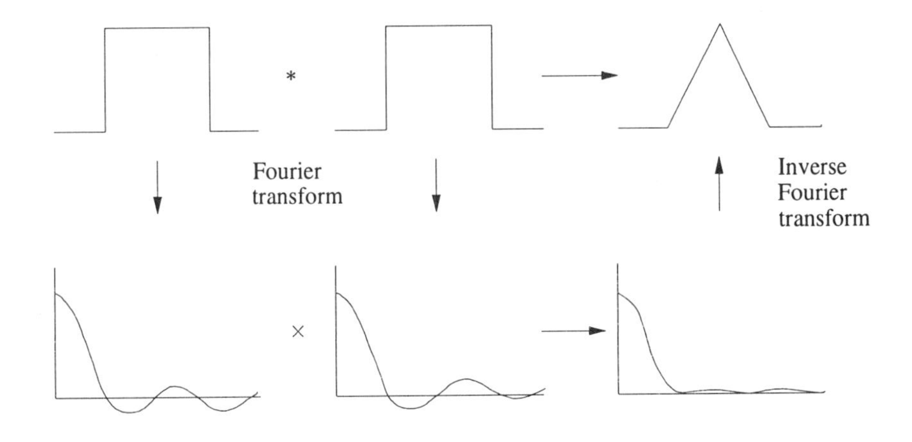

Figure 9–22. Fourier transform representation of correlation. The asterisk indicates correlation, and × indicates multiplication. The autocorrelation of a rectangular signal yields a triangular waveform (top path). The same result can be obtained by multiplying the Fourier transforms together and taking the inverse transform of the product (bottom path).

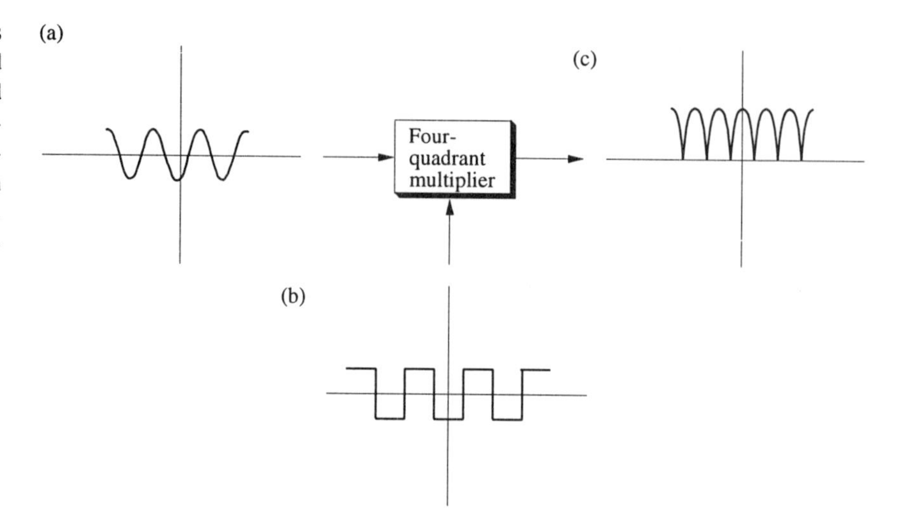

Figure 9–23. Synchronous demodulation. The signal (a) and reference (b) are multiplied together in a four-quadrant multiplier. The reference wave is adjusted to be exactly in phase with the signal (modulated carrier). The multiplier output for this condition is shown in part c.

specifically as a correlation technique when the complete correlation function of two waveforms is evaluated.

Lock-In Amplification

The **lock-in amplifier** is useful for S/N enhancement of signals that can have their maximum power density at a frequency other than 0 Hz. The lock-in amplifier is an example of an S/N enhancement system that uses a cross-correlation technique. The basic steps in a complete lock-in amplifier system include modulation, selective amplification (often tuned amplification), synchronous demodulation, and low-pass filtering. The demodulation step of the lock-in amplifier provides the phase discrimination. Synchronous demodulation can be carried out with a four-quadrant multiplier, as shown in Figure 9–23. The result of the synchronous demodulation is a full-wave rectified output of those signal components of the same frequency *and phase* as the reference. The final step in the recovery of the signal is to send the multiplier output through a low-pass filter. This step simply decreases the fluctuations of the synchronously rectified carrier and produces an output voltage proportional to the amplitude of the carrier wave. Note that the cross-correlation operation is present in the synchronous demodulation and low-pass filtering step.

Hardware lock-in amplifiers are available from several manufacturers. It is also possible to simulate lock-in amplifier performance in real time with an interfaced microcomputer system. The reference waveform can be used to trigger data acquisition of the carrier wave. If the frequency of the carrier is not too high, say 1 kHz or less, several samples can be acquired during each half-cycle of the carrier. Alternate half-cycles are then added together with the correct polarity so as to carry out synchronous demodulation. The values for several hundred cycles are then averaged to provide a "low-pass-filtered" output via a DAC.

Boxcar Integration

The boxcar integrator is a versatile gated integrator for measuring repetitive signals. The boxcar technique involves gating out a particular section of a waveform and

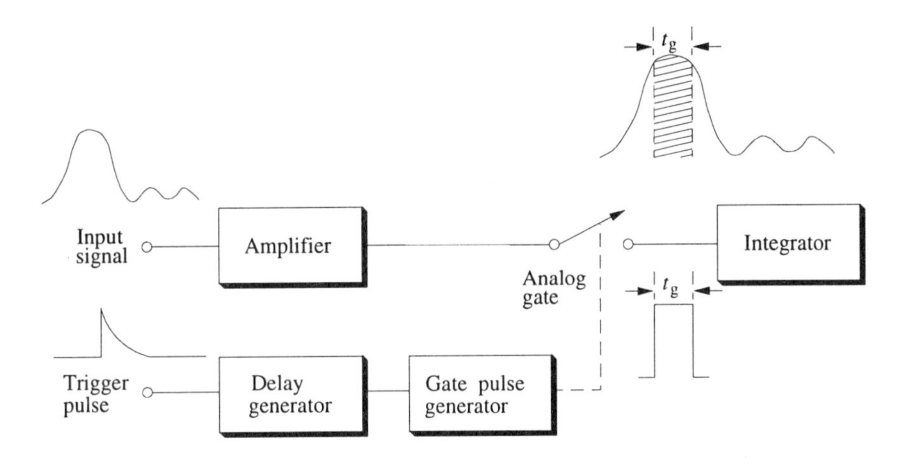

Figure 9–24. Boxcar integrator. The amplified input signal is gated to an integrator for a short gate time t_g controlled by the gate pulse generator. The gate pulse occurs at a fixed time delay relative to the trigger signal. The same time slice of the input waveform can be integrated for multiple repetitions to improve the S/N, or the delay generator can slowly scan the gate time across the input waveform. This allows examination of multiple sections of the waveform or of the entire waveform.

integrating successive gated signals to improve the S/N. It is particularly useful for measuring repetitive short-pulse signals and signals that have a slow repetition rate or a low duty cycle. A block diagram of a boxcar integrator is shown in Figure 9–24.

The analog gating operation can be thought of as a multiplication operation in which the input waveform is multiplied by a normalized rectangular pulse (amplitude = 1). The analog gating and integration operation is then a form of cross-correlation in which the signal waveform is cross-correlated with a rectangular gating pulse at a relative displacement *t* set by the delay generator. The boxcar gate delay can also be slowly scanned in time so that the cross-correlation is carried out across the complete waveform. With a sufficiently narrow gate pulse and a relatively slow sweep on the delay time, the shape of a pulse as brief as 2 ns can be recorded on a strip chart recorder. Gate pulse widths as low as 100 ps are available on commercial boxcar integrators.

The key step in the boxcar integrator system is the analog gating. This step is a sampling operation, and all the criteria for accurate sampling discussed in Chapter 2 must be satisfied. The gate pulse width t_g is the aperture time of the sampling operation. If the gate pulse is scanned across the waveform, the increments in the delay time between samples must be small enough to satisfy the Nyquist sampling theorem.

The boxcar integrator function is also readily implemented with a computer data acquisition system, as illustrated in Figure 9–25. In the example shown, the time of sampling is controlled by a real-time clock. Multiple-point averaging and scanning are readily accomplished with quite simple programming. A computer boxcar integrator system, however, cannot achieve the very fast aperture times of a hardware system.

The boxcar integrator is thus a very versatile measurement system. It can be thought of as a lock-in amplifier for pulsed signals. As a gated integrator, the boxcar is a single-channel averager for repetitive waveforms. In its scanning mode, it is a cross-correlation computer. The boxcar integrator can measure very fast waveforms and narrow pulses. However, for many signals, multichannel averagers offer significantly faster measurement times because hundreds of channels can be averaged simultaneously. Also, more general cross-correlation analysis requires more complex cross-correlations than those between a simple rectangular pulse and a signal waveform.

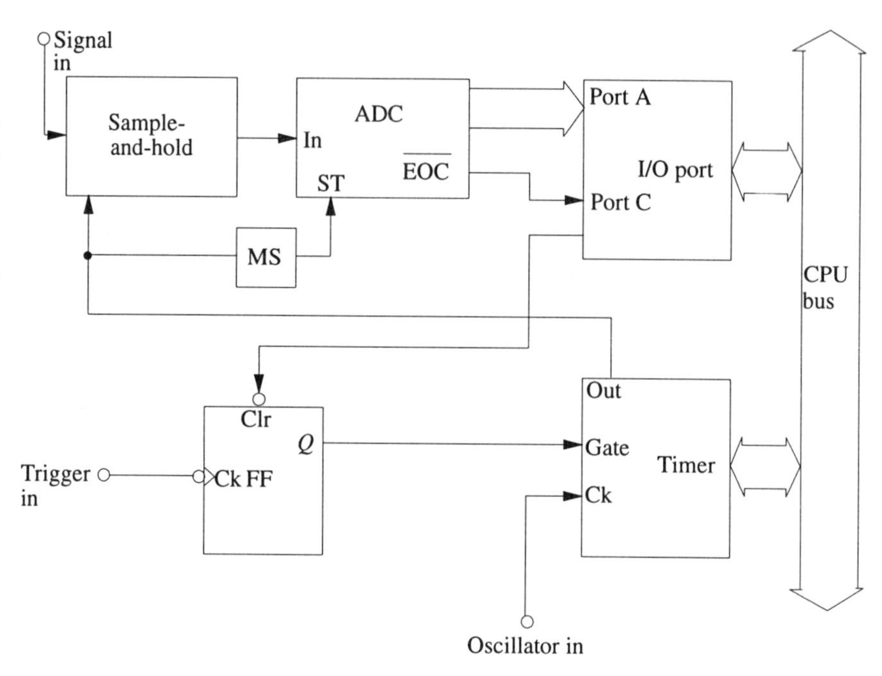

Figure 9–25. Computer boxcar integrator. A programmable timer is used to generate a variable delay between the trigger signal and the acquisition of the sample. Through program control, a specified number of samples can be taken at any given delay, the delay can be scanned at the desired rate, and the time increment between points can be varied.

Multichannel Averaging

In many experimental measurements it is necessary to recover a complete repetitive signal from a noisy waveform. A **multichannel averager** acquires a large number of evenly spaced samples across the complete waveform on *each* repetition instead of one sample per repetition as in the boxcar integrator. However, because the multichannel averager takes N samples across a waveform, the complete waveform can be recovered N times as fast. As with any averaging or integrating technique, the signal builds up as the number of scans, and the noise builds up as the square root of the number of scans. Thus, the S/N improves as $N^{1/2}$.

Most multichannel averagers are digital instruments based on the general block diagram of Figure 9–26. Each repetition of the waveform is digitized at a desired number of points, and these values are added to previous values and stored in memory. The general structure of the digital multichannel averager may be based

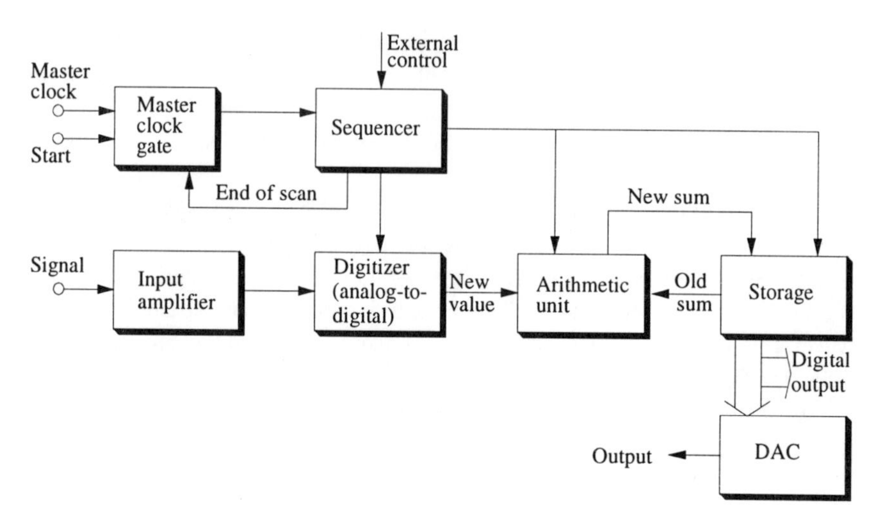

Figure 9–26. Multichannel averager block diagram. The start pulse that initiates the repetition of the signal waveform gates a master clock to a sequencer (hardware or computer) that controls the sampling, digitization, addition, and storage of the new values. A DAC provides an analog output for display.

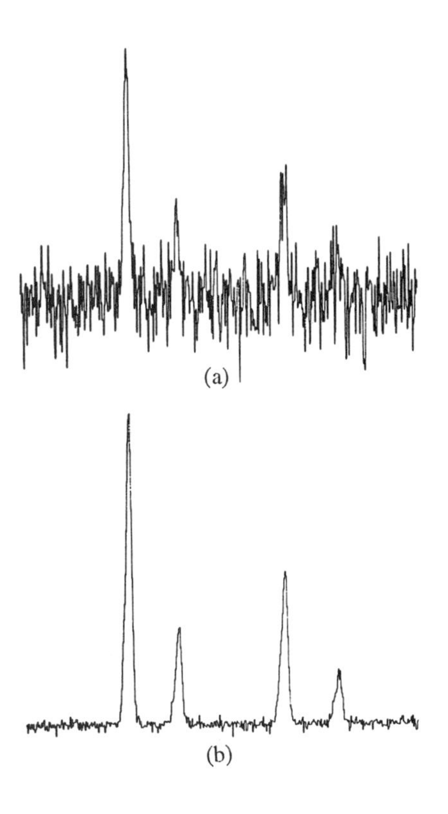

Figure 9–27. Effects of multichannel averaging. (a) One scan of a noisy waveform; (b) the result of 100 scans.

on a stand-alone hardware system, or it may be just a program in a microcomputer system. Hardware averagers are usually faster than the software-based averagers, but multichannel averagers are not, in general, extremely fast. Sampling rates are normally less than 200 kHz for hardware averagers and are often less than 50 kHz for computer-based systems. Typical input circuitry includes a fast sample-and-hold and a fast ADC. Precision can be traded off for sampling rate if desired.

If averaging is not required, considerably faster sampling rates can be achieved. Instruments called **transient recorders** can digitize and store a complete single trace of a waveform. Sampling rates for transient recorders can be as high as 1 GHz with 8-bit resolution. The transient recorder block diagram is similar to that shown in Figure 9–26 except that the adder is not used.

In all multichannel averagers, the sampling operation described above is an extremely important step. All the criteria for accurate sampling with respect to sampling rate, sampling duration, aperture time, and quantizing time must be satisfied.

Multichannel averaging can achieve impressive S/N enhancement for certain types of signals, as illustrated in Figure 9–27. One hundred scans should improve the S/N by a factor of 10 compared with a single scan. Quantizing noise can limit the measurement precision when the ADC does not have enough resolution or, more often, when two large signals are subtracted to yield a small result. If the noise in the system (ADC and signal) is much less than the quantization interval Q, quantizing noise cannot be reduced by averaging. This is true because the noise is a round-off error, which is identically reproduced on each scan. However, a level of system noise of Q or greater randomizes the quantizing error and allows quantizing noise reduction by averaging. When the total system noise is very small, an S/N enhancement can be realized by adding random noise to the original signal prior to multichannel averaging. As long as the random noise is not too intense, its final level can be reduced to an acceptable point by averaging.

Correlating with Complex Waveforms

The basic correlation operation applies to all S/N enhancement techniques discussed in this section. However, except for the scanning boxcar integrator, the correlation was evaluated at only one relative displacement and even then with only a rectangular reference pulse. Cross-correlation and autocorrelation of complete waveforms are, not surprisingly, also very useful S/N enhancement techniques.

In cross-correlation it is useful to distinguish the signal waveform and the reference waveform and to think in terms of the reference waveform moving with respect to the signal waveform. For example, in the scanning boxcar integrator, the rectangular gate pulse is the reference waveform that is shifted or slid across the signal. In more general cross-correlations, any desired shape could be used for the reference waveform. The optimal shape of the reference cross-correlation waveform for maximum S/N enhancement has been thoroughly investigated. For repetitive single-pulse signals, a noise-free version of the signal itself provides a close-to-optimal reference pulse shape. An example of the cross-correlation operation for S/N enhancement is shown in Figure 9–28.

The goal of cross-correlation with pulse waveforms need not be simply S/N enhancement. Useful modifications of the signal waveform such as differentiation and resolution enhancement can be carried out by cross-correlation with bipolar pulses. In general, cross-correlations carried out with complex reference waveforms can be considered a measure of the degree of similarity between the signal and the reference waveforms. The value of the cross-correlation function at zero displacement ($\tau = 0$) is the sum or time average of the product of the two waveforms, and its magnitude is a measure of the common features shared by the two waveforms. As such, cross-correlation is highly useful in automated signal and pattern recognition systems. As with simple pulse cross-correlations, more complex correlations can be carried out under conditions of poor S/N.

Correlation carried out between a signal waveform and a time-shifted version of the same signal waveform is called autocorrelation. This type of correlation can be useful in recovering a periodic signal from noise when no reference waveform is available. The signal becomes its own reference. However, any phase information in the original periodic signal is lost in the autocorrelation operation. The autocorrelation patterns of several periodic waveforms are illustrated in Figure 9–29. Note that the pattern for random noise is quite different from that of the

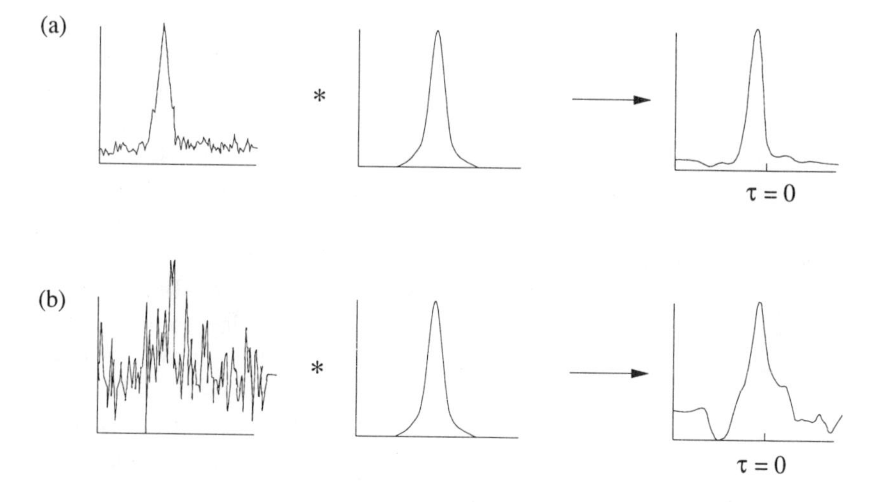

Figure 9–28. S/N enhancement by cross-correlation. The noisy signal on the left is cross-correlated (indicated by *) with a noise-free version of itself; the result is the smoothed signal on the right. Note that some distortion in signal shape is introduced as the signals are broadened and rounded by the cross-correlation. The noise level in signal b greatly exceeds that in a.

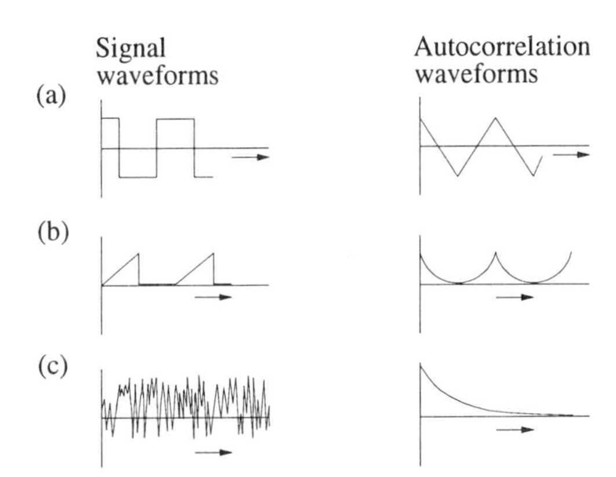

Figure 9–29. Autocorrelation patterns of several waveforms. When it is autocorrelated, random noise of limited bandwidth (c) gives rise to an exponential decay. Periodic signals (a and b) give patterns that persist after the noise has decayed.

signals shown. Thus, if a periodic waveform is present in a noisy signal, its autocorrelation pattern persists after that for the noise has decayed.

Although a few manufacturers supply stand-alone correlation instrumentation, it is not widely available at this time. However, correlation analysis can be readily performed with computer-based systems. This provides great flexibility in pretreating the data, carrying out the correlation operation, and displaying the data for easier interpretation. The correlation may be carried out on stored waveforms by multiplication of the Fourier transforms of the waveforms and inverse transformation or by time averaging of the product of the two waveforms as a function of displacement.

9–4 Measuring the Rate of Events

It is often necessary to measure the rates of processes or events, and some aspects of rate measurements are distinct from direct measurements of magnitude. This section builds on the concept of frequency measurement developed in Chapter 2 and goes on to explore the measurement of the rates of irregularly or randomly spaced events.

Rates of Randomly Spaced Events

The rate of a regularly recurring event is its frequency. The normal measurement of frequency is accomplished by a counter that accumulates the number of events or cycles occurring during a specific time interval. It is helpful here to remember that a frequency measurement is an integrating measurement. All events during the gate time are counted, and the result, events per time, is the average rate of events over that time. Many events of interest in nature—radioactive decay of unstable nuclei and the arrival of photons from an ordinary light source, for example—occur at irregular intervals. These events occur randomly in time, and this irregularity of occurrence affects the measurement process.

At low rates of incidence, photons, decay particles, and ions are often detected by conversion to emitted electrons followed by electron multiplication. The electron multiplier is the charge amplification dynode chain used in the photomultiplier tube of Figure 8–28. Each incident particle or photon that causes electron emission produces a cascade of 10^5 to 10^8 electrons at the last dynode. This burst of electrons has a charge of 10^{-11} to 10^{-14} C, which causes a current pulse in the

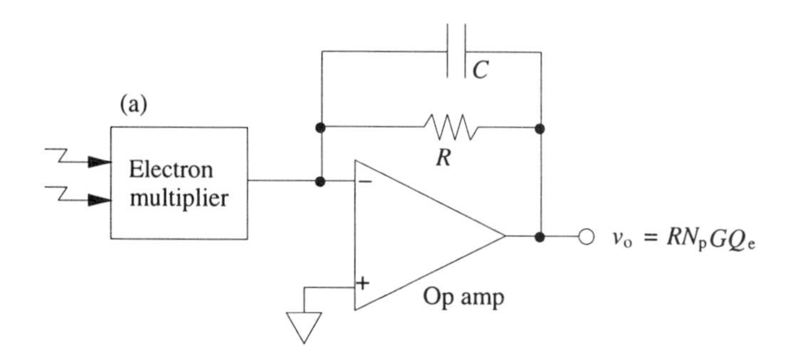

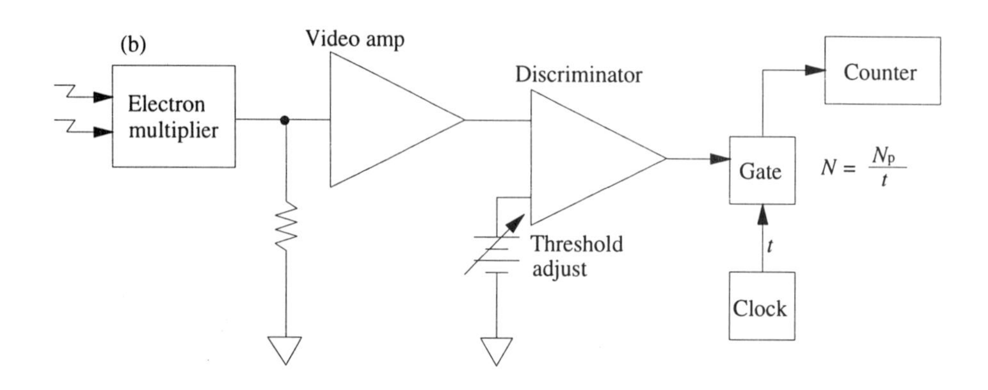

Figure 9–30. Analog and digital techniques for measuring the event rate at an electron multiplier detector. In the analog circuit (a), a current follower converts the electron multiplier current to a proportional output voltage. The current is equal to the charge per second, which is the number of pulses per second N_p multiplied by the electron multiplier gain G multiplied by the electron charge Q_e. The capacitor with capacitance C produces a low-pass filter with an RC time constant much longer than the interval between pulses. The digital circuit uses a high-frequency amplifier (amp) and comparator to produce a logic pulse for each current pulse. The counter reading N is the number of pulses that occur over the gate time interval t.

connection to the last dynode. All the cascading electrons produced by a single particle arrive at the last dynode within a time span of 3 to 10 ns. The peak pulse current (pulse charge divided by pulse width) is thus in the range of 1 µA to 3 mA, a readily detectable pulse amplitude. Two systems are used to measure rates with electron-multiplying detectors: analog circuits that measure the average current produced by the packets of charge and digital circuits that count the number of current pulses in a given time. The digital technique is called **pulse counting** or, in optical systems, **photon counting**. Both of these techniques are illustrated in Figure 9–30.

The randomness of the pulse times causes an uncertainty in the measured rate because there is a variation in the number of events that occur in a given period. The larger the number of pulses measured, the smaller the relative error in the measurement. For the digital technique, the standard deviation σ for the measurement is equal to the reciprocal of the square root of the count value, $(N_p)^{-1/2}$. For analog measurement, the standard deviation of the current follower output voltage σ_v is $(iRGQ_e/2C)^{1/2}$, where i is the average output current of the electron multiplier. The relative standard deviation σ_v/v_o of the analog output increases as the signal decreases, as can be seen by dividing the expression for σ_v by $v_o = iR$ to obtain

$$\frac{\sigma_v}{v_o} = \left(\frac{GQ_e}{2iRC}\right)^{1/2}$$

This same expression shows, as expected, that the relative standard deviation improves (decreases) as the *RC* time constant increases.

At low count rates, the pulse-counting technique has a number of distinct advantages. First, at very low count rates, the average analog current is extremely low. (At 1000 pulses per s, the current is 10^{-8} to 10^{-11} A depending on *G*.) Such low currents require a large (and therefore noisy) feedback resistor in the current follower, a long time constant to smooth out the pulses, and a low-drift amplifier to minimize $1/f$ noise. Other error sources are seen from the counting circuit waveforms shown in Figure 9–31. At low count rates, the current from the low-level pulses can become significant in the analog technique, as can the pulse-to-pulse charge variation. Conversely, with the pulse-counting technique, long integration times can be used without drift to obtain high measurement precision even for very low count rates.

The pulse-counting technique would be ideal if it were not for its problems at high count rates. These problems are due to an inability to distinguish pulses that occur at nearly the same time. The **resolving time**, or **dead time**, of the counting system is defined as the minimum distinguishable time interval between two pulses. The dead time for most systems is longer than the pulse width from the electron multiplier. It results from pulse broadening in the amplifier, the limited rise and fall times of the discriminator, and the maximum clock rate of the counter. The dead time for a good counting system is about 10 ns. If more than one pulse occurs in any 10-ns interval, **pulse pileup** occurs, causing a count error. The probability P_n that *n* pulses will occur within the resolving time t_d of a pulse counter is given by the Poisson distribution as

$$P_n = \frac{(\overline{R}t_d)^n}{n!}\exp(-\overline{R}t_d)$$

where $\overline{R}$ is the average pulse rate. If no pulse or one pulse arrives during t_d, no pulse overlap occurs, but if two or more arrive during t_d, count loss occurs. The

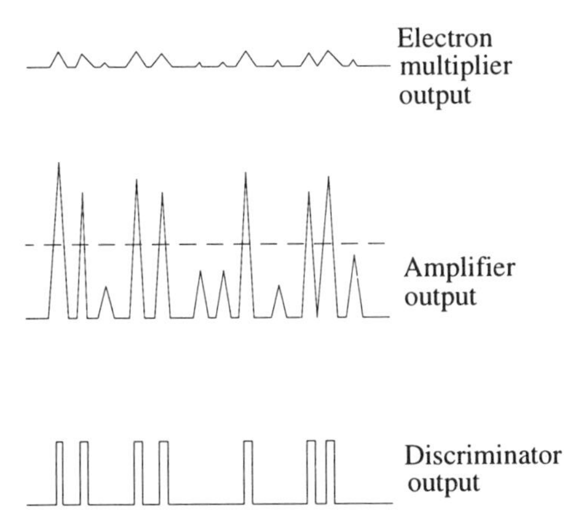

Figure 9–31. Waveforms in a digital pulse-counting circuit. The amplified signal from the electron multiplier shows low-amplitude noise pulses that arise from electron emission unrelated to the measured events that start partway down the dynode chain. The heights of pulses from true events are unequal because of variations in multiplier gain from pulse to pulse. When the discriminator level is appropriately set, multiplier noise and gain variations do not affect the pulse-counting measurement.

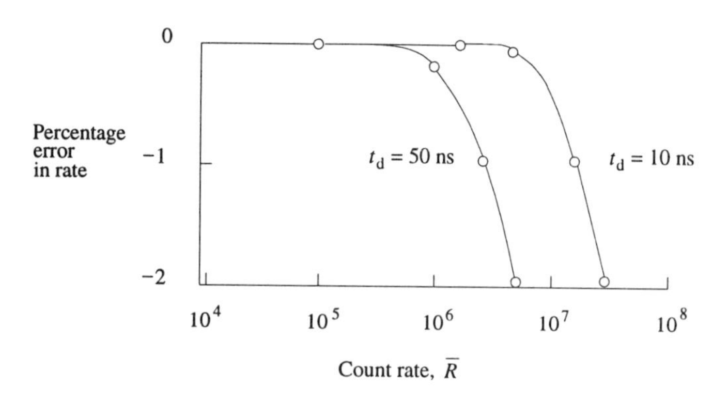

Figure 9–32. Pulse-counting error due to pulse overlap. $\overline{R}$ is the average count rate, and t_d is the counter dead time. Errors for other values of t_d can be calculated from the values of $\overline{R}t_d$ for 2, 1, 0.1, 0.01, and 0.001% error; these are 0.22, 0.15, 0.045, 0.015, and 0.005, respectively.

probability P_l that overlap will occur is the total probability (unity) less each of the other probabilities:

$$P_l = 1 - P_0 - P_1 = 1 - \exp(-\overline{R}t_d) - \overline{R}t_d \exp(-\overline{R}t_d)$$

Figure 9–32 shows the error due to pulse pileup as a function of frequency for two values of t_d. A counting system with a t_d of 10 ns has a count error of 0.1% at an average count frequency of 4.5 MHz. This same counter can count regularly spaced events up to 100 MHz without error. The average current at a pulse rate of 900 kHz, where the pulse-counting error begins to be significant for a 50-ns dead time, is between 0.014 and 14 μA, depending on the multiplier gain. Because this current level is easily measured with a simple op amp circuit, the widest-range random-event rate-measuring system would combine a pulse counter for low rates and a simple analog circuit for average currents above 10^{-8} A.

Commercial pulse-counting systems sometimes offer several additional operating modes. In one, rate is measured by determining the time required to achieve a predetermined count value. This mode maintains a constant standard deviation for each measurement. If the source of the random events is chopped, a kind of lock-in detection can be used by controlling the counting mode with the chopper drive. Counts obtained during a source OFF state can be subtracted from counts obtained during an equal-duration ON state. An up–down counter is convenient for this application. If this cycle is repeated many times without clearing the counter, the noise in the ON and OFF rates averages out, and the difference between the ON and OFF counts accumulates.

Related Topics in Other Media

Video, Segment 9

- Using sampling and integration to improve signal-to-noise ratio

Laboratory Electronics Kit and Manual

- Section J–2
  - ¤ Wiring and testing an analog-to-digital converter
  - ¤ Demonstrating an electronic digital servo system

Chapter 10

Troubleshooting Your Instruments

10–1 Understanding the Process

Anyone who drives a car knows that there are occasional malfunctions that need repair. Likewise, users of scientific instrumentation are familiar with equipment malfunctions that result in downtime and often serious loss of data, time, and productivity. These troubles require prompt attention and decisions. Is it something simple that I can fix myself? Should I call the instrument manufacturer? Or a local instrument repair person? Or . . . ? The answer, of course, depends on one's ability to define the problem and organize a logical approach.

Defining the Problem

Identifying the problem and deciding on a logical approach to correct a malfunction are the first steps in what is often called "troubleshooting." In general, we refer here to electronic malfunctions, but the principles of troubleshooting are quite similar for other types of equipment.

There are equipment malfunctions that only a skilled troubleshooter can locate and repair. Fortunately, however, an instrument user who understands the concepts and principles presented in the previous chapters of this book already has the basis for developing troubleshooting skill. By analyzing the instrument data flow and then following a systematic troubleshooting procedure, the malfunction can often be located and even corrected. Learning to do this yourself can save much time and money. Thus, a troubleshooting approach is presented here, and subsequent sections focus on some of the most common problems that are readily corrected.

Organizing the Approach

To start with, it is important for the troubleshooter to know something about the equipment. The more information the better for organizing the approach. The function of each major subsystem should be understood. If not known, the information can usually be found in the service manual, which should also contain a list of common malfunctions with typical symptoms. By observing symptoms and making a few simple measurements, it is often possible to define the problem and even identify a faulty subsystem. The types of test instruments required for these basic measurements are outlined in this section.

With preliminary information in hand, several approaches are possible. If it appears that a specific subsystem is the culprit and a spare unit is available and easily substituted, it might be expeditious to substitute the spare module and get your equipment back in operation. However, before substituting a spare module, it might be prudent to check your observations with the service department of the instrument manufacturer. A telephone call could help you check your analysis of the problem and provide useful recommendations based on the manufacturer's

2861-2/94/0329$08.40/1 Published 1994 American Chemical Society

expertise. Also, many computer-based instruments come with elegant diagnostic routines that point to particular malfunctioning elements and list probable causes.

Getting Equipped

Various types of sophisticated troubleshooting instruments that are important for the tough problems are available. These are essential for the professional repair shops. Here, we are interested in making rather straightforward measurements by using common, readily available test instruments such as the digital multimeter (DMM), oscilloscope, signal generator, timer/frequency counter, and logic probe. A review of these basic test devices follows.

Digital multimeter (DMM). The general-purpose DMM is first introduced in Chapter 2, and its functional block diagram is shown in Figure 2–14. A fuller description of the DMM, including illustrations of the DMM input circuits for voltage, current, and resistance measurements, is presented in Chapter 3, Section 3–2. The DMM measurement of power supply and circuit voltages and tests for electrical continuity or short circuits are often sufficient to locate typical malfunctions described later in this chapter. Thus, the compact and inexpensive DMM is a very versatile test instrument for troubleshooting.

Oscilloscope. The basic principles and applications of an oscilloscope are presented in Chapter 3, Section 3–4, *Using the Oscilloscope To Observe Signals.* Various types of oscilloscopes are described in that section. A rather compact and convenient oscilloscope for general-purpose troubleshooting is shown in Figure 10–1.

The oscilloscope is a powerful troubleshooting instrument. Because it provides visual displays of signal waveforms, it enables the user to see shapes, magnitudes, and frequency and phase relationships of signals. Also, critical time relationships between two signals are readily observed with a dual-trace oscilloscope. The observed signals can be compared with the expected signals, which are often shown in an instrument manual.

Signal (Function) Generator. Signal sources for which the output varies in specific ways as a function of time have many measurement and control applications. Also, they can be very useful in troubleshooting because they provide known signals that can be observed and measured at successive points in the circuits. Application of this *signal tracing* procedure is illustrated in the next section.

Both microcomputer and hardware function generators have been introduced in previous chapters. The former is described in Chapter 1, Section 1–4, and Chapter 2, Section 2–5, under *Emulating Standard Test Instruments* and *Generating Analog Signals,* respectively. It has the advantage of being able to generate many types of complex signals. The basic circuits of hardware function generators are described in Chapter 7, Section 7–1, entitled *Generating Waveforms.* These circuits are frequently combined into a versatile test instrument, as illustrated by the block diagram in Figure 10–2. Commercial function generators have many standard or optional features, of which only a few are illustrated in Figure 10–2. The direct current (dc) level control enables the user to add a dc component to the output waveforms. In addition to manual frequency control, most generators

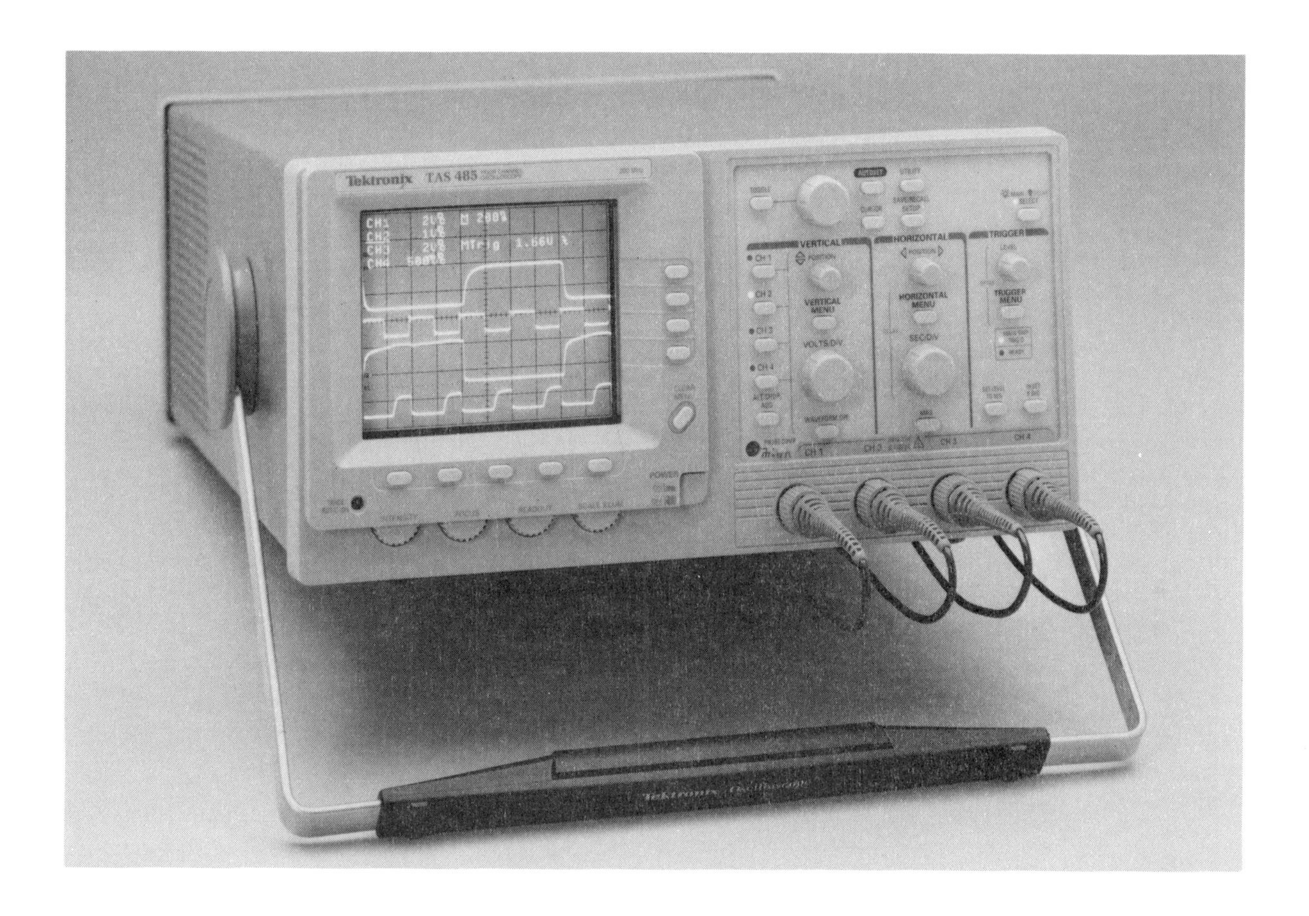

Figure 10–1. Oscilloscope with four input channels and 200-MHz frequency response. (Photo supplied by and used with permission of Tektronix, Inc.)

provide an external control input. This allows control of the frequency by a variable applied voltage. Thus, the generator operates as a voltage-controlled oscillator and can be used to generate frequency-modulated (FM) waveforms and as a swept-frequency oscillator.

Time interval/period/frequency counter. Measurements of frequency, period, or time interval of events can provide helpful clues in troubleshooting. Principles of these systems are introduced in Chapter 2, Sections 2–1 and 2–4. A block diagram of an electronic counting system is shown in Figure 2–2. Switching of events and clock outputs into the scaler and the gate inputs provides time interval,

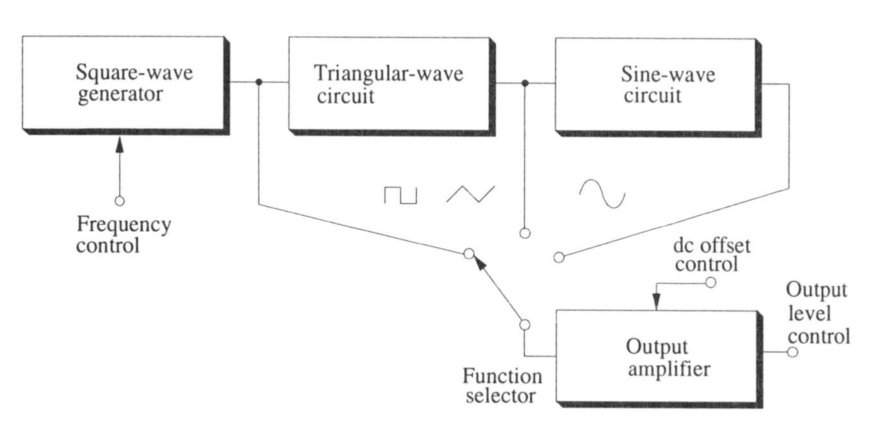

Figure 10–2. Block diagram of function generator.

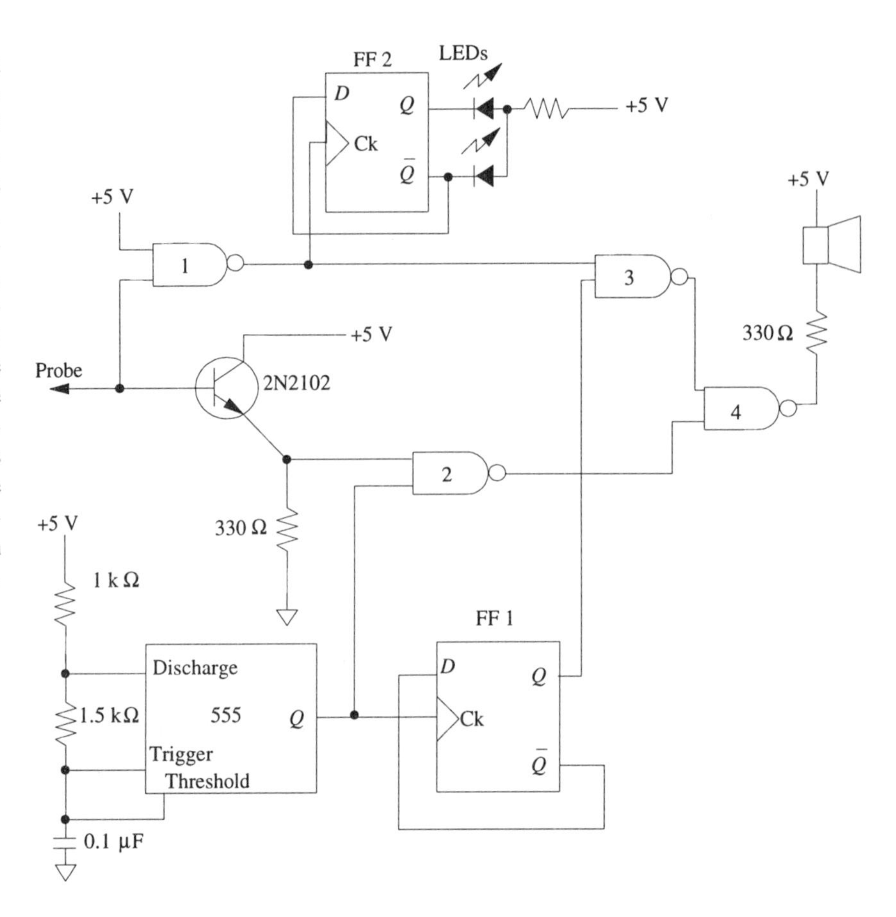

Figure 10–3. Logic probe with audible output. The high-frequency tone is generated by the 555 astable oscillator. A tone one octave lower (half the frequency) is obtained from the output of FF 1. The high and low tones are multiplexed to the speaker by gates 2, 3, and 4. The multiplexer is controlled by gate 1 and the transistor. A LO at the probe input turns gate 3 OFF and gate 2 ON to produce the high tone. When not connected, the control inputs to gates 2 and 3 are both LO, and no tone is produced. The FF 2 circuit alternates the illuminated LED on each LO–HI edge at the probe input. Ck, clock.

period, frequency, and frequency ratio measurements, as illustrated in Figures 2–19 through 2–23.

Logic probe. A very convenient device for testing digital circuits is the **logic probe.** It is shaped like a thick pencil with a needlelike probe at the tip. Light-emitting diode (LED) indicators on the probe and/or audible high and low frequencies can indicate the logic level at the point in the circuit contacted by the probe.

The logic probe shown in Figure 10–3 has a convenient feature: a HI logic level gives a high-frequency tone, and a LO logic level gives a tone one octave lower (half the frequency). The flip-flop 2 (FF 2) circuit alternates the illuminated LEDs on each LO–HI edge at the probe input.

It is possible to perform complex digital data analysis by using logic and signature analyzers. These are rather sophisticated test instruments that require considerable troubleshooting experience and are not discussed here.

Instrument manual. A good repair manual can greatly facilitate troubleshooting. The manufacturer should supply an instrument service manual at the time of purchase. It is recommended that repair manuals be listed in the purchase specifications. If the service manual is not shipped with the instrument, it can probably be ordered at extra cost.

In the absence of repair manual documentation, instrument troubleshooting might consist primarily of the basic observations described in Section 10–2 and interchanging modules or circuit boards (called board swapping) with those of a similar instrument that is functioning correctly.

10–2 Following the Data Flow

The emphasis from the beginning of this book has been on data and its flow in instrumental systems. It is stated on the first page of Chapter 1 that there is a timeless approach to understanding modern instrumentation that *concentrates on the data and not on the devices;* this approach focuses on the form the data can take in various parts of the instrument and how the data are processed in achieving the desired results. This concept also applies to troubleshooting. We must locate the system functional module that is interrupting the data flow or is processing the data incorrectly.

In previous chapters of this book we have investigated various functional units for acquiring, encoding, storing, processing, optimizing, manipulating, presenting, measuring, and controlling data. In this chapter we want to find which one of these units has stopped doing its job. The overall map for locating the faulty unit is the **system functional block diagram.** More-detailed maps are often necessary to locate specific problems within a malfunctioning unit or subsystem. These are the **schematic and subsystem block diagrams** found in some instrument manuals.

Studying Functional Block Diagrams

The first diagrams presented in this book, Figures 1–1 and 1–2, are system block diagrams of computer-controlled instruments. In the previous section of this chapter, there is a system block diagram of a basic square-triangular-sine-wave function generator (Figure 10–2). Throughout the other chapters of this book are many system block diagrams for instruments such as a counter (Chapter 2, Figure 2–2), a DMM (Chapter 2, Figure 2–14), a digital storage oscilloscope (Chapter 3, Figure 3–28), linear and switch-mode regulated power supplies (Chapter 4, Figures 4–30 and 4–31), null comparator control and measurement systems (Chapter 5, Figures 5–1 and 5–2), and a general automated system (Chapter 8, Figure 8–2). There are also dozens of subsystem functional block diagrams throughout the book. The block diagrams focus on the main functional units and data flow for the system. Therefore, a study of block diagrams is important in preparing to search for a faulty unit.

Locating Functional Units within the Instrument

The physical locations of major functional units for some systems are relatively easy to find. For example, those who have a modular stereo audio system can readily identify the functional units because several are in separate boxes or sections of the chassis, as illustrated in Figure 10–4. The audio signal sources consisting of tape deck, compact disk drive, phonograph, and FM tuner are all shown in separate units, although some might be combined in one box. They all feed into a preamplifier unit, where the desired signal source is switch selected. The preamplified signal source is frequently sent into one or more units (such as equalizer, expander, etc.), which can modify and process the signal to provide optimum sound based on personal taste in given situations. The processed signals

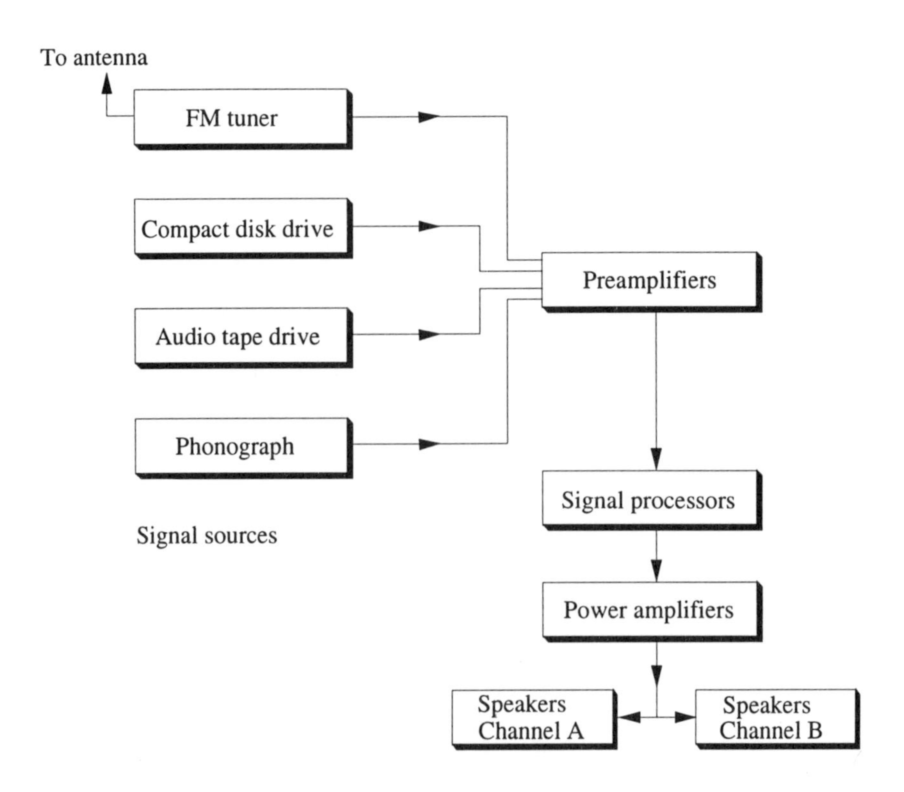

Figure 10–4. Stereo system functional modules.

feed into another unit containing the power amplifiers. The outputs of the power amplifiers then feed into two or more speaker boxes strategically located in the room.

The physical separation of the functional units also exists in some modular scientific instruments. This is especially desirable for some research instruments that undergo frequent modifications. More often, however, most of the functional modules are enclosed in one instrument case. Fortunately, some manufacturers show with drawings or photographs the physical locations of functional modules in the system. Others might describe the general locations of various units. Also, separate circuit boards are sometimes used for the functional modules. In all instruments the functional units can be located and identified by relating the equipment to schematic diagrams and various instrument inputs and outputs. The circuit schematic diagrams are usually drawn to show boundaries of functional units. Signals between functional circuits are often easy to access and identify.

Identifying Possible Problem Areas

Various *common sense* observations can be quickly made to identify possible problem areas. A frequent problem is **interconnections.** Are the line cords all plugged in? Is there a loose, broken, or missing connection between modules? Inspect the equipment for obvious overheated or burned parts, smelly transformers, leaking capacitors, broken circuit boards or wires, corrosion, blown fuses or circuit breakers, or anything else that might indicate a problem.

If there are multiple input signal sources and the system works well except for one data channel, the obvious approach is to focus attention on the faulty input channel. If the stereo system illustrated in Figure 10–4 works perfectly with audio tapes, compact disks, and phonograph records but not with the FM tuner, then attention can be focused on the FM tuner, its input and output connections, and the antenna.

If there are multiple methods of observing instrument output data, such as video terminals, printers, and recorders, and all work except one, it is necessary to focus attention on the unit not operating and its connections.

Modern equipment that has self-test and diagnostic routines can identify common malfunctions and display the problem area(s) on a terminal. This is certainly the trend for the future. Automatic calibration and diagnosis of troubles are already built into some complex systems and should become increasingly available in the future. Therefore, the instrument user will be able to identify a common problem area through automatic equipment diagnosis.

Testing for Essential Responses

If the above common sense observations need confirmation or have failed to identify the likely problem area, the next troubleshooting approach is to test the instrument for essential responses. There are certain logical steps to take in making these tests. They are described in general terms in this section. A more specific interpretation of these measurement tests, as related to fixing power supplies, fixing analog and digital circuits, and solving microcomputer problems, is described in the next section (Section 10–3).

Checking power conditions. If there is one statement that needs repeating over and over again to a novice troubleshooter, it is "You must have power—and it must be the right power." Therefore, the first simple electrical test is to use a voltmeter (or DMM) to determine whether all of the voltage supplies are producing the required voltages under normal loads. If the output voltage from a supply is zero or too low, it will be helpful to know whether the output voltage is correct without the load connected. It is also useful to determine whether the necessary voltages exist at the appropriate circuit boards. For analog circuits, the voltages to be expected at certain points under static conditions are usually indicated on the schematic diagrams. For digital circuits, the correct initial voltages can be checked with the circuits in the reset state.

The dc power supplies must provide "clean" dc voltages that remain constant within acceptable limits. Any alternating current (ac) components or spikes on the dc lines must be sufficiently small for the application. Use of the oscilloscope to observe and measure superimposed signals on the dc lines can provide helpful information when one is trying to solve certain malfunctions described in the next section.

Testing with standard sources. Instruments that operate with voltage or current input transducers can be checked for correct responses by applying standard voltage or current sources. This is especially useful for instruments like pH/pIon meters, which utilize somewhat fragile or easily fouled electrode pair input transducers. For example, for pH instruments it is possible to substitute for the input electrodes and standard solutions a specialized standard voltage source called a "pH test box." It gives accurate millivolt outputs corresponding to increments of 1 pH unit from 0 to 14 pH at dial-selected temperatures. Therefore, the instrument response can be checked without depending on the reliability of the input transducer and various pH standard solutions over a range of temperatures.

Other specialized test devices can be conveniently substituted for input transducers when one is troubleshooting instruments. The standard variable sources can be used for instrument calibration as well as for checking overall response.

Finding the Faulty Unit

Thus far we have emphasized the importance of studying the functional block diagram, locating functional units in the instrument, using common sense visual inspection of the malfunction, and testing for essential responses. In many situations these rather simple procedures will enable you to find the faulty unit. If you have not yet found it, some general troubleshooting methods provide efficient approaches to the problem.

Starting at the output and moving to the input. Since the troubleshooting started because there was either an incorrect or no output, a basic procedure is to put the instrument in the normal operating mode, start at the output unit, and sequentially test the units while moving toward the instrument input. That is, first determine whether the final output module has the correct input signal. If not, move to the module that precedes it. If a correct signal appears at the input but not at the output of a functional unit, then the problem has been isolated to this unit. It is important, of course, to know the proper output for each sequential module operating under dynamic conditions.

Knowing the proper signals. Comparison of the output signals of a faulty system either to the correct signals often shown in an instrument manual or to those observed by measurements in a properly functioning similar instrument will make it easy to recognize erroneous waveforms, timing sequences, and digital-bit patterns.

Injecting test signals. When the normal instrument signals are not readily available for testing purposes because of complex operating conditions, various signals that simulate dynamic conditions can be generated and made available at the inputs of functional modules. The basic function generator described in Section 10–1 is appropriate in many situations. In more complex cases, the digital or computer-generated test signals described in Chapter 2, Section 2–5, are more applicable.

Checking at the midpoints. The most efficient way of checking a complex system is to test the output signal at about the midpoint of the functional modules and then sequentially test the output signal at about the midpoints of the remaining modules in the section not working. Keep testing at about the midpoint of the section that is not producing a proper response until the faulty module is found. For example, in Figure 10–5, the system has eight functional units. If the output is checked at the midpoint (unit 4 output) of the system and the signal is not correct, then the output is checked halfway between units 1 and 4, that is, the output of unit 2. If the output is correct at unit 2, the signal is checked halfway between units 3 and 4. If the signal is correct at the output of unit 3, then the malfunction is in unit 4. Thus, the faulty module has been found by making signal measurements at only

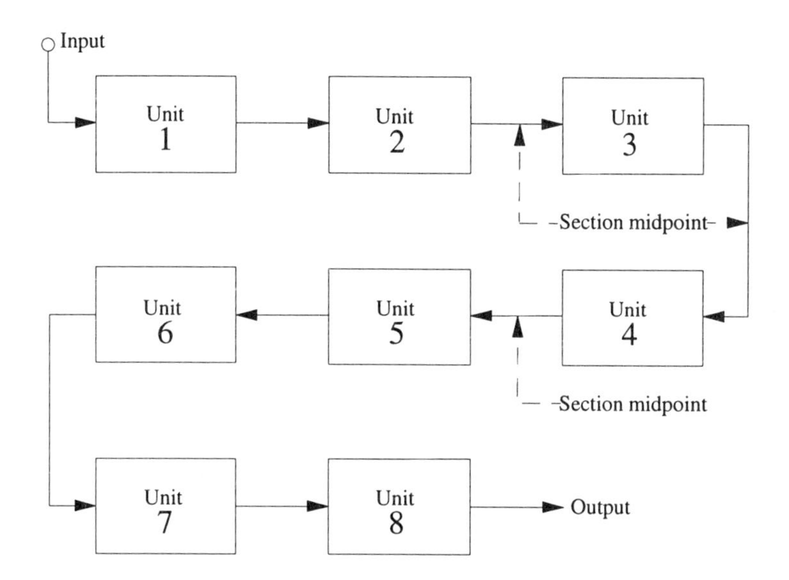

Figure 10–5. Locating a faulty functional unit by checking output signals at successive midpoints of the system and sections. For an instrument with eight functional modules in series, the midpoint of the system is at the output of unit 4. If the signal at unit 4 is incorrect, the output of unit 2 is checked, and if unit 2 output is correct, the next check is at the midpoint of the two remaining units or in this case at the output of unit 3.

three points in the system. This technique is very effective and is often referred to as the *divide and conquer method.*

Complex instruments could have dozens of functional modules, and the midpoint technique (sometimes called half-splitting) is especially efficient in these cases. Very complex systems are conveniently broken into subsystems, and each subsystem can be checked independently of the overall system by using the technique just described.

Opening feedback loops. When there are feedback loops in various parts of instrument systems, it is often necessary to open (break) the feedback path to troubleshoot the units in the feedback loop. Appropriate test signals must then be fed into the units where the loop is broken. Observing the system response while varying the test signals will indicate whether the system is working correctly.

Deciding on the Solution

One of the troubleshooting techniques described in the previous sections of this chapter should have enabled you to find the faulty functional module. Possibly the specific problem in or between modules was identified through the simple observations described above. Replacing a broken lead, cleaning poor interconnections (a very frequent problem), or some other simple, specific solution may have put your instrument back in working order. If not, it might be necessary to select an optimum solution among several alternatives.

One of the common procedures is to completely replace a faulty module if a good module is readily available. Then the faulty unit can be turned over to a repair shop, returned to the manufacturer, or seen by a repair person called in for on-site service. A phone call to the manufacturer's service department could help with this decision. This module substitution practice is especially good when the equipment needs to be operating again quickly. However, the user might not have a spare unit readily available to swap for the faulty one or might enjoy the challenge and fulfillment of fixing it. Also, each situation gives the troubleshooter a greater understanding of the instrument and a reservoir of experience for quickly solving future problems. When a faulty module has been identified and the decision is to "fix it yourself," specific approaches are recommended depending on the types of devices and electronic circuits within the module. These suggestions are presented next.

10–3 Fixing It Yourself

The most frequent instrument malfunctions are in the power supplies. Therefore, specific suggestions on troubleshooting power supplies are presented in the first subsection. The finding of faults in analog circuits and the diagnosis of digital circuits and microcomputer problems are discussed in subsequent subsections.

Troubleshooting Power Supplies

It is not surprising that power supplies are one of the most frequent causes of instrument malfunctions. They are subjected to various nonideal conditions such as overheating (poor ventilation), transient voltage spikes on the line, overloads, and large swings in the line voltage. Power supplies are generally designed for good ventilation and protection against typical overloads and other anticipated problems. However, excessively nonideal conditions and aging of components cause failures. The linear and switch-mode ac-to-dc regulated power supplies are most frequently found in modern instruments, and typical overall problems of both types of supplies are discussed here. However, the various protection devices are considered first, because they might shut down the supply before any damage occurs. It is assumed here that Chapter 4 on power supplies has been read and understood.

Input/output protection. It is now common practice to connect a noise suppression unit between the ac line power outlet and the line input to microcomputers or other equipment in which transient voltage spikes and noise can cause damage or loss of data. Excessive transients in some of these units can cause the line voltage input to be interrupted, and the protection unit will need to be reset after the source of transients no longer exists.

The ways in which protection devices in the power supply influence troubleshooting should be considered. The devices and circuits are described in Chapter 4, Section 4–7. A fast-acting crowbar circuit is typically connected at the output of a regulator and thus directly across the load, as shown in Chapter 4, Figure 4–37. If the regulator circuit fails and the voltage increases significantly, the crowbar silicon-controlled rectifier fires and the output of the power supply is essentially shorted. Thus, the load is protected. If current-limiting circuits (as shown in Chapter 4, Figure 4–36) are in operation, the supply current is restricted to safe levels even when the output is shorted. Either a fuse or a circuit breaker in the power supply circuit is obviously important if the current-restrictive circuit should become disabled.

Thermal sensing is also incorporated in regulator integrated circuits. If the regulator gets too hot, it shuts down and provides both voltage regulator and load protection. Hysteresis is designed into a thermal circuit within the integrated circuit. Therefore, the regulator will not turn back ON until the device cools. Cooling time might require several minutes, and the overall supply would not be ready for testing until sufficient time had elapsed.

Linear ac-to-dc regulated supply. Troubleshooting a power supply module should proceed as described in Section 10–2. Become acquainted with the functional block diagram, such as Figure 4–30 for a linear regulated supply, and follow with a thorough visual check. Replace a fuse or reset a circuit breaker if necessary. Look for parts that are obviously burned or discolored and replace them. Then proceed with voltage and oscilloscope measurements.

A dual, linear regulated ±15-V supply is shown in Figure 10–6. If the loads are disconnected and both power supply outputs are not at correct voltages, then the input unregulated voltages to the regulators are probably not correct for operation. By measuring voltages moving from outputs to inputs and understanding the functions of each section of the supply (as described in Chapter 4), the faulty part(s) can be determined. **CAUTION!** ***Dangerous power exists in many power supplies, including the input line. Use extreme care in making measurements so that the body does not contact the circuits.***

Correct dc voltages can be measured even though there is troublesome oscillation at the power supply outputs. This oscillation can be readily observed with oscilloscope measurements. It is generally caused by bad regulator bypass capacitors. If the capacitors are polarized (such as tantalum), it is essential to connect the replacements with correct polarities.

The large-capacitance filter capacitors tend to deteriorate with age and non-ideal operating conditions such as high temperatures. Therefore, they are typically suspect. Oscilloscope measurements will indicate a large ripple voltage across an unregulated supply with a bad filter capacitor.

The ON–OFF switches do go bad and should not be ignored if observations and measurements point in their direction. With the power supply unplugged, resistance measurements across the switch can help confirm whether these switches are a problem.

Components must be replaced only with those of correct values. Power ratings as well as resistance values for resistors, type and values for fuses, voltage ratings, and capacitance for capacitors are some of the parameters to compare. Selection of wrong-valued replacements could be damaging to the equipment and can even present a fire hazard if a fuse has a current rating that is much too high.

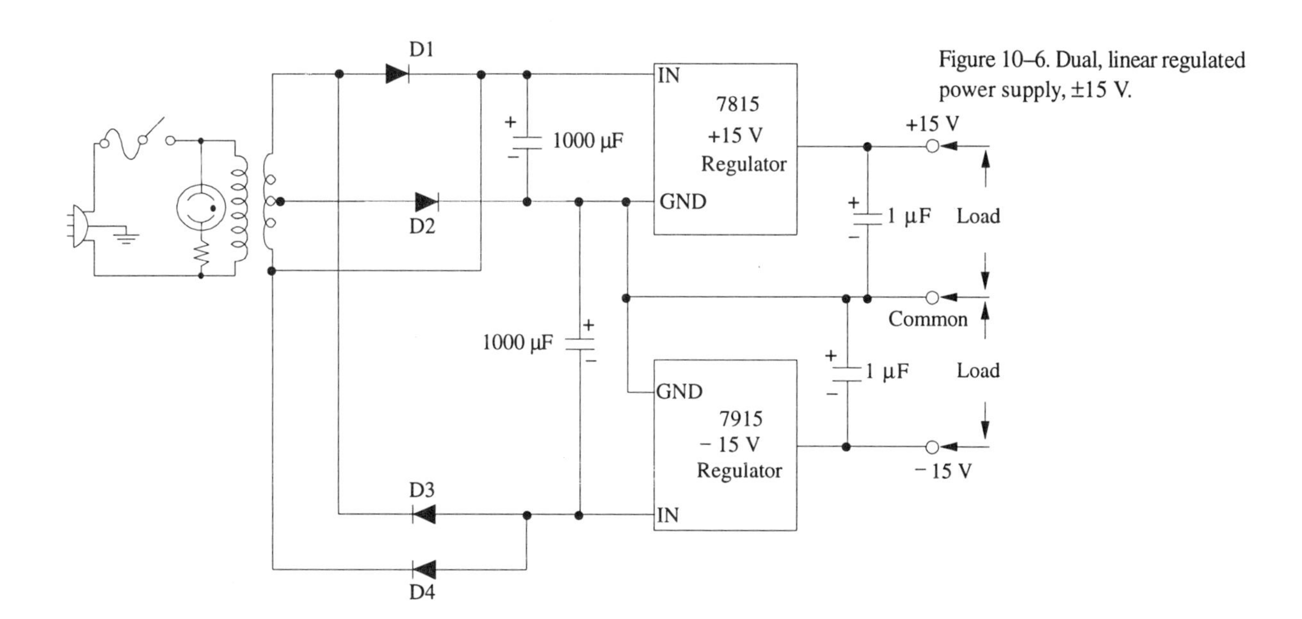

Figure 10–6. Dual, linear regulated power supply, ±15 V.

After repairs have been made on a power supply module, it should be reconnected to the normal load and checked to ensure proper operation under load.

Switch-mode ac-to-dc regulated supply. The switch-mode-regulated supply has gained widespread application as the power source in millions of microcomputers and other electronic equipment because of its efficiency and minimum size. It eliminates much heat, and the bulky 60-Hz transformer in the linear regulated supply and the filter capacitor can also be much smaller. For many applications, these advantages of switch mode compared with linear supplies offset the disadvantages of more complexity and more line noise and electromagnetic interference (EMI), which need to be filtered out. As with all power supplies, problems can occur during operation, especially under nonideal conditions. These problems include development of too much ripple voltage, EMI or voltage spikes, oscillations, loss of output voltage, or loss of regulation.

The troubleshooting approach is quite similar to that described for linear supplies. However, the switch-mode supplies are difficult to repair unless you can visually see the problem. Refer to the instrument service manual and circuit diagrams of the power supply for information. The simplified circuit for one type of switch-mode off-the-line regulated supply, shown in Figure 10–7 can be used here for considering the general types of troubles referred to above.

An overload on the power supply can cause the protection circuits to effectively shut off the output voltage. Therefore, before starting to troubleshoot a power supply, reduce the load and try the supply again after allowing time for cooling if the supply has thermal protection.

If there is no output voltage with a minimum or "dummy" load, check the output of the off-the-line rectifier circuit. If there is no output, the line input should be checked. If there is line input, then the filter capacitor could be shorted or the rectifier diodes blown. If the line is checked with an instrument with a grounded input lead such as an oscilloscope, be careful so that the line is not shorted.

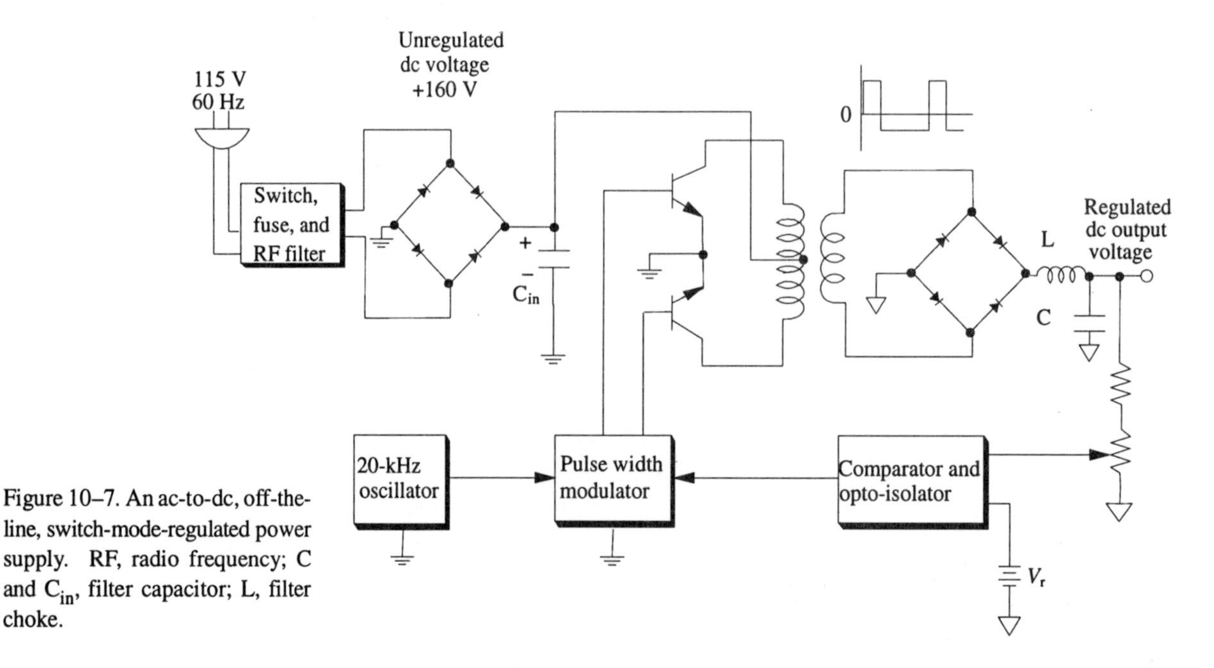

Figure 10–7. An ac-to-dc, off-the-line, switch-mode-regulated power supply. RF, radio frequency; C and C_{in}, filter capacitor; L, filter choke.

With unregulated voltage available but no regulated output voltage, it is important to refer to the specific power supply circuit diagram. The output regulator integrated circuit or filter capacitor might be bad. It might also be necessary to open the feedback loop and insert signals to help in finding the fault.

Large ripple or switching spikes indicate that one of the filter capacitors is bad. A replacement capacitor must have a low equivalent series resistance to filter the ripple and spikes effectively.

In the past, EMI from switching power supplies caused serious problems. However, modern switch-mode supplies are built with short interconnections and careful noise suppression so that EMI is usually minimal. The residual noise is larger than in a good linear power supply but definitely acceptable. If EMI increases with power supply use, it is necessary to check the filter circuits.

Finding Faults in Analog Circuits

The distinctive features of analog and digital signals have been described and applied throughout the preceding chapters. Most so-called "real world" information (such as position, pressure, strain, volume, mass, acidity, light intensity, and hundreds of other parameters) is readily converted into analog electrical information in the form of voltages and currents, or vice versa. The analog electrical data can, of course, be converted into digital data for processing and manipulating and then converted back to analog data for necessary analog action by output transducers. For practical reasons, such as cost or simplicity, analog circuits are often used throughout a system. In either case, whether the system is all analog or a hybrid of analog and digital circuits, it is the purpose of this subsection to focus on general ways to find faults in the analog circuits.

Universal analog system. The importance to troubleshooting of understanding functional block diagrams was emphasized in Section 10–2. The overall operation of analog systems can be illustrated by the general block diagram in Figure 10–8. The physical or chemical parameter to be measured or controlled is converted into analog electrical information (charge, current, or voltage) by the input transducer. The electrical analog is then processed and/or manipulated by various circuits so as to give a suitable electrical analog for operating the output transducer, such as a motor or an analog meter. The output transducer converts the electrical information into the desired measurement or control operations.

For the all-analog system, the processor/manipulator consists entirely of analog circuits. However, in modern instruments it is more probable that the processor/manipulator will be a hybrid of analog and digital circuits. For most instruments this section of the instrument consists of several functional modules, and a more detailed block diagram is required to illustrate its overall operation.

Troubleshooting approach. The general troubleshooting methodology presented in Section 10–2 is directly applicable for analog systems. If the typical common sense observations mentioned in that section have not indicated the probable problem area, check the power conditions and other essential responses. If the input transducer is not rugged or is difficult to operate repeatedly, it is a good idea to substitute a standard test source for ease in testing the system. From the universal block diagram in Figure 10–8 it can be seen that if the output transducer is checked and has no problems and if the system is not working correctly with a

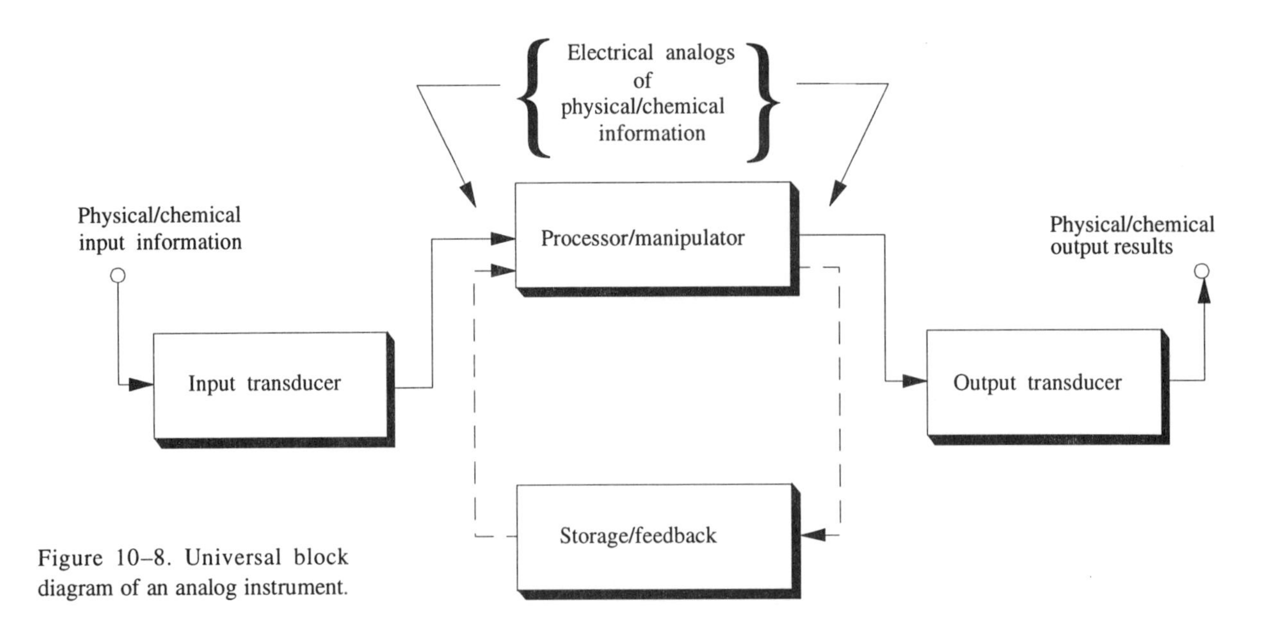

Figure 10–8. Universal block diagram of an analog instrument.

reliable test source connected, then the fault must be with the processor/manipulator circuits or with interconnections to the input and output transducers. The following examples illustrate some typical considerations in troubleshooting analog systems.

Recording pH/pIon instrument. The block and pictorial diagram in Figure 10–9 shows the functional units of a recording pH/pIon instrument. The pH/pIon of a solution is transduced into a voltage by selective electrodes, and the voltage is processed by various analog circuits to give an output voltage v_u equal to 100 mV/pH or pIon as shown in Figure 10–9. This analog voltage output is utilized as the input signal to a digital or analog readout. An analog-to-digital converter and LED display circuitry can give a direct digital readout of pH. Also, a digital printer can record the digital value of the pH or pIon, as illustrated in Figure 10–9.

A classic analog system for continuously indicating the value of pH or pIon is the potentiometric strip chart recorder. This type of recorder is seldom used now because inexpensive microcomputer/printer systems are readily available and more versatile. However, the analog potentiometric readout system shown in Figure 10–9 is included here because it illustrates the operation of a generally applicable electromechanical servo system. The pen's position is continuously controlled to be the analog of the pH or pIon.

A brief description of the characteristics of the pH/pIon voltage conditioning circuits and the recording system should provide an understanding of the overall system before some typical considerations in troubleshooting the units are described.

The operation of the circuits for processing the electrode signals related to solution pH or pIon is as follows.

1. The output of either a pH or pIon electrode pair is fed into a high-quality operational amplifier that is connected as a voltage follower. This provides a very high input impedance and a low input current to accommodate electrodes with extremely high resistance (such as 100 to 10 000 MΩ).

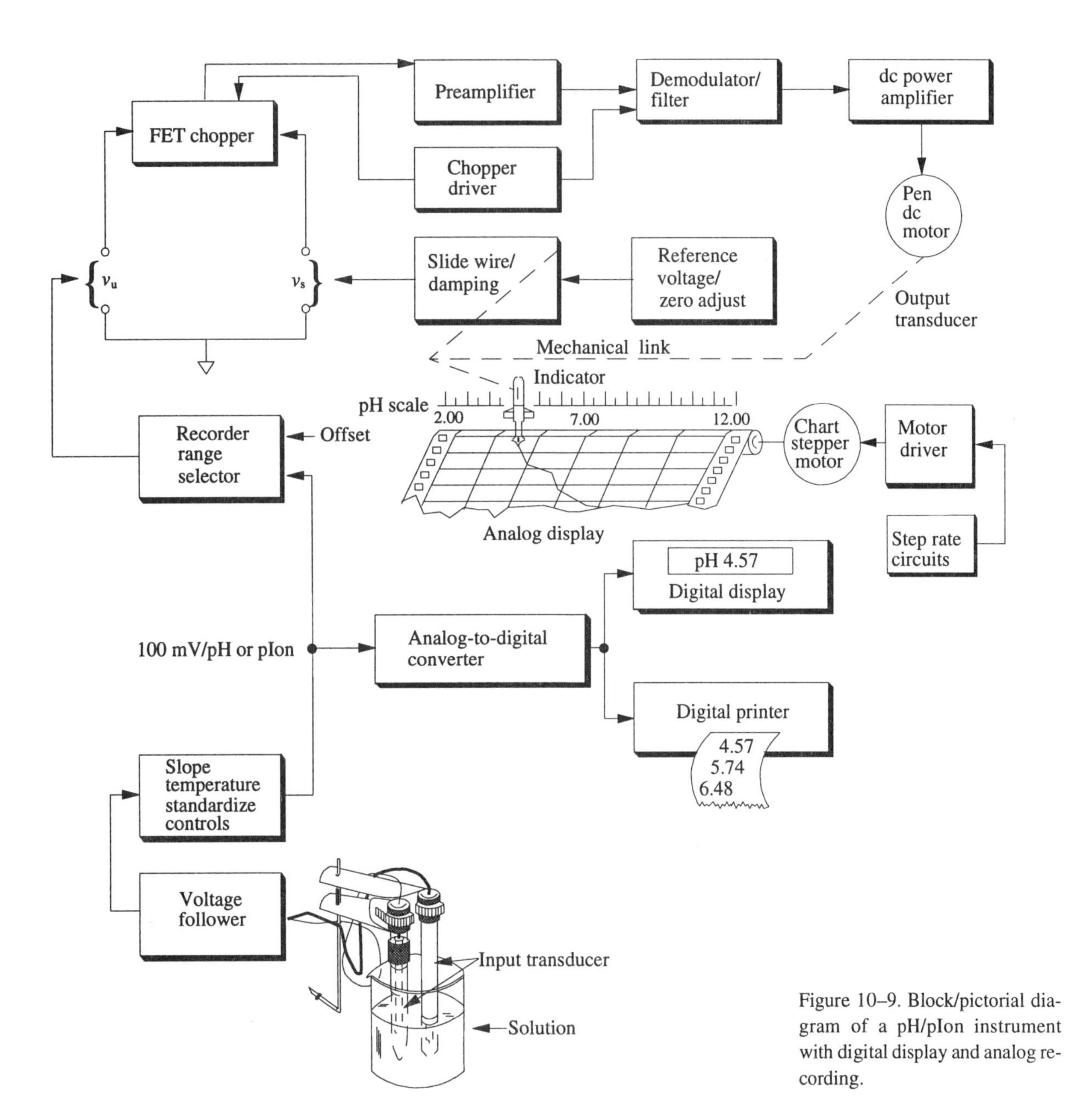

Figure 10–9. Block/pictorial diagram of a pH/pIon instrument with digital display and analog recording.

2. The output of the follower is fed to a circuit that adjusts the voltage to accommodate the response characteristics of specific electrodes. A standardize control shifts the instrument response laterally, and a slope control rotates the response about the zero-crossing point (for pH electrodes, it is pH 7). A temperature control corrects the slope for changes in temperature. The circuit provides an output voltage of 100 mV/p*X*.

The recorder system illustrated at the top of Figure 10–9 operates as follows.

1. A solid state field effect transistor (FET) chopper is used to alternately compare the input voltage v_u from the pH/pIon processing circuits to a variable reference voltage v_s, which is controlled in magnitude by an electromechanical feedback circuit.
2. A preamplifier amplifies the difference voltage between v_u and v_s.

3. The amplified chopped signal is demodulated and filtered. This provides a dc voltage whose polarity depends on whether v_u or v_s is the larger.
4. The dc voltage is amplified by the dc power amplifier to provide sufficient current to operate a dc motor, which is mechanically linked to a precision slide wire and an indicator pen.
5. The voltage across the slide wire is provided by an accurate Zener reference voltage. A zero adjust enables v_s to equal zero volt anywhere on the slidewire, and zero can thus be moved to any desired position on the chart.
6. The damping circuit provides an output signal that leads the slide wire output voltage v_s. This prevents overshoot or oscillation as the servo comes to balance.
7. At balance, when $v_u = v_s$, the potentiometric input impedance is "infinite."
8. A recorder range selector circuit can provide switch-selected outputs such as 100, 200, 500, and 1000 mV/p*X*. Therefore, if the chart recorder is set to respond full scale for a 1-V input and the range selector is switched to 100 mV/p*X*, a change of 10.00 pH in the solution will cause the pen to travel the full width of the chart. The scale in Figure 10–9 is shown to vary from 2.00 to 12.00 because the standardize control was set with the indicator pen in the center of the chart for a pH 7 solution.
9. An offset voltage can be applied to the recorder range selector circuit so that any change of 5, 2, or 1 pH or pIon can be displayed across the full chart.

With this understanding of the functional units, troubleshooting becomes clear. The system breaks readily into sections, that is, the display circuits and the electrode conditioning circuits, and each section can be checked separately. If the lead between the sections is disconnected, a variable standard millivolt source can be connected to the digital display or analog recorder inputs and varied from 0 to 1000 mV. If they respond correctly, then the difficulty must be in the input pH/pIon.

If a pH test box is available, it can be substituted for the electrodes and solutions to facilitate testing of the circuits. Again, by using the general methodology described in Section 10–2, the faulty functional block can be isolated. Each of the blocks contains circuits that have been discussed in previous chapters of this book as follows.

1. The electrodes are discussed in Chapter 8, Section 8–3.
2. The voltage follower module is basically the same as that presented in Chapter 5, Section 5–3.
3. The slope, temperature, and standardize controls are manual potentiometers that vary the gain of inverting operational amplifiers (op amps) such as those presented in Chapter 5, Section 5–4.
4. Interaction of controls is provided by a summing amplifier (*see* Chapter 5, Section 5–5).
5. The recorder range selector is an inverting op amp (*see* Chapter 5, Section 5–4, and Chapter 7, Section 7–2) with four accurate switch-selected gains of 1, 2, 5, and 10.

These circuits are relatively simple and can easily be checked for appropriate operation. The most likely problems are associated with the input transducer electrodes or the connectors. Thus, the use of the pH/pIon test box can be very helpful in isolating the fault.

The principles of the potentiometric recorder are described in Chapter 5, Section 5–3, and the individual devices and analog circuits are also presented in previous sections. The general troubleshooting techniques described in Section 10–2 should enable you to find a faulty functional module. By understanding the block diagram and principles described in previous chapters and referring to the schematic diagram in the instrument manual, it should be possible to locate and repair most circuit problems within a faulty module.

Dual-channel boxcar integrator. The lower portion of Figure 10–10 is a functional block diagram of an analog signal processor/manipulator that is used as part of an automated selective system for measuring concentrations of atomic species in sample solutions. The upper portion of Figure 10–10 shows the functional units of a rather complex but very selective input transducer. The output stage of this transducer is a photomultiplier (PM) tube (described in Chapter 8, Section 8–5). The intensity of the emitted light at a wavelength characteristic of a given element, such as calcium or iron, is converted into a current by the PM tube. The intensity of the measured atomic fluorescence is directly related to the concentration of the selected element in the sample. The desired flourescence signal generated by the vapor source that is illuminated by an intermittently operated hollow-cathode lamp is measured during the ON periods. This is illustrated by the PM current output signal plotted as a function of time. The control circuits must provide synchronization signals for the analog measurement system and the hollow-cathode lamp. The fluorescence signal is often small, buried in noise, and also superimposed on the background.

It is the overall function of the analog measurement system to do the following.

1. It integrates the signal plus background for the period when the hollow-cathode lamp is ON.
2. It integrates the background information for an equal time when the hollow-cathode lamp is OFF.
3. It subtracts the two integrated values to give an output that is the net fluorescence signal for a specific species in the sample solution.
4. The ON–OFF cycles are repeated many times, and the system measures the sum of the signals for the multiple cycles. This improves the signal-to-noise ratio and the precision in measuring the concentration of the selected element.

Referring to the block diagram of the analog processor/manipulator circuits shown in the lower portion of Figure 10–10, the operation is as follows.

1. The conditioned signal from the PM tube is alternately switched through channel 1 or 2 in synchronization with the ON–OFF periods of the hollow-cathode lamp.
2. The information sent into channel 2 is inverted so that its integrated value for the same period as for channel 1 is effectively subtracted.
3. The integrator output is shown for a couple ON–OFF cycles by the oscilloscopic trace in Figure 10–10.
4. A sample-and-hold circuit is included at the output of the integrator so that it can sample and hold the output signal when the net signal is available at the end of the desired series of n pulses. Therefore,

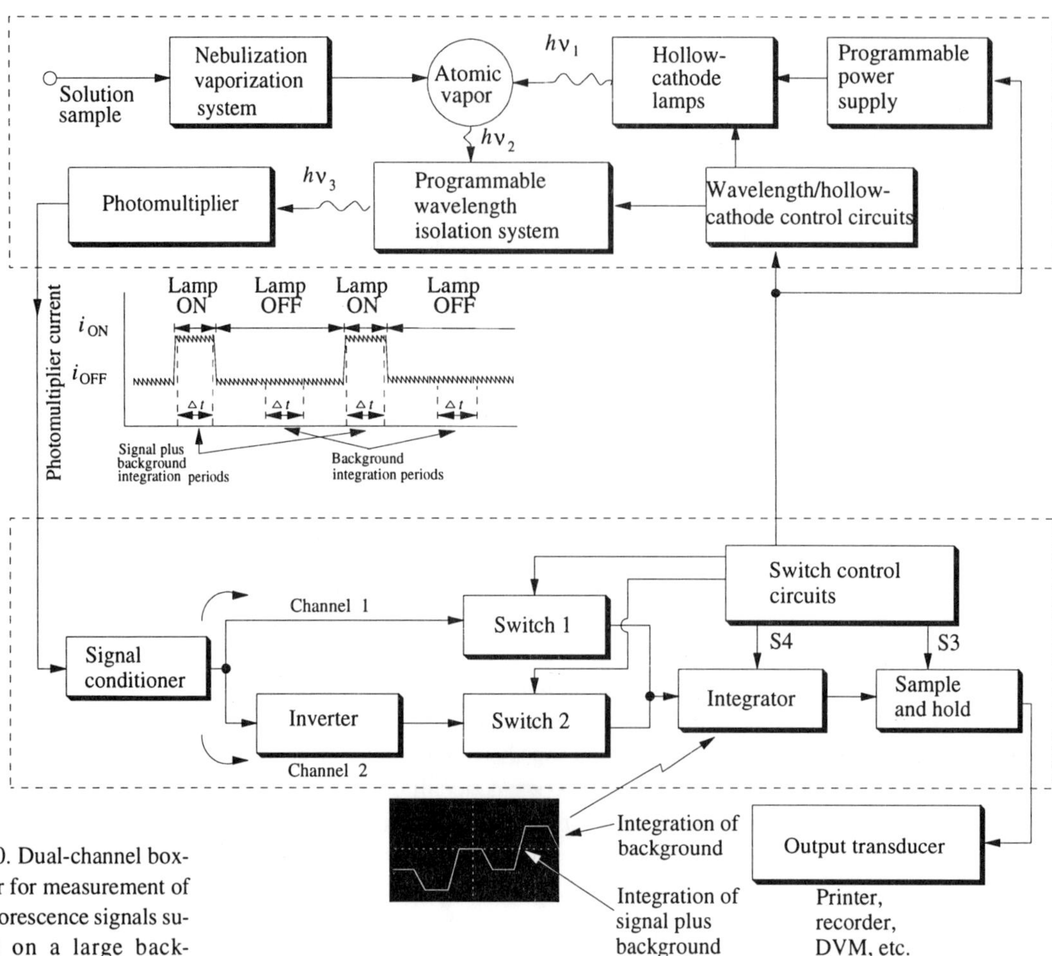

Figure 10–10. Dual-channel boxcar integrator for measurement of low-level fluorescence signals superimposed on a large background and noise.

the integrator can be reset for the start of the next measurement cycle without changing the signal to the output transducer (recorder, printer, digital voltmeter, etc.).

The specific analog circuits to accomplish the above are shown in the schematic diagram in Figure 10–11 and include the following.

1. A current-to-voltage op amp circuit (OA1) converts the PM current signals to voltages that are directed to channel 1 or 2 by analog field effect transistor switches (S1 and S2).
2. An op amp inverter (OA2) of unity gain inverts the polarity of the channel 2 voltage.
3. The voltages result in two currents that are alternately connected to the summing point of an integrator op amp (OA3) for exactly the same length of time. Since the signal in channel 2 is inverted by OA2, the integrated current for this channel is subtracted from the integrated current in channel 1.
4. The sample-and-hold op amp circuit (OA4) is operated to hold the net signal at the end of the desired *n* pulses by operation of switch S3.
5. The integrator is then cleared for the next measurement by operation of switch S4.

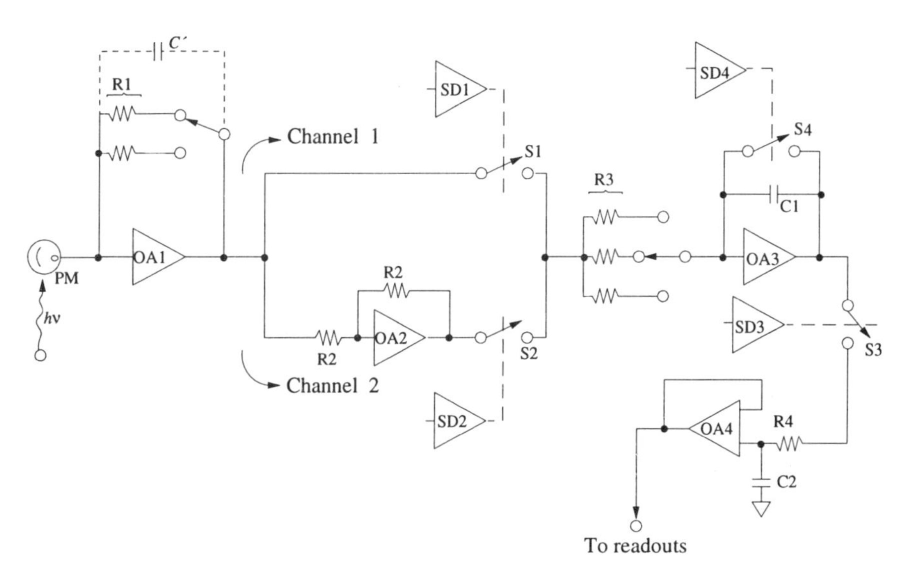

Figure 10–11. Schematic diagram of the analog dual-channel boxcar integrator. OA, op amp; SD, switch driver.

The analog circuits in this system are the basic op amp circuits described in Chapter 5. The troubleshooting approach is straightforward. Test signals can be provided by a voltage reference source that can be gated ON and OFF by a solid-state switch controlled by a waveform generator. This eliminates the need to operate the complex chemical-to-electrical input transducer while checking circuit responses of the analog measurement system. Individual op amp stages are readily checked for expected responses by voltage measurements and oscilloscope observations. When op amp circuits are not giving the expected responses, check for faulty feedback, which can cause oscillations (unstable op amp circuits); op amp outputs that are at limit; and summing points (for inverting op amp circuits) that are not at virtual common.

Conclusions. The analog examples presented above illustrate how sections of rather complex instruments can be isolated for simplification of troubleshooting. Test signals that simulate the performance of the separated section of the real instrument can facilitate the troubleshooting and can often be used for calibration or standardization of a module. This was illustrated with recommendations to use the pH/pIon test box to replace the electrodes when checking the analog processor/manipulator section of the recording pH/pIon instrument. A standard variable reference voltage source can be used when checking the recorder section.

For the atomic fluorescence instrument, the complex input transducer that converts chemical information into an electrical current can be replaced with a signal that simulates the PM current signal shown in Figure 10–10. Both of the instrument examples also have some digital circuits. These will be considered in the next section.

Diagnosing Digital Circuits and Microcomputer Instrumentation

Microcomputers and other digital instrumentation have been discussed throughout this book, especially in Chapters 1, 2, 6, and 9. Basic digital circuits such as logic gates, flip-flops, and counters were described in detail in Chapter 6. The principles of various digital circuits were investigated in several chapters. If the operation and characteristics of these circuits were understood from those studies, the troubleshooting of the equipment should follow the general procedures already described in this chapter. For example, the digital voltmeter is often used as a primary or auxiliary readout, as in the boxcar integrator or pH instrument. Its circuits have been described in Chapter 9. These basic circuits can be diagnosed by using the standard troubleshooting procedures already presented. The same is true for the control circuits that use basic gates, flip-flops, and counters such as in the example of the analog boxcar integrator described in the previous section.

Although the basic digital circuits are readily diagnosed by traditional troubleshooting methodology and test instruments, when it comes to the more complex microcomputer instrumentation problems, additional approaches are important. Some systematic approaches, other than using traditional equipment and methodology, for diagnosing faults in complex microcomputer instrumentation follow.

Using self-test diagnosis. Most equipment is becoming more and more user friendly. The increasing sophistication of the hardware has allowed elegant programs, including reliable self-test diagnostic programs, to be developed. Thus, an instrument user can run the program that tests the hardware of the instrument system and automatically identifies the faulty units. Little or no troubleshooting experience is necessary with this approach.

Unfortunately, good self-test programs do not exist for older instrumentation. Also, present programs are often not very thorough and can fail to find defects. Future instruments should have greatly improved self-test programs that encompass all functional units.

Analyzing the signature. For analyzing the signature, a test program is repeated continuously in a loop. The "signature" or waveforms at the connections or nodes in the instrument are observed. They are compared with the signatures when the system is working correctly, as determined from tables of correct signatures. It is thus possible to trace signatures until the faulty unit or component is found.

A signature analyzer is generally used for this type of diagnosis. Although an oscilloscope can be used, it is inconvenient because the waveforms are usually too long. This type of analysis is relatively slow at isolating a functional module. However, signature diagnosis works effectively for isolating the defect to a specific microchip or connection.

Single stepping and logic analyzer. These troubleshooting procedures for digital systems are effective but have several disadvantages. They require considerable digital troubleshooting experience to use successfully. Also, a complete schematic of the systems and an assembly code listing of test software are required, and unfortunately, many manufacturers do not make these elements available for proprietary reasons. Therefore, these methods are not considered in this chapter.

10–4 Making the Right Connections

A subtheme for this book has been "making the right connections." It has been shown how various functional units or modules are used for acquiring, encoding, storing, processing, optimizing, manipulating, displaying, measuring, and controlling data. For these and other operations to be performed correctly by an instrument, it is necessary to make the right connections between and within functional modules. Also, it has been shown in the previous sections of this chapter that troubleshooting requires making connection of the right test equipment. Then, by making the right measurements, it is possible to locate the faulty module and component or the poor or wrong connections.

The concept of following the data flow has been emphasized throughout this book. Therefore we complete this chapter in the same way. Following the data flow is essential for understanding how an instrument operates and likewise for troubleshooting. We believe that this timeless approach to instrumentation will make studies of topics in each chapter of the book a worthwhile investment for many years to come. Even though devices improve or change in character, the process of following data through the right connections will endure indefinitely.

Related Topics in Other Media

Video, Segment 10

- Data flow through an instrument's functional modules
- Common causes of malfunctions (connections and power supplies)
- Use of diagnostic software to test computer components
- Divide and conquer to locate a faulty functional unit

Laboratory Electronics Kit and Manual

- Supplement 2
  - ¤ Summarizes specific experiments and concepts related to troubleshooting
- Section A
  - ¤ Observing data flow and operating a major test instrument (digital multimeter)
  - ¤ Checking electrical continuity
- Section B
  - ¤ Measuring resistance, voltage, and current
  - ¤ Measuring voltage to locate a "break" in the circuit
- Section C
  - ¤ Checking power supplies
- Sections D and E
  - ¤ Connecting op amps and reading schematic diagrams

- Sections F and G
  - ¤ Testing logic gates and flip-flop circuits
  - ¤ Reading schematic diagrams
  - ¤ Correcting connection errors
- Sections H, I, and J
  - ¤ Connecting more complex circuits
  - ¤ Testing ability to correct errors in wiring
- Appendix 1
  - ¤ Demonstrating the principle of a dual-trace oscilloscope
  - ¤ Using the oscilloscope as a test instrument

Supplement 1

Generating and Controlling Analog Electrical Quantities

The analog electrical quantities that are used to convey information are charge, current, voltage, and power. In this supplement, charge is shown to be the basis for the other quantities. All electrical phenomena result from the action of charged species on each other and on their environment. The separation of positive from negative charges leads to a potential difference (voltage), and the flow of charged particles is called a current.

S1–1 Electrical Quantities

The atoms that make up the molecules of matter are composed of negatively charged electrons, equally but positively charged protons, and electrically neutral neutrons. In a neutral atom or molecule, the number of electrons is equal to the number of protons. Atoms or molecules sometimes have extra electrons and carry a negative charge, or they may have a deficiency of electrons and carry a positive charge. The charge must be some multiple of 1.603×10^{-19} C, the charge on an electron. Materials that have mobile positive or negative charges are called **conductors**. Materials with no mobile charge carriers are called **insulators**. They can be used to isolate conductors that have different voltages, because there will be no path for charge carriers to move between them.

Voltage

Because unlike charges attract each other, energy is required to separate species with opposite charges. When charged species are separated to create regions in matter with different charges, potential energy differences occur between these regions. The energy required to separate the charges is converted to an electrical potential energy. An electrical potential energy difference is called a **voltage**, and its magnitude is expressed in **volts** (V). One volt of electrical potential results when 1 J of energy is required to separate 1 C (6.23×10^{18} electron charges) of charge; that is, volts equals joules per coulomb. Electrical voltage is given the symbol v or V. As with all electrical quantities, it is common practice to use a lowercase symbol (v) when the variation in that quantity is of interest and an uppercase symbol (V) when the quantity is considered constant in value.

Electrical potential differences can be created in many different ways. A few of these are shown in Figure S1–1. The Van de Graff generator (Figure S1–1a) is a particularly graphic example of charge separation that can be used to produce quite high voltages. The electrical battery (Figure S1–1b) is a good example of charge separation produced by electrochemical processes. Many input transducers such as photovoltaic cells (Figure S1–1c) and ion-selective electrodes (Figure S1–1d) are voltage sources that convert an input quantity into a related voltage (*see* Chapter 8). Symbols for voltage sources are given in Figure S1–2.

2861-2/94/0351$08.00/1 Published 1994 American Chemical Society

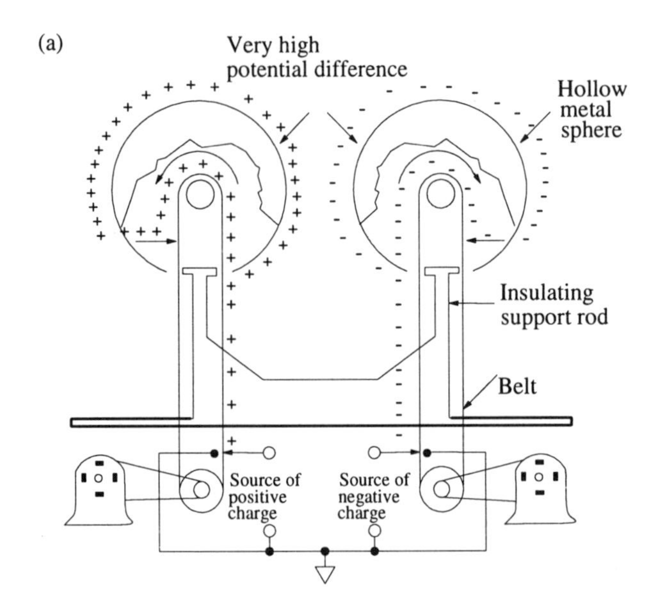

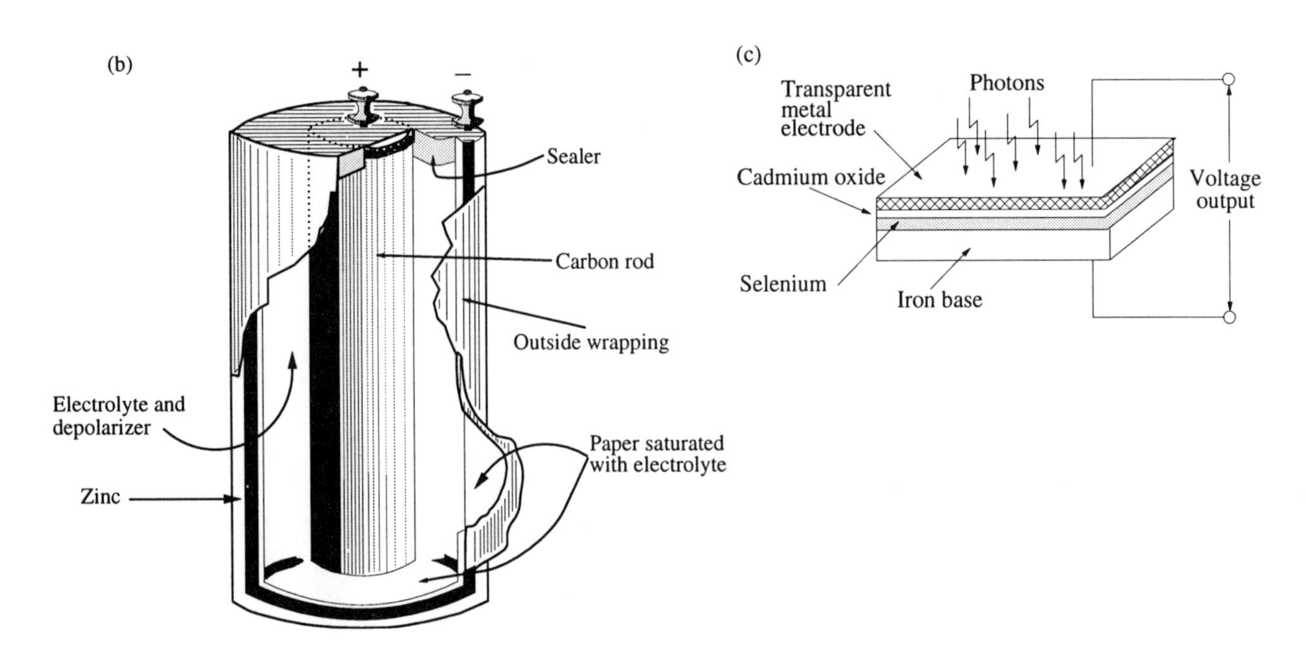

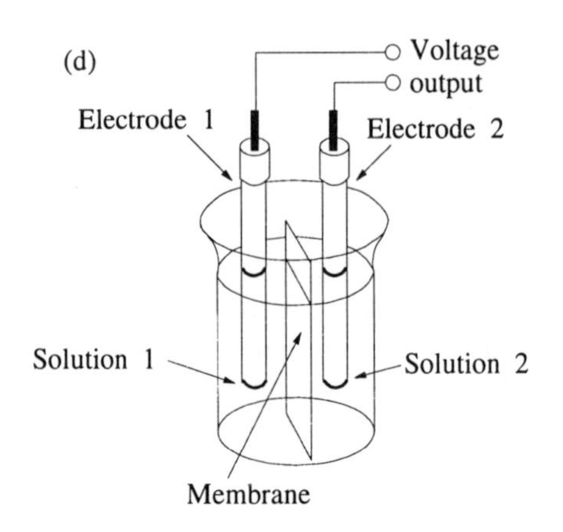

Figure S1–1. Sources of electrical potential difference. (a) Van de Graff generator; (b) battery; (c) photovoltaic cell; (d) ion-selective electrode. These devices convert various forms of energy into electrical potential energy (voltage).

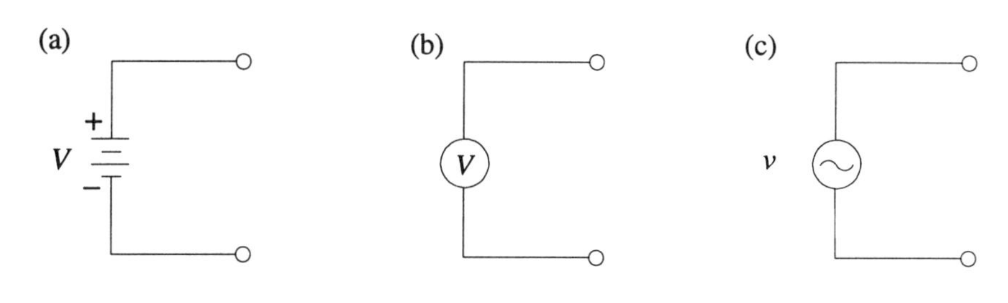

Figure S1–2. Symbols for voltage sources. The symbol for a battery (a) is often used as a generic symbol for a dc voltage source. The symbol in part b is more general and can be used for an ac or dc voltage source. The symbol in part c is used for a sine-wave source.

When charges are separated to produce an electrical potential difference, or voltage, the magnitude of the voltage is proportional to the amount of charge separated Q, so that

$$V = kQ \qquad \text{(S1–1)}$$

From equation S1–1, the amount of charge Q (coulombs) that must be transferred to cause a 1-V potential difference is $1/k$. This constant is called the **capacitance** C. Thus,

$$C = Q/V \qquad \text{(S1–2)}$$

From equation S1–2 we see that C has the units coulombs/volt (C/V), which is called the **farad** (F). A device with a capacitance of 10^{-6} F (1 μF) and a potential difference of 1 V across it must contain (by storing) 1 μC of separated charge. Devices called **capacitors** are made and used specifically for their property of capacitance. However, any pair of conductors has the property of capacitance. The capacitance of a capacitor (or pair of conductors) increases with increasing area of the conductors, decreasing separation between the conductors, and increasing dielectric constant of the material between the conductors. Below are the symbols for capacitors.

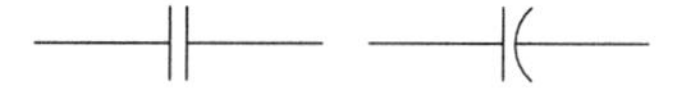

Current

If a conducting material (e.g., a metal, a semiconductor, or an electrolytic [ionic] solution) has an electrical potential difference across it, positive charges are attracted toward the more negative region, while negative charges are attracted toward the more positive region. The rate of charge flow that results from the movement of charge carriers is the electrical **current** I or i. The magnitude of the current is expressed in **amperes** (A). A rate of charge motion of 1 C/s is 1 A of current.

Current sources are very commonly encountered in instrumentation, and many input transducers can be considered current sources (*see* Chapter 8). Several of these devices are shown in schematic form in Figure S1–3 along with the symbols that are commonly used to indicate current sources. For each of the sources shown, it is the output current and not the voltage that is of interest. Such current sources are measured and treated differently from the voltage sources shown in Figure S1–1.

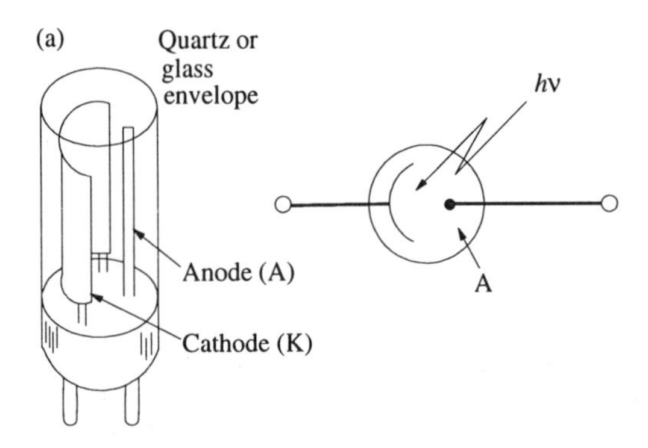

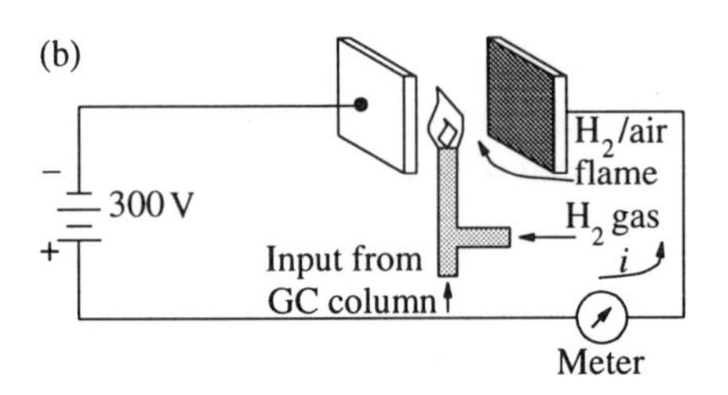

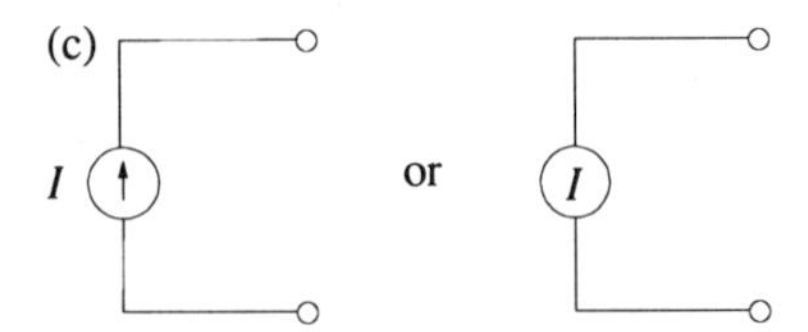

Figure S1–3. Current sources and their symbols. (a) Phototube and schematic symbol; (b) flame ionization detector; (c) generic circuit symbols for current sources. GC, gas chromatography.

Ohm's Law

Electrical voltage, sometimes called electrical pressure, is analogous to pressure in a fluid flow system. Electrical current is analogous to flow rate. Just as pressure and flow rate are related in a flow system, current and voltage are related in an electrical system. For many conductors, the current through the conductor is proportional to the voltage across it. That is,

$$I = GV \tag{S1–3}$$

where G is the **conductance** of the conductor. The conductance depends on the concentration of mobile charge carriers, their mobilities, and the geometry of the conductor and its contacts. The current–voltage relationship in a conductor can also be expressed in terms of the conductor's **resistance** R to the flow of charge. Resistance is just the reciprocal of conductance: $R = 1/G$. Hence, Ohm's law can be written as

$$I = V/R \tag{S1–4}$$

The unit of resistance is the **ohm** (Ω). If a conductor with a resistance of 1 Ω has a voltage of 1 V across it, 1 A of current flows through the device (ohms equals volts per ampere). The unit of conductance is the **mho** (Ω^{-1}), which can be defined as amperes per volt.

Resistors are devices made specifically for their resistance properties. They are used extensively to control the amount of current in a circuit and to direct the flow of charges to specific devices. According to Ohm's law, when a resistor with a resistance of R ohms is connected between two conductors that have a potential difference of V volts, charge flows through the resistor at a rate of V/R amperes. Unless the charge difference between the conductors is replenished, the current will diminish the stored charge and thus the potential. This in turn reduces the current. However, the current will exist as long as $V \neq 0$. Thus, there can never be a potential difference between regions of a conductor without an immediate flow of charge acting to reduce the potential difference to zero. However, if a source of energy that can maintain a potential difference between the two points (such as at the terminals of a battery) is present, then the electrical current in the conductor will be continuous.

A Complete Circuit

A diagram of a complete conducting path between the two terminals of a battery is shown in Figure S1–4. Such a continuous conducting path between the points of potential difference (e.g., the positive and negative terminals of the battery) is called a **complete circuit**. The conducting path here includes the switch contacts, the light-emitting diode (LED), the resistor, and the connecting wires. The connecting wires offer little resistance and are usually assumed to be ideal conductors. Many switches are also nearly ideal conductors, but in some cases the resistance offered by switches must be considered.

In a circuit containing only one conducting path, such as that shown in Figure S1–4, *the current must be the same through every element in the circuit at any instant.* If the battery voltage and the circuit resistance are known, the current in the circuit can be calculated by applying Ohm's law. In the circuit of Figure S1–4, for example, if the LED resistance is assumed to be negligible, the current I is

$$I = V/R = 5\ \text{V}/1\ \text{k}\Omega = 0.005\ \text{A}\ (5\ \text{mA})$$

When the switch is changed to the open position, charge cannot flow through the switch. The circuit is then incomplete and is often referred to as an **open circuit**.

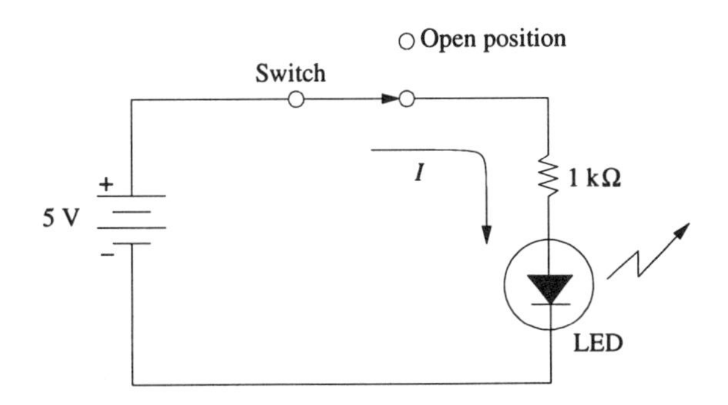

Figure S1–4. Simple complete circuit. When the switch is closed, the conducting path between the battery terminals is completed and there is a steady current I through the conductor LED and current-limiting resistor). The current arrow indicates the direction of the flow of *positive* charges.

The current in an open circuit is zero. The full battery voltage must therefore appear across the contacts of the open switch. This is, in fact, a good rule of thumb to use in locating a nonfunctional component. If the full power supply voltage appears across it, the element behaves as if it had infinite resistance (i.e., an open circuit).

Power

Many voltage sources can be described as charged capacitors with a means of maintaining or replenishing their charge as needed. We will now consider the connection of a **load** (any conducting device) of resistance R to a steady voltage source of V volts. When the load is connected, the current supplied by the voltage source or the rate of its discharge is $I = V/R$ coulombs per second. Clearly, the source cannot maintain its initial or no-load output voltage if the rate of discharge exceeds the rate at which charge separation can be produced within the source. *There is therefore a limit to the current that a voltage source can supply before its output voltage is significantly decreased.* If the load is connected for t seconds, the total charge Q discharged is It coulombs. The **work** W performed on the load during discharge is

$$W = QV = ItV \tag{S1–5}$$

where the units of W are joules. **Power** is defined as the rate of doing work, that is, work per time, or W/t. A work rate of 1 J/s is equal to 1 W (**watt**) of power (P). Therefore,

$$P = W/t = QV/t = ItV/t$$

from which we obtain

$$P = IV \tag{S1–6}$$

According to equation S1–6, the power in watts supplied by the voltage source and dissipated in the load at any instant is equal to the voltage applied to the load multiplied by the current in the load at that instant. If the load is a resistance, substitution of Ohm's law (equation S1–4) gives two additional equations for the power dissipated in the load. (In a resistor, the dissipated power appears as heat.) These equations are

$$P = I^2R \text{ and } P = V^2/R \tag{S1–7}$$

As an example, we can calculate the power dissipated in the resistor of Figure S1–4. We have already found that the current I is equal to 0.005 A with R = 1 kΩ. Hence, the power dissipated is $P = (0.005\ \text{A})^2 \times 1\ \text{k}\Omega = 0.025\ \text{W}$ (25.0 mW). If the resistance is 100 Ω, the current I is 0.05 A and the power P is $(0.05\text{A})^2 \times 100\ \Omega = 0.250\ \text{W}$. In this case we would need a resistor with a power dissipation rating of greater than 0.25 W to avoid overheating. Resistive components should

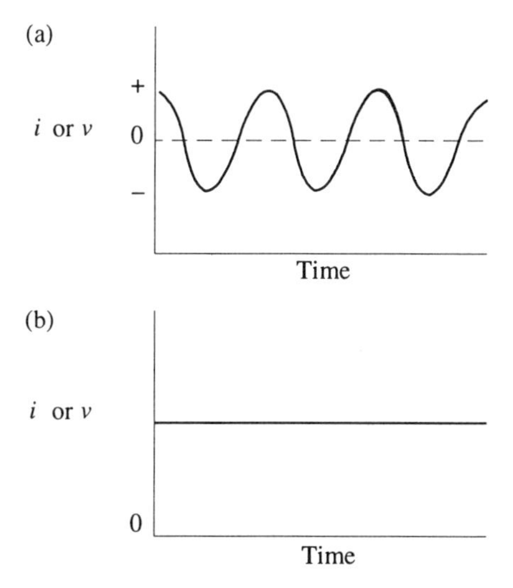

Figure S1–5. Plots of ac and dc quantities. (a) In an ac, or alternating current or voltage, the sign of the charge separation or the direction of the current reverses periodically. (b) In contrast, dc, or direct current or voltage, maintains the same direction and sign.

be selected with a knowledge of the maximum power that will be dissipated in them.

Voltage sources that are specifically constructed to provide power for circuits, transducers, and other devices are called **power supplies**. Batteries and supplies that operate from the alternating current power lines are examples (*see* Chapter 4). Some voltage sources, such as electromagnetic generators, periodically reverse the direction of the charge separation. This results in a voltage of alternating polarity called an **alternating voltage**, as shown in Figure S1–5a. When an alternating voltage is connected to a load, the current direction reverses with every reversal of the polarity of the voltage source. Such a current is called an **alternating current**. The abbreviation **ac** is commonly used for alternating current, and the corresponding abbreviation **dc** denotes unidirectional or **direct current**, as shown in Figure S1–5b.

The terms ac and dc are also used to indicate power sources, circuits, and components that are designed for alternating voltage and/or ac operation or for unipolar operation, respectively. For example, a battery for a portable instrument may be referred to as a "dc voltage source," whereas an "ac instrument" is usually designed to operate from the ac power lines. Some systems, such as laptop computers, can be operated either from an internal battery or from a dc power supply derived from the ac power lines. The characteristics of ac quantities and circuits are introduced in Supplement 2.

S1–2 Series dc Circuits

A current exists whenever there is a conducting path between two points of different potential, regardless of how circuitous or complex the path may be. The series circuit of Figure S1–6 is composed of two resistors and a battery. Because there is only a single conducting path for charge carriers, the current must be the same everywhere in a series circuit. The current (rate of charge flow) in the circuit depends on both resistors. Because there are two elements in series in this circuit, we cannot simply apply Ohm's law to obtain the circuit current. As shown below, we must rely on a conservation law, known as Kirchhoff's voltage law, to determine the current in this circuit.

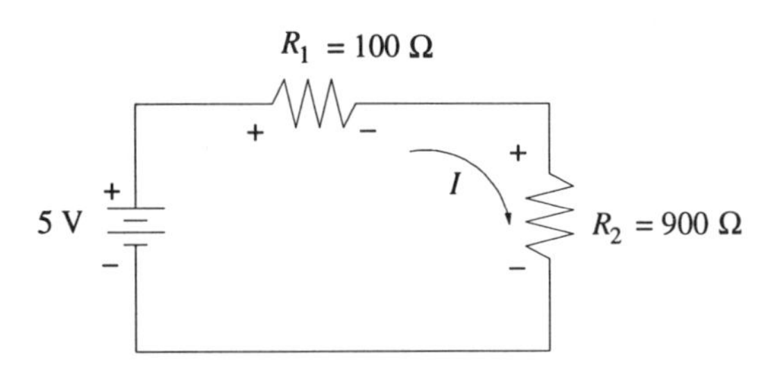

Figure S1–6. Series circuit. In a series circuit there is only one path for current, and this path includes every circuit element in sequence. Each terminal of each element is in contact with only one other element. For example, the two resistors have only one point in common (right contact at R_1 connected to top contact at R_2).

Kirchhoff's Voltage Law

Ohm's law can be used to obtain the potential difference across each of the series resistors. Whenever a current I exists through a resistor of resistance R, a rearrangement of Ohm's law tells us that the potential difference V is equal to the product IR. This potential difference is often called the ***IR* voltage drop** or just ***IR* drop**, because the voltage decreases in going across the resistor in the direction of the current, as shown in Figure S1–6.

For the circuit of Figure S1–6, Kirchhoff's voltage law tells us that the *sum of the IR drops in the series circuit must equal the potential difference at the battery terminals*, because there are no other elements in the circuit across which a potential difference can exist. Thus,

$$V = IR_1 + IR_2$$

Hence, for the series circuit of Figure S1–6, the current I is given by

$$I = V/(R_1 + R_2) = 5 \text{ V}/(100 \ \Omega + 900 \ \Omega) = 0.005 \text{ A (5 mA)}$$

The IR drops across the resistors are $IR_1 = 0.005 \text{ A} \times 100 \ \Omega = 0.5$ V and $IR_2 = 0.005 \text{ A} \times 900 \ \Omega = 4.5$ V. Note that the same current (0.005 A) would be obtained in the circuit if the two resistors were replaced by a single resistance R_s with a value equal to the sum of the two resistances, $R_s = R_1 + R_2 = 1000 \ \Omega$.

In general, Kirchhoff's voltage law states that *the algebraic sum of the voltages from all voltage sources and IR drops encountered at any instant as a path is traced out around a complete circuit (closed-loop path) is zero.* For a series circuit composed of N resistors,

$$V = IR_1 + IR_2 + IR_3 + \ldots + IR_N \qquad \text{(S1–8)}$$

The total series resistance is the sum of the resistances of the N resistors in series:

$$R_s = R_1 + R_2 + R_3 + \ldots + R_N \qquad \text{(S1–9)}$$

In other words, a series combination of resistances, R_1, R_2, R_3, . . . R_N, is *equivalent* to a single resistance R_s.

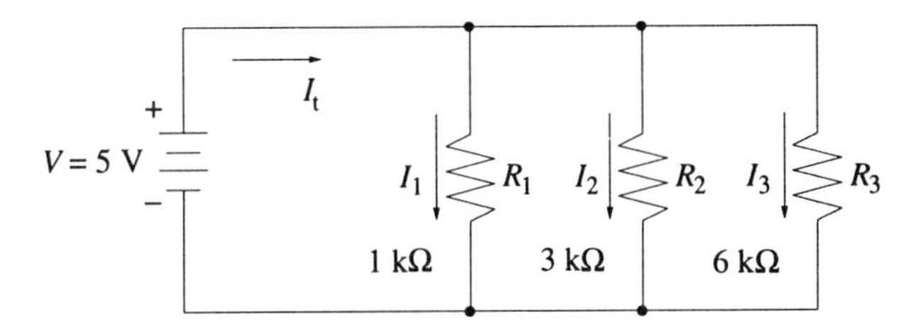

Figure S1–7. Parallel circuit. Multiple paths exist for completing the circuit. Current exists in these separate paths simultaneously. Both ends of each parallel resistance R_1, R_2, and R_3 are connected. The current through R_1 is $V/R_1 = 5$ V/1 kΩ = 5 mA. The current through R_2 is 1.4 mA, while that through R_3 is 0.83 mA. The total current is 7.23 mA.

Parallel Circuits

Often, several conducting paths between two points in a circuit provide independent alternate routes for the transport of charge. Such separate conducting paths between the two points are said to be connected **in parallel**. Note that *the same voltage is applied to each conducting path in parallel.* We will analyze the parallel circuit of Figure S1–7 by using another conservation law, Kirchhoff's current law.

Kirchhoff's Current Law

The current through the individual resistors in Figure S1–7 can be calculated from Ohm's law: $I_1 = V/R_1$, $I_2 = V/R_2$, $I_3 = V/R_3$. Kirchhoff's current law states that the sum of the currents in the parallel conductors must be equal to the total current I_t supplied by the battery in order to avoid a net accumulation of charge at some point in the circuit. For the circuit of Figure S1–7,

$$I_t = I_1 + I_2 + I_3 = V/R_1 + V/R_2 + V/R_3 = V(1/R_1 + 1/R_2 + 1/R_3)$$

Because by Ohm's law $I_t = V/R_p$, the same total current in the parallel circuit would be obtained if the three resistors were replaced by a single resistor of value $1/R_p = 1/R_1 + 1/R_2 + 1/R_3$. Thus, in a circuit consisting of three parallel resistors, the total parallel conductance G_p is the sum of the conductances of the separate paths:

$$G_p = G_1 + G_2 + G_3 = 1/R_p = 1/R_1 + 1/R_2 + 1/R_3 \qquad \text{(S1–10)}$$

In general, if N resistors are in parallel, the **equivalent resistance** is

$$R_p = \frac{1}{\frac{1}{R_1} + \frac{1}{R_2} + \ldots + \frac{1}{R_N}} \qquad \text{(S1–11)}$$

For two resistors in parallel,

$$R_p = \frac{1}{\frac{1}{R_1} + \frac{1}{R_2}} = \frac{R_1 R_2}{R_1 + R_2} \qquad \text{(S1–12)}$$

S1–3 Ground and Circuit Common

Electronic circuits generally have a stable reference point from which all voltages are measured. This point of stable potential is called the **circuit common**. It is often provided in actual circuits by a very low resistance conductor, such as a large copper wire or foil, to which one terminal of power supplies, signal sources, and many other circuit components are connected. This common point is indicated in circuit diagrams by a downward-pointing triangular symbol, as shown in Figure S1–8.

When a circuit is linked to other circuits in a system, the commons of those circuits are often connected together to provide the same reference potential for the entire system. In circuit diagrams, the complete circuit is often simplified by indicating connections to common by downward-pointing triangular symbols. Hence, the addition of a voltmeter to the parallel circuit of Figure S1–7 to measure the voltage across the parallel network might be indicated by the simplified diagram of Figure S1–8a. Here, the voltage source, the parallel network, and the voltmeter are shown connected to the common point. This is equivalent to but simpler than showing a wire running from the negative terminal of the voltage source to the bottom terminal of the resistors to the common terminal of the voltmeter.

When a circuit common is not connected to a system common point or to earth ground, the circuit is said to be **floating**. Some circuits must be operated in this manner to carry out the circuit function or to isolate one part of the circuit from another. For example, to measure the voltage drop across R_1 in the series circuit of Figure S1–6, the voltmeter common cannot be connected to the voltage source common. Instead, the meter must be connected across the resistor as shown in Figure S1–8b. Here, the voltmeter is floating with respect to the series circuit common. In another example, a difference amplifier is used to amplify voltage differences between two voltage sources, which are themselves referenced to common. Both sources have varying voltages, and the amplifier inputs must be

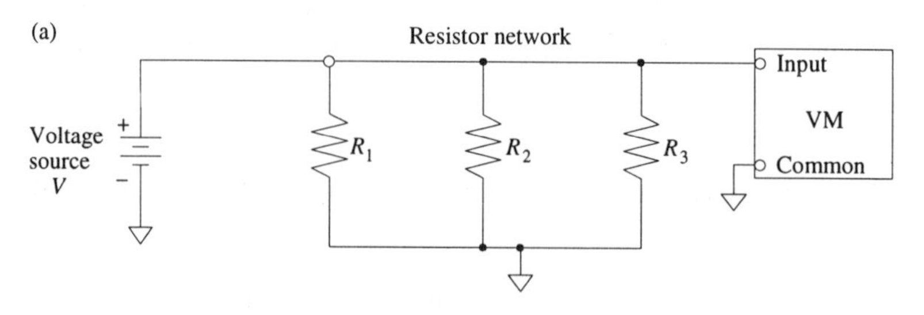

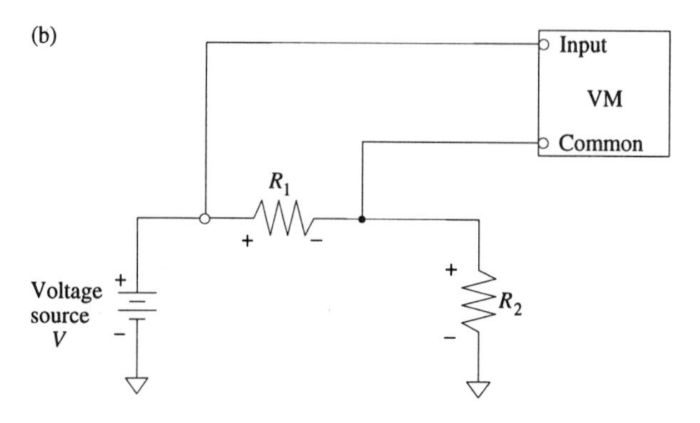

Figure S1–8. Use of common connections in a circuit diagram. (a) A voltmeter (VM) is connected to the parallel network of Figure S1–7. Now the voltage source, parallel resistors, and voltmeter are conveniently shown connected to the same common point, indicated by the downward-pointing triangular symbol. Elements connected to the same point are of course connected to each other. (b) A voltmeter is connected to the series circuit of Figure S1–6 to measure the voltage drop across R_1. The voltmeter must be floating and cannot be connected to the circuit common in this case.

floated in order to amplify the difference voltage and avoid shorting out one of the sources.

The term *ground*, precisely used, refers to earth ground, the universal reference point for potential measurements, power distribution, and radio transmissions. Connections to earth ground are made at a ground rod, metal water pipe, or the third wire of the ac power line. "Ground" is often loosely used to mean the common potential of a circuit even when the circuit is not connected to earth ground. It is better practice to use the term common unless an actual earth ground is intended. Appendix A discusses in detail grounding and shielding techniques for circuits.

Supplement 2

Describing Waveforms and Reactive Circuits

Many different types of repetitive or **periodic** waveforms are useful in instrumentation or result from natural phenomena. For simple waveforms such as sinusoidal waveforms, the entire amplitude–time function can be described by a simple equation that includes the frequency of the waveform, the time, and the peak amplitude. In general, signals that result from complex phenomena are not simple sine waves composed of single frequencies, and their description is necessarily more complex (recall the frequency spectrum described in Chapter 2, Section 2–2). However, all periodic signals, no matter how complex, can be represented as a combination of simple sine waves.

S2–1 Sine-Wave Signals

A sine-wave signal is a voltage or current that varies sinusoidally with time. Objects whose displacements are sinusoidal functions are said to be undergoing *simple harmonic motion* and are very common in nature. Sinusoidal current is produced by rotating a wire loop in a uniform magnetic field as in a power generator. The sine-wave signal is the only true single-frequency waveform.

The generation of a sine wave from a rotating vector of magnitude A_p is illustrated in Figure S2–1. The sine wave results from the projection of the vector on the vertical axis as the vector rotates counterclockwise with a uniform **angular velocity** ω. If the vector is rotating at the rate of 1 rps, the sine wave repeats itself periodically once every second. Each revolution of the vector produces one cycle. The period is the time interval t_{per} required to produce one cycle. Because a cycle is generated each time the vector sweeps through 360°, or 2π rad, the time axis is conveniently expressed as the angle of the vector. Because the vector sweeps one cycle every t_{per} seconds, the frequency f is

$$f = 1/t_{per} \tag{S2–1}$$

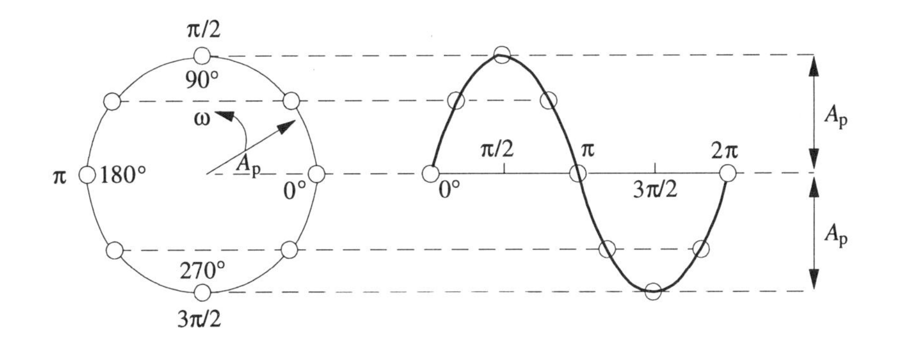

Figure S2–1. Development of a sine wave from the projection of a rotating vector. The vertical displacement of the point of the rotating vector is plotted against the angular rotation in radians.

2861-2/94/0363$08.00/1

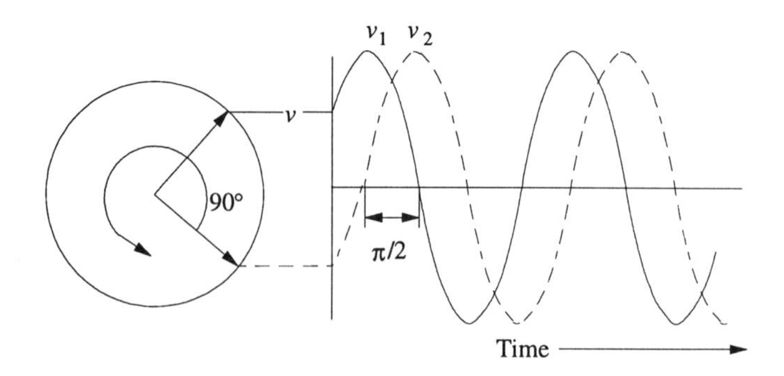

Figure S2–2. Two sine waves with a 90° phase difference. Both vectors must rotate at the same rate (i.e., have the same frequency) to maintain a fixed angle between them. The signal v_2 lags (follows after) signal v_1 by 90°.

The vector is rotating at a rate ω or 2π rad every t_{per} seconds; thus,

$$\omega = 2\pi/t_{per} = 2\pi f \tag{S2–2}$$

If the vector represents a current or voltage, the instantaneous current i or voltage v can be given as

$$i = I_p \sin \omega t = I_p \sin 2\pi ft \tag{S2–3}$$

$$v = V_p \sin \omega t = V_p \sin 2\pi ft \tag{S2–4}$$

where I_p and V_p are the peak values of the current and voltage, respectively.

If two voltages have the same frequency but cross zero at different times, as in Figure S2–2, they are said to be **out of phase**. The difference in time can be conveniently expressed in terms of an angle (fraction of a complete 360° angular rotation) called the **phase angle** ϕ.

If the two sine waves are shown as rotating vectors with the same origin (Figure S2–2), the phase angle can be seen to arise from the fact that one vector leads the other. A generalized voltage sine-wave signal can be expressed as

$$v = V_p \sin (\omega t + \phi) \tag{S2–5}$$

where ϕ is the phase angle between the sine wave under consideration and a reference sine wave.

Because the sine wave is such a simple waveform, *only two pieces of information are obtainable from a sine-wave signal: the frequency* ω *and the amplitude* V_p. If another signal of the same frequency is available as a time reference, the phase angle ϕ between them may also carry information.

S2–2 ac Signals

An ac voltage or current has been defined as one that changes sign periodically. The voltage and current are almost always of this kind in ac power circuits. Electrical signals, however, are often combinations of a dc signal level, or average

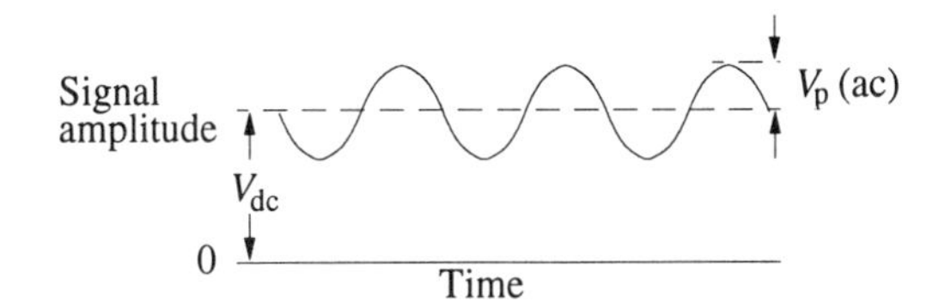

Figure S2–3. The ac and dc components of a signal. A varying signal has both ac and dc components if the average value of the signal over a long time is not zero. The average value is the dc component, and the variation about that average is the ac component.

value, and an ac component that alternates about the dc level. In such cases it is useful to define the **ac signal** as that part of the signal that varies with time. Figure S2–3 shows a signal with a sinusoidal variation of V_p volts about an average dc signal level of V_{dc}. The total signal may be considered the sum of the steady, or dc, voltage and the varying ac voltage. An ac signal is readily isolated from a dc component in an electronic circuit.

Nonsinusoidal Periodic Signals

The sine wave is just one example of a periodic waveform. The **rectangular wave** shown in Figure S2–4a is a series of regularly occurring pulses. If the widths of the positive and negative half-cycles are identical, as in Figure S2–4b, the waveform is called a **square wave. Ramp** and **sawtooth** waves are also shown (Figures S2–4c and S2–4d, respectively). The total variation from the most positive to the most negative voltage is the **peak-to-peak** voltage. The dc level of each of the waveforms in Figure S2–4 is different. The average, or dc, level is that value above and below which the voltage–time or current–time *areas* are equal for one complete cycle. If the waveform is asymmetric, that is, it has dissimilar shapes for the positive and negative alternations, the dc level is often *not* the midpoint between the peak values.

Fourier Series Waveform Analysis

Any periodic waveform, such as those shown in Figure S2–4, has a fundamental frequency, or repetition rate, and contains information at other frequencies. These waveforms can be represented by a **Fourier series expansion**. A Fourier series of sine waves is a special combination that is a summation of multiples of a single fundamental frequency. The multiples are known as **harmonics**, and they may have various amplitudes and phase angles. The Fourier series is a special case of

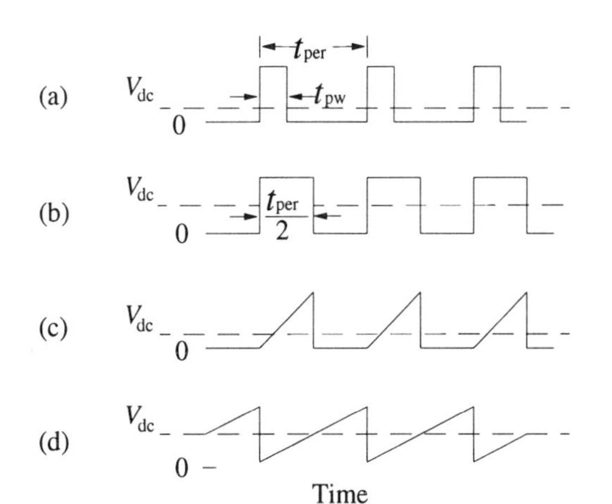

Figure S2–4. Nonsinusoidal periodic signals. (a) Rectangular wave; (b) square wave; (c) ramp; (d) sawtooth. The waveforms have the same period, repetition rate, and peak-to-peak amplitude.

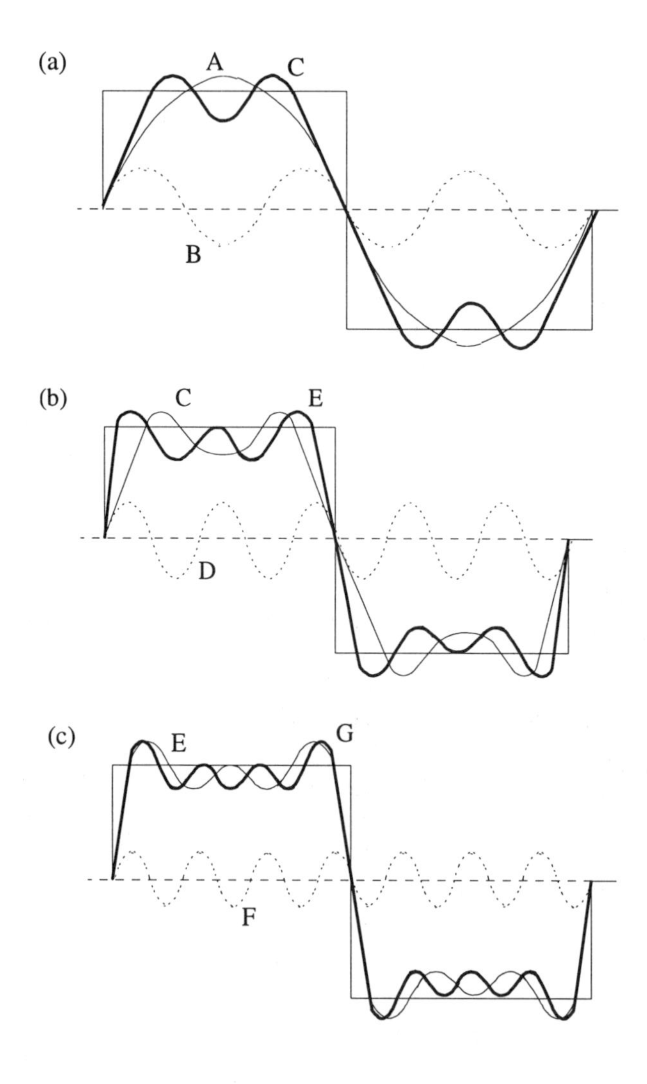

Figure S2–5. Harmonic composition of a square wave. (a) Fundamental A and third harmonic B are added to give curve C. (b) Adding the fifth harmonic D gives E, the fundamental plus the third and fifth harmonics. (c) The seventh harmonic F is added to the fundamental plus the third and fifth harmonics to give G.

the Fourier transform discussed in Chapter 2. Any single-valued, periodic waveform $v(t)$ can be represented by a Fourier series expansion.

As an example of how a Fourier series can be used to represent the waveform of a periodic signal, a square wave can be constructed as shown in Figure S2–5. Note that curve G, which is the graphical sum of the fundamental and the third, fifth, and seventh harmonics, is a fairly good approximation to the square wave. The complete Fourier series would include higher odd harmonics, each in decreasing amplitude. A frequency spectrum (plot of the relative amplitudes at each frequency versus the frequency) of a square wave is shown in Figure S2–6. For

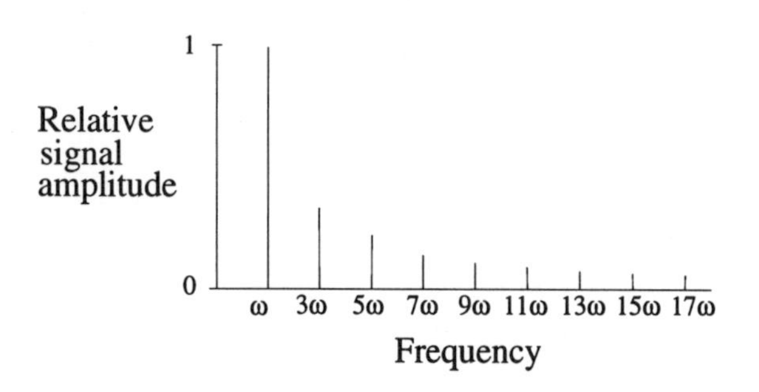

Figure S2–6. Frequency spectrum of a square wave. Only the odd harmonics occur in the square wave, but these have substantial amplitudes even at very high frequencies.

the square wave, the harmonic at 3ω has an amplitude one-third that of the fundamental, while that at 5ω is one-fifth that of the fundamental, etc. From Figure S2–5 it can be seen that the higher-frequency components are needed to produce the abrupt changes in the signal at the leading and trailing edges. Lower-frequency components produce the flat top and bottom portions. Because the square wave contains low- and high-frequency constituents, it is a very useful waveform in testing the frequency response of measurement and control systems. A triangular wave can also be represented by a Fourier series. In contrast to the square wave, a triangular wave has frequency components at both even and odd harmonics.

S2–3 Measures of Periodic Signal Amplitude

There are four measures in common use to describe the amplitudes of periodic signals. The peak value (e.g., peak voltage) is useful where the maximum positive or negative excursions are of principal interest. Where the total magnitude of the variation is of interest, the peak-to-peak value is used. The root-mean-square (rms) value is useful in power calculations. Some ac meters, such as moving-coil meters, respond to the average value of the waveform and calculate the rms value. Other meters may respond directly to the rms value. Digital multimeters (Chapter 3, Section 3-2) and oscilloscopes (Chapter 3, Section 3–4) respond to the peak-to-peak value. For sine-wave signals the relationships among these four measures are straightforward, and conversion factors may be incorporated in the measuring device. However, for more complex waveforms it is important to be aware that the method of calculation may be inappropriate.

Peak Values

The **peak value** is the voltage or current at the maximum point in the waveform. If the signal has both positive and negative polarities during a cycle, there is a positive peak value V_{p+} and a negative peak value V_{p-}, as shown in Figure S2–7. If the waveform is centered about 0 V, V_{p+} and V_{p-} are equal in magnitude. The total variation in voltage from the positive peak to the negative peak is the **peak-to-peak** value V_{p-p}. For a waveform that is symmetrical about the zero point, $V_{p-p} = 2V_p$.

Average Values

The **average value** of a waveform is the average of the *absolute magnitude* of the signal during an entire period. It thus depends on whether a signal is entirely ac or has a dc level associated with it. The average ac value is illustrated in Figure S2–8 for several waveforms that are symmetrical about 0 V. For the sine wave in Figure

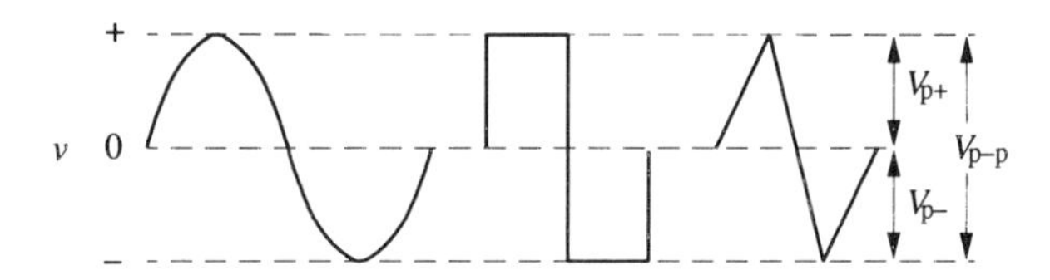

Figure S2–7. Peak measures of common ac waveforms. The peak-to-peak voltage V_{p-p} is the sum of the maximum positive and maximum negative peak values (V_{p+} and V_{p-}, respectively).

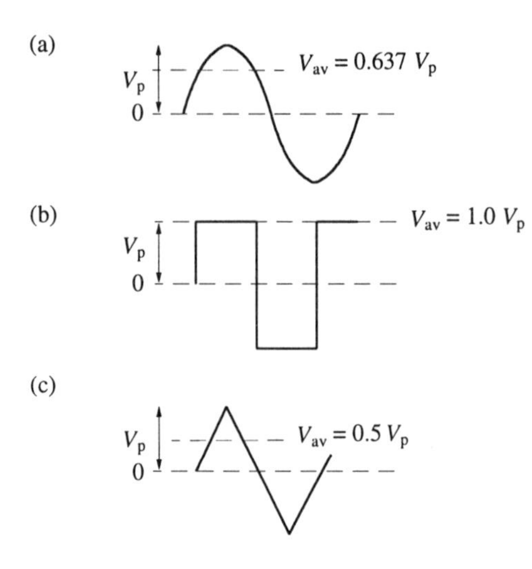

Figure S2–8. Average measures of common ac waveforms. For the sine wave (a), the square wave (b), and the triangular wave (c), the average absolute amplitude is a different fraction of the peak value.

S2–8a, the average value is obtained by determining the area under the waveform during a half-cycle and dividing by the time of one half-cycle. The result is

$$V_{av} = 0.637\ V_p \qquad \text{(S2–6)}$$

The average value of the symmetrical square wave shown in Figure S2–8b is equal to V_p. On the other hand, the average value of the triangular waveform in Figure S2–8c is 0.5 V_p, as shown.

Root-Mean-Square (rms) Values

The rms value of a waveform is the effective power value. The rms current produces the same heating (power dissipation) in a resistor as an identical dc current. In other words, 1 A rms produces the same amount of heat in a resistor as 1 A dc. The power dissipated in a resistor at any time is proportional to the square of the current through the resistor ($P = I^2R$) or of the voltage across the resistor ($P = V^2/R$). Therefore, the average power dissipated by a varying signal is proportional to the average of the square of the current or voltage amplitude over one complete cycle. The steady voltage or current that is equivalent in power dissipation to the varying voltage or current is the square root of the average of the squared amplitude values. This is called the **root-mean-square (rms) value.** The operation of taking the rms value for a sine wave is shown graphically in Figure S2–9. The square of the sine

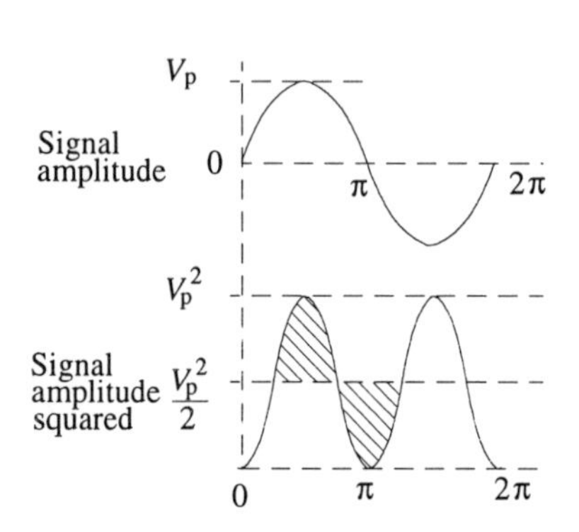

Figure S2–9. Graphical determination of the rms value of a sinusoidal signal. The squared signal is a sine wave that is positive over the entire waveform. The average is $V_p^2/2$, as demonstrated by the equal areas of the two hatched sections of the waveform.

Table S2–1. Peak, Average, and rms Values for ac Signals

Waveform	V_p	V_{av}	V_{rms}
Sine	1	0.637	0.707
Square	1	1	1
Triangle	1	0.500	0.576

wave ($V_p^2 \sin^2 \omega t$), plotted directly below the sine wave, indicates the heating effect of the current, which is directly proportional to the square of the current. Taking the square root of the mean of the square gives

$$V_{rms} = (V_p^2/2)^{1/2} = V_p/2^{1/2} = 0.707\ V_p \qquad \text{(S2–7)}$$

The relation between the effective (rms) and peak currents for a sinusoidal signal is the same as for the effective and peak voltages. Thus, $I_{rms} = 0.707\ I_p$. Calculations of voltages and currents in ac circuits provide self-consistent values. For example, if rms voltages are used, the currents will have rms values. Note that peak values for sine waves are obtained by multiplying the rms values by the reciprocal of 0.707 (1.414), so that $I_p = 1.414\ I_{rms}$ and $V_p = 1.414\ V_{rms}$.

The rms values for the other waveforms shown in Figure S2–8 are given in Table S2–1. The bipolar square wave has a mean square voltage of V_p^2, which means that for this waveform, $V_p = V_{av} = V_{rms}$. The rms value of a triangular wave is obtained by taking the rms of the linear voltage function $v = kt$ over the time from 0 to V_p/k. Thus, we see that relationships among the several useful measures of periodic signal amplitude are highly dependent on the wave shape. The various measures of sine-wave amplitude are summarized graphically in Figure S2–10 for a sinusoidal current waveform.

S2–4 Reactive Circuits

Whenever we desire to change signal levels rapidly in a circuit or system, we must be aware that circuit components can react against these changes and affect the rate at which the changes take place. Reaction to signal variations can be classified as **inductive reactance** or **capacitive reactance**. Capacitors and inductors are often intentionally used for their reactive effects in reactive circuits.

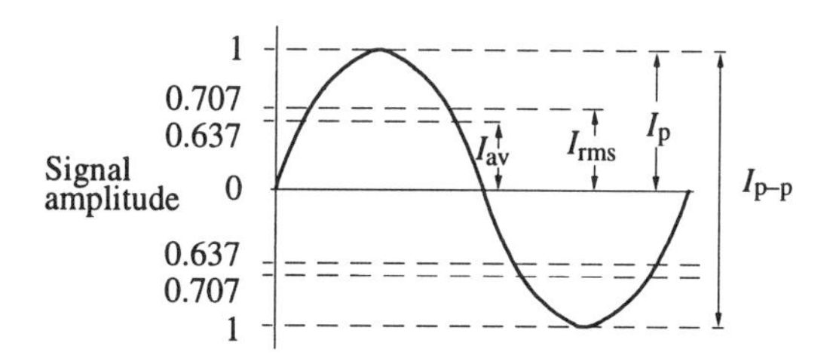

Figure S2–10. Common measures of amplitude for a sine wave.

Two types of signal level changes are commonly encountered. In one type, the signal levels change rapidly. This type of change is experienced in a computer data acquisition system, for example, where we might desire to sample a new data point every few microseconds. Reactive components can seriously affect the time response and thus the data acquisition rate of such a system as discussed below. In the other type, a continuous waveform such as a sine wave is applied to a reactive circuit. In this case the presence of capacitance and inductance in the circuit can affect the steady-state current and the time relationship between the current and the voltage in the circuit. The exact effects depend on the frequency of the waveform as well as the reactance of the circuit. These reactive effects may be turned to our advantage. For example, high- and low-pass filters are reactive circuits that are designed to pass a particular frequency range and exclude all others.

Let us consider first the application of a continuous sinusoidal signal to a circuit and later come back to the case in which a step function signal is applied. For ac signals, the **impedance** Z (in ohms) accounts for both the resistive effect of a circuit component and its reactance to change. The magnitude of the impedance is $Z = V_p / I_p$, where V_p and I_p are the peak values of the voltage and current, respectively. As the rate of signal variation (its frequency) approaches zero (dc), the impedance of a device approaches its resistance.

For simple resistors the current is always proportional to the instantaneous applied voltage as given by Ohm's law. For purely resistive devices, $Z = R$ at all frequencies. Application of a sinusoidal voltage $v = V_p \sin(\omega t + \phi)$ to a pure resistor gives rise to a sinusoidal current $i = I_p \sin(\omega t + \phi)$. The phase angle ϕ does not change; for pure resistance, the current and voltage sine waves are in phase for all frequencies. Pure resistance is an ideal that is closely approached by many practical resistors.

Capacitive Reactance

A capacitor exhibits reactance because it stores charge. The quantity of charge q stored in a capacitor depends on the voltage across it v_C:

$$q = Cv_C \tag{S2–8}$$

If the voltage v_C varies with time, the changing voltage across the capacitor must correspond to a change in charge, and therefore a current is required. Because current is the rate of charge flow, the current required is directly proportional to the rate of change of the capacitor voltage [$i = dq/dt = C(dv_C/dt)$]. The voltage across a capacitor cannot change instantaneously since this would require an infinitely large current. The capacitor thus reacts against a change in the voltage across it. If the capacitor is connected to a sine-wave voltage source of voltage $v = V_p \sin \omega t$, the current is $i = \omega C V_p \cos \omega t$. The peak current is $I_p = \omega C V_p$; thus, the instantaneous current is $i = I_p \cos \omega t$. The impedance of a device is $Z = V_p / I_p$, and for a pure capacitor the impedance, called the capacitive reactance X_C, is $V_p/(\omega C V_p)$. Thus,

$$X_C = 1/(2\pi f C) = 1/(\omega C) \tag{S2–9}$$

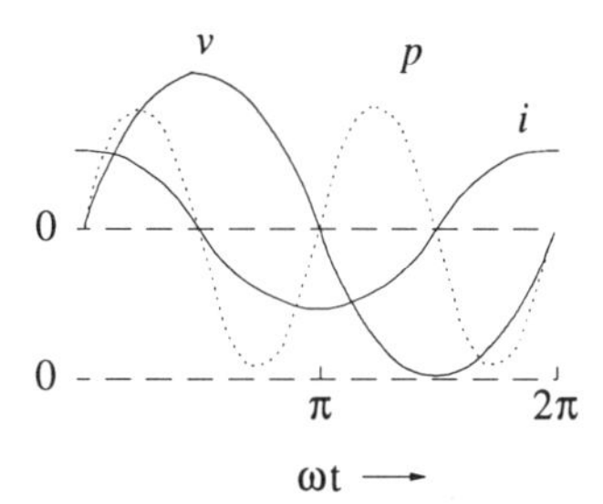

Figure S2–11. Current, voltage, and power waveforms in a circuit consisting only of a sine-wave voltage source and a pure capacitor. The current wave precedes the voltage by $\pi/2$ rad, or 90°. The current thus leads the voltage. The instantaneous power p is $p = vi$.

The capacitive reactance X_C has the units ohms and is dependent on frequency, as shown by equation S2–9. The higher the frequency, the less reactance a capacitor offers to the flow of charge. Because of the way in which its reactance depends on frequency, a capacitor behaves like an open circuit to a constant, that is, dc, voltage ($X_C \rightarrow \infty$ as $f \rightarrow 0$).

When a sine-wave voltage source is connected to a capacitor, the resulting current is a cosine wave. There is thus a 90° phase difference between the current and voltage waveforms in a capacitor, as shown in Figure S2–11.

Series Resistor–Capacitor (RC) Circuits

A **series RC circuit** is extremely useful as a **filter**. This simple combination can be used to reduce the amplitudes of low-frequency signal components, in which case it is called a **high-pass filter**, or to attenuate high-frequency components (a **low-pass filter**). Low-pass filters are often used in dc measurements to attenuate higher-frequency noise components such as the 60-Hz noise from the ac power distribution system.

Two arrangements of the series RC circuit with a sinusoidal voltage source are shown in Figure S2–12. To understand the frequency dependence of the voltages across the two components, remember that the resistance of a pure resistor is frequency independent and that the reactance of the capacitor increases with decreasing frequency. At low frequencies, therefore, the capacitive reactance is very high and a large fraction of the source voltage appears across the capacitor ($X_C >> R$). In fact, for dc voltages ($\omega = 0$) the capacitive reactance is infinite and all the source voltage appears across the capacitor. As the frequency increases, X_C decreases and a larger fraction of v_s appears across the resistor. When the output of the series RC circuit is taken across the capacitor (Figure S-12a), the circuit is called a low-pass filter, because dc voltages are unattenuated.

The series RC network has a total impedance magnitude Z and a phase angle ϕ given by

$$Z = \sqrt{R^2 + X_C^2} \qquad \text{(S2–10)}$$

$$\tan \phi = \frac{X_C}{R}$$

For a low-pass filter, the frequency at which the capacitive reactance equals the resistance, called the **upper cutoff frequency** f_2, is readily derived from

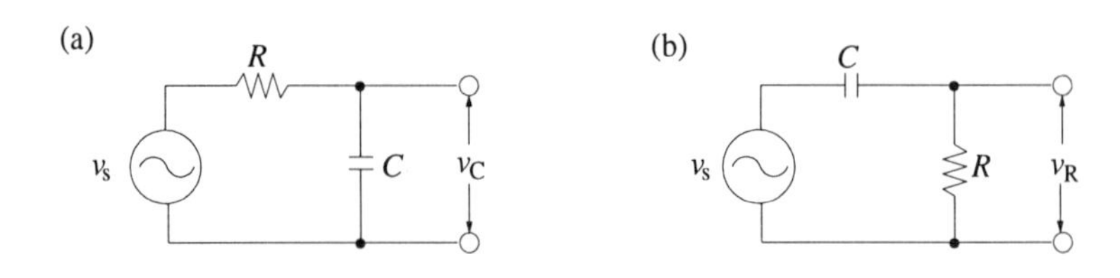

Figure S2–12. Series RC circuit. This series arrangement of a resistor and a capacitor is an ac voltage divider. Unlike a resistive voltage divider, the divider fraction is frequency dependent for a series RC circuit. The arrangement in part a is a low-pass filter, and that in part b is a high-pass filter.

equations S2–9 and S2–10 for the conditions $X_C = R$ and $f = f_2$. The result is

$$f_2 = 1/(2\pi RC) \tag{S2–11}$$

At the upper cutoff frequency, the peak voltage across the capacitor V_{pC} equals 0.707 V_p, where V_p is the peak voltage of the source. The ratio V_{pC}/V_p, which is called the **amplitude transfer function**, is essentially unity at frequencies much lower than f_2, because all the source voltage appears across C at very low frequencies. It is convenient to plot the ratio V_{pC}/V_p against frequency in a log–log plot as shown in Figure S2–13. Usually, the transfer function is expressed in **decibels** (dB), where the decibel is defined in terms of the circuit power gain [dB = 10 log (P_{out}/P_{in})]. The term has been extended through the relationship $P = V^2/R$ to express the circuit voltage gain in decibels as dB = 20 log (V_{out}/V_{in}). (Technically, the voltage ratio relationship should be used only when the impedances across which the voltages appear are equal.) At the upper cutoff frequency, the filter output is 3 dB down from its unattenuated value (low-frequency value). For this reason the upper cutoff frequency is often called the upper **3-dB point**. At frequencies much higher than f_2, the network transfer function falls by 20 dB for each decade (10-fold) increase in frequency. The plot shown in Figure S2–13a is known as a **Bode diagram**. It is readily constructed from two limiting regions: a low-frequency V_{pC}/V_p value of 0 dB and a slope of ≈0, and a high-frequency slope of

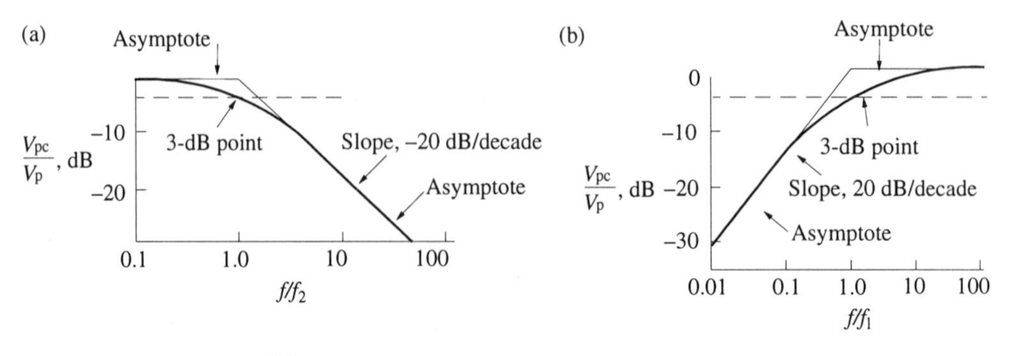

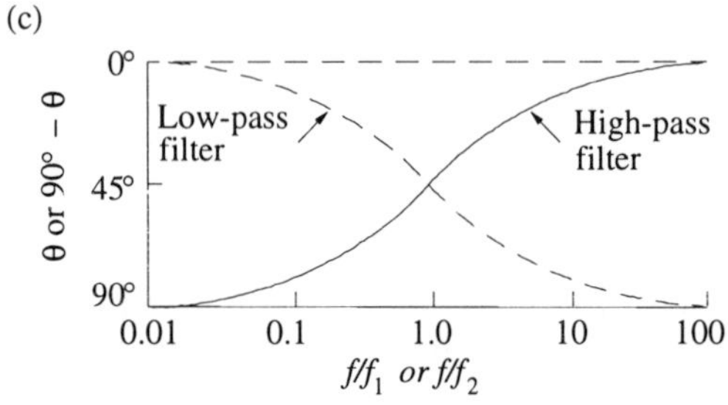

Figure S2–13. Frequency response plots for low- and high-pass filters. (a) Bode diagram of a low-pass filter. The asymptotes are extensions of the linear portions of the curve. (b) Bode diagram of a high-pass filter. (c) Phase shift of low- and high-pass filters versus frequency.

-20 dB/decade. The two limiting slopes meet at the 3-dB point. These asymptotes represent the true response quite accurately except for the region very near f_2.

If the output of the filter is taken across the resistor (Figure S2–12b), the circuit is a high-pass filter, because the attenuation at the output increases at low frequencies. The point at which $X_C = R$ in this case is known as the **lower cutoff frequency** or lower 3-dB point f_1 and is given by $f_1 = 1/(2\pi RC)$. The Bode diagram for the high-pass filter is shown in Figure S2–13b. Again, the Bode diagram can be closely approximated by two limiting slopes that meet at the lower 3-dB point.

High- and low-pass filters also shift the phase of the output signal with respect to the input signal. Recall that with a pure resistor, current and voltage are always in phase, and with a capacitor, the current leads the voltage by 90°. Hence, V_{pR} always leads V_{pC} by 90°. Plots of ϕ versus f/f_1 and of $(90° - \phi)$ versus f/f_2 for V_{pC} are shown in Figure S2–13c for the high- and low-pass filters. At the 3-dB point, the phase shift ϕ is 45°.

For nonsinusoidal waveforms, frequency-selective networks affect the shape of the waveform. Consider a square wave applied to an RC low-pass filter. Because the square wave contains high-frequency components in its rising and falling edges, severe distortions can occur in these regions of abrupt changes. The step response of reactive circuits is discussed in detail in a later section of this supplement.

Frequency-Compensated Voltage Divider

High-resistance voltage dividers are required in many measurement situations and many instruments. Oscilloscopes, computer-based instruments, ac voltmeters, and amplifiers are a few examples of units that may use high-resistance dividers. The frequency responses of such systems can often be limited by unintentional stray capacitance, as shown by the voltage divider of Figure S2–14a. The stray capacitance C_2 is shown in dotted lines. Often by adding a second capacitance C_1 in parallel with resistance R_1, as shown in Figure S2–14b, the effect of stray capacitance can be nearly eliminated. Such compensation capacitors are often used in the attenuator probes of oscilloscopes and the attenuators of amplifiers to compensate for the input capacitance.

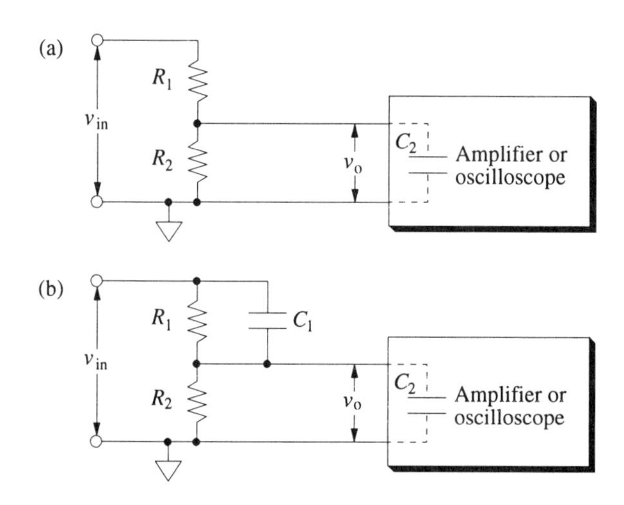

Figure S2–14. Reactive voltage dividers. (a) Uncompensated voltage divider. The capacitance shown in dotted lines is stray capacitance or the input capacitance of the scope or amplifier. (b) Frequency-compensated voltage divider. If $R_1C_1 = R_2C_2$, the divider fraction is independent of frequency.

The compensation circuit consists of two parallel RC networks in series. The impedance Z_2 of components C_2 and R_2 is given by

$$Z_2 = \frac{R_2}{1 + j\omega C_2 R_2}$$

and that of components C_1 and R_1 is

$$Z_1 = \frac{R_1}{1 + j\omega C_1 R_1}$$

The fraction v_o/v_{in} appearing across the parallel combination of R_2 and C_2 is

$$\frac{v_o}{v_{in}} = \frac{Z_2}{Z_1 + Z_2} = \frac{R_2/(1 + j\omega R_2 C_2)}{R_1/(1 + j\omega R_1 C_1) + R_2/(1 + j\omega R_2 C_2)} \qquad \text{(S2–12)}$$

If the compensating capacitance C_1 is chosen such that $R_1C_1 = R_2C_2$, equation S2–12 reduces to

$$v_o/v_{in} = R_2/(R_1 + R_2)$$

and the voltage divider fraction is independent of frequency.

Inductance

A device with **inductance** reacts against a change in current by developing a voltage that opposes the current change. All electrical components have some inductance, but **inductors** are purposely used in circuits to react against changes in current. They can serve important control functions.

An inductor is based on the principle that a varying magnetic field induces a voltage in any conductor in that field. A practical inductor may simply be a coil of wire, as shown in Figure S2–15a. A schematic diagram of an inductor connected to an ac voltage source is shown in Figure S2–15b. The current in each loop of wire generates a magnetic field that passes through neighboring loops. If the current through the coil is constant, the magnetic field is constant and no action takes place. If the current increases, however, the magnetic field lines expand and the changing magnetic flux generates a voltage in each loop. This induced voltage

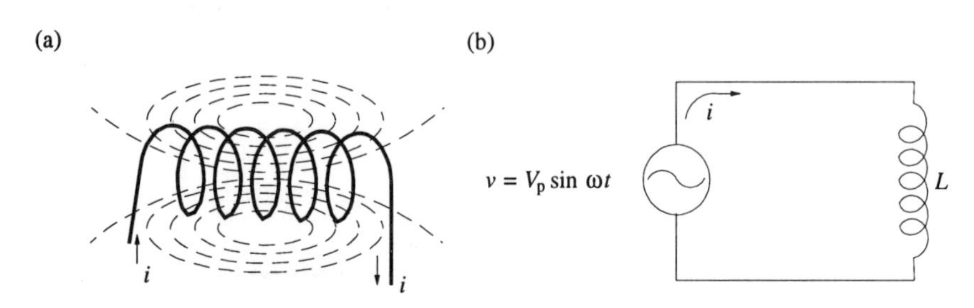

Figure S2–15. Inductance. Current in a coil produces a magnetic field (a). The energy stored in the field is proportional to the current. The energy absorbed or released when the current is changed reacts against the change in current. The inductor is then an impedance in an ac circuit (b).

(**electromotive force**, or **emf**) is counter (of opposite polarity) to the change in the applied voltage. This counter emf thus tends to impede, or react against, any change in current through the coil. The counter emf is directly proportional to the rate of change of current through the coil. The proportionality constant is the inductance L, which has the unit **henrys** (H). An inductance of 1 H induces a counter emf of 1 V when the current is changing at a rate of 1 A/s. Inductance can be important even when dealing with wires and cables. A coaxial cable, for example, typically has an inductance of 212 nH/m.

Inductive Reactance

In an ac circuit, like that of Figure S2–15b, the inductor offers reactance to ac. The voltage across the inductor v_L is directly proportional to the rate of change of current and the inductance L ($v_L = L\,di/dt$). When the inductor is connected to a sine-wave voltage source of voltage $v = V_p \sin \omega t$, the current is given by $i = (-V_p/\omega L) \cos \omega t$. For a pure inductor the impedance ($Z = V_p/I_p$) is the **inductive reactance** X_L ($X_L = V_p/V_p/\omega L$) with the units ohms. It is given by equation S2–13.

$$X_L = \omega L = 2\pi f L \qquad \text{(S2–13)}$$

Inductive reactance, like capacitive reactance, is frequency dependent. For inductors, however, the reactance *increases* with increasing frequency. An inductor is said to be a short circuit to dc, because $X_L \rightarrow 0$ as $f \rightarrow 0$.

When a sine-wave voltage source is connected to an inductor as described above, the current through the inductor lags the voltage across it by 90°. Note that this is opposite to the phase relationships of the current and voltage waveforms in a capacitive circuit.

Series Resistor–Inductor (RL) Circuits

Like series RC circuits, series RL circuits are used as filters. The impedance Z is equal to $(R^2 + X_L^2)^{1/2}$, and the phase angle ϕ is given by $\tan \phi = X_L/R$. In the low-pass filter of Figure S2–16a, the output is taken across the resistor. At very low frequencies, where $X_L << R$, the full source output appears across the resistor. The

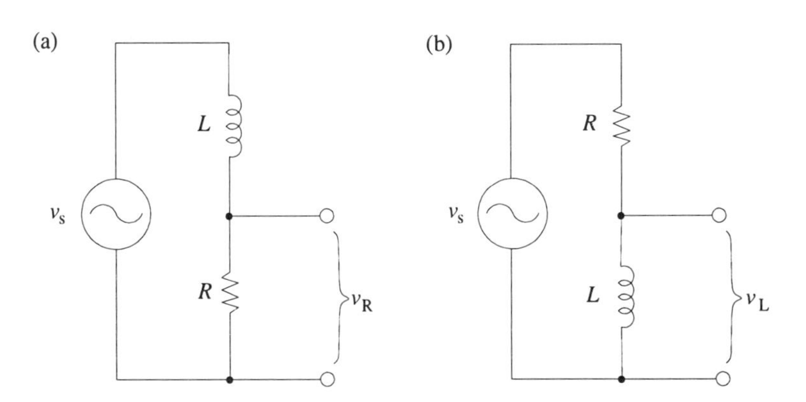

Figure S2–16. RL voltage dividers. The increase in the impedance L with increasing frequency makes the circuit in part a a low-pass filter and that in part b a high-pass filter.

upper cutoff frequency f_2 is reached when $R = X_L$ and is given by

$$f_2 = R/(2\pi L) \tag{S2–14}$$

The Bode diagram is the same as for an RC low-pass filter (Figure S2–13a) except that V_{pR}/V_p is plotted against f/f_2, and f_2 is given by equation S2–14.

The RL high-pass filter of Figure S2–16b produces the full source voltage at the output only at high frequencies, where $X_L >> R$. At frequencies lower than the lower cutoff frequency f_1, the transfer function V_{pL}/V_p decreases by 20 dB for each decade decrease in frequency. Again, the Bode diagram is the same as that for the RC high-pass filter in Figure S2–13b except that V_{pL}/V_p is plotted against f/f_1 and $f_1 = R/(2\pi L)$.

Resistor–Inductor–Capacitor (RLC) Resonant Circuits

A **series RLC circuit** can exhibit the phenomenon of resonance. Resonant circuits are used in tuned amplifiers, frequency-selective filters, and oscillators. For the series resonant circuit there is a particular frequency, called the resonance frequency, at which the capacitive reactance X_C equals the inductive reactance X_L. Because the current leads the voltage in a capacitor and lags it in an inductor, the reactances of these devices cancel at the resonance frequency. The total circuit impedance is at a minimum at the resonance frequency and equal to the resistance. Because the impedance is at a minimum at the resonance frequency, the current in the circuit reaches a maximum that decreases rapidly at higher or lower frequencies.

A circuit consisting of a resistor, a capacitor, and an inductor in parallel also displays resonance. Here, the total parallel impedance and the voltage pass through a maximum at the resonance frequency.

Step Response

In addition to the steady-state responses of reactive circuits described above, the transient responses of such circuits are extremely important. In some cases, step function changes are the result of applying a voltage pulse or series of pulses to a circuit. For example, a "start" pulse is used to signal the beginning of serial data transmission to and/or from a computer. Alternatively, a transient signal can be the result of switching a steady voltage to a circuit, as would occur with a digital-to-analog converter. In either case, the transition of the signal values from their initial (before-the-pulse) values to their final (steady-state) values is not instantaneous because of reactances in the circuit. Every component of a circuit has some inductance or capacitance, and these reactances, along with the circuit resistances, limit the response time.

We shall illustrate the step response of a series RC circuit with the switching circuit of Figure S2–17. The series circuit is still a voltage divider with the applied voltage divided between the capacitor voltage v_C and the resistor voltage v_R. Here, the applied voltage is V in switch position 2 and 0 V in position 1. Suppose that the switch has been in position 1 long enough for the capacitor to be completely discharged. Because there is no applied voltage in switch position 1 and the capacitor is completely discharged, the voltage across the capacitor v_C and the

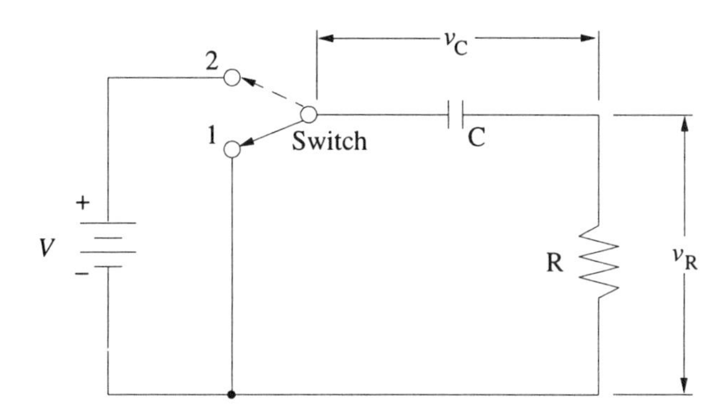

Figure S2–17. Series RC circuit and step signal source. In switch position 2, the capacitor charges toward the supply voltage. In switch position 1, the capacitor discharges through the resistor.

voltage across the resistor v_R are both zero. Now the switch is changed to position 2, and the applied voltage is V. At the instant of closing the switch, the charge on the capacitor is still zero, so that $v_C = 0$ V. Because Kirchhoff's voltage law applies (i.e., $v_C + v_R = V$), the entire source voltage V appears across the resistor at the instant the switch is changed to position 2. The current at this instant is v_R/R. The current in the circuit begins to charge the capacitor. At any time,

$$V = v_R + v_C = iR + q/C \tag{S2–15}$$

Thus as q and v_C increase with time, i and v_R must decrease. All four of these quantities change exponentially with time, as shown in Figure S2–18.

Unlike the resistive voltage dividers we have previously encountered, the series RC circuit is a time-dependent voltage divider. In Figure S2–18, the time scale is calibrated in RC units. The product RC has the units seconds and is called the **time constant**. The capacitor charge q at any time t after the switch has been moved to position 2 can be found by substituting dq/dt for i in equation S2–15 and solving the resulting differential equation. The current i at any time t is

$$i = Ve^{-t/RC}/R \tag{S2–16}$$

After a time $t = RC$ (one time constant), the current i, equal to $(V/R) \times e^{-1}$, is $(V/R) \times 0.368$. Hence, after one time constant, the current is 36.8% of its value at the instant of impressing the source voltage V on the circuit. At this same time, the

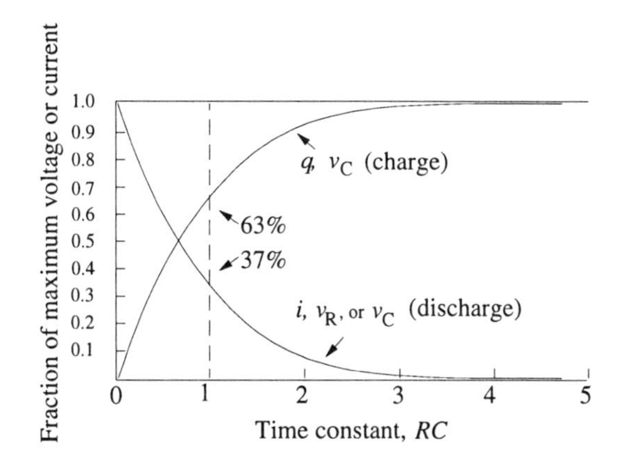

Figure S2–18. Charge and discharge curves for a series RC circuit. On charging, the capacitor voltage v_C reaches 63% of the supply voltage in one time constant. On discharging, the capacitor voltage falls to 37% of its fully charged value in one time constant.

Table S2–2. Voltages across Capacitor and Resistor in Series RC Circuit

Time (*RC*)	Capacitor Charging (% *V* Applied)		Capacitor Discharging (% *V* Initial)	
	v_C	v_R	v_C	v_R
τ	63.2	36.8	36.8	36.8
2τ	86.5	13.5	13.5	13.5
2.3τ	90.0	10.0	10.0	10.0
3τ	95.0	5.0	5.0	5.0
4τ	98.2	1.8	1.8	1.8
4.6τ	99.0	1.0	1.0	1.0

voltage across the resistor is 36.8% of its initial value and the capacitor is charged to 63.2% of the source voltage. The time constant of the RC circuit is often given the symbol τ. Table S2–2 gives values of v_C and v_R for different multiples of τ both for charging the capacitor and for discharging it after it has been charged to voltage V. When $t = 4.6\tau$, the capacitor is charged to 99.0% of the impressed voltage and v_R is only 1% of its initial value. For practical purposes the capacitor is often considered to be fully charged when $t = 5\tau$. It is important to keep in mind that the voltage across a capacitor (or any element with capacitance) cannot change instantly; instead, it changes exponentially with time.

When an oscilloscope is connected to the series RC circuit first across the resistor and then across the capacitor, various wave shapes similar to those shown in Figure S2–19 can be observed when the switch of Figure S2–17 is turned on and off or when a rectangular pulse train is used as the input. The output waveform depends on the ratio of the *RC* time constant τ to the pulse width T_p. The leading

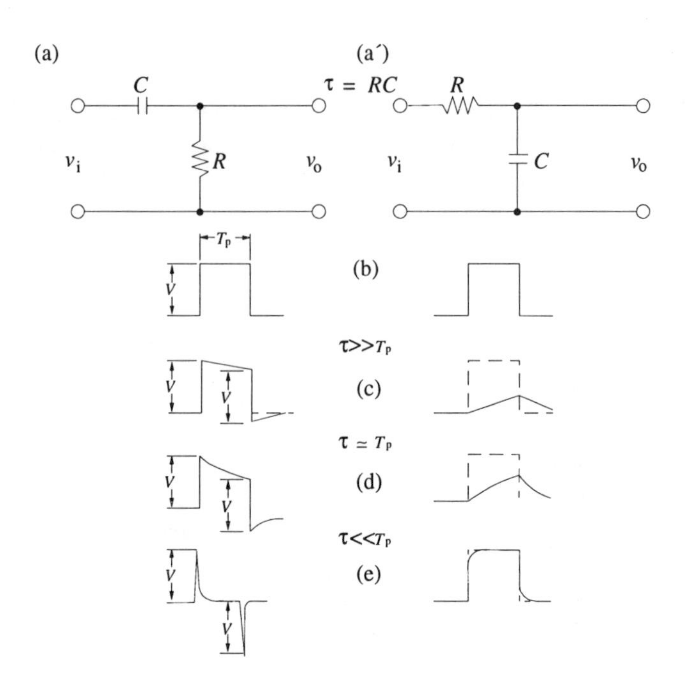

Figure S2–19. Waveforms in RC series circuits for the output taken across the resistor (a) and the capacitor (a′). The input voltage v_i is shown in part b. The waveforms in parts c, d, and e result from various ratios of the time constant τ to the pulse width T_p.

edge of the output across the resistor (circuit a) is always steep as long as the input voltage has a steep leading edge. In contrast, the leading edge of the capacitor output (circuit a′) always changes exponentially.

Sharp positive and negative pulses can be obtained across the resistor when the *RC* time constant is much shorter than T_p, as can be seen in Figure S2–19e. The circuit in this case is called a **differentiator**, because the sharp spikes are approximately the derivative of the input pulse shape. When the time constant is greater than the pulse width (Figure S2– 19c), the voltage across the capacitor is nearly a linear ramp voltage. In this case the RC circuit is referred to as an **integrator**, because the capacitor charging is nearly proportional to v_i as long as $v_C << v_i$. More nearly ideal differentiators and integrators can be constructed with operational amplifier circuits as described in Chapter 5, Section 5–6.

The presence of inductance in a circuit makes it impossible to change the current value instantaneously, just as capacitance makes it impossible to change the voltage instantaneously. In either case there is a time lag between the attempt to change the voltage or current level and the attainment of that change. When a voltage *V* is switched rapidly to a series RL circuit, the voltage across the resistor is initially zero at the instant of switching because the inductor prevents an instantaneous change in current. Because $v_L = L(di/dt)$, the current increases at a rate of *V*/*L* amperes per second. As the current increases, some of the voltage *V* appears across *R* ($v_R = iR$). This decreases both the voltage across the inductor v_L and the rate of increase of the current. The result is an exponential decrease in v_L according to $v_L = Ve^{-tR/L}$ and an exponential increase in v_R according to $v_R = V(1 - e^{-tR/L})$. The quantity *L*/*R* is the time constant for the RL circuit. After several time constants, v_L approaches zero, while v_R approaches the applied voltage *V*, and the current *i* approaches *V*/*R*. When the switch is opened, the current suddenly becomes zero. The inductor reacts against this sudden change in current by developing a large negative value of v_L. Because *di*/*dt* is very large when the switch is opened, the value of v_L when the current is interrupted in real inductors can be thousands of volts. This can cause arcing across switch contacts and eventual destruction of the switch. This same effect is used to generate high-voltage sparks for internal combustion engines and other applications.

Supplement 3

Understanding Diodes, Transistors, and Other Solid-State Devices

The advent of solid-state devices for switching and amplification has been an essential part of the revolution in electronics and modern instrumentation. Solid-state switches enable digital control functions to be implemented on a nanosecond to microsecond time scale. These switches are the basic building blocks of digital circuits. Solid-state amplifiers have been an integral part of the recent development of extremely high quality, inexpensive operational amplifiers (op amps), instrumentation amplifiers, and automatic feedback control systems.

Despite dramatic differences in their application, switching and amplification circuits have much in common. In each case the active device controls the current in one part of the circuit in response to a controlling signal applied through another part. In switch applications the active device is clearly either ON or OFF; in amplifier applications the active device operates throughout the region between the maximum conducting and nonconducting limits.

Two main classes of transistors are used as active devices for both switching and amplification: the bipolar junction transistor (BJT) and the field effect transistor (FET). In bipolar devices the controlled current path includes one or more pn junctions; in field effect devices the current through a single type of semiconductor (n or p) is controlled by the influence of an electric field. This supplement begins with a description of pn junctions and both types of transistors.

The analog switch packages, considered functionally in Chapter 7, Section 7–2, are described here in terms of their internal elements and circuits. The principles and power-switching applications of four-layer devices such as silicon-controlled rectifiers (SCRs) and triacs are also considered. Basic bipolar and FET amplifiers are introduced and then combined to form a representative example of an op amp.

S3–1 Bipolar Devices

Most solid-state switches and amplifiers depend upon the characteristics of the pn junction for their control capabilities. Such pn junctions and pn junction diodes are reviewed briefly here before the bipolar transistor is introduced.

A pn junction is formed in a semiconductor at every interface between p-doped and n-doped regions, as illustrated in Figure S3–1a. Under conditions of zero bias, electrons from the n region and bonding electron vacancies (holes) from the p region neutralize each other in the junction region. The concentration profiles for holes and electrons are shown in Figure S3–1b. In the example shown, the dopant concentration is slightly less in the n region than in the p region, and the total concentration of charge carriers is depleted in the junction region by many orders of magnitude. This region of reduced charge carrier concentration is the **depletion region**, and the contact junction potential appears across this region as shown in Figure S3–1c. This contact potential is about 0.3 V for a germanium pn junction and about 0.6 V for silicon.

2861-2/94/0381$08.00/1 Published by American Chemical Society

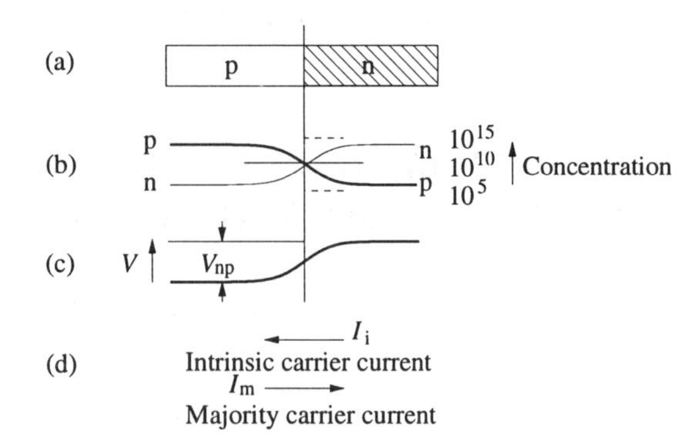

Figure S3–1. The pn junction. (a) Block representation; (b) concentration profile of charge carriers in carriers per cubic centimeter; (c) potential profile; (d) junction current directions.

The concentration of charge carriers in the depletion region is essentially that of the intrinsic semiconductor. These charge carriers are electron–hole pairs generated by thermal excitation. The contact potential difference across the depletion region causes the generated holes to move to the p region and the electrons to move to the n region. The current due to carriers in the intrinsic region thus moves from right to left as shown in Figure S3–1d. The magnitude of the intrinsic carrier current I_i depends on the rate of carrier generation in the depletion region, which is a function of temperature only. The net flow of positive charge from right to left due to the intrinsic carrier current tends to reduce the contact potential across the junction. However, as the potential barrier is decreased, the number of majority carriers on each side with enough energy to cross the junction increases. The current due to majority carriers I_m moves from left to right and increases exponentially with increasing temperature and decreasing junction potential:

$$I_m = Ke^{-Q_e V_{np}/kT} \qquad \text{(S3–1)}$$

where k is the Boltzmann constant (J/K), Q_e is the electron charge (C), and T is the temperature (K). When no external current source is applied to the pn junction, the net current across the junction must be zero; that is, the intrinsic and majority currents must cancel exactly. For any given temperature there is only one value of the junction potential for which this is true. One can see qualitatively how the higher intrinsic carrier generation rate of germanium over silicon results in the lower junction potential for germanium.

When an external bias is applied, essentially all of the bias voltage appears as a change in the voltage across the pn junction. This is shown by the family of potential profiles in Figure S3–2. The change in V_{np} has a great effect on the majority carrier current I_m but little effect on the intrinsic current I_i. At zero applied bias, I_m is equal to I_i, so that

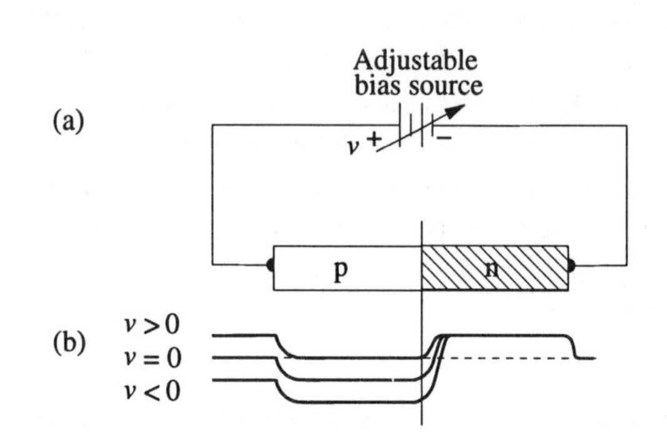

Figure S3–2. (a) A pn junction with bias voltage source connected; (b) potential profiles showing the effect of bias voltage on the pn junction potential.

$$I_i = \mathring{I}_m = Ke^{-Q_e \mathring{V}_{np}/kT} \tag{S3–2}$$

where $\mathring{I}_m$ and $\mathring{V}_{np}$ are the values for the majority current and junction potential, respectively, at zero bias. At nonzero bias the junction potential is $\mathring{V}_{np} - v$, and a net forward current i can cross the junction. The net current is the difference between I_m and I_i:

$$i = I_m - I_i = Ke^{-Q_e(\mathring{V}_{np} - v)/kT} - I_i \tag{S3–3}$$

Rearranging equation S3–3, we obtain the Shockley equation:

$$i = I_i(e^{Q_e v/kT} - 1) \tag{S3–4}$$

The Shockley equation describes the familiar exponential current–voltage behavior of the pn junction under forward-bias conditions (*see* Chapter 4, Figure 4–6e) and shows that under reverse-bias conditions the current *is* essentially I_i, often called the **reverse-bias saturation current**. The reverse-bias current depends on temperature, as shown in equation S3–2, and upon other sources of energy at the junction (e.g., light), but it is essentially independent of the reverse-bias voltage applied as long as this voltage is less than the reverse breakdown voltage.

The Bipolar Junction Transistor (BJT)

Seldom has a device had such a dramatic impact on our leisure and work as has the transistor. This remarkable device has brought about a complete revolution in electronic circuitry and has made commonplace items of products that were science fiction visions not so many years ago.

The BJT is composed of two n- or p-type semiconductor regions separated by a thin region of the opposite type of semiconductor. This arrangement is illustrated in Figure S3–3 with the standard symbols for BJTs. Two types of transistors result,

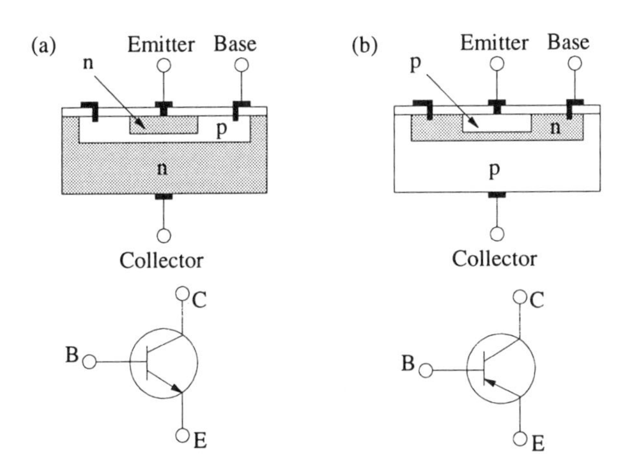

Figure S3–3. Basic transistor types. Structure and symbol of an npn transistor (a) and a pnp transistor (b). B, base; C, collector; E, emitter.

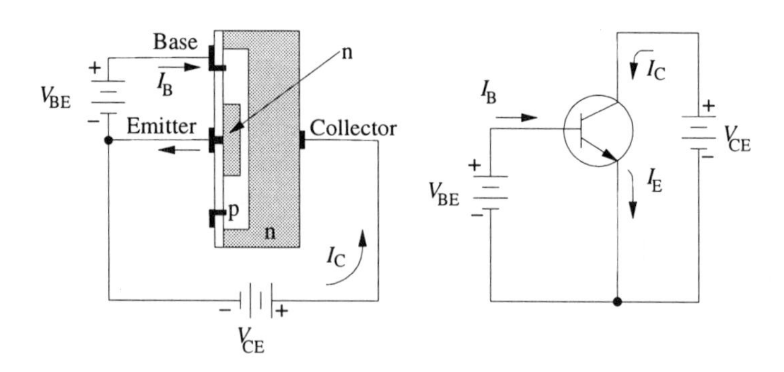

Figure S3–4. The npn transistor biased in the ON state. The transistor is in the common-emitter configuration.

the npn and the pnp, named for the order of the semiconductor types used in their construction. The thin center region is called the **base**, and the other two regions are called the **emitter** and the **collector**. In three-terminal-switch operation, an actuating signal that is applied between the base and emitter controls conduction between the collector and emitter.

An npn transistor in the ON state is diagrammed in Figure S3–4. The base–emitter junction is forward-biased by voltage V_{BE}. Under the influence of the forward bias, emitter electrons are accelerated into the base region. Some of these electrons encounter holes in the base and produce the base current I_B. However, the base region is so thin and so lightly doped that there is a high probability that the electrons from the emitter will traverse the base region without being neutralized by a hole. These electrons are collected by placing a positive voltage on the collector, and the collector current I_C results. Because only a small fraction of the electrons from the emitter combine with holes in the base, the collector current is much larger than the base current. Therefore, a small base current controls a much larger collector current. When the transistor conducts, the collector voltage V_{CE} is larger than the base voltage V_{BE} and the collector–base junction is reverse-biased. When the base–emitter junction is reverse-biased, both pn junctions are reverse-biased and the collector–emitter conduction path is turned off. The operation of the pnp transistor is completely analogous: holes rather than electrons are emitted and collected, and the current arrows and battery polarities are reversed.

A current–voltage curve for a representative transistor is shown in Figure S3–5. There is actually a whole family of current–voltage curves, one for each value of base current. For values of V_{CE} greater than 1 V, the collector current is relatively constant and is approximately proportional to the base current. The **forward**

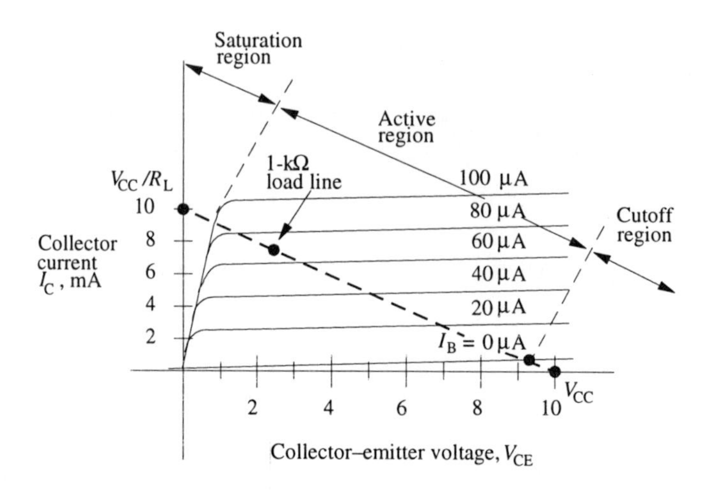

Figure S3–5. Collector characteristic curves for the common-emitter connection of an npn transistor. The dashed load line connects V_{CC} and V_{CC}/R_L. The collector voltage and current are given by the point on the load line that is intersected by the characteristic curve for the applied base current. For I_B = 68 μA, V_{CE} and I_C are given by the point in the middle of the load line.

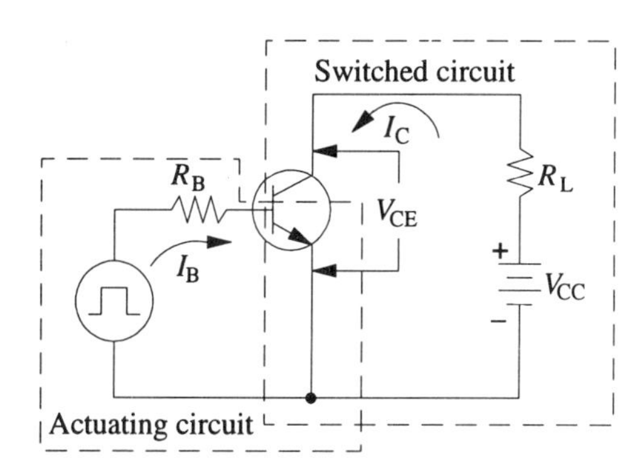

Figure S3–6. Common-emitter npn transistor switching circuit. The actuating circuit provides base current I_B, which controls the conducting state of the transistor. For the analysis of Figure S3–5, the actuating signal is assumed to vary from 0 and +4.0 V, R_B = 50 kΩ, and V_{CC} = 10 V.

current gain characteristic β is defined as the ratio of the collector current to the base current, or $\beta = I_C/I_B$. The value of β is approximately constant in the linear amplification region and is about 100 for the transistor type shown in Figure S3–5. The value for β is roughly the probability of an emitted charge carrier passing through the base region and arriving at the collector (100:1 in this case). Values of β for commercially available transistors vary from ~5 to >1000. From the flatness of the current–voltage curve, it can be seen that once V_{CE} is high enough to collect the emitted charge carriers, a further increase in V_{CE} does not give a significantly higher current.

A simple npn transistor switch circuit is shown in Figure S3–6. The actuating circuit provides the base current I_B that controls the conducting state of the transistor. The switched circuit consists of the collector supply voltage V_{CC} and the load R_L. When the base current is high enough to force the transistor into conduction, the much larger collector current I_C exists in the load. When the transistor is turned off by the actuating signal, the current through R_L is greatly reduced.

The switching circuit of Figure S3–6 can be analyzed to find I_C for the two conduction states of the transistor by a simple graphical method using the characteristic curves of Figure S3–5 and a linear equation called the **load line** for the switched circuit. This method is illustrated in Figure S3–5.

For best operation as a switch, I_B should have values that allow only for the maximum or minimum values of I_C. For the load line shown in Figure S3–5, the maximum value of I_C is V_{CC}/R_L = 10 V/1 kΩ = 10 mA. The value of β can be used to estimate the value of I_B required: I_B(ON) >> I_C(max)/β ≈ 10 mA/100 ≈ 100 μA. The operating point on the load line for I_B = 100 μA is readily identified in Figure S3–5. For this point, V_{CE} is approximately equal to 0.2 V. Because V_{BE} is equal to 0.6 V (for a conducting junction), the base–collector voltage is about +0.4 V. In other words, the base–collector junction is actually forward biased by 0.4 V. In this condition the transistor is said to be in **saturation**. The saturation, or maximum ON condition, is a state achieved by applying enough base current to control a collector current that is greater than the circuit allows.

Light-Activated Switches

In a **light-activated switch**, the conductivity or the current between the two switch contacts depends upon the amount of light falling on a photosensitive device. These devices are used as light detectors, as described in Chapter 8. The light can also be used as the actuating signal in switching applications of the photodetectors. This application has found extensive use in position indication, where mechanical motion is used to interrupt the light path from source to detector. The combination

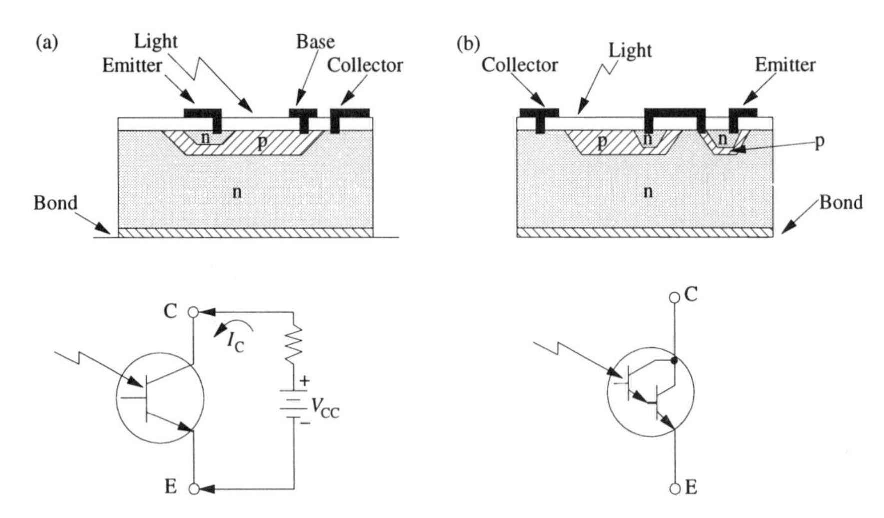

Figure S3–7. (a) Phototransistor structure and circuit with a typical load and bias; (b) photodarlington structure and circuit symbol.

of light and detector used as actuator and switch has the advantage of excellent electrical isolation between the actuating and controlled circuits, as described in Chapter 8, Section 8–6. This allows transmission of data between systems that are highly noise sensitive or that may differ in voltage by thousands of volts.

The photoconductive cell and the photodiode described in Chapter 8 have some application as light-activated switches. These applications are generally limited to low signal levels. Although the OFF (no-light) conductivity or current can be very low, the ON (full-light) conductivity or current is often not very high (that is, R_{ON} is hundreds of ohms, or I_{ON} is tens of microamperes).

A very popular photodetector for switch applications is the **phototransistor** shown in Figure S3–7a. As in normal transistors, the collector–base junction is reverse-biased. Light strikes this region through a window and creates electron–hole pairs that give rise to a positive "leakage" current to the base. The positive charge, collecting in the doped base region, forward biases the base–emitter junction to produce a collector current I_C. The collector current is β times the light-induced base current, where β is the forward current gain of the transistor. Collector ON currents of several milliamperes are obtained with this type of device.

Still larger current capacity is obtained with a **photodarlington**, an integrated two-stage amplifier device shown in Figure S3–7b. The emitter of the phototransistor is connected directly to the base of another transistor that shares a common collector with the first. The collector–emitter current of the phototransistor is the base current for the second transistor. Thus, the light-induced current is effectively amplified by β^2. With this device the second transistor can be driven into saturation with moderate levels of actuating light.

S3–2 Field Effect Transistor (FET) Devices

FETs find a variety of applications in switching and in linear circuits. As switches, they have OFF/ON resistance ratios superior to those of bipolar transistors. The high input impedance of FET amplifiers makes them nearly ideal as voltage amplification elements. This section considers the basic operating principles of the junction FET (JFET) and the metal oxide semiconductor FET (MOSFET).

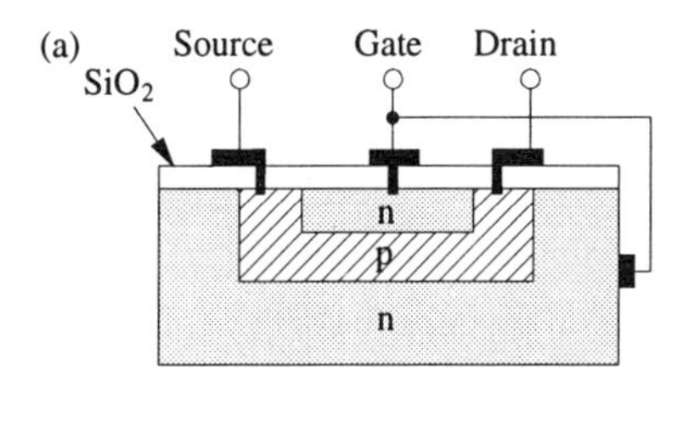

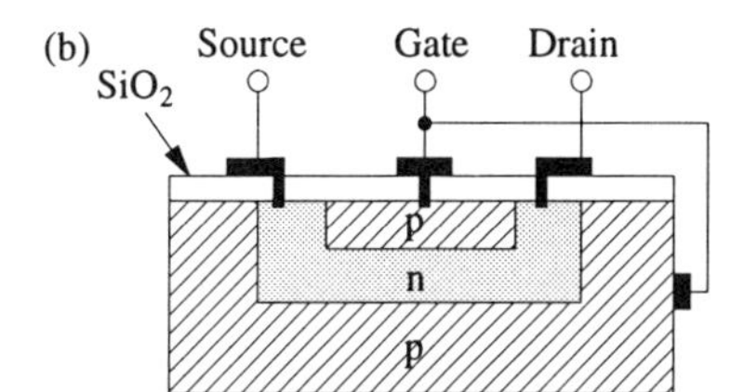

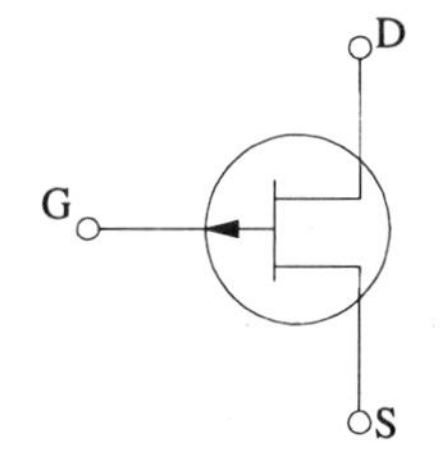

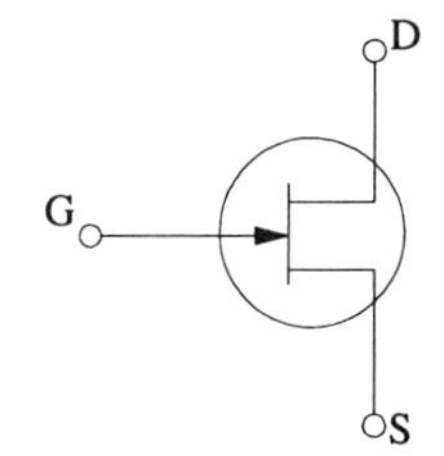

Figure S3–8. JFET structures and symbols. (a) p-channel JFET; (b) n-channel JFET. D, drain; S, source; G, gate.

The JFET

The **JFET** is made of a channel of unipolar (only one type of charge carrier) semiconductor with a contact at each end (the **source** and the **drain**). Conductivity between the source and the drain depends upon the charge carrier density and the dimensions of the channel, which are, in turn, affected by the depletion layer thickness of one or more reversed-biased pn junctions placed along the channel. This is shown in Figure S3–8 for JFETs made of both p-type and n-type channels. The channel dimension control region is called the **gate**.

The bias on an n-channel JFET is shown in Figure S3–9. The junction of gate and channel is reverse-biased by the gate source voltage V_{GS} to create a depletion region in the channel. The drain source voltage V_{DS} must then be of opposite polarity to the gate to maintain this reverse bias. Increasing the reverse bias on the gate narrows the effective source drain channel and causes a decrease in I_D. For the p-channel JFET, the voltage polarity and current directions are of course reversed. In switch and amplifier applications, the control or actuating signal is connected to the gate, and the source and drain connections are placed in the circuit to be controlled.

The current–voltage characteristic curves are shown in Figure S3–10 for a typical n-channel JFET. As expected, the drain current is smaller for increasingly negative values of V_{GS}. In the ohmic region ($V_{DS} < V_p$), an increase in V_{DS} increases I_D proportionally. The effect of the gate bias voltage is just to change the channel resistance. At higher values of drain voltage, the drain current becomes self-limiting

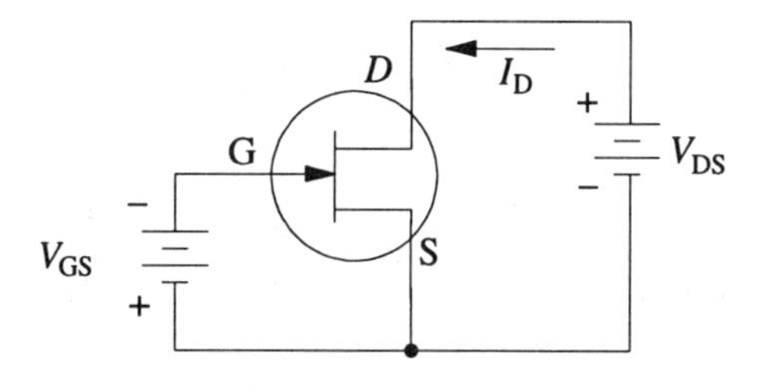

Figure S3–9. Voltages and current in an n-channel JFET.

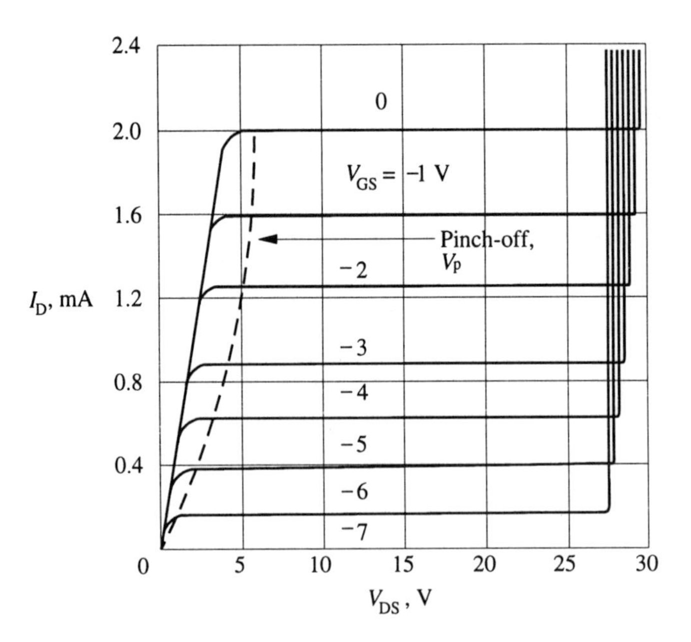

Figure S3–10. Characteristic curves for the n-channel JFET.

at a value determined by V_{GS}. This is called the "pinch-off" region because the *IR* drop in the channel affects the bias on the junction. The result is to keep the *IR* drop constant. From the current–voltage curves, the JFET can be seen to make an excellent switch. The ON condition is achieved at $V_{GS} = 0$ V, and the OFF condition occurs when V_{GS} is a large negative value. The ON resistance is typically 25 to 100 Ω, and the OFF resistance is 10^9 Ω or greater.

The MOSFET

In the JFET, the gate–channel junction serves primarily to impose a reverse-bias electric field on the channel. Because the junction is rarely used in forward-bias conduction, a normal pn junction is not required. In the MOSFET, an insulating layer of metal oxide is interposed between the gate electrode and the channel (Figure S3–11). The channel is an extremely thin layer of semiconductor of a type opposite to that of the **substrate** (the semiconductor piece on which the device is fabricated). On the surface of the channel are successively deposited a few micrometers of metal oxide insulator and metallic gate connection. In Figure S3–11a, a p-type substrate is shown; the channel and the drain and source contacts

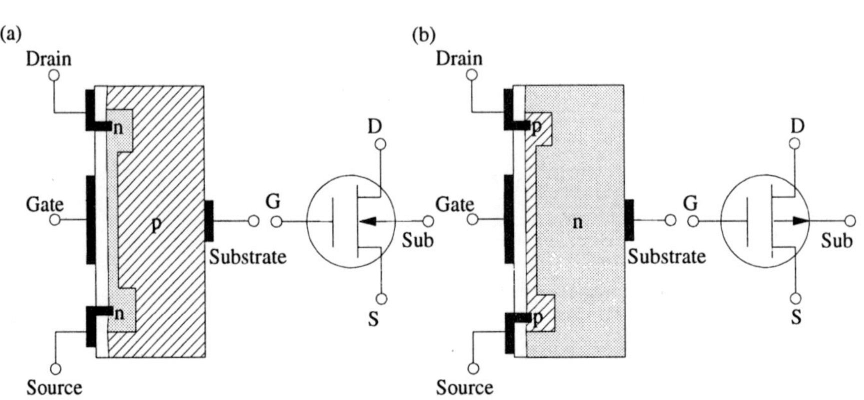

Figure S3–11. Basic structures and symbols for depletion-mode MOSFETs. (a) n-channel device; (b) p-channel device.

are therefore n-type semiconductors. Just as in the JFET, a reverse bias (in this case positive) applied to the gate reduces the number of charge carriers in the channel. This is called the **depletion mode**, because increasing the gate bias depletes the channel conductivity. A p-channel MOSFET is similar to the n-channel device, as can be seen in Figure S3–11b, except that the p and n regions are switched. The current–voltage curves of MOSFET devices have the same features as those of the JFET shown in Figure S3–10. An advantage of the MOSFET is its extremely high gate input impedance of $10^{12}\ \Omega$.

The presence of the insulating layer prevents conduction between the gate and channel even for what would have been a forward bias in the JFET. Because of this feature, it has been possible to build MOSFETs in which the gate bias is used to enhance the channel conductivity. In these **enhancement mode** MOSFETs, no channel is diffused at all. The application of a forward bias actually creates a channel of the opposite type on the surface of the substrate. The surface channel provides the conducting link between the source and drain contacts. The characteristic curves are not as flat as those of the depletion mode FETs. An advantage of the enhancement mode FET is that the gate signal is of the same polarity as the supply for the drain circuit.

All types of FETs are used extensively as analog switches. They have low noise, no distortion due to junction characteristics, high impedance in the actuating circuit, and fast response. The relative simplicity and low power requirements of MOSFETs have fostered today's microcomputer revolution. As loose components, MOSFETs must be handled with care, because the gate insulation can be permanently damaged by relatively small static charges.

S3–3 Analog Switches

The integrated circuit analog switch packages discussed in Chapter 7 are made from bipolar and field effect devices. In many switches, combinations of devices are used to provide the various driver and switch functions. In this section, several analog switching circuits that use the BJT, JFET, and MOSFET are described.

Transistor Analog Switches

Several BJT analog switch circuits are illustrated in Figure S3–12. In the series switch circuit (Figure S3–12a) the gate is closed when the transistor base is positive with respect to v_s. The reverse-biased base–collector and base–emitter junctions make an open circuit output when the gate is closed. A negative pulse is used to open the gate (and turn the transistor on). When negative signal levels are to be switched, an npn transistor with a positive-going actuation signal is used. Note in Figure S3–12a that the base of the transistor is connected to the common point of the actuating and controlled signals. Accordingly, this arrangement is called the **common–base configuration**.

The shunt switch shown in Figure S3–12b uses an npn transistor to shunt the output terminals for positive signals when the switch base is driven positive. The current-limiting resistance R_e avoids short-circuiting the source when the transistor is on. This switch gives better transmission accuracy for the open gate. The shunt switch is in the **common–emitter configuration**.

The series–shunt combination shown in Figure S3–12c is used for precision analog switching. BJTs are frequently used in an inverted or **common–collector**

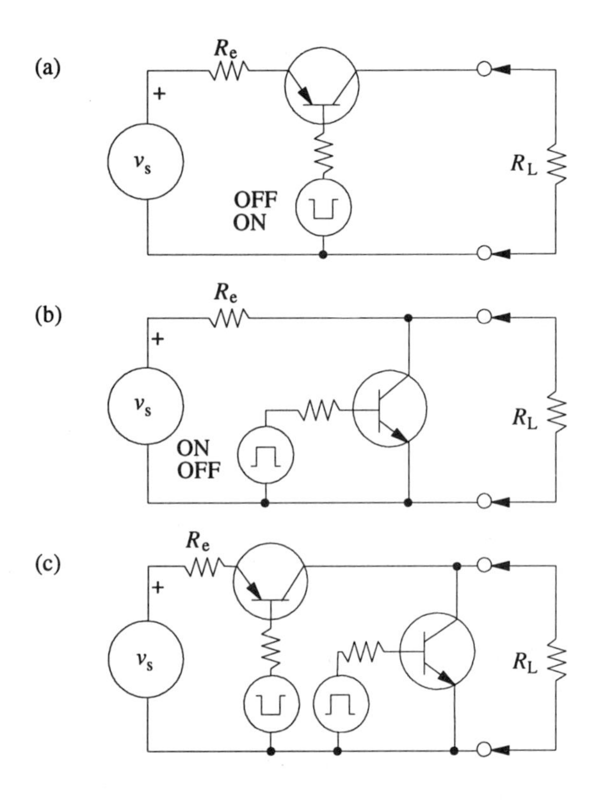

Figure S3–12. Transistor analog switch circuits. (a) Series switch; (b) shunt switch; (c) series–shunt switch.

configuration, with the emitter and collector connections reversed from those of Figure S3–12b. The voltage offset is smaller when a BJT is operated in the inverse mode, the ON resistance can be as low as 2 Ω, and the switching speed can be as short as 10 ns.

FET Analog Switches

As might be expected, the FET family (JFET and MOSFET) makes excellent analog switches. The FET switch offers high analog transmission accuracy because it does not introduce a junction potential in series with the signal. It also offers the versatility of being able to conduct current in either direction. Its disadvantages with respect to the BJT switch are higher resistance and generally slower switching speed.

When the FET is used as a voltage switch (i.e., to switch a signal that might have an amplitude of several volts), a switch driver circuit is needed to convert HI/LO logic-level signals into signals suitable for turning the FET on and off. Figure S3–13 illustrates an FET voltage switch and its associated two-transistor driver. The driver circuit must provide a gate voltage that is enough more positive than the switched voltage in one state and enough more negative than the switched voltage in the other state to turn the FET completely on and off. A voltage swing larger than the maximum switched voltage is generally required.

When a JFET is used as a current switch (i.e., to switch a current source to a point at common), it can be turned on and off by ordinary transistor–transistor logic (TTL) logic-level signals. Figure S3–14 illustrates use of a JFET current switch at the input of an op amp current follower. When the JFET is OFF, the input diode limits the drain–source voltage to 0.6 V. Because the drain and source voltages are always near zero, the driver voltage needs to vary only between zero and the

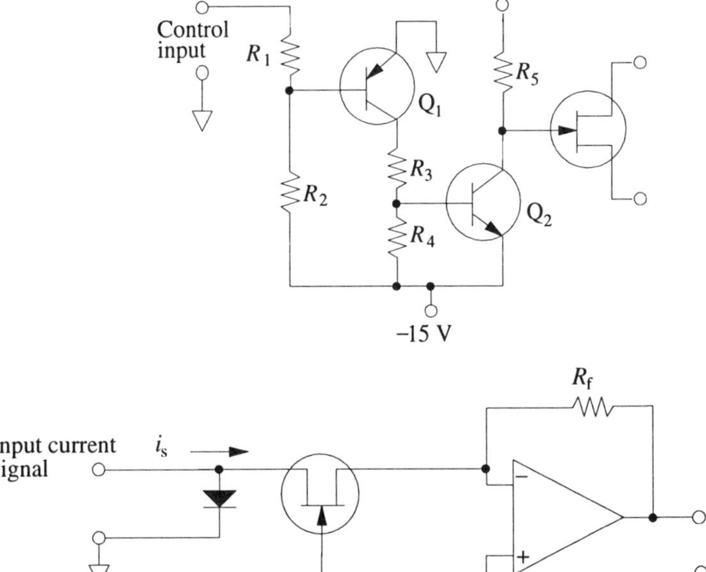

Figure S3–13. FET voltage switch and two-transistor driver. The drive circuit provides gate voltages sufficient to turn the FET completely on in one state and completely off in the other. Q_1 and Q_2 are driver transistors.

R_f

Input current signal i_s

Logic-level signal

$v_o = -i_s R_f$

Figure S3–14. JFET current switch. When used as a current switch, the JFET can be turned on and off by ordinary logic-level signals.

pinch-off voltage (<5 V). The small voltage drop across the JFET and the corresponding small driver voltage makes the current switch faster than the voltage switch.

S3–4 Power Control Switches

Several semiconductor switches called **thyristors** have at least four layers, such as pnpn. Thyristors can be triggered into conduction, but they do not require any control current to maintain conduction. Because of this, the circuitry for controlling a thyristor switch is usually simple and consumes very little power. Thyristors find many applications in the control of ac and dc power, including the zero-crossing switch described at the end of this section.

The Silicon-Controlled Rectifier (SCR)

The SCR is a four-layer pnpn device. Its structure and symbol are illustrated in Figure S3–15. The three adjacent layers starting at the anode form a pnp transistor,

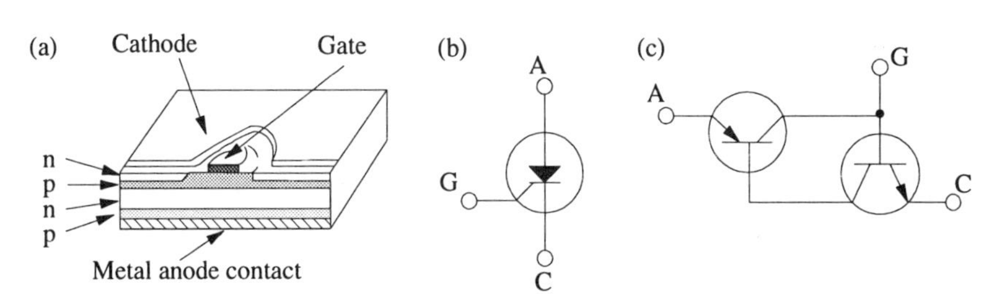

Figure S3–15. Silicon-controlled rectifier (SCR). (a) Structure; (b) circuit symbol; (c) two-transistor equivalent. G, gate; A, anode; C, cathode.

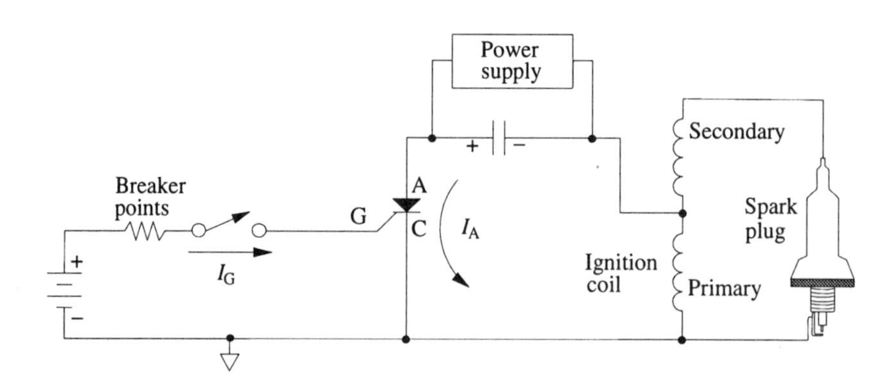

Figure S3–16. Solid-state automobile ignition system. A small gate current I_G switched to the SCR by the breaker points causes the SCR to conduct and discharges the capacitor through the primary turns of the ignition coil. The high voltage produced in the secondary causes the spark.

and the three starting at the cathode constitute an npn transistor. In the two-transistor representation of the SCR shown in Figure S3–15c, the pnpn structure is considered a complementary npn–pnp transistor pair.

If the anode of the SCR is connected to the positive terminal and the cathode is connected to the negative terminal of the power supply, the center np junction is reverse-biased, both transistors are off, and the SCR does not conduct. If base current is then supplied to the npn transistor by way of the gate terminal, the npn transistor turns on. The collector of the npn transistor draws current from the base of the pnp transistor and turns it on. The on pnp transistor provides current through its collector to the base of the npn transistor; this current keeps the npn transistor on even after the original gate current is removed. Thus, all that is needed to initiate conduction is a pulse of gate current, and all that is needed to maintain conduction is a minimum anode–cathode current called the **holding current**. To stop conduction in the SCR, the current is reduced below the holding current by removing the anode-to-cathode supply voltage, or if the supply is ac, the SCR turns off when the supply reverses polarity. A few thyristors known as **gate turn-off switches** are turned off by withdrawing current from the gate, but for normal SCRs this is an inefficient method of stopping conduction. Because the SCR conducts in only one direction, it is classified as a **reverse-blocking triode thyristor**.

Figure S3–16 illustrates the use of an SCR in a solid-state automobile ignition system. The capacitor is charged from the power supply. When the spark plug is to fire, the breaker points close momentarily and a small gate current is applied to the SCR. The SCR turns on, and the charged capacitor discharges through the few primary turns of the ignition coil. The sudden surge of current in the ignition primary generates the very high voltage in the secondary of the coil that produces the spark. As soon as the capacitor has discharged, the current through the SCR falls below the holding current, the SCR turns off, and the charging cycle begins again. In an ordinary ignition system, the points wear out because they are required to switch the full ignition coil current.

The Triac

A **bidirectional triode thyristor**, unlike the SCR, can conduct in both directions. The **triac**, an example of such a device, behaves much like two SCRs connected in a head-to-toe manner, as the triac circuit symbol and circuit model in Figure S3–17 show. Because the triac can conduct in both directions, the terminals are labeled MT1 and MT2 (main terminals 1 and 2) rather than anode and cathode. The gate signal is connected to both SCRs to allow conduction to be triggered by either positive or negative gate currents.

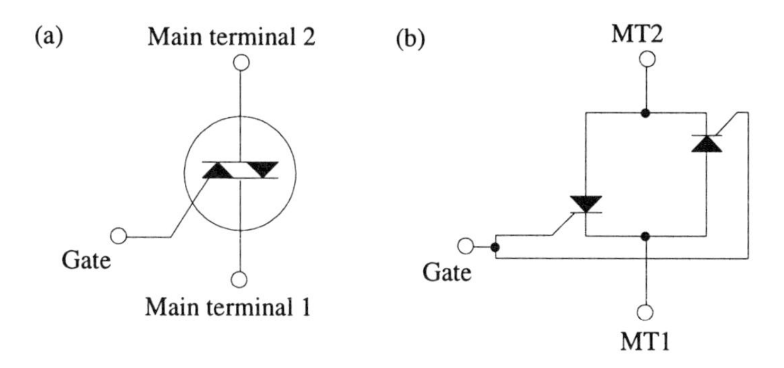

Figure S3–17. Triac. (a) Circuit symbol; (b) circuit model.

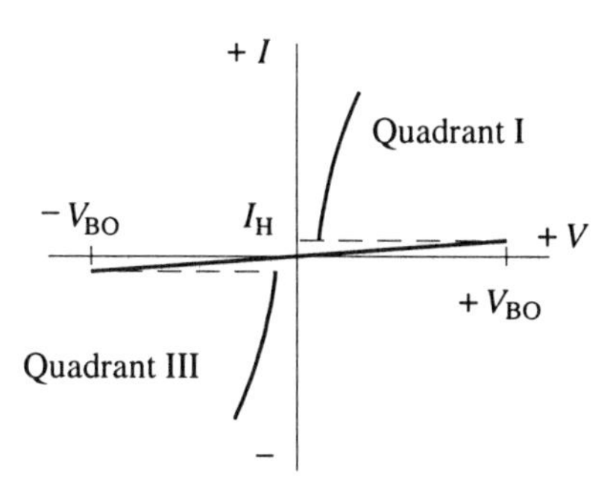

Figure S3–18. Current–voltage curve of the triac.

The current–voltage curve of the triac is illustrated in Figure S3–18. In quadrant I, the voltage at MT2 is positive with respect to that at MT1. The triac can be triggered to conduct forward current at voltages less than the break-over voltage V_{BO} by a control current at the gate. In quadrant III, the voltage at MT2 is negative with respect to that at MT1, and conduction can occur in the reverse direction when the triac is triggered. In normal triac use, the ac voltage to be switched should always be less than V_{BO}, because a voltage that exceeds V_{BO} will cause conduction that is not controlled by the gate. The triac remains in its conducting state until the main terminal current falls below the holding current I_H.

The triac is an extremely versatile switch. It can be triggered by either positive or negative dc currents, by ac currents, or by pulses. Triacs are used in power control applications when full-wave control is desirable. Figure S3–19 shows a triac used as the control element of a solid-state lamp dimmer. A trigger circuit gates the triac into conduction at a time during each half-cycle determined by the setting of the control dial. This method of power control is known as **phase control**, because the triac is triggered at a specific point in the cycle (phase) of the ac waveform. Here,

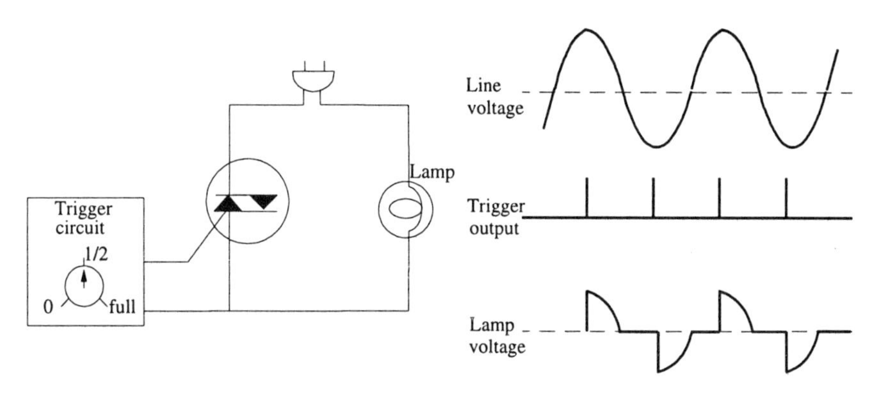

Figure S3–19. Triac lamp dimmer.

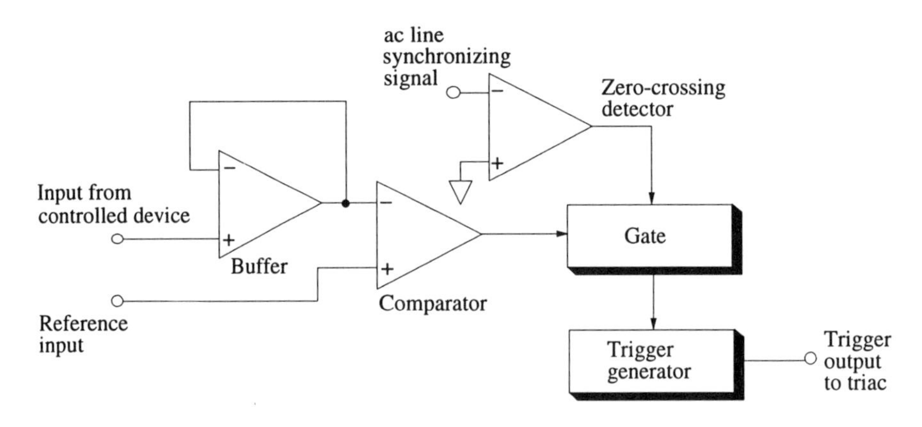

Figure S3–20. Zero-crossing switch. The trigger generator produces a triac trigger pulse the first time that the ac line voltage crosses zero after the control signal falls below the reference level.

for the half-power setting the triac is triggered into conduction in the middle of each cycle. This method of regulation is called **full-wave phase control** switching. Phase control regulation is much more efficient than current control by resistive devices that convert the unused power (I^2R) into heat.

Zero-Voltage Switches

Whenever a significant current is switched on or off very suddenly, as it can be with solid-state switches, the inductance of the conducting path acts as an antenna generating a broad-band, high-frequency electromagnetic noise called radio frequency noise or **RF noise**. The generation of RF noise in ac power switching can be minimized by turning the switch on or off when the ac current is zero. The combination of a circuit to sense when the switched ac waveform crosses zero and an SCR or triac power switch can provide an almost ideal zero-crossing switch. A typical integrated circuit zero-crossing triac controller is shown in Figure S3–20. In addition to the zero-voltage sensing and trigger circuits, it contains an input comparator and a buffer. Here, a voltage signal from the device being controlled is compared to a reference signal. When the control input falls below the reference input, the comparator output changes state and enables the gate to pass the zero-crossing detector output to the trigger generator. Thus, the triac is triggered into conduction on the first zero crossing after the control signal falls below the desired level. Note that the zero-crossing detector is a voltage comparator that senses when the ac *voltage* crosses 0 V. For this zero crossing to correspond to the time of zero current, the load controlled by the triac must be resistive, that is, nonreactive. Encapsulated modules that include a zero-voltage switch controller and a triac (often optically isolated) are commercially available. These **solid-state ac relays** are rugged and easy to use and respond to standard TTL control signals.

S3–5 Transistor Amplifiers

In amplification, an active electronic device such as a BJT or an FET is used to control the voltages and currents from the power supply that are applied to the load,

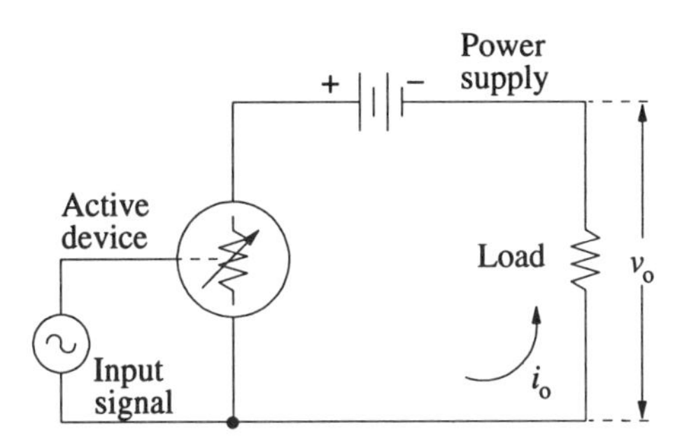

Figure S3–21. Principle of amplification. The active device (BJT or FET) controls a large voltage or current from a power supply in response to the small input signal.

as shown in Figure S3–21. The BJT or FET control element is actuated by the input signal. The amplifier gain is a result of the small voltage or current required by the control element to control larger voltages or currents from the power supply.

Basic JFET Amplifiers

In Figure S3–22, a silicon n-channel JFET is shown in a basic common-source configuration. The term **common source** indicates that the source terminal of the JFET is common to both the input and the output terminals of the amplifier.

Majority carriers (electrons for the n-channel silicon FET) flow from the source S to the drain D because of the drain supply voltage V_{DD}. Because the gate–channel junction is reverse-biased (in this example by a voltage source V_{GG}), I_G is very small (10^{-8} to 10^{-13} A). Therefore, the gate–source voltage v_{GS} is equal to $v_{in} + V_{GG}$. The current i_D through the channel of the JFET is controlled by v_{GS}, and v_{GS} thus controls the voltage drop $i_D R_D$ across the drain load resistance R_D.

The output voltage v_o is equal to the difference between the supply voltage V_{DD} and the voltage drop across R_D caused by the drain current. Thus,

$$v_{DS} = v_o = V_{DD} - i_D R_D \qquad \text{(S3–5)}$$

The JFET amplifier behaves like a voltage generator controlled by the input voltage. The input voltage v_{in} can be much smaller than the output voltage v_o. Hence, the JFET amplifier can provide voltage gain. Because the output voltage changes are opposite in polarity to the input voltage changes, the JFET common-source amplifier is an inverting amplifier. The JFET amplifier provides high input resistance.

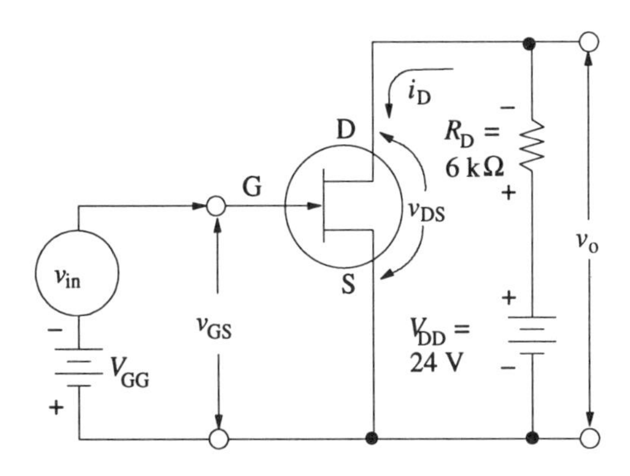

Figure S3–22. Common-source JFET amplifier.

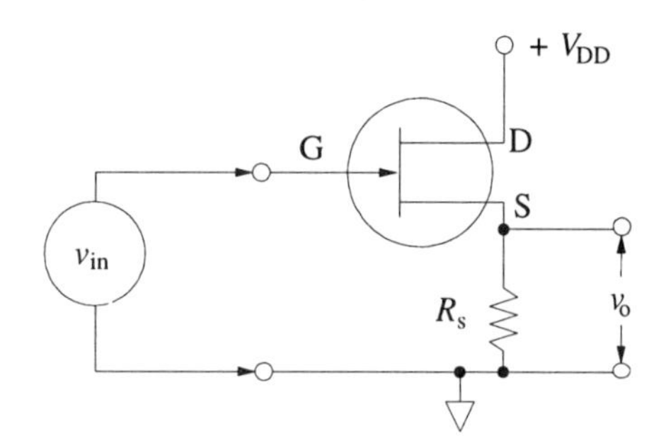

Figure S3–23. JFET source follower. The amplitude is noninverting, and the gain is nearly unity.

In the **source follower** circuit shown in Figure S3–23, the load resistor is connected to the source rather than to the drain. The output voltage v_o is equal to $i_D R_s$, and the input voltage v_{in} is equal to $v_{GS} + i_D R_s$. Because $i_D R_s$ is much larger than v_{GS} for reasonable values of R_s, most of the input variation appears across R_s. This amplifier circuit is noninverting and has a gain of approximately unity.

Basic BJT Amplifiers

A basic BJT common–emitter amplifier is shown in Figure S3–24. The principle of amplification is analogous to that in the JFET amplifier, but one great difference between the BJT and the FET amplifiers is the appreciable base current required by the BJT for control of the emitter–collector current. The base current i_B and the base–emitter voltage v_{BE} influence the charge distribution and thus the operating characteristics of the BJT. As a result, the characteristics of the output circuit (emitter to collector) depend on certain parameters of the input circuit. Similarly, conditions in the output circuit have a small effect on the characteristics of the input circuit. To a first approximation, this latter complication can be neglected. In circuit operation the BJT amplifier behaves as if it were a current generator controlled by the input current. The basic BJT common–emitter amplifier thus provides current gain. It also provides voltage gain, phase inversion of the input signal, and relatively low input resistance.

The other BJT amplifier configurations are the common–collector and the common–base. The common–base configuration provides low input resistance and very high output resistance and is noninverting. It has almost unity current gain and a potentially high voltage gain and finds use as a constant-current source and as an impedance-matching stage.

The common–collector amplifier is usually called an **emitter follower** amplifier. Its operation is very similar to that of the JFET source follower. The

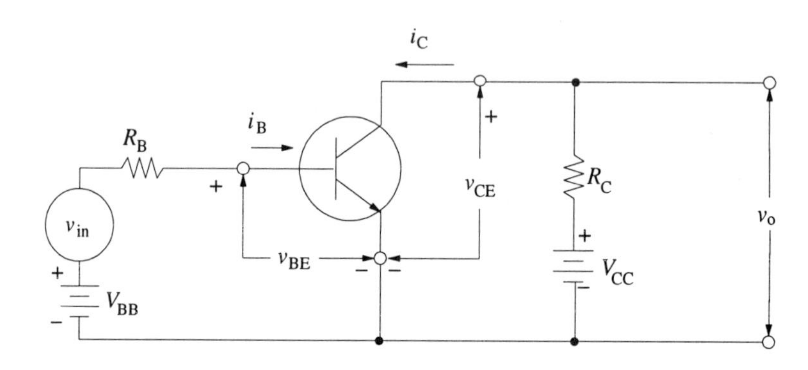

Figure S3–24. BJT common–emitter amplifier.

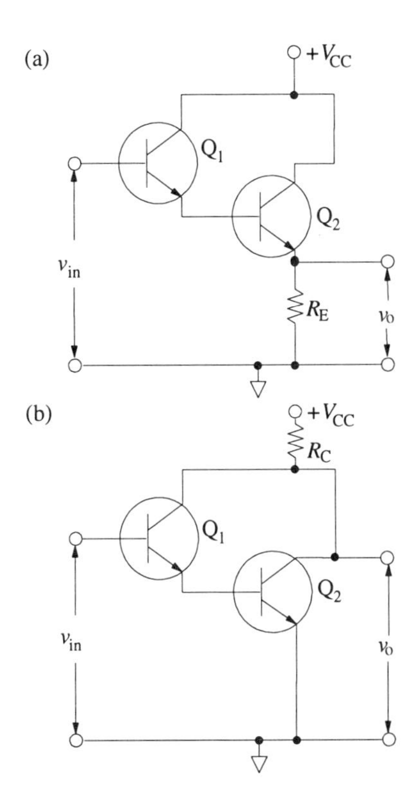

Figure S3–25. Darlington pairs. (a) Emitter follower Darlington amplifier; (b) Darlington voltage amplifier.

emitter follower circuit has the desirable characteristics of relatively high input impedance, low output impedance, and nearly unity voltage gain. It is a noninverting configuration and is commonly used as a buffer amplifier or an output stage.

The transistor pair connected as shown in Figure S3–25 is known as a **Darlington pair**. In this configuration the collectors of the pair are connected together, and the emitter of the input transistor Q_1 is connected to the base of the second transistor Q_2. The Darlington pair can function either as an emitter follower (Figure S3–25a) or as a voltage amplifier (Figure S3–25b). The Darlington emitter follower provides a higher input impedance, lower output impedance, and more nearly unity voltage gain than its single-transistor equivalent. The voltage amplifier configuration provides the voltage gain of the common–emitter amplifier while retaining the high input impedance of the emitter follower amplifier. Darlington pairs are commercially available in single packages.

Difference Amplifier

An amplifier whose output is a function of the difference between two input voltages is a **difference amplifier**. Ideally, the output of a difference amplifier is not responsive to common-mode voltages, temperature variations, and supply voltage fluctuations.

A basic FET difference amplifier is shown in Figure S3–26. The sources of the two FETs are connected through source resistances R_{s1} and R_{s2} to a constant-current source made from resistance R_{cm} and the negative supply voltage $-V$. The differential output v_{od} is taken from drain to drain as illustrated. An effective differential amplifier should have high gain to a difference signal but low gain to any common-mode signal (high common-mode rejection ratio). The common-mode

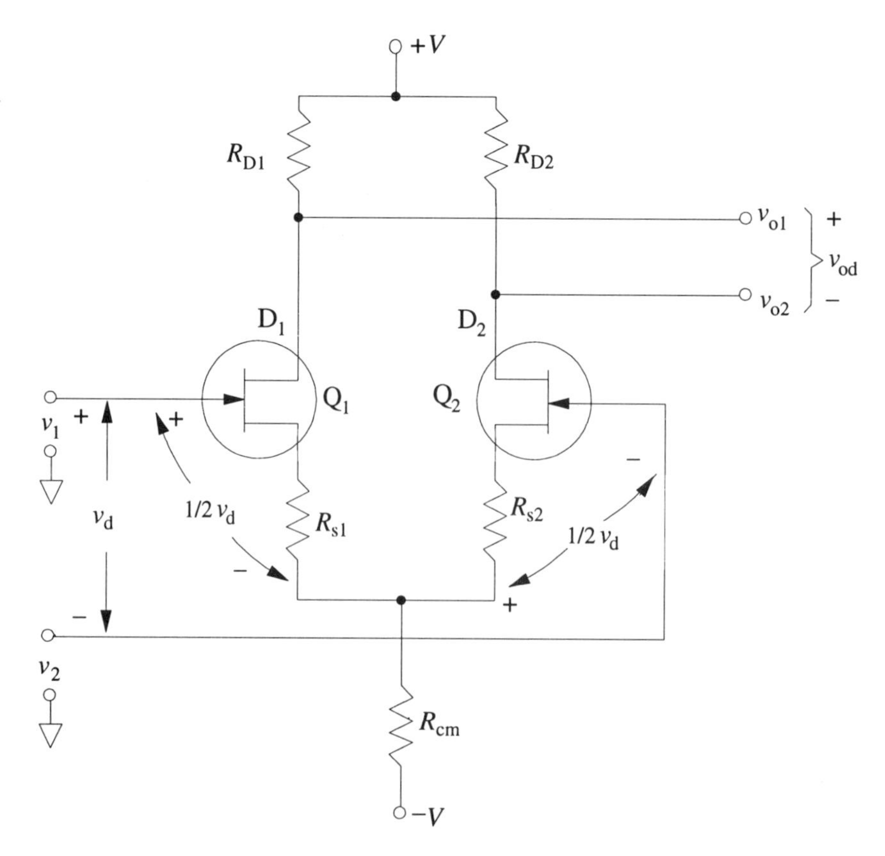

Figure S3–26. JFET difference amplifier. The constant-current source composed of R_{cm} and $-V$ makes the output insensitive to a common-mode signal. The difference voltage v_d is divided equally across each stage and produces equal and opposite current changes in the drain load resistors. This produces equal and opposite output voltages from each FET.

response is determined largely by the quality of the constant-current source. A common-mode signal should produce a negligible change in the currents through the drain resistors and thus produce a correspondingly low change in output voltage. To obtain a bias current that is independent of the common-mode signal, R_{cm} should be very large. Alternatively, the current source can be controlled to be constant by an active circuit.

When a difference signal v_d is applied between the two inputs, one-half of v_d appears across each stage when the resistances and FETs are matched (Figure S3–26). This equal division of the differential input signal produces equal and opposite current changes in the two source resistors and no change in the overall current supplied by the constant current source. The resulting current changes in the drain load resistors produce equal and opposite output voltages. The differential output voltage is then

$$v_{od} = v_{o1} - v_{o2} = A_1(v_d/2) - A_2(v_d/2)$$

where A_1 and A_2 are the voltage gains of FETs Q_1 and Q_2, respectively. If Q_1 and Q_2 are well-matched, A_1 and A_2 are identical and equal to A. The difference output v_{od} is equal to Av_d.

The difference amplifier is widely used as an input stage because of its ability to reject common-mode signals, its relative insensitivity to environmental changes, and its amplification of true difference signals. A difference amplifier is illustrated in the next section as the input stage for a general-purpose operational amplifier.

S3–6 The Operational Amplifier (op amp)

The op amp introduced in Chapter 5 is used in hundreds of different ways to provide elegant solutions to measurement and control problems. It is, of course, not necessary to know what is inside the op amp block (represented simply by a triangular symbol) in order to use an op amp profitably. However, most users feel more comfortable when they have some idea of how op amps are designed, and certainly the user can appreciate specifications and limitations better after investigating the internal operation of a typical op amp. Therefore, this section is devoted to a look inside the triangular symbol. Op amp specifications are summarized in Chapter 7, Section 7–4.

The op amp that we have chosen to investigate is constructed on a semiconductor microchip. This integration of several basic amplifier circuits into a single chip has resulted in units no larger than and costing no more than a single transistor. The general-purpose op amp is to function as a gain block with dc integrity, high input impedance, low output impedance, and wide-band frequency response. The ways in which these characteristics are introduced can be understood by studying Figure S3–27, a simplified circuit diagram of an integrated circuit op amp.

The input circuit is a differential amplifier using matched p-channel JFETs as source followers. An external 10-kΩ potentiometer can be used to balance the differential amplifier and compensate for any offset. The differential input provides for the inverting and noninverting inputs and for an excellent common-mode

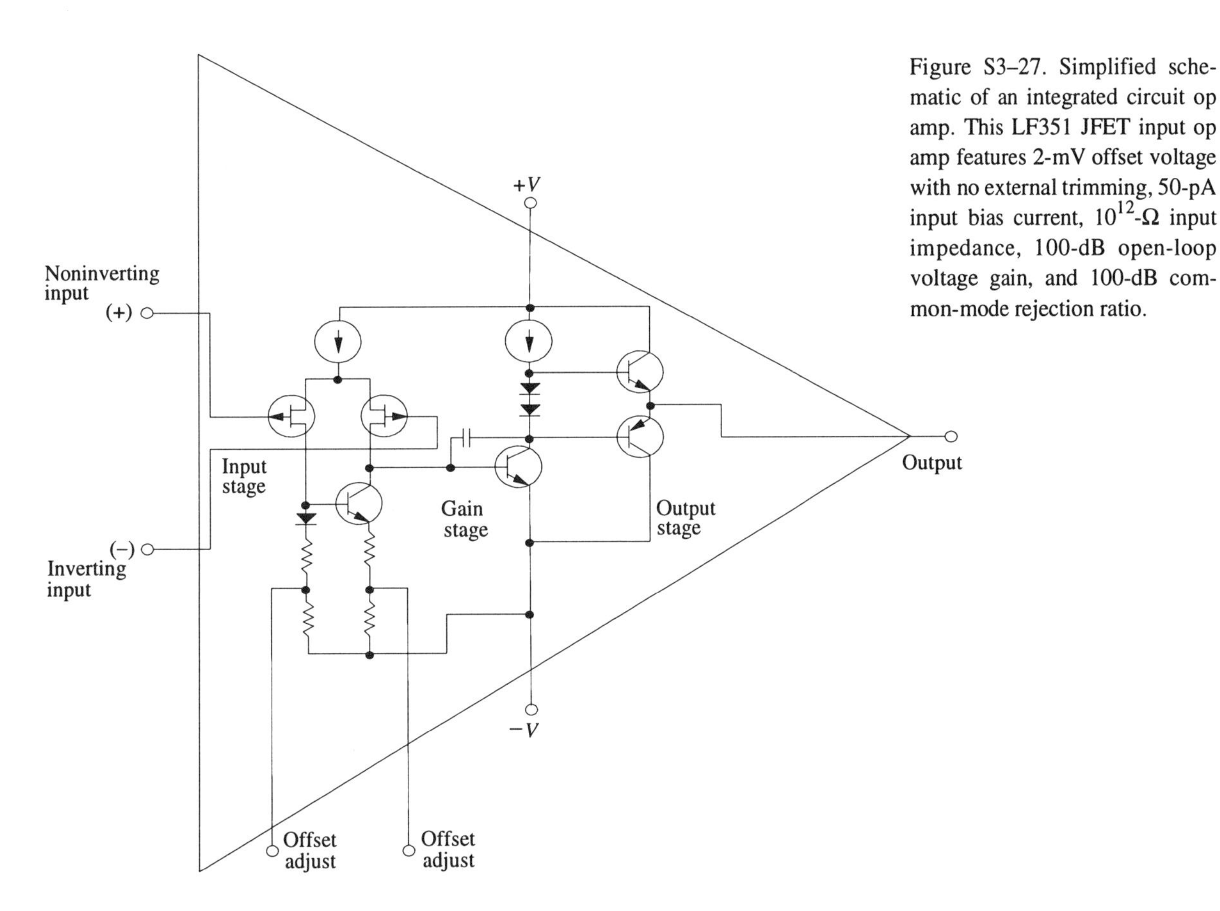

Figure S3–27. Simplified schematic of an integrated circuit op amp. This LF351 JFET input op amp features 2-mV offset voltage with no external trimming, 50-pA input bias current, 10^{12}-Ω input impedance, 100-dB open-loop voltage gain, and 100-dB common-mode rejection ratio.

rejection ratio. The JFETs also provide a very high input impedance and a very low input bias current.

BJTs (shown as a single transistor in Figure S3–27) are coupled together to produce a second stage with high current gain. This stage is usually a Darlington pair with a current gain approximately equal to the product of the β values of the two transistors.

The output stage is a cascaded complementary emitter follower circuit. It has the necessary low output impedance (<1 Ω with negative feedback) for driving external loads.

Appendix A

Grounding and Shielding

The quality of electronic measurement and control systems often depends directly on the care taken by the designer and the user in minimizing unwanted noise and pickup. Interference noise problems are often difficult to assess, and their elimination remains something of an art; however, following the basic guidelines developed in this appendix when designing and interconnecting system components can solve a large fraction of all noise problems and eliminate much of the frustration and time involved in tracking down an interference problem that appears during use.

A–1 Grounding

Voltage is not an absolute quantity but is the potential difference between two points. To establish and maintain reproducible and safe voltages in a circuit, a stable reference point from which all voltages are measured must be established. This single stable reference point is called the circuit common. When a circuit is linked to other circuits in a measurement system, the commons of the circuits are often connected together to provide the same common for the entire system. The circuit or system common may also be connected to the universal common, earth ground, by connection to a ground rod, water pipe, or power line common (*see* Supplement 1, Section 1–3).

The term "ground" has come to be a general term for the system common whether it is connected to earth ground or not. Thus, the verb "to ground" may mean to connect either to common or to earth ground. The more specific terms "common" and "earth ground" are preferable, but because they lack convenient verb forms, they are not often used. Compound terms such as "ground loops" and "ground plane" also apply to a system common as well as to an earth ground. Thus, whenever the term "ground" is used, the context must be studied to determine exactly what is meant.

Safety Grounds

For safety reasons the chassis of electronic equipment must be grounded. If it is not, stray impedances between a voltage source and the chassis and between the chassis and ground can cause a fraction of the source voltage to appear on the chassis, presenting a shock hazard. An even more dangerous situation could arise if the insulation between the ac power line and the chassis were to break down and the ac line were to contact the chassis. If the ac line can be fused, the chassis is at the line voltage and capable of supplying an amount of current limited only by the fuse. Grounding the chassis, however, causes the fuse to blow when a breakdown of the power line insulation occurs, removing the danger. Safety grounds are always connected to earth ground; signal commons may or may not be earth grounded.

2861-2/94/0401$08.00/1 Published 1994 American Chemical Society

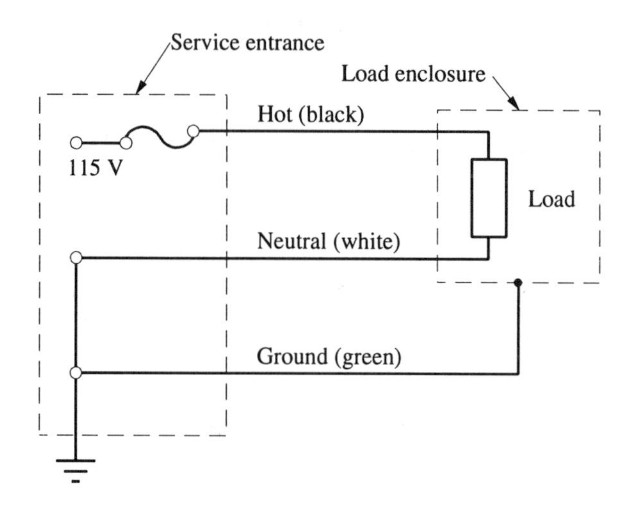

Figure A–1. Standard three-wire U.S. 115-V ac power distribution. The hot wire (black) is fused, and any load current is returned through the neutral wire (white). The ground wire (green) only carries current during a fault and then only until the fuse or a circuit breaker interrupts the circuit.

In the United States, the National Electrical Code requires the three-wire 115-V ac power distribution system illustrated in Figure A–1. Enclosures should be connected to the safety ground (green), because it carries no current and is always at ground potential. The neutral and safety ground are connected together only at the point where the ac power enters the building.

Signal Commons

When signal commons are considered, it is important to remember that no conductor is perfect—that is, all conductors have inductance and resistance—and that two physically separate commons or earth grounds are seldom, if ever, at the same voltage.

The commons of low-frequency analog circuits are usually best interconnected at a single point. Use of a single-point common to eliminate ground loops is shown in Figure A–2 for an op amp inverting amplifier. A ground loop is particularly troublesome if the two commons are unstable with respect to each other. It is important that the connections to the single common point have very low resistance and high current-carrying capacity so that ohmic drops along the connections are minimized. Typically, a large copper wire or foil is used. This is particularly important when several connections are made to a single common and when some of the connections are long, as they would be when the signal source must be remote from the measurement circuits. Even so, at radio frequencies the resistance is increased by the "skin effect," and inductive reactance can be very large.

When several circuits share a single-point common, a series or parallel connection can be made, as illustrated in Figure A–3. In both cases the resistance to common must be very low, because ultimately, a single conductor must carry the sum of all the currents from every component in the system. It may in fact become impractical to have a single common point, because the current-carrying capacity cannot be provided. In this case it may be safer to have several stable common points and tolerate some ground loops. This sort of compromise is often necessary in solving the grounding problems associated with large installations and buildings, such as a computer center, or when the circuitry is subjected to interference that may cause large currents, such as that from electrical storms.

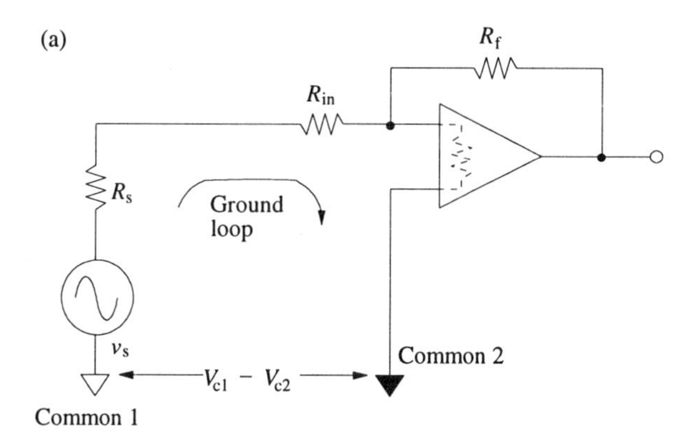

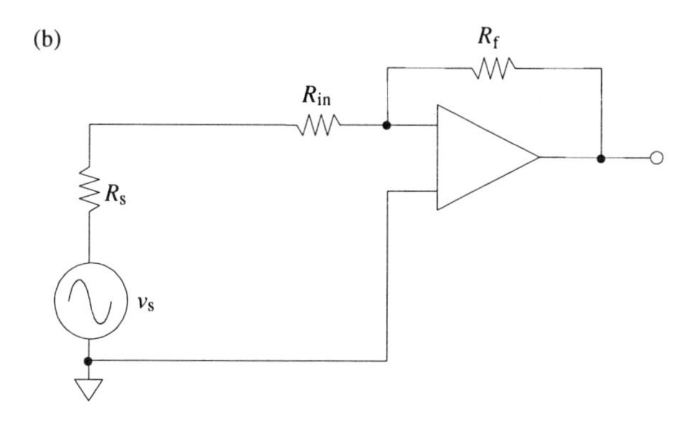

Figure A–2. Ground loops. In part a the signal common (common 1) is at a voltage different from that of the amplifier common (common 2). This gives rise to a ground loop and an erroneous signal. Ground loops can be eliminated by establishing a single common point as in part b.

In many laboratory measurement situations, it is not possible or practical to have a single common point, particularly if the signal source is remote from the measurement system. In these cases it is advantageous to use a differential or instrumentation amplifier, as discussed in Chapter 7, Section 7–2. Even though a potential difference exists between the signal common and the amplifier common, the erroneous signals generated by ground loops are common mode and are rejected by the difference amplifier (see Chapter 7, Figure 7–20). Therefore, it is unnecessary for the two common voltages to be stable with respect to each other. The input impedance of the difference amplifier should be large compared with the source impedance to keep the lines identical and thus retain high common-mode rejection.

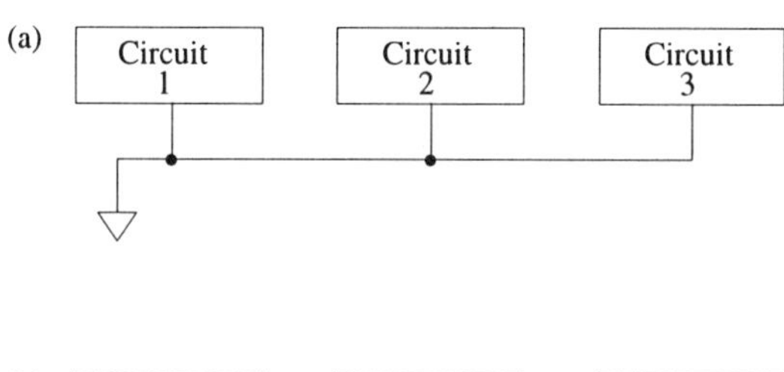

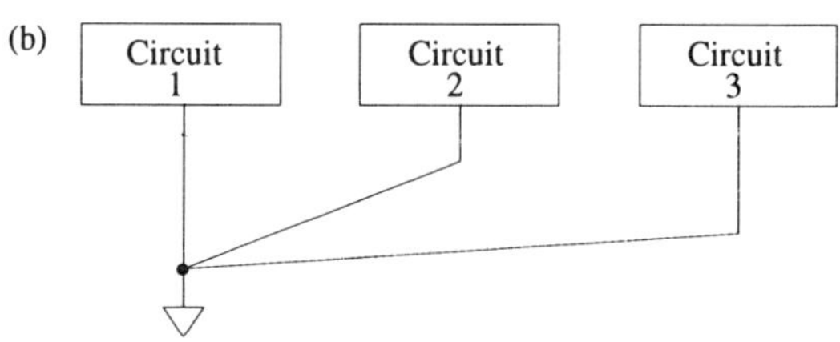

Figure A–3. Single-point commons. The series connection (a) is simpler from a wiring standpoint than the parallel connection (b). However, the parallel connection is less noisy, because currents to common from different circuits do not cross couple.

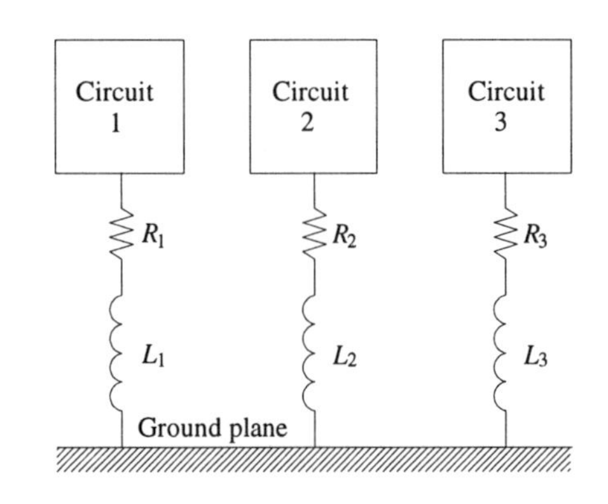

Figure A–4. Multipoint-common system for high frequencies. Circuits are connected to the closest low-impedance ground plane. Because a ground plane can be made with very low inductance, its impedance can be low compared with that of separate conductors connected to a single point. Each connection to the ground plane should be kept as short as possible (only a fraction of an inch in very high frequency circuits).

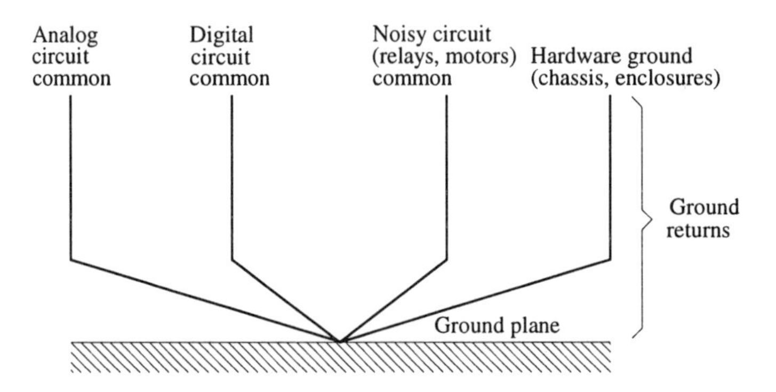

Figure A–5. Single-point connection for several circuit types. Noise coupling between circuits can be minimized by using separate ground returns connected at only one point. The ac power ground (green wire) is connected to the hardware ground.

Single-point-common systems suffer serious limitations at high frequencies (>1 MHz), because the inductive reactance of lengthy conductors to common can be large. Therefore, high-frequency circuits, including digital circuits, use a multipoint-common system, as illustrated in Figure A–4, to minimize the impedance. Such multipoint commons should not be used at low frequencies (<1 MHz), because the common currents from all circuits go through the same impedance of the ground plane and cross couple.

A high-quality common is necessary on circuit boards that contain digital integrated circuits (ICs) because of their rapid switching speeds. Either a low-impedance ground bus or a ground plane covering a large percentage (60% or more) of the board is satisfactory. If power buses are used, inductances can be kept low by making the bus as wide as possible (0.1 in. [ca. 0.3 cm] or more). Unused gate inputs should be connected to common or through a series resistor to the power supply voltage V_{CC}, whichever is appropriate.

Most complex systems require separate ground returns (connections to common) for circuits of widely different power and noise levels. Thus, several low-level analog circuits can share a common ground return that is different from the digital ground return. Electromechanical components that tend to generate noise, like relays and motors, can share a ground return line separate from the signal grounds, as illustrated in Figure A–5. Connecting the separate ground returns at a single point greatly minimizes most low-frequency grounding problems.

Power Supply Decoupling

The dc power supply and its distribution system are usually shared by a variety of circuits; thus, noise generated by one circuit can couple to other circuits through the power system. To minimize this coupling, it is good practice to place a capacitor across the power bus at each circuit. Decoupling analog circuits is very important, because any noise on the supply lines can add noise to the signal, influence amplifier gains, and possibly cause oscillations.

Digital circuits with totem pole output stages (*see* Chapter 6, Section 6–5) are particularly strong sources of power supply voltage transients. When a totem pole gate switches states, there is a short time in which both output transistors are on and a low-impedance path exists between the supply and the ground. This can result in power supply current spikes as large as 100 mA each time the gate changes state. A high-speed decoupling capacitor, typically 0.01 to 0.1 μF, at the IC package acts as an extra current source and helps prevent supply voltage transients during switching. Decoupling capacitors (10 to 100 μF) should also be used on each printed circuit board at the point of power entrance to the board. For capacitor characteristics, *see* Appendix B. Additional information on the avoidance of transmission line effects with TTL circuits is given in Appendix C, Table C–1.

A–2 Shielding

Another very important method for minimizing the pickup of unwanted interference noise is shielding. Shielding involves surrounding signal-carrying wires, components, circuits, or complete systems with a conducting material that is connected to common. Although shielding of cables is stressed here, the same principles apply to shielding amplifiers, components, or systems.

Shielded Cables

Several kinds of shielded cables are available for external signal connections and critical internal connections. Coaxial cable (Figure A–6) is useful from zero frequency (dc) to several hundred megahertz, because its characteristic impedance

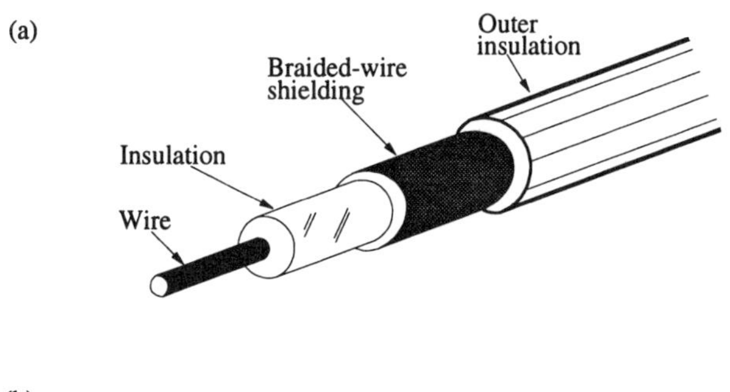

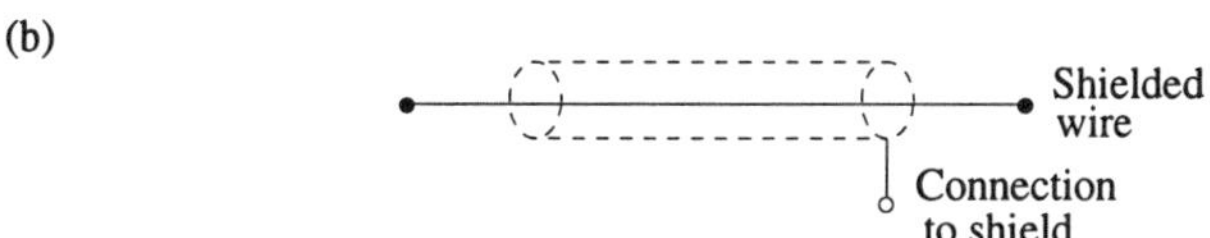

Figure A–6. Coaxial-cable construction (a) and schematic (b). The braided shield provides excellent electric field protection but only fair magnetic shielding. The shield should be terminated uniformly by a coaxial [BNC-, ultrahigh frequency (UHF)-, or N-type] connector.

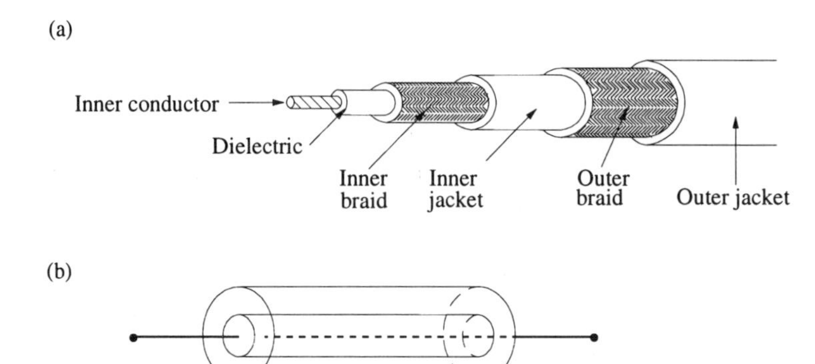

Figure A–7. Triaxial-cable construction (a) and schematic (b). The inner conductor is surrounded by an inner braided shield as in normal coaxial cable and by an outer shield. The outer braid is grounded and bypasses both ground loop currents and capacitive pickup.

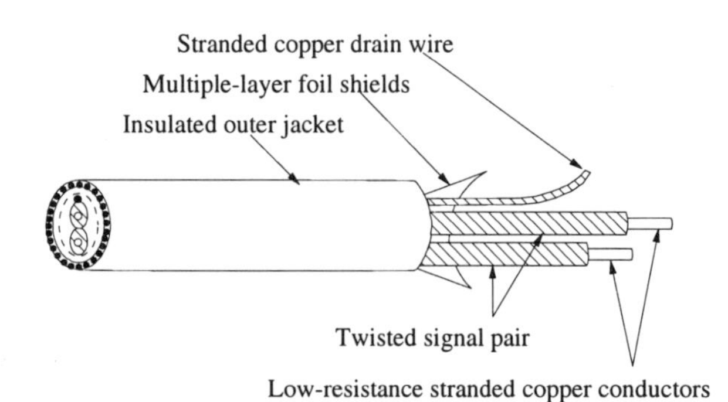

Figure A–8. Shielded twisted-pair cable. The foil shield provides better coverage and more effective magnetic shielding but less flexibility and durability than a braided shield.

is uniform over this range (*see* Appendix C). The coaxial cable provides good protection against other electric fields and capacitive pickup if the shield is grounded. However, the shield is part of the signal path, and grounding on both ends can cause ground loops. The triaxial (or triax) cable shown in Figure A–7 provides an additional outer copper braid that is insulated from the signal conductors and acts as a true shield. Triax cables are, unfortunately, expensive and rather awkward to use.

Shielded twisted pairs (Figure A–8) have characteristics similar to those of triaxial cables, but they are normally limited to signal frequencies from dc to 1 MHz. The inner conductors carry the signal current, and the shield carries the noise current. Twisting the two signal-carrying wires cancels randomly induced noise pickup and protects against capacitive coupling and magnetic fields. Any noise current in the shield couples equally to the two conductors by mutual inductance, and the induced voltages thus cancel.

Grounding of Cable Shields

A very important question that invariably arises when shielded cables are used is where to ground the cable shield so as to avoid ground loops and capacitive pickup. For low-frequency circuits where single-point commons are used, the shield should be grounded at only one point. The exact placement of the shield-to-ground connection depends upon whether the signal source or the receiver is grounded.

For illustrative purposes it will be assumed that a low-frequency (<1-MHz) signal source is being connected to an amplifier. If the signal source has one

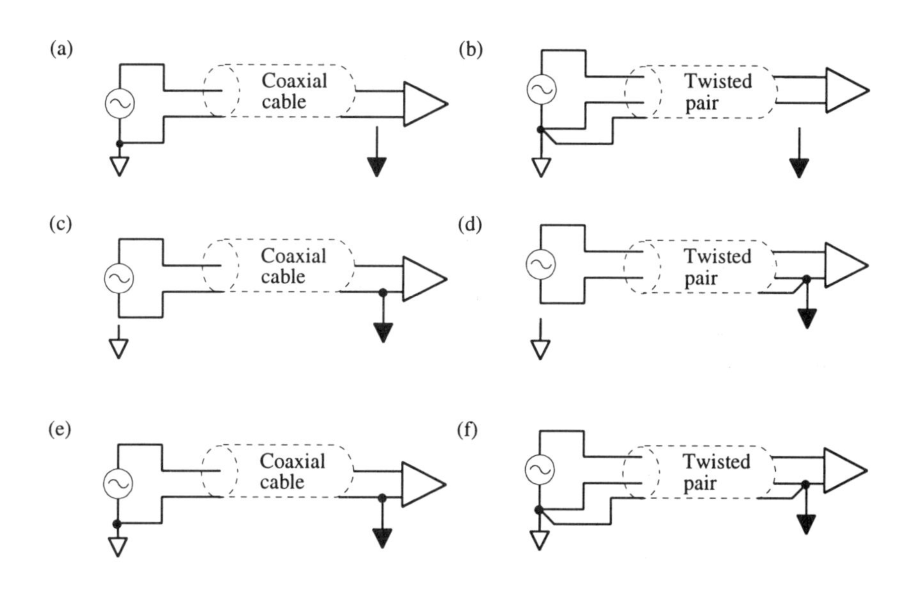

Figure A–9. Preferred connections of cable shields for grounded source (a, b), grounded amplifier (c, d), and grounded source and amplifier (e, f).

grounded lead but neither amplifier input is grounded, the shield should always be connected to the source common even if this is not at earth ground. Such a connection is shown in Figure A–9a for a coaxial cable and in Figure A–9b for a twisted-pair cable. The signal common is the only connection point that produces no noise voltage between the amplifier input terminals.

If the signal source is floating but one amplifier lead is grounded, the only shield connection that precludes a noise voltage between the amplifier input terminals is the amplifier common as shown in Figures A–9c and A–9d. If the signal circuit is grounded both at the source and at the amplifier, a ground loop results, and the amount of noise pickup depends on the susceptibility of the loop to electric and magnetic fields. The preferred connection of the shield ground in this case is shown in Figures A–9e and A–9f. In each case, grounding the shield at both ends forces some of the ground loop current through the shield rather than the center conductor or twisted-wire pair. To provide better noise immunity in this case, it is necessary to break the ground loop by an isolation transformer, an optical coupler, or a differential amplifier, as discussed next.

A–3 Isolation

Ground loops can be broken by transformer coupling between amplifier stages. For ac signals below about 5 MHz, a simple isolation transformer as shown in Figure A–10 effectively isolates the input circuitry from the output. Optical

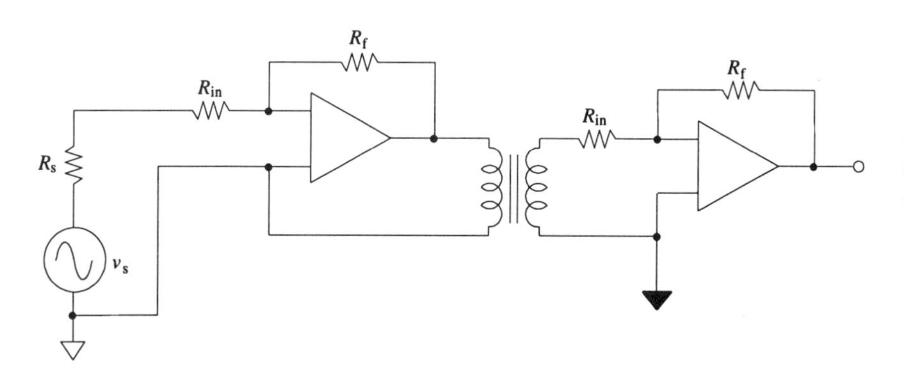

Figure A–10. Isolation transformer for ac signals. The two amplifier stages are transformer coupled to provide excellent isolation between the input and output circuits.

isolators as discussed in Chapter 8, Section 8–6, can provide excellent isolation between circuits. Optical isolators contain a light-emitting diode and a phototransistor or photodiode in the same package. They are especially useful for isolation of digital circuits, where linearity through the coupler is unnecessary.

Many analog manufacturers produce isolation amplifiers that are transformer or optically coupled. Isolation amplifiers can be used simply as op amps and yet provide excellent ohmic isolation between the signal source and the output. Transformer-coupled isolation amplifiers use modulation techniques so that their low-frequency response extends to dc. Many contain dc-to-dc converters (*see* Chapter 4, Section 4–6) so that the input amplifier power supply is also isolated from the output stage supply. When gain accuracy and linearity are most important, transformer-coupled amplifiers provide excellent characteristics; optically coupled amplifiers provide higher speed.

A–4 Radio Frequency (RF) Shielding

High-frequency interference in circuits is frequently referred to as **RF interference**. Many sources of RF interference can be found in laboratory environments. Spark sources, flash lamps, and gaseous discharges for lasers are but a few. RF interference can be quite serious, rendering many digital circuits completely inoperable.

Enclosing the sensitive circuit in a metal shield and using shielded cable can provide RF shielding. A conductor that has a large surface area (mesh or braid) makes an excellent RF ground. The shield should be terminated at both ends, as is a signal cable for high-frequency signals (*see* Appendix C). For best shielding, two separate shields such as are provided by a triaxial cable should be used.

A–5 Guarding

Many available differential amplifiers and measurement devices have an additional shield called a **guard shield.** The guard shield, which surrounds the amplifier or meter, should be held at a potential that prevents current through any unbalanced source impedance. This eliminates the differential input noise voltage. The rule to follow when connecting a guard shield is to ensure that no common-mode current occurs in any of the input resistances. This usually means that the guard should be connected to the source terminal with the lowest impedance to common. In addition, a shield around a high-gain amplifier should always be connected to the amplifier common.

Consider, for example, the measurement of the IR drop across the resistance R_s with a guarded digital voltmeter as shown in the circuit of Figure A–11. Connecting the guard to the low-impedance source terminal eliminates the noise current in the input circuitry of the meter.

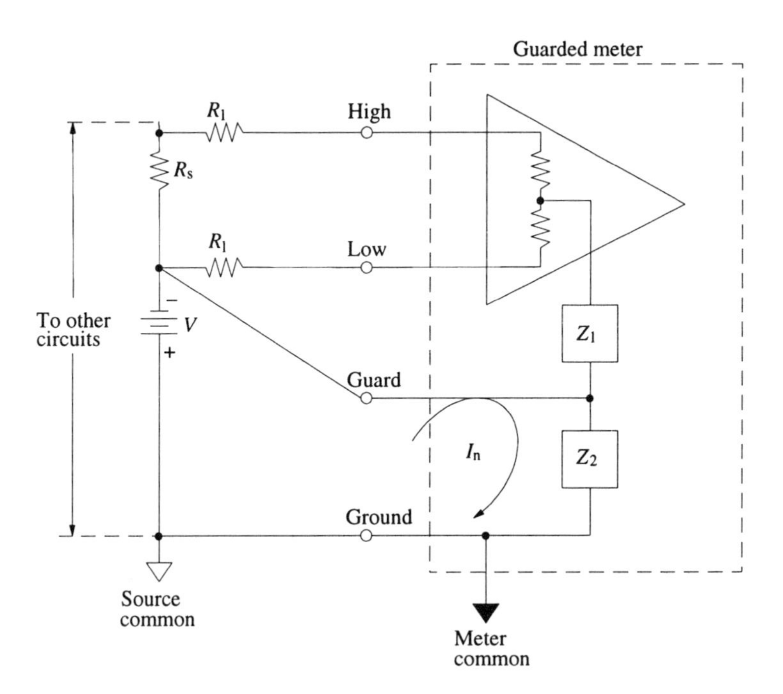

Figure A–11. Use of a guarded meter. Connect the guard to the low-impedance terminal of the source to avoid noise currents at the meter input. If the guard is connected to the source common or meter common, any noise in V or any induced noise causes a noise current I_n in the low meter lead. If the guard is connected to the low meter terminal, a noise current is produced in the line resistance R_l.

Appendix B

Components

B–1 Resistors

Commercial resistors are wire-wound, carbon composition, or film. As the name suggests, **wire-wound resistors** are made by winding resistance wire around an insulating form and providing contact at each end. Three types of wire-wound resistors are shown in Figure B–1. Wire-wound resistors can have high precision and high power dissipation capabilities.

A **carbon composition resistor** is made by forming a mixture of graphite powder, silica, and a binder into a solid cylinder pressed between two conductors, as shown in Figure B–2. The resistance is changed by varying the ratio of carbon to silica in the resistive element.

Film resistors are formed by depositing a thin conducting film on a cylindrical insulating ceramic substrate. Contact is then made to each end. For precision resistors, a spiral groove can be cut through the film to increase the resistance, as shown in Figure B–3. The automated cutter ceases when the desired resistance is attained. The conducting film in a film resistor can be metal, metal oxide, carbon, or cermet, which is a mixture of glass and metal alloys that is glazed onto an alumina substrate. The cermet film is relatively thick and very durable.

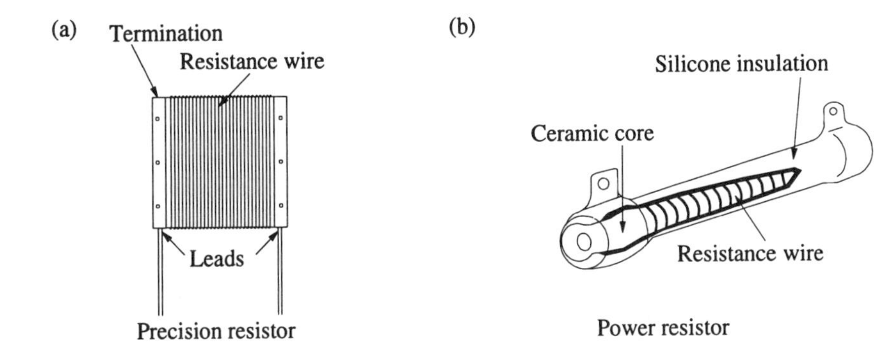

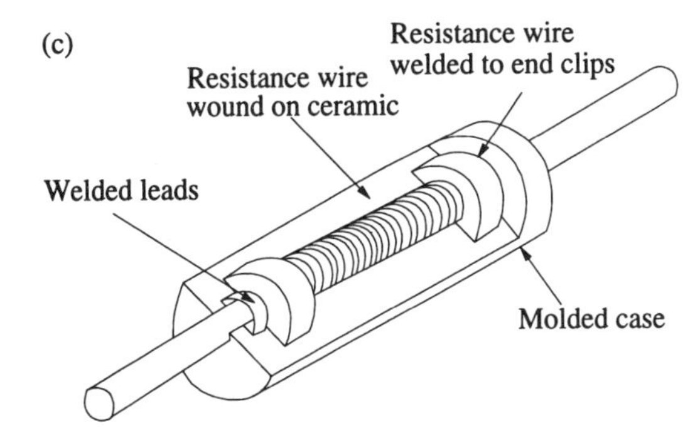

Figure B–1. Wire-wound resistors. (a) High precision; (b) high power with radial leads; (c) medium power or precision with axial leads.

2861-2/94/0411$08.40/1 Published 1994 American Chemical Society

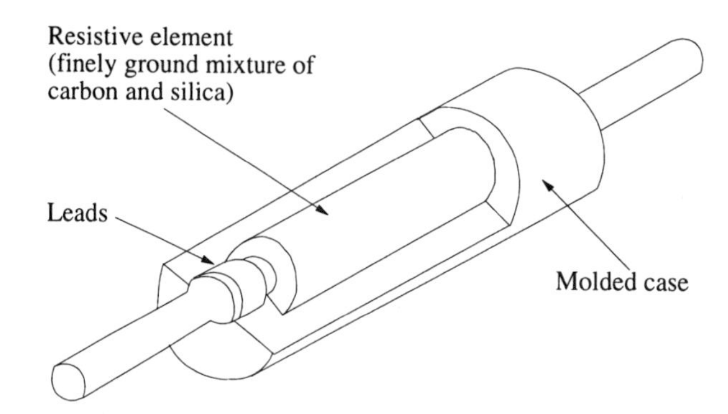

Figure B–2. Carbon composition resistor.

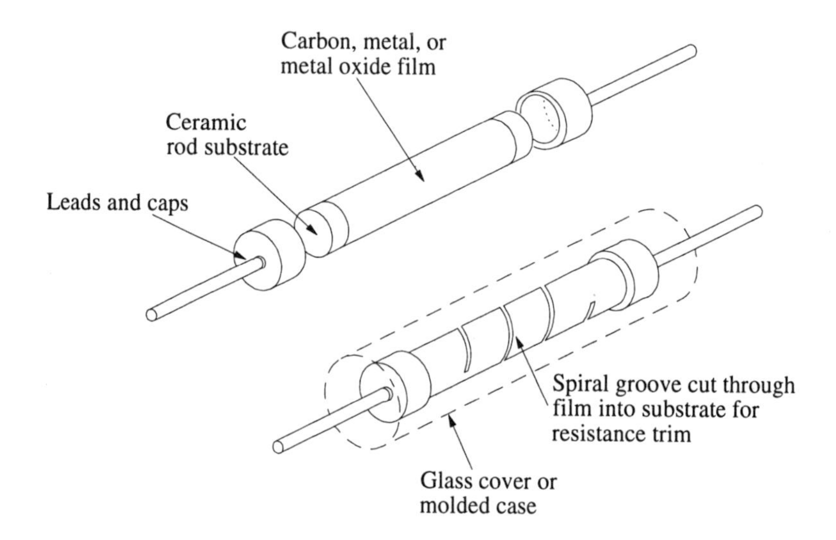

Figure B–3. Film resistor.

Tolerances

The tolerance of a resistor is the expected agreement of its actual and nominal resistances when purchased, expressed as a percentage of the nominal value. Resistor tolerances can be divided into four categories: general purpose, ±5 to ±20%; semiprecision, ±1 to ±5%; precision, ±0.1 to ±1%; and ultraprecision, ±0.01 to ±0.1%. Carbon composition resistors are semiprecision. Metal film resistors can achieve accuracies in the precision range, but only wire-wound resistors are capable of ultraprecision applications.

General-purpose resistors are available in ±5, ±10, and ±20% tolerances. The available values of these resistors are shown in Table B–1. The resistor values listed increase in increments of approximately 5%. This accounts for the nonuniform intervals and the lack of round numbers. Note that the two significant digits are the same in each decade. The resistances of general-purpose resistors are marked on them in a code of colored bands (Figure B–4). Each number from 0 to 9 is assigned a color. The color of the band closest to the end of the resistor represents the first figure of the resistance; the second band represents the second figure; and the third band represents the power of 10 by which to multiply the first two figures to get the total resistance. Thus, a resistor coded with yellow, violet, and red bands is $47 \times 100\ \Omega$ or 4.7 kΩ. A sequence of blue, gray, green is $68 \times 10^5\ \Omega$, or 6.8 MΩ; and a sequence of green, blue, black is 56 Ω. For resistances between 1 and

Table B–1. General-Purpose Resistor Values

1.0	**10**	**100**	**1000**	**10 k**	**100 k**	**1.0 M**	**10 M**
1.1	11	110	1100	11 k	110 k	1.1 M	11 M
1.2	**12**	**120**	**1200**	**12 k**	**120 k**	**1.2 M**	**12 M**
1.3	13	130	1300	13 k	130 k	1.3 M	13 M
1.5	**15**	**150**	**1500**	**15 k**	**150 k**	**1.5 M**	**15 M**
1.6	16	160	1600	16 k	160 k	1.6 M	16 M
1.8	**18**	**180**	**1800**	**18 k**	**180 k**	**1.8 M**	**18 M**
2.0	20	200	2000	20 k	200 k	2.0 M	20 M
2.2	**22**	**220**	**2200**	**22 k**	**220 k**	**2.2 M**	**22 M**
2.4	24	240	2400	24 k	240 k	2.4 M	
2.7	**27**	**270**	**2700**	**27 k**	**270 k**	**2.7 M**	
3.0	30	300	3000	30 k	300 k	3.0 M	
3.3	**33**	**330**	**3300**	**33 k**	**330 k**	**3.3 M**	
3.6	36	360	3600	36 k	360 k	3.6 M	
3.9	**39**	**390**	**3900**	**39 k**	**390 k**	**3.9 M**	
4.3	43	430	4300	43 k	430 k	4.3 M	
4.7	**47**	**470**	**4700**	**47 k**	**470 k**	**4.7 M**	
5.1	51	510	5100	51 k	510 k	5.1 M	
5.6	**56**	**560**	**5600**	**56 k**	**560 k**	**5.6 M**	
6.2	62	620	6200	62 k	620 k	6.2 M	
6.8	**68**	**680**	**6800**	**68 k**	**680 k**	**6.8 M**	
7.5	75	750	7500	75 k	750 k	7.5 M	
8.2	**82**	**820**	**8200**	**82 k**	**820 k**	**8.2 M**	
9.1	91	910	9100	91 k	910 k	9.1 M	

NOTE: Values for ±10 and ±20% are in **boldface** type. All values are available in ±5% tolerance.

Color	Figure	Multiplier	Tolerance
Black	0	1	
Brown	1	10	±1
Red	2	10^2	±2
Orange	3	10^3	
Yellow	4	10^4	
Green	5	10^5	±0.5
Blue	6	10^6	±0.25
Violet	7		±0.1
Gray	8		±0.05
White	9		
Gold		0.1	±5
Silver			±10
No band			±20

Figure B–4. General-purpose resistor color code.

Tolerance marking
Multiplier figure
2nd figure
1st figure

10 Ω, gold is used for the third band. Thus a sequence of orange, white, and gold is 3.9 Ω.

Precision resistors have standard values that are within 1% or less of each other. This results in many more values per decade, as shown in Table B–2. The color code markings for precision resistors are shown in Figure B–5. Four bands are used to give three significant digits and a multiplier. The wide band at the far end of the resistor indicates a precision resistor and gives the tolerance according to the table shown. The resistance of a precision resistor is most often printed on the body along with an alphanumeric code for other data about the resistor. The resistance is generally given as the digits and a multiplier, just as in the color code. Thus, 103 is 10×10^3 Ω = 10.0 kΩ, and 3322 is 332 $\times 10^2$ Ω = 33.2 kΩ.

Factors that can change the value of a resistor in applications include temperature variation, hours operated at high temperature, soldering, shock, overload, and moisture. In close-tolerance situations, these should be considered. Manufacturer's installation and handling recommendations should be followed, and excessive electrical, thermal, or mechanical stress should be avoided. The temperature coefficient of resistance may be quite significant. This is generally given as α_R, the relative change in R per degree Celsius from the value at 25 °C. For a composition resistor, α_R may be 0.005, which means that the resistance would change 0.5% per degree Celsius. A resistor that is 10 kΩ at 25 °C might be 11.25 kΩ at 50 °C. Metal film and wire-wound resistors have the lowest temperature coefficients of resistance.

Table B-2. Precision Resistor Values

1.00	**1.33**	**1.78**	**2.37**	**3.16**	**4.22**	**5.62**	**7.50**
1.01	1.35	1.80	2.40	3.20	4.27	5.69	7.59
1.02	**1.37**	**1.82**	**2.43**	**3.24**	**4.32**	**5.76**	**7.68**
1.04	1.38	1.84	2.46	3.28	4.37	5.83	7.77
1.05	**1.40**	**1.87**	**2.49**	**3.32**	**4.42**	**5.90**	**7.87**
1.06	1.42	1.89	2.52	3.36	4.48	5.97	7.96
1.07	**1.43**	**1.91**	**2.55**	**3.40**	**4.53**	**6.04**	**8.06**
1.09	1.45	1.93	2.58	3.44	4.59	6.12	8.16
1.10	**1.47**	**1.96**	**2.61**	**3.48**	**4.64**	**6.19**	**8.25**
1.11	1.49	1.98	2.64	3.52	4.70	6.26	8.35
1.13	**1.50**	**2.00**	**2.67**	**3.57**	**4.75**	**6.34**	**8.45**
1.14	1.52	2.03	2.71	3.61	4.81	6.42	8.56
1.15	**1.54**	**2.05**	**2.74**	**3.65**	**4.87**	**6.49**	**8.66**
1.17	1.56	2.08	2.77	3.70	4.93	6.57	8.76
1.18	**1.58**	**2.10**	**2.80**	**3.74**	**4.99**	**6.65**	**8.87**
1.20	1.60	2.13	2.84	3.79	5.05	6.73	8.98
1.21	**1.62**	**2.15**	**2.87**	**3.83**	**5.11**	**6.81**	**9.09**
1.23	1.64	2.18	2.91	3.88	5.17	6.90	9.20
1.24	**1.65**	**2.21**	**2.94**	**3.92**	**5.23**	**6.98**	**9.31**
1.26	1.67	2.23	2.98	3.97	5.30	7.06	9.42
1.27	**1.69**	**2.26**	**3.01**	**4.02**	**5.36**	**7.15**	**9.53**
1.29	1.72	2.29	3.05	4.07	5.42	7.23	9.65
1.30	**1.74**	**2.32**	**3.09**	**4.12**	**5.49**	**7.32**	**9.76**
1.32	1.76	2.34	3.12	4.17	5.56	7.41	9.88

NOTE: Values are given for only 1 decade. All other decades have the same significant digits. Values for ±1% values are in **boldface** type. All values are available in ±0.1, ±0.25, and ±0.5%.

Color	Tolerance, %
Black	
Brown	±1
Red	±2
Orange	
Yellow	
Green	±0.5
Blue	±0.25
Violet	±0.1
Gray	±0.05
White	
Gold	±5
Silver	±10

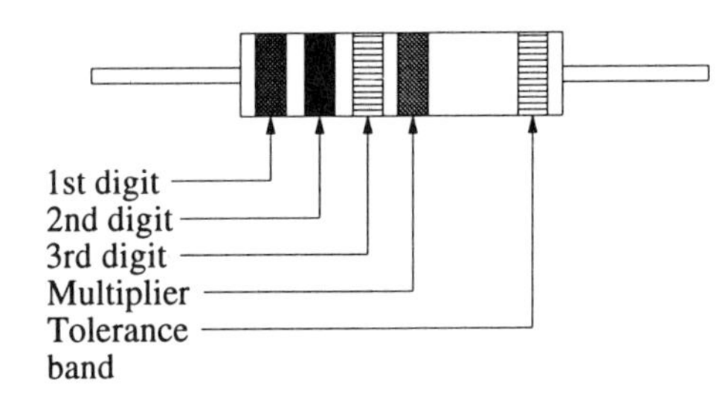

Figure B–5. Color code position for precision resistors.

Power

A resistor converts electrical power to heat. The amount of power thus converted can be calculated from

$$P = IV = I^2R = \frac{V^2}{R}$$

where P is the power in watts and I and V are, respectively, the current in and the voltage across the resistance R. When power is dissipated in a resistor, the resistor heats up until its rate of heat loss to the surroundings equals the rate of electrical heat conversion. Resistors differ in their abilities to withstand high temperatures and to lose heat to the surroundings. Increasing the size of the resistive element aids in both respects. The approximate sizes of composition resistors in four power ratings are shown in Figure B–6. For dissipation greater than 2 W, power resistors are used. Rather low gauge resistance wire is wound on a ceramic form that withstands high temperatures. Power resistors should be mounted so that they can readily dissipate heat to the air or to the instrument case.

Table B–3 summarizes the ratings of power dissipation, size, tolerance, temperature coefficient, and maximum voltage for a number of common resistor

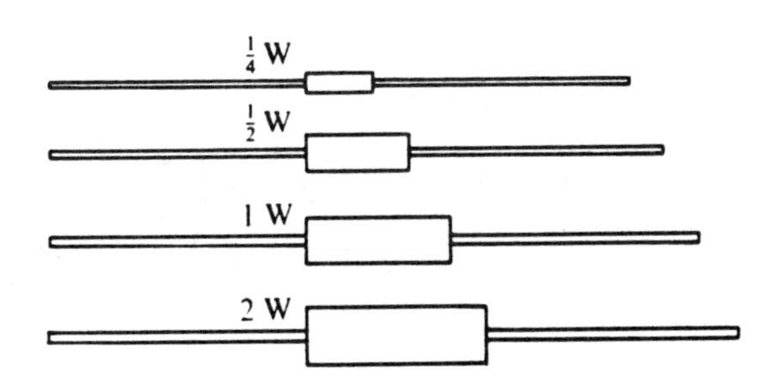

Figure B–6. Composition resistors of various power ratings drawn to actual size.

Table B–3. Resistor Properties

Resistive Material	Power Rating at $T_A = P_D$, W	Ambient Temperature T_A, °C	Body Dimensions, in. X_L	Body Dimensions, in. D	Resistance Range, Ω	Temperature Coefficient, $\alpha_R \times 10^6$, K^{-1}[a]	Resistance Tolerance, ±%	Maximum Working Voltage V_M, V
Metal film	1/20	125	0.150	0.065	10 to 2×10^5	−100 to +25	0.1, 0.5, 1	
Deposited carbon film	1/10	70	0.250	0.090	1 to 4×10^5	−250 to +500	1, 2, 5, 10	
Metal film	1/10	125	0.250	0.110	10 to 3×10^5	−100 to + 25	0.1, 0.5, 1	
Carbon composition	1/8	70	0.145	0.062	2.7 to 1×10^8	−5000 to +5000	5, 10, 20	
Deposited carbon film	1/8	70	0.375	0.125	1 to 3×10^6	−1500 to −250	1, 2, 5, 10	150
Cermet film	1/8	70	0.150	0.065	4.7 to 1.5×10^5	−200 to +100	1, 2, 5	
Metal film	1/8	125	0.375	0.125	25 to 1.5×10^6	−100 to +25	0.1, 0.5, 1	
Wire-wound precision	1/8	125	0.375	0.125	1 to 3×10^5	+20	0.05, 0.1, 0.5, 1	
Carbon composition	1/4	70	0.250	0.090	2.7 to 1×10^8	−5000 to +5000	5, 10, 20	250
Deposited carbon film	1/4	70	0.625	0.188	1 to 5×10^6	−1500 to −250	1, 2, 5, 10	250
Cermet film	1/4	70	0.250	0.090	10 to 1.6×10^5	−200 to +100	1, 2, 5	
Metal film	1/4	125	0.625	0.188	25 to 3×10^6	−100 to +25	0.1, 0.5, 1	
Wire-wound precision	1/4	125	0.750	0.250	10 to 4.5×10^5	+10	0.01, 0.05, 0.1, 1	
Carbon composition	1/2	70	0.375	0.140	1 to 1×10^8	−5000 to +5000	5, 10, 20	350
Deposited carbon film	1/2	70	0.750	0.250	1 to 1×10^7	−1500 to −250	1, 2, 5, 10	350
Cermet film	1/2	70	0.375	0.140	4.3 to 1×10^6	−200 to +100	1, 2, 5	
Metal film	1/2	125	0.750	0.250	10 to 4×10^6	−100 to +25	0.1, 0.5, 1	700
Wire-wound precision	1/2	125	1.000	0.375	10 to 1.2×10^6	+10	0.01, 0.05, 0.1, 1	
Carbon composition	1	70	0.562	0.225	2.7 to 1×10^8	−5000 to +5000	5, 10, 20	500
Deposited carbon film	1	70	1.062	0.375	1 to 1.5×10^7	−1500 to −250	1, 2, 5, 10	500
Cermet film	1	70	0.562	0.190	10 to 1×10^6	−200 to +100	1, 2, 5	
Metal film	1	125	1.062	0.375	25 to 4×10^6	−100 to +25	0.1, 0.5, 1	1000
Wire-wound power	1	25	0.250	0.085	0.5 to 1×10^3	+20 to +100	0.1, 0.5, 1	
Carbon composition	2	70	0.688	0.318	10 to 1×10^8	−5000 to +5000	5, 10, 20	1500
Deposited carbon film	2	70	2.188	0.375	2 to 1×10^8	−1500 to −250	1, 2	1500
Cermet film	2	70	0.688	0.318	10 to 1.5×10^6	−200 to +100	1, 2, 5	
Metal film	2	125	2.188	0.375	100 to 6×10^6	−100 to +25	0.1, 0.5, 1	
Wire-wound power	2	25	0.406	0.094	0.5 to 2.5×10^3	+20 to +100	0.1, 0.5, 1	
Cermet film	2–115	25			10 to 1×10^6	−500 to +500	1, 2, 5, 10	
Wire-wound power	2–250	25			0.1 to 2.7×10^5	+50	0.1, 0.5, 1, 3	

SOURCE: Reprinted with permission from Giacoletto, L. J. *Electronics Designers' Handbook*, 2nd ed.; McGraw-Hill: New York, 1977; p 3–38.

[a] $\alpha_R = \frac{1}{R} \left| \frac{\delta R}{\delta T} \right|_{T = 25°C}$

types. Metal film resistors withstand higher temperatures than composition and other film types. The rating is the maximum temperature of the air surrounding the resistor in which it can dissipate its rated power.

Series and Parallel Resistors

The resistance R_s of N resistors in series is $R_s = R_1 + R_2 + R_3 + \ldots + R_N$. The actual value of the resistance can differ from the nominal value by the error tolerance, which is generally a constant percentage of the nominal value. The error tolerance of resistors in series is then a combination of the individual error tolerances. If the resistances and tolerances are all equal, then the tolerance of the combination is the square root of the sum of the squares of the individual tolerances, because the variances are additive. If the resistances are unequal, the tolerance of the largest R often dominates. If all the tolerances are equal percentages, the tolerances of all smaller resistors may be negligible. The power dissipated in resistors in series is in direct proportion to the resistance, because P is equal to I^2R and the current is the same in each. Awareness of these facts can allow use of lower accuracy and lower power for the lower-valued resistors in a series circuit.

The combined resistance R_p of N parallel resistors is

$$R_p = \frac{1}{1/R_1 + 1/R_2 + 1/R_3 + \ldots + 1/R_N}$$

In the parallel case, the tolerance of the resistor of lowest resistance has the largest influence on the combined tolerance. The power dissipated in the resistors is in inverse proportion to their value, because P is equal to V^2/R and the same voltage is applied to all.

Noise

Two types of electrical noise, Johnson noise (*see also* Chapter 9, Section 9–2) and current noise, are generated in resistors. Johnson noise is caused by the random thermal motion of charge carriers in a conductor and is unavoidable. The root-mean-square amplitude of Johnson noise is $\bar{v}_{rms} = (4kTR\,\Delta f)^{1/2}$ where k is Boltzmann's constant, T is the temperature in degrees Kelvin, R is the resistance, and Δf is the bandwidth of the system influenced by the noise. This noise is clearly reduced at lower temperatures, lower resistance, and narrower bandwidth.

Current noise is a low-frequency noise caused by a current in a non-homogeneous conductor such as the carbon composition material. Like $1/f$ noise, current noise is inversely proportional to frequency. Current noise is considerable in carbon composition resistors, where it is generally the dominant noise source. For low-noise applications, metal film or wire-wound resistors should be used. The noise in these resistors approaches the theoretical lower limit of the Johnson noise equation.

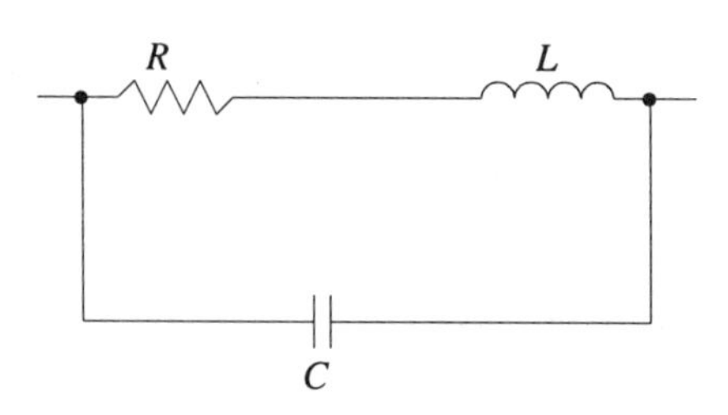

Figure B–7. Equivalent circuit of a practical resistor.

Frequency Characteristics

A practical resistor is not a purely resistive device. The inductance of its leads or windings and the capacitance of its end terminals result in the approximate equivalent circuit shown in Figure B–7. The consequence of the reactances is that the overall impedance is a function of the frequency of the current or voltage applied. The capacitance and inductance associated with the resistance vary considerably from one type of resistor to another. The carbon composition resistor has negligible inductance and a capacitance of 0.25 to 0.5 pF. The 3-dB point (*see* Supplement 2, Section S2–4) for a parallel *RC* circuit is $f_0 = 1/(2\pi RC)$. *If the* capacitance does not depend on the value of *R*, Rf_0 is a constant:

$$Rf_0 = 1/(2\pi C)$$

For a typical carbon composition resistor, the value of Rf_0 is about 0.5 MHz·MΩ. This means that the effective resistance of a 1-MΩ resistor is "down 3 dB"—that is, it is 707 kΩ—at 0.5 MHz. At 10 MHz, the highest value of *R* that can be used with less than 30% error is 50 kΩ. Carbon and metal film resistors have an Rf_0 constant of generally 3 to 5 MHz·MΩ. The slight series inductance increases the impedance at high frequencies and partially compensates for the capacitive decrease. Wire-wound resistors can have very high values of series inductance because of the coil-type winding of the resistance wire. Ultrahigh-precision wire-wound resistors are thus often truly precise only for dc circuits. Some wire-wound resistors use noninductive windings (two parallel coils wound in opposite directions). Such windings can extend the useful frequency range of wire-wound resistors into the low audio frequency range.

Variable Resistors

Some resistors are made to have an adjustable or variable resistance by exposing the resistive element so that a movable contact, or wiper, can contact the element at any point along the resistor. Such resistors are called rheostats or potentiometers. The value of the total or maximum resistance is usually stamped on the case. The resistive element is usually either a coil of resistance wire or a strip of resistive film. Several types of potentiometers are shown in Figure B–8. The word **rheostat** is used for high-wattage potentiometers, and a **trimmer** is a potentiometer used for an adjustment that is required only occasionally.

Potentiometers are made of the same materials as fixed resistors and so have similar qualities. In addition, they have several other characteristics. The linearity of a potentiometer often affects its accuracy in a given application. The deviation of the linear relationship between output resistance and degree of rotation is usually given as the maximum resistance deviation from a straight line relative to the total

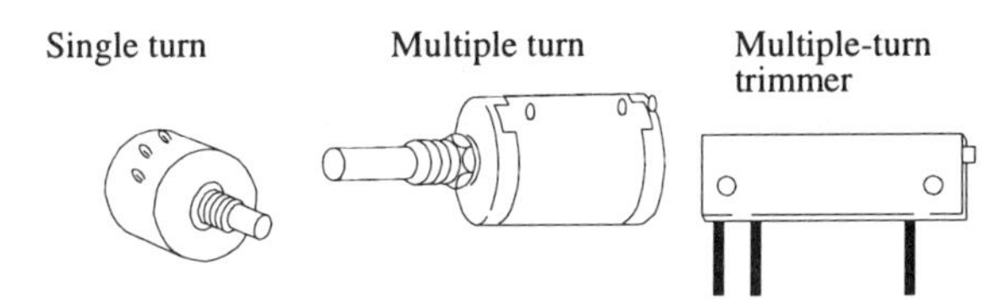

Figure B–8. Variable resistors.

(end-to-end) resistance. Potentiometers may require anywhere from a 270° turn to 25 complete turns to traverse the entire resistance range. Their lifetime is measured in the number of cycles (end-to-end rotations) they are designed to endure. Trimmers have a low cycle lifetime (in the hundreds); some cermet and wire-wound potentiometers give very long service rated in hundreds of thousands of revolutions.

The power rating of a potentiometer determines the power it can dissipate, assuming the entire element is involved in the dissipation. If the current is passing through only part of the element, for example, between the wiper and one end, the power rating must be reduced proportionately. A safe approach is to calculate the maximum current for the full resistance at the rated power and then not exceed this same current through any part of the resistance. For example, a 1-W, 10-kΩ potentiometer can withstand a current of $I = (P/R)^{1/2} = \sqrt{10^{-4}} = 10^{-2}$ A. Thus, the current through any part should be limited to 10 mA or less.

B–2 Conductors

Metallic conductors are used to connect electronic components in the desired circuit. The assumption of negligible resistance and reactance in the connecting wires and contacts is often valid. However, conductors do have finite resistances and therefore limited current-carrying capacity. Conducting wires also exhibit a small inductance and capacitance that can be significant at high frequencies. Two types of conductors, wires and printed circuit foil, are considered in this section.

Wire

Metallic wire is an essential part of all electronic systems. Its low resistance can provide a nearly ideal electrical connection between components. Wire is also used to fabricate inductors, transformers, and wire-wound resistors. The usual electrical hookup wire is made of copper because of copper's very high conductivity and good flexibility at moderate costs. The copper wire is often plated or "tinned" with a thin layer of another metal such as silver or tin to make it easier to solder other components to it, and then the wire is covered with an insulator, usually plastic.

Copper wire is available in many diameters, called **gauges**. The larger the diameter, the lower the resistance per unit length. Table B–4 gives the resistances of several sizes of copper wire. A conductor made of several collected small wires is called **stranded** wire. It offers improved flexibility and is less likely to break under repeated flexing. To provide some perspective, 22-gauge wire is the normal hookup wire size, 18-gauge wire is used for household lamp cords, and 12- and 14-gauge wires are used for house wiring.

A single wire of diameter d centimeters that is h centimeters above a ground plane exhibits a capacitance, an inductance, and therefore a characteristic impedance (*see* Appendix C, Section C–1) according to the following relationships:

Table B–4. Copper Wire Characteristics

A.W.G.[a]	Number of Strands	Diameter per Strand, mm	Resistance, Ω/m	Current Capacity, A[b]
30	1	0.255	0.346	0.144
24	1	0.511	0.0804	0.577
24	7	0.022	0.0804	0.577
22	1	0.644	0.0501	0.918 (5)
22	7	0.025	0.053	0.918 (5)
20	1	0.812	0.0316	1.46 (7.5)
20	7	0.032	0.0332	1.46 (7.5)
18	1	1.024	0.0198	2.32 (10)
18	7	0.040	0.0203	2.32 (10)
16	1	1.291	0.0125	3.69 (13)
16	19	0.029	0.0130	3.69 (13)
14	1	1.628	0.0078	5.87 (17)
12	1	2.053	0.0049	9.33 (23)

[a]American Wire Gauge (A.W.G.) is a means of specifying relative wire diameter. The lower the A.W.G. number, the larger the diameter and the lower the resistance per unit length.
[b]The values in parentheses would bring the temperature of the wire to 100 °C if the wire were bundled or enclosed and the ambient temperature were 57 °C (135 °F).

$$C = \frac{24.12}{\log \frac{4h}{d}} \text{ (pF/m)}$$

$$L = 0.46 \log \frac{4h}{d} \text{ (}\mu\text{H/m)}$$

$$Z_o = 138 \log \frac{4h}{d} \text{ (}\Omega\text{)}$$

Printed Circuit Foil

Most component-to-component connections in modern circuits are accomplished by patterns of copper foil on an insulating epoxy circuit board. This construction technique is simple and reliable, because all the connections are correctly completed once the components have been properly soldered in place on the board. Most printed circuit (PC) boards are 1/16 or 3/32 in. (1.59 or 2.38 mm) thick. The copper foil pattern may be on one or both sides or in additional layers sandwiched between the outside layers. The copper foil is very thin, from 0.001 to 0.0025 in.

Table B–5. Recommended Maximum Currents for Various PC Trace Widths[a]

Trace Width, in. (cm)	Maximum Current, A
0.015 (0.038)	1
0.03 (0.08)	2
0.08 (0.2)	3
0.12 (0.30)	4
0.156 (0.40)	5

[a]Assumes a 0.00125-in. (0.00318-cm) foil, continuous current, and adequate ventilation.

(0.04 to 0.1 mm), but it is often coated with a thicker layer of tin/lead solder. The width of the conducting foil trace between two contacts determines the maximum current that should pass in that connection, as shown in Table B–5. Exceeding the maximum recommended current can cause overheating and destruction of the circuit board. The traces on the PC boards are often quite close together. The minimum practical spacing between traces depends upon the maximum difference in voltage of the traces. Table B–6 gives the minimum spacing recommended for various voltages.

In the design of the PC board pattern, it is often necessary to make a connection between the patterns on the two sides of the board by a feedthrough. In some boards, the holes are copper plated after drilling. These plated-through boards automatically connect top and bottom patterns wherever a hole is drilled. Only one side of a plated-through board needs to be soldered when components are mounted (and demounted). If the board is not plated through, component leads or wires soldered on both sides of the board make the top-to-bottom connections.

High-frequency or low-noise circuit boards often use a nearly solid foil on one conductor layer as a ground plane. This limits the connecting traces (other than ground connections) to the remaining layers. The use of one or more ground planes increases shielding and reduces cross talk between the traces (*see* Appendix A). The trace and ground plane combination also produces a transmission line effect for the connection with capacitance, inductance, and a characteristic impedance.

Table B–6. Minimum Spacing between PC Traces for Various Maximum Voltages

Voltage, V	Minimum Separation, in. (cm)
5	0.010 (0.025)
25	0.013 (0.033)
50	0.025 (0.064)
120	0.060 (0.15)
300	0.120 (0.30)

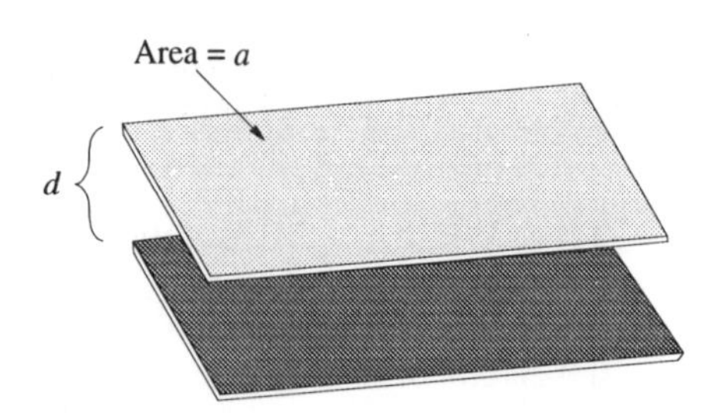

Figure B–9. Pictorial representation of a capacitor.

B–3 Capacitors

A capacitor consists of two conductors separated by an insulator. The pictorial representation of a simple capacitor in Figure B–9 shows two metal plates of area a separated by a distance d. The capacitance is directly proportional to the overlapping area of the plates and inversely proportional to the distance between them. Thus, $C = \varepsilon a/d$, where ε is the proportionality constant. If a vacuum occupies the space between the capacitor plates, the value of ε is 8.854×10^{-12} F/m. This constant is called ε_0, the permittivity of free space. If an insulating material with polarizable molecules (a dielectric) is placed between the plates, the molecules of the dielectric tend to align under the influence of an applied voltage. This process takes energy from the field and increases the capacitance (the charge required to attain a particular voltage). The ratio of the capacitance with the dielectric to that with a vacuum is the dielectric constant K_d for that material. Table B–7 lists the

Table B–7. Dielectric Constants for Common Materials

Material	K_d
Vacuum	1.0000
Air	1.0001
Paper (impregnated)	3.7
Polyester	3.0–4.5
Polystyrene	2.5
Polycarbonate	3.2
Polypropylene	2.1
Polysulfone	3.1
Teflon	2.1
Mica	5.4
Glass	4.5–9.1
Quartz	3.8
Steatite	6.0
Titanium dioxide	80–120
Barium titanate	200–16 000
Aluminum oxide	8.4
Tantalum oxide	27.6
Oil	2.2

K_d values for a number of commonly used dielectrics. The capacitance of a parallel plate capacitor with a dielectric insulator is

$$C = \frac{\varepsilon_0 K_d a}{d}$$

Capacitor Types

Practical capacitors are made with various combinations of conductors and dielectrics. Families of capacitors are based on the type of dielectric used: film such as paper or plastic, rigid material such as mica or ceramic, metal oxide such as Al_2O_3 or Ta_2O_5, and fluid such as oil or gas.

Film capacitors are made by rolling two strips of foil and dielectric into a cylinder as shown in Figure B–10. The unit is sealed with wax or plastic. The larger the area of the foil and the thinner the dielectric film, the higher the capacitance. However, if too high a voltage is applied to the capacitor, the dielectric breaks down and a discharge occurs between the two foils. Thus, capacitors are rated by both capacitance and breakdown voltage. Higher breakdown voltages require thicker dielectric films and thus greater size for a given capacitance. Tubular film capacitors are made with films of impregnated paper or of various plastics such as polyester, polystyrene, polycarbonate, polypropylene, polysulfone, and polytetrafluoroethylene (Teflon). Paper has good high-voltage characteristics but is being replaced by polyester. The plastic dielectrics differ in dielectric constant, breakdown voltage, resistance, and degree of signal loss (*see Frequency Characteristics* below). Polycarbonate combines high quality with small size. The premium film capacitors are made with Teflon.

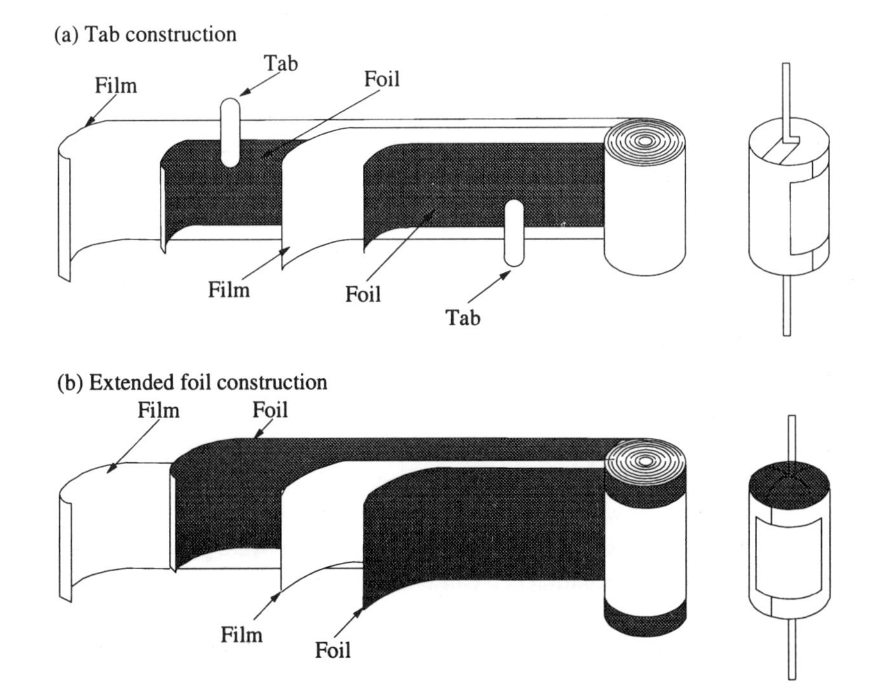

Figure B–10. Construction of tubular film capacitors. The tab construction (a) is more economical, but current must pass through the coiled foil to get to the capacitance of the inner windings. The coiled path adds inductance to the device. This is avoided by extending the foils (b) to opposite sides to make contact along the long edge of the foil. When finished, one foil is the outermost conductor, and the other is shielded by it.

Figure B–11. Construction of mica capacitor.

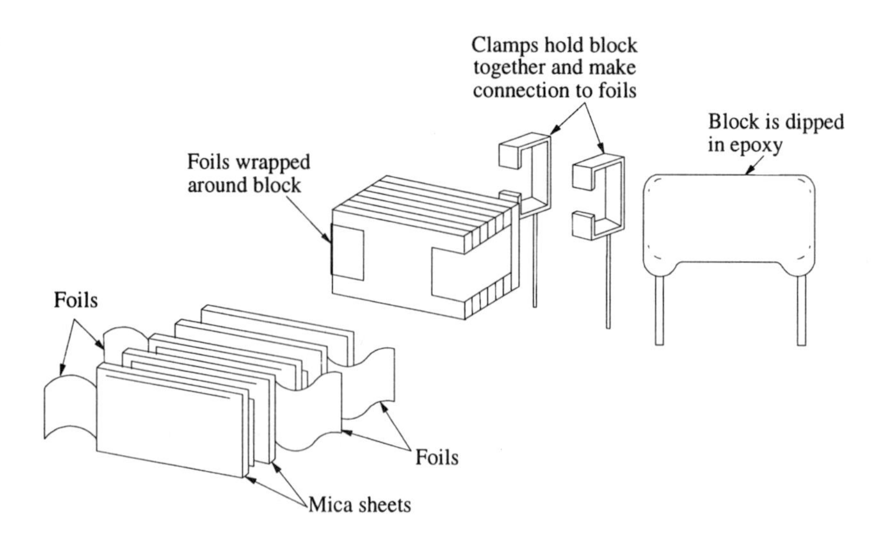

The foil in the film capacitor can be replaced by vapor deposition of a metal coating on one side of the film. Such metallized film capacitors are very compact, because the metal coating is much thinner than the foil. Because the current-carrying capability of the conductor is also greatly reduced, metallized film capacitors should not be used in situations where high ac currents (*see* Supplement 2, Section S2–2) are encountered. Metallized capacitors are more resistant to breakdown, because an arc occurring at a weak spot in the dielectric vaporizes the metal and breaks the connection to the area.

Capacitors made from rigid dielectrics such as mica, glass, or ceramic are stacked rather than rolled, as shown in Figure B–11. Metallized mica is also used. When glass is used, 0.001-in. (0.04-mm)-thick films of glass are interleaved with metal foil. After the contacts are attached, the assembly is surrounded with glass and fired to produce an extremely durable device. Glass capacitors are stable to very high temperatures and are quite insensitive to shock. Ceramic dielectric sheets are often stacked with conducting layers of silver paste. The stack is then fired and encapsulated. The very high K_d of some ceramics makes these capacitors extremely compact, and single-thickness units in disk or tube shapes such as those shown in Figure B–12 are often practical. Manufacturers distinguish two types of ceramic capacitors. Type I capacitors are made from ceramic that is largely titanium dioxide (TiO_2). They have reasonable temperature stability and good high-frequency characteristics. Type II capacitors are made largely from barium titanate ($BaTiO_3$). Their capacitance is highly dependent on temperature and voltage, and they have poor high-frequency characteristics. The extremely compact Type II ceramics thus have limited application.

Electrolytic capacitors are made of aluminum foil, tantalum foil, or tantalum sponge with a surface that has an anodic formation of metal oxide film. The anodized metal foil is in an electrolytic solution. The oxide film is the dielectric between the metal and the conducting solution. Because the dielectric is so thin, a high capacitance can be obtained in a small space. Most electrolytic capacitors must be used in a circuit where the polarity is always in one direction. If the polarity is reversed, the oxide is reduced, and the dielectric is destroyed; gas evolved at the electrode contacting the solution can cause the capacitor to explode. Some electrolytic capacitors are made of two anodized metal electrodes connected by the electrolyte. Such capacitors can be used with bipolar signals or connected either

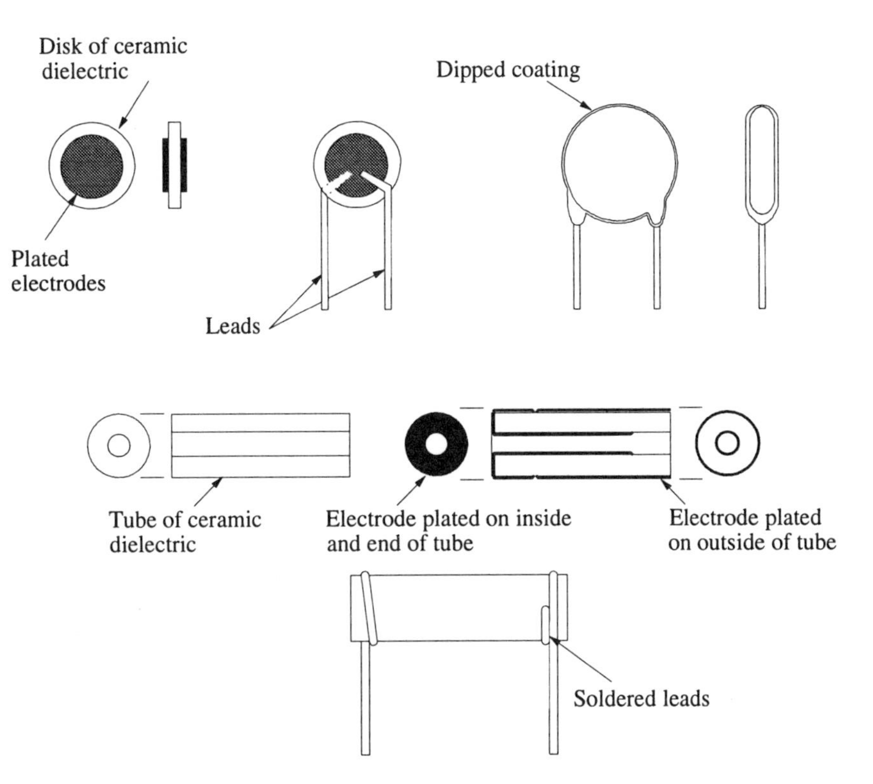

Figure B–12. Disk and tubular ceramic capacitors.

way in the circuit. The fact that the oxide film is a relatively low resistance dielectric results in significant leakage current. The oxide in aluminum electrolytics tends to deteriorate with time even when it is not used. Some tantalum electrolytics use a dry electrolyte and can thus be more completely encapsulated. Electrolytic capacitors are generally used for filtering applications when a leaky dielectric can be tolerated and a large capacitance in a small space is essential.

Oil, air, and gas dielectric capacitors are used for specialized applications that are generally not electronic in nature. Various types of oil capacitors are used for heavy ac such as in circuits with large line-operated motors, for heavy dc filtering such as in a power supply for arc welding, and for high-energy discharges such as in a flash lamp supply. Air and gas dielectrics are used when very high precision and stability or very high breakdown voltage is required.

Values and Tolerance

Capacitors can be obtained in various tolerance ratings from ±20 to ±0.5%. Because of dimensional changes, capacitors have a high temperature coefficient of capacitance. Among film capacitors, polycarbonate and Teflon have the lowest temperature dependence and polyester has the highest. Mica and glass capacitors with temperature coefficients of less than 100 ppm/°C can be obtained. Among ceramics, only very low K_d dielectric materials have reasonable temperature stability for critical applications.

Humidity can also affect capacitance values by as much as 20% in paper capacitors. Teflon and polypropylene have the lowest water absorption among film types, and polyester has relatively high absorption. Mica and glass capacitors are unaffected by humidity.

Table B-8. Tolerance Code for Capacitors

Letter	Tolerance, %
M	±20
K	±10
J	±5
G	±2
F	±1
E	±0.5

The capacitance of an ideal capacitor is not affected by the voltage applied. This is generally true of all types of capacitors except electrolytics and Type II ceramics. Fairly serious distortion can result when these latter types are used for coupling or in high-pass or low-pass filters.

Capacitor values range from a few tenths of a picofarad to thousands of microfarads. The capacitance value is generally marked on the body of capacitor. If the value is followed by MF, μF, or MFD or if the value marked is less than 1, the value is in microfarads (examples: 0.1 MFD, 0.22 MF, 0.01 ± 20%). If the value is greater than 1 and microfarads are not indicated, the value is in picofarads. This value can be expressed directly in picofarads (examples: 7.5, 27, 470, 1600) or in a three- or four-digit code. The code digits are followed by a letter that gives the tolerance according to Table B–8. As in precision resistor marking, the first digits in the code are the significant digits, and the last digit is a multiplier. Thus, capacitors marked 220 K, 471 J, and 1003 F have values of 22 pF ± 10%, 470 pF ± 5%, and 0.1 μF ± 1%, respectively. The contact to the outside foil of tubular film capacitors is indicated by a band encircling that end of the capacitor. For the lowest amount of noise, this end should be connected to the circuit point with the lower impedance to common. Some mica and tubular ceramic capacitors are coded as shown in Figure B–13.

A capacitor does not hold a charge indefinitely, because the dielectric is never a perfect insulator. Capacitors are rated for **leakage**, the conduction through the dielectric, by the leakage resistance–capacitance product MΩ·μF. The lowest-leakage capacitors are those of Teflon or polystyrene film, with 10^6 MΩ·μF at 25 °C. (A 0.01-μF capacitor has a leakage resistance of 10^{14} Ω.) Other dielectrics have less than 1/10 this resistance. High temperature increases leakage. The resistance of polystyrene capacitors falls to 10^5 MΩ·μF at 85 °C. Electrolytic capacitors show high leakage.

Some of the charge on a capacitor cannot be recovered immediately because of **dielectric absorption**, in which long-term polarizations or interfacial effects absorb energy that cannot be quickly returned. Thus, a charged capacitor can be momentarily discharged to 0 V but have a small fraction of its original voltage on it several moments later. Dielectric absorption is measured as the fraction of the charge that is not available in a short time. Teflon and polystyrene have the lowest values, at about 0.02 to 0.5%. Values for polypropylene and polycarbonate are 0.1 and 0.5%, respectively, and polyester and paper are much worse. Some applications such as analog integrators require capacitors with high charge accuracy. For such applications, low leakage and low dielectric absorption are essential.

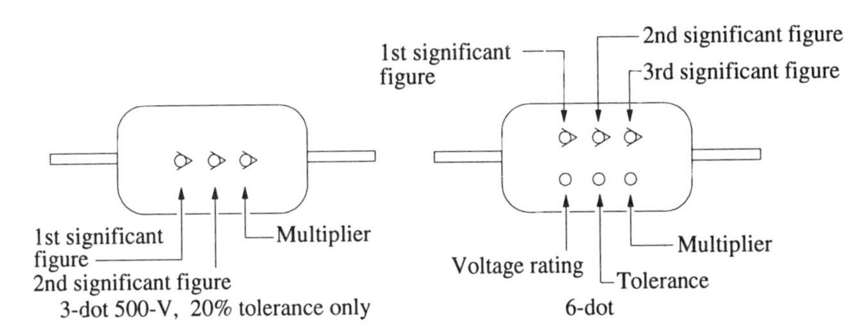

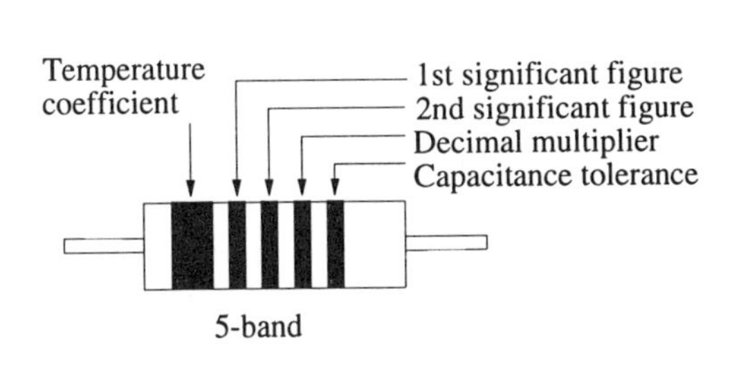

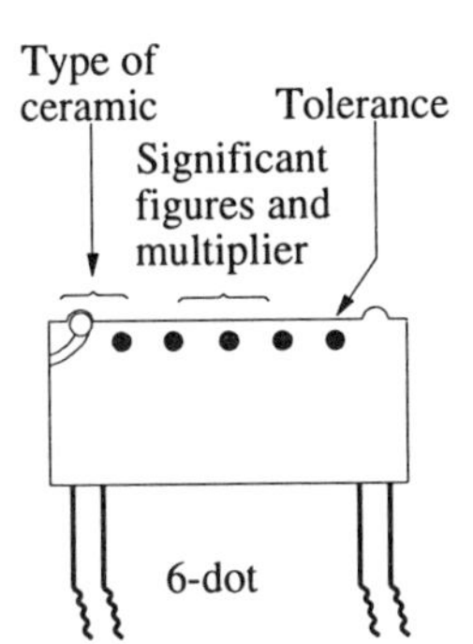

Figure B–13. Codes for mica and ceramic capacitors.

Color Code for Mica and Ceramic Capacitors

			Capacitance Tolerance			
Color	Significant Figure	Multiplier	>10 pF	<10 pF	Temperature Coefficient	Voltage Rating
Black	0	1	±20%	2.0 pF	0	
Brown	1	10	±1%		30	100
Red	2	100	±2%		80	200
Orange	3	1000			150	300
Yellow	4	10 000			220	400
Green	5		±5%	0.5 pF	330	500
Blue	6				470	600
Violet	7				750	700
Gray	8	0.01		0.25 pF	30	800
White	9	0.1	±10%	1.0 pF	500	900
Gold			±0.5%			1000
Silver			±10%			2000

Series and Parallel Capacitance

The capacitance of parallel capacitors is additive so that

$$C_p = C_1 + C_2 + C_3 + \ldots + C_N$$

When capacitors are in series, their reactances (inversely proportional to capacitance) are additive. Thus, $X_{C_s} = X_{C_1} + X_{C_2} + \ldots + X_{C_N}$. When $1/(\omega C)$ is substituted for X_C and the ω's are canceled,

$$C_s = \frac{1}{1/C_1 + 1/C_2 + \ldots + 1/C_N}$$

When unequal capacitances are in parallel, the largest values influence the combined capacitance most, but for capacitors in series, the combined capacitance is always less than the smallest value. The breakdown voltages of parallel capacitors must all be at least equal to the applied voltage; in series combinations of capacitors, the total applied voltage is divided among the capacitors in inverse proportion to their value. Two 0.01-μF, 200-V capacitors in series would give a combined capacitance of 0.005 μF and a breakdown voltage of 400 V.

Frequency Characteristics

The equivalent circuit of a practical capacitor is shown in Figure B–14. Assume that the parallel resistance R_p is so much larger than X_C that it can be neglected. The series resistance R_s includes the lead resistance, but the power loss on charging and discharging due to polarizing the dielectric is generally much larger. This loss is proportional to the current and shows up as an effective series resistance. The inductance is due to the capacitor leads and foil. The magnitude of the total impedance of the capacitor is not then simply X_C; it is $Z = [R_s2 + (X_C - X_L)^2]^{1/2}$. A vector plot of the complex impedance is shown in Figure B–15. The effect of R_s is indicated by three different measures, as shown by the equations in Figure B–15. The **power factor** (PF) is the ratio of the resistive loss to the total impedance. The **dissipation factor** (DF) is the ratio of the resistance to the net reactance. The ideal PF and DF are zero, and the ideal Q is infinite. Because the DF varies somewhat with frequency, the value at 1 kHz is generally specified. The lowest values of DF for film capacitors are 2×10^{-4} for Teflon, 3×10^{-4} for polypropylene, and 5×10^{-4} for polystyrene. Polyester and Type I ceramics have DF values of about 0.005. At high frequencies (>10 kHz for a 0.01-μF capacitor), where the value of X_C becomes quite low, the value of DF increases rapidly.

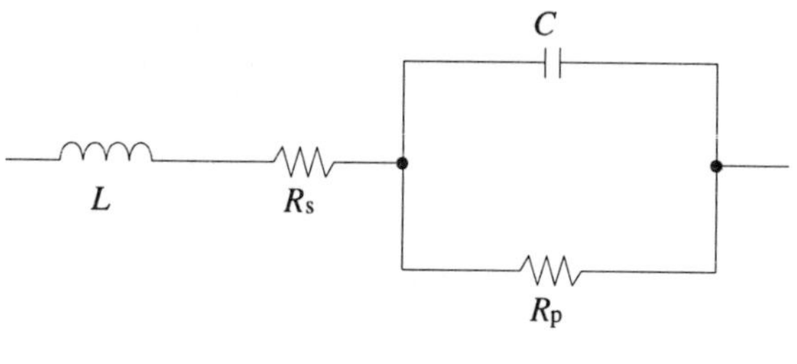

Figure B–14. Equivalent circuit of a practical capacitor. R_p is the leakage resistance, and R_s includes lead resistance and losses in the polarization of the dielectric.

C
L
Rs
Rp

R_s

$X_C - X_L$ θ Z

$$\text{Power factor, PF} = \frac{R_s}{Z} = \cos\theta$$

$$\text{Dissipation factor, DF} = \frac{R_s}{X_C - X_L} = \cot\theta$$

$$\text{Quality factor, } Q = \frac{1}{\text{DF}}$$

Figure B–15. Vector plot of the impedance of a practical capacitor.

The impedance of an ideal capacitor decreases linearly as the frequency increases. Plots of the impedance against frequency for some actual capacitors are shown in Figure B–16. The impedance decreases with increasing frequency as expected until the increasing impedance due to the inductance is of comparable value. At higher frequencies the impedance increases, and the device actually exhibits a net inductance at these frequencies. The frequency at which the capacitive and inductive reactances are equal is the resonant frequency for the capacitor. Obviously, a capacitor should not be used above its resonant frequency if capacitive reactance is desired. Note that the resonant frequency increases with decreasing nominal capacitance for a tubular film capacitor. For very high frequency applications, rigid dielectric capacitors are recommended. The stacked construction eliminates much of the inductance. Type I ceramics have negligible frequency effects past 100 MHz.

Variable Capacitors

Adjustable capacitors with values to a few hundred picofarads are available. They are made by adjusting the fraction of the area of the conducting plates that overlaps to produce the capacitance. The geometries used are rotating semicircles and sliding cylinders. The dielectrics are air, mica, and ceramic. Some variable capacitors are trimmers, designed to be adjusted only occasionally; others, like radio tuning capacitors, are made for frequent use. Varactor diodes (pn junction diodes in which the capacitance across the junction depends on the reverse-bias voltage) are finding increasing use as adjustable capacitors. An adjustable resistor or a digital-to-analog converter is used to vary the diode bias voltage. The characteristics of several types of variable capacitors are summarized in Table B–9.

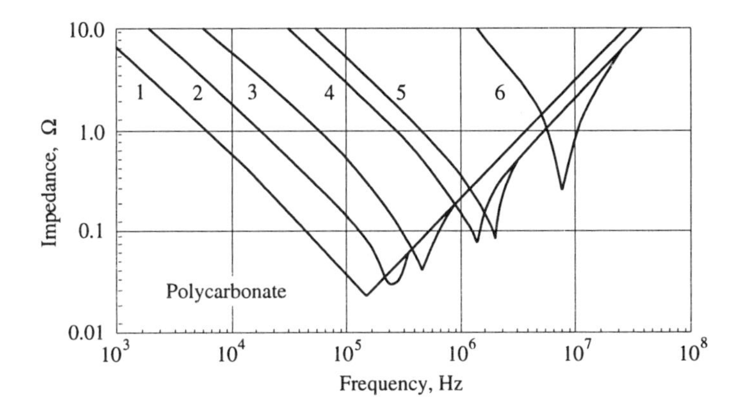

Figure B–16. Impedance versus frequency curves for several values of metallized polycarbonate capacitors. Curve 1, 22.44 μF; curve 2, 7.71 μF; curve 3, 2.37 μF; curve 4, 0.55 μF; curve 5, 0.29 μF; curve 6, 0.013 μF.

Table B–9. Characteristics of Selected Variable Capacitors

Type	Capacitance Range, pF	dc Working Voltage	Typical Q at 1 MHz	Maximum Tempera-ture, °C	Relative Cost
Trimmer					
Ceramic	7–45	500	500	85	Medium
Quartz	0.5–30	1250	1500	125	High
Mica	7.5–50	200	250	70	Medium
	1400–3000				
Film	5–50	100			Low
Air	9–145	500	1500	85	Medium
Multigang air					
General purpose	15–100	500	250	125	Medium
	350–550				
Precision	25–115	250	2000	125	High
	100–1150				
Voltage-controlled diode	2–6	20–100	750	125	Medium
	30–100				

SOURCE: Reproduced with permission from Giacoletto, L. J. *Electronics Designers' Handbook*, 2nd ed.; McGraw-Hill: New York, 1977; p 3–37.

B–4 Inductors and Transformers

Inductors

Inductors are never purely inductive devices, because there is always some resistance in and some capacitance between the coil windings. When choosing an inductor (occasionally called a choke) for a specific application, it is necessary to consider the value of the inductance, the dc resistance of the coil, the current-carrying capacity of the coil windings, the breakdown voltage between the coil and the frame, and the frequency range in which the coil is designed to operate.

Inductors with inductance values ranging from several hundred henrys down to a few microhenrys are available. To obtain a very high inductance, it is necessary to have a coil with many turns. The inductance can be further increased by winding the coil on a closed-loop iron or ferrite core. To obtain as pure an inductance as possible, the dc resistance of the windings should be reduced to a minimum. This can be done by increasing the wire size, which of course increases the size of the choke. The size of the wire also determines the current-handling capacity of the choke, because the work done in forcing a current through a resistance is converted to heat in the resistance. Magnetic losses in an iron core also account for some heating, and this heating restricts any choke to a certain safe operating current. The

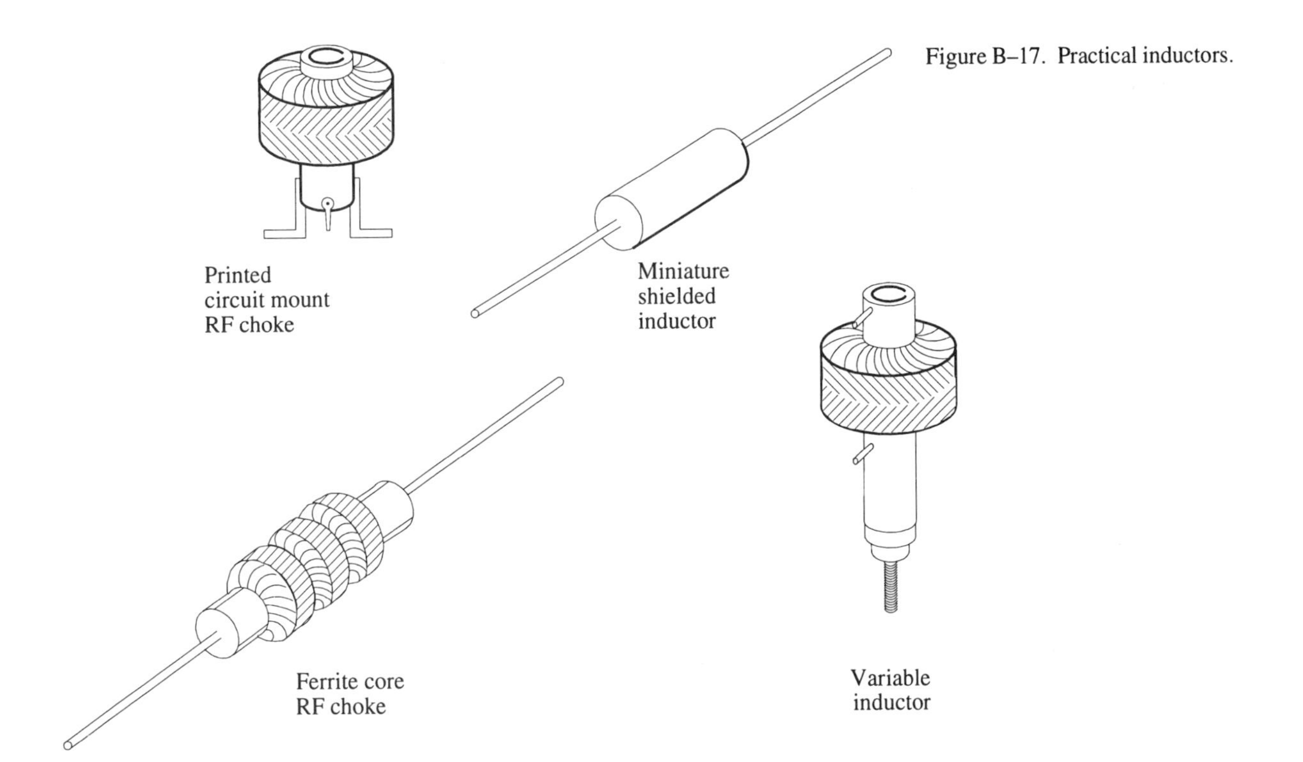

Figure B–17. Practical inductors.

windings of the coil must be insulated from the frame as well as from each other. Heavier insulation, which necessarily makes the choke more bulky, is used in applications where there will be a high voltage between the frame and the winding. The losses sustained in the iron core increase as the frequency increases. At about 15 kHz they become so large that the iron core must be abandoned. This results in coils of reduced coupling efficiency, but fortunately very large inductance values are not used at high frequencies. The iron core chokes are restricted to low-frequency applications, but ferrite core RF (radio frequency) chokes are good at high frequencies. Several practical inductors are illustrated in Figure B–17. In a variable inductor the magnetic coupling between the windings of the coil or the effective number of turns is varied in order to change the total inductance value. The magnetic coupling can be changed by varying the orientation of part of the coil winding, or the effective number of turns can be varied by positioning a silver-plated brass core.

Another type of variable choke is called the swinging choke. The inductance of this choke increases with decreasing current. This type of choke is used to advantage in certain power supply circuits.

Large inductors, rated in henrys, are used principally in power applications. The frequency in these circuits is relatively low, generally 60 Hz or low multiples thereof. In high-frequency circuits, such as those found in FM radios and television sets, very small inductors (of the order of microhenrys) are frequently used.

Transformers

A transformer is two or more inductors arranged so that the coils are inductively coupled to each other. As in simple inductors, the coupling can be through air, ferrite, or iron, depending on the size of the inductance and its desired frequency

Figure B–18. Connection of dual primary-power transformers for 115- or 230-V power sources.

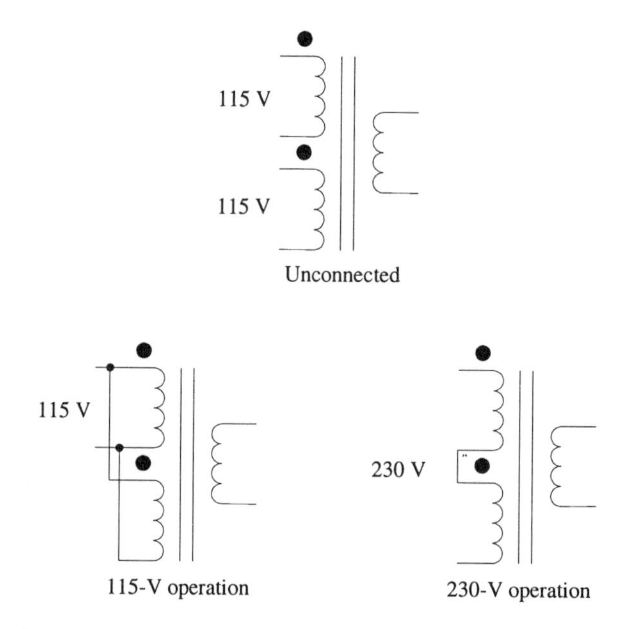

of operation. Two kinds of transformers are most commonly available as standard products. High-frequency pulse transformers are used for coupling ac signals while isolating the dc levels of primary and secondary circuits. Power transformers are used to provide various 60-Hz voltages from the power lines. Power transformers now usually have two primary windings, which allow them to be used with either 115- or 230-V power, as shown in Figure B–18. A single power transformer can provide several voltages for different circuit needs by including multiple secondaries in the same transformer.

Appendix C

Digital Data Transmission

In modern digital systems it is often necessary to interconnect circuits that may be far apart. Even though digital signals are much less susceptible to noise and distortion than analog signals, precautions are necessary when signal leads are more than a few inches long. Because digital signals have transition times of only a few nanoseconds, it is important to understand the high-frequency characteristics of the lines used to transmit such signals. For that reason this section begins with a brief consideration of the impedance characteristics of transmission lines. Then line drivers and receivers that permit accurate transmission of digital signals over long distances are presented. The section concludes with a discussion of serial data transmission standards. Additional practical information concerning grounding and shielding techniques can be found in Appendix A.

C–1 Transmission Lines

The lines used to transmit electrical signals from one point to another in electronic systems are known as **transmission lines**. Some of the common types of lines, parallel-wire lines, single-wire lines above ground planes, and coaxial cables, are illustrated in Figure C–1. The parallel-wire configuration usually consists of equal-diameter wires held apart by an insulating material. The familiar antenna lead-in for a television set is a parallel-wire transmission line. An example of a single-wire line is the printed circuit board foil that interconnects various circuits

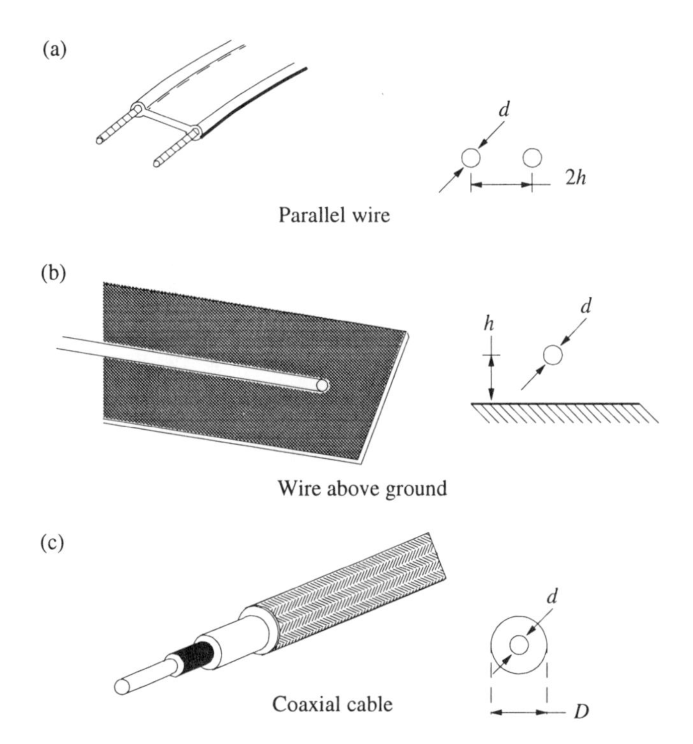

Figure C–1. Transmission lines. (a) Parallel wire; (b) wire above ground; (c) coaxial cable. The corresponding capacitance expressions are

$$(a)\ C \approx \frac{\pi\varepsilon}{\ln \frac{4h}{d}}$$

$$(b)\ C \approx \frac{2\pi\varepsilon}{\ln \frac{4h}{d}}$$

$$(c)\ C \approx \frac{2\pi\varepsilon}{\ln \frac{D}{d}}$$

The capacitance expressions are correct to about 5% if $h/d > 1$. ε, dielectric constant of the material between the wires.

2861-2/94/0433$08.00/1 Published 1994 American Chemical Society

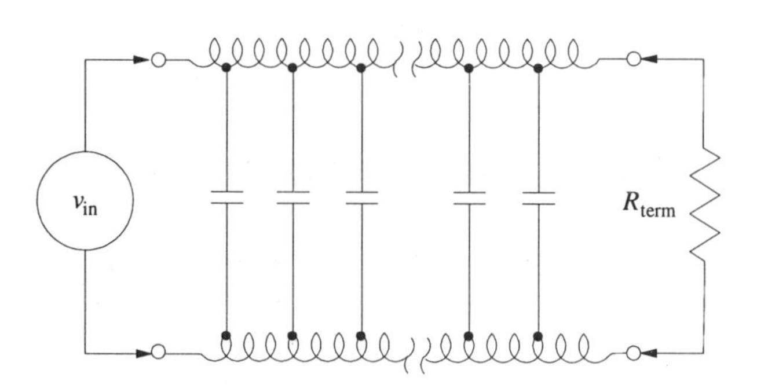

Figure C–2. Equivalent circuit of a transmission line. The line has a characteristic impedance given by $Z_o = (L/C)^{1/2}$, where L is the inductance per unit length and C is the capacitance per unit length. R_{term}, terminating resistance.

on a board. The coaxial line is often used for input and output connections between laboratory instruments. It has the advantage that the signal on the cable is shielded from external radiation sources. Likewise, external sensors are shielded from the signal transmitted by the line.

Transmission lines are characterized by a series inductance L per unit length and a parallel capacitance C per unit length, both of which act to impede changes in the voltage applied to the transmission line. The load that the transmission line presents to a varying signal source is determined by the inductance and capacitance of a short section at the input end. As the equivalent circuit of Figure C–2 suggests, the effect on the input signal of the sections of series and parallel reactance farther from the input is increasingly less, so that a limiting **characteristic impedance** Z_o, given by $(L/C)^{1/2}$, is soon reached. Note that Z_o is independent of the length of the line and the frequency of the signal. For lines in which losses due to the resistance of the wire and leakages between the conductors can be neglected, Z_o is a pure resistance, but it is clearly not the dc resistance of the line. Because L and C are functions of the line geometry type and dimensions, lines with various characteristic impedances are available. For parallel-wire lines, Z_o is typically about 300 Ω; for coaxial and twisted-pair lines, values from 50 to 100 Ω are most common. The popular RG58/U coaxial cable has a Z_o of 50 Ω, an L of 212 nH/m, and a C of 85 pF/m.

A change in v_{in} applied to the transmission line travels down the line at a rate given by $t_d = \sqrt{LC}$, where t_d is the transmission delay time per unit length of line (*see* Note C–1). Consider the application of a step function to a transmission line as shown in Figure C–3. The transmission line appears to have a pure resistance $R_o = Z_o$ to the source V with source resistance R_s. The initial voltage v_i at the transmission line input is thus $V[R_o/(R_s + R_o)]$ as the step is applied. The current applied to the transmission line is a steady v_i/R_o as the voltage step moves down the line, as shown in Figure C–3b. When the edge reaches the end of the line, it simply terminates if the IR drop across R_t is exactly equal to the step voltage v_i, that is, if $(v_i/R_o) R_t$ is equal to v_i. For this to be true, R_t must be equal to R_o. If R_t is larger than R_o, the IR drop across R_t is larger than v_i and a step edge is generated that is reflected back down the transmission line toward the source. The amplitude of the reflected step is a fraction ρ of the original step, where $\rho = (R_t - R_o)/(R_t + R_o)$. The quantity ρ is called the **reflection coefficient** (*see* Note C–2). The voltage at the termination after the reflection is $v_i(1 + \rho)$. A value of R_t less than R_o causes the IR drop across R_t, and thus the reflected voltage, to be less than v_i. The equation for ρ indicates that it is negative for values of R_t less than R_o. Figure C–3b shows the reflected steps for values of R_t that are not equal to R_o. In the extremes, if R_t is very large (open circuit), ρ is equal to +1, but if R_t is very low (short circuit), ρ

Note C–1. Transmission Line as a Delay Line

Because transmission lines have characteristic delay times, $t_d = (LC)^{1/2}$, they are sometimes used as intentional delay elements. For example, the RG58/U cable has a delay time of

$$t_d = (212 \text{ nH/m} \times 85 \text{ pF/m})^{1/2} = 4.2 \text{ ns/m}$$

Note C–2. Derivation of Reflection Coefficient

For a step voltage of v_i, the current step is v_i/R_o. The fraction of the voltage step reflected is defined as ρ. The voltage at termination is therefore $v_i(1 + \rho)$. The current in R_t is thus $v_i(1 + \rho)/R_t$. The current in the reflected step is $\rho v_i/R_o$. The step current is equal to the sum of the terminating and reflected currents:

$$\frac{v_i}{R_o} = \frac{v_i(1+\rho)}{R_t} + \frac{v_i\rho}{R_o}$$

This equation can be solved for ρ to give

$$\rho = \frac{R_t - R_o}{R_t + R_o}$$

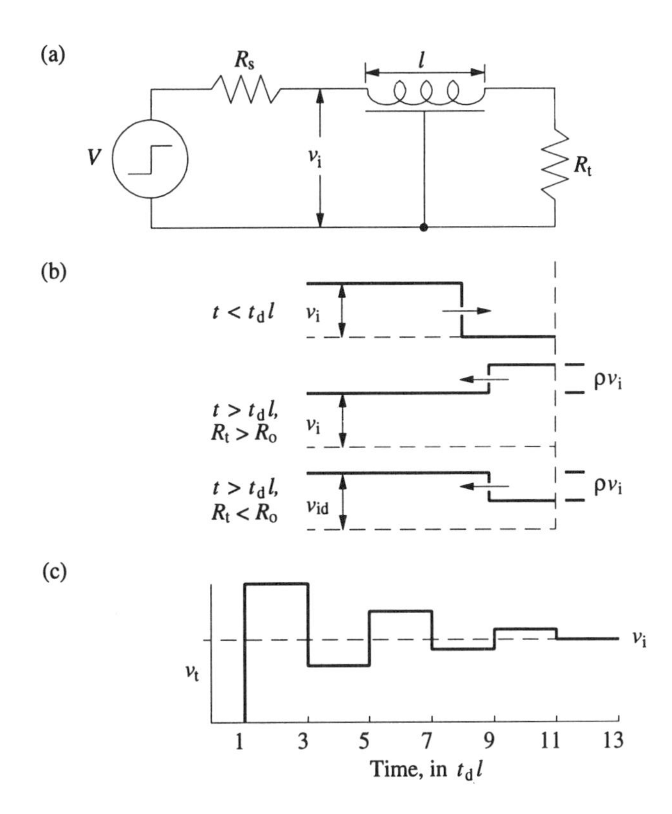

Figure C–3. Step function response of transmission line. (a) A source with internal resistance R_s applies a step voltage v_i to a transmission line of impedance R_o terminated with R_t. (b) The voltage step moves along the line before $t_d l$, the delay time of the line. At the termination, a positive, negative, or zero reflection occurs, depending upon whether R_t is, respectively, greater than, less than, or equal to R_o. (c) If $R_t > R_o$ and $R_s < R_o$, multiple reflections can occur before the output voltage reaches a steady state.

is equal to −1. To avoid reflections altogether, it is necessary for R_t to be equal to R_o, for which $\rho = 0$.

A common case occurs when the source impedance is lower than R_o ($0 > \rho' > -1$) and the transmission line is connected only to a high-impedance input ($R_t > R_o$ and $1 > \rho > 0$). A step voltage v_i applied to this line produces a reflection of $+\rho v_i$ to give a terminal voltage of $v_i + \rho v_i$. When this reflection reaches the source, it sees a low terminating resistance, and a reflection edge of ρ' times ρv_i is generated to give a net signal value of $v_i(1 + \rho + \rho'\rho)$, where ρ' is the reflection coefficient for the source termination R_s. This multiple reflection process continues until the reflected edges become immeasurably small, as shown in Figure C–3c. If R_s were 0 and R_t were ∞, ρ and ρ' would be 1 and −1 and a steady voltage would never be attained. In practice, resistive losses in the line eliminate the possibility of infinite oscillation, but the example shows that when neither end of the line is terminated properly, oscillations can be extensive. It should also be noted that when R_s is not zero and R_t is not equal to R_o, the final voltage approaches $VR_t/(R_s + R_t)$, which is different from v_i. Therefore, to accurately transmit signal voltage changes that are short compared to $t_d l$ (fast signals or long lines), proper termination is essential. For example, with emitter-coupled logic gates, termination is advised when a connecting wire is longer than about 10 cm. For slow signals or short connections, termination is often unnecessary.

C–2 Line Drivers and Receivers

In addition to termination precautions, special **line drivers** and **line receivers** should be used when signals are to be transmitted over distances of meters. For

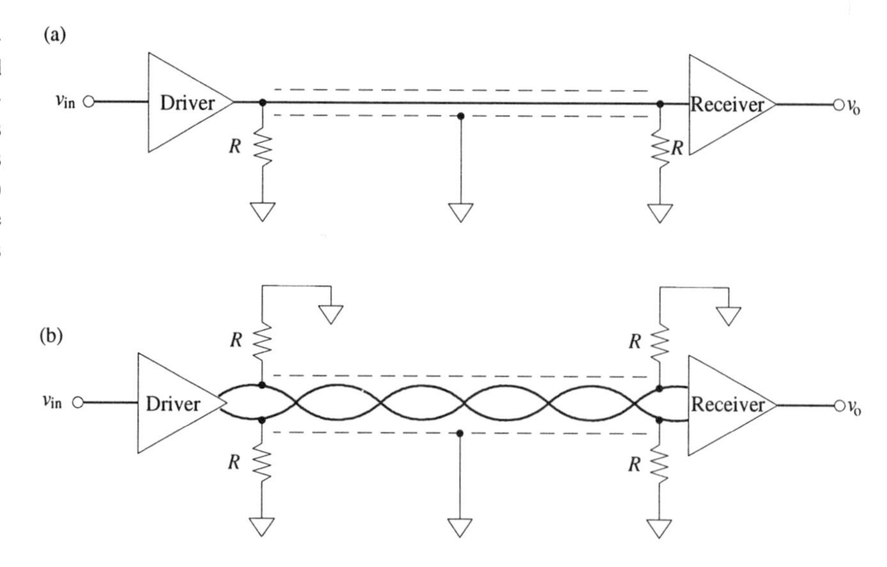

Figure C–4. Digital data transmission techniques. (a) Single-ended line with an input driver and output receiver. If possible, the line is terminated on both ends with its characteristic impedance. (b) Balanced transmission line (twisted pair). Note that both lines are terminated.

moderate distances the driving and receiving circuits are generally single-ended and the line is a wire, printed circuit board foil, coaxial cable, or ribbon cable with possible commons at either end, as shown in Figure C–4a. Driver type permitting, the cable is terminated at its receiving end in its characteristic impedance (for hookup wire or printed circuit board foil, about 200 Ω) to avoid reflections. The terminating resistor on the driver end cannot always be used, but it is safest to do so. In the special case of open collector driver gates, the terminating resistor on the receiver end is connected to the supply voltage and serves the dual purpose of pull-up resistor and line termination resistor.

For long distances, balanced output drivers and differential input receivers are recommended. These are for use with balanced lines such as the shielded twisted pair shown in Figure C–4b. The balanced line transmission is less susceptible to induced noise (*see* Appendix A), because both lines are driven identically, exposed to the same influences, and detected differentially. Both lines in a balanced pair must be terminated.

Integrated circuit line drivers and receivers as well as two-way communication devices called **transceivers** are available in many logic families. Many of these have tristate outputs and can therefore be used in shared line situations.

For TTL systems, the guidelines given in Table C–1 have been established to minimize transmission line effects (*see also* Appendix A).

C–3 Serial Communication Standards

Digital data can be transmitted between locations in several ways. Where very high speed is required, parallel transmission lines with the substantial expense of multiwire cables are necessary. Many digital instruments and computer peripherals do not require extremely high speed transmission and thus communicate via serial (single-channel) transmission lines. Terminals, printers, plotters, teletypes, and data loggers are but a few of the many devices that rely on serial communications.

The serial communication standards for the electrical characteristics of the transmitted signals, data formats, and transmission rates are summarized in Table C–1.

Table C–1. TTL Transmission Line Considerations

1. Use direct wire interconnections that have no specific ground return for lengths up to about 10 in. (25.4 cm) only. A ground plane is always desirable.
2. Direct wire interconnections must be routed close to a ground plane if they are longer than 10 in. and should never be longer than 20 in. (ca. 51 cm).
3. When using coaxial or twisted-pair cables, design around approximately 100-Ω characteristic impedance. Coaxial cable of 93 Ω is recommended. For a twisted pair, no. 26 or no. 28 wire with thin insulation twisted about 30 turns per ft. (ca. 30 cm) works well. Higher impedances increase cross talk, and lower impedances are difficult to drive.
4. Ensure that transmission line ground returns are carried through at both transmitting and receiving ends.
5. Connect reverse termination at driver output to prevent negative overshoot.
6. Decouple line-driving and line-receiving gates as close to the package V_{CC} and ground pins as practical, with a 0.1-μF capacitor.
7. Gates used as line drivers should be used for that purpose only. Gate inputs connected directly to a line-driving output could receive erroneous inputs because of line reflections, long delay times, or excessive loading on the driving gate.
8. Gates used as line receivers should have all inputs tied together to the line. Other logic inputs to the receiving gate should be avoided, and a single gate should be used as the termination of a line.
9. Flip-flops are generally unsatisfactory line drivers because of the possibility of collector commutation from reflected signals.

SOURCE: Adapted from Morris, R. L.; Miller, J. R. *Designing with TTL Integrated Circuits*; McGraw-Hill: New York, 1971; p 105.

Electrical Standards

Teletype communications occur via **current loops**, where a logic 1 is the presence of a current (~20 mA) in the loop and a logic 0 is the absence of current. Data rates are typically 110, 150, or 300 bits per s for teletype communication. Current loops are very low impedance lines that are quite insensitive to noise. Digital signals can be transmitted over distances approaching 1 mi. (ca. 1.609 km) without information loss.

The Electronics Industry Association (EIA) standard RS-232C covers electrical characteristics and physical specifications for serial transmission. The standard also defines control signals for standard telephone connection equipment and modulator–demodulators (modems). Electrically, nominal +12- and –12-V signals are used for data and control signals. The bit rates may be any of the following standard rates: 19 200, 9600, 4800, 2400, 1200, 600, 300, 150, 110, 75, or 50 bits per s. In Table C–2, the RS-232C standard is compared with two more recent long-distance standards, RS-422 and RS-423. The advantage of these last two

Table C–2. Serial Communication Standards

Attribute	RS–232C	RS–422[a]	RS–423[a]
Logic 1	–1.5 to –36 V	$V_A > V_B$	$V_A = +$
Logic 0	+1.5 to +36 V	$V_A < V_B$	$V_B = +$
Maximum data rate	20 kilobits/s	10^6 bits/s	10^5 bits/s
Receiver input minimum	1.5 V (single ended)	100 mV (differential)	100 mV (differential)
Maximum line length	100 ft.[b]	5000 ft.	5000 ft.

[a] V_A is the voltage on wire A; V_B is that on wire B.
[b] 1 ft. = ca. 30 cm.

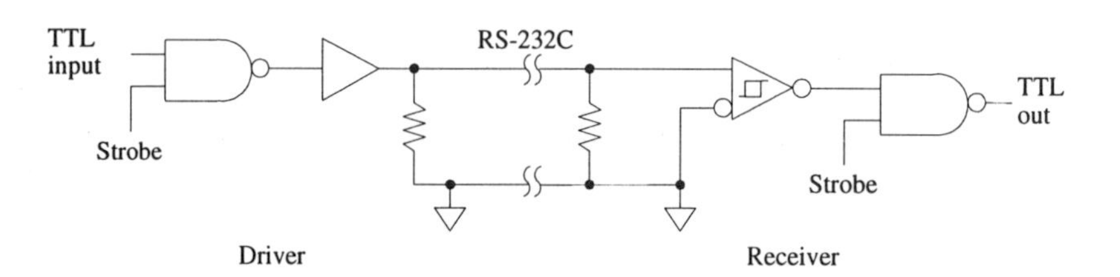

Figure C–5. Typical TTL RS-232C driver receiver.

standards is that they are differential. This allows longer distances and higher data rates.

Line drivers and receivers for conversion between TTL voltages and the EIA standard voltages are readily available. Figure C–5 illustrates a typical RS-232C communication line.

Data Formats

Serial data can be encoded in several formats. One format, the American Standard Code for Information Interchange (ASCII), uses 7 binary bits to encode 128 possible characters. An eighth bit may be used for parity. The ASCII code is very popular for encoding characters for CRT (cathode ray tube) displays, teletypes, printers, etc.

Synchronous and Asynchronous Serial Communication

For the receiver to decode serial data, the beginning and end of each byte or word transmitted must be indicated. With asynchronous serial transmission, this is done by sending a **start bit** to delineate the beginning of an 8-bit character and one or more **stop bits** to signal the end. Figure C–6 illustrates the transmission of an 8-bit ASCII character by the asynchronous serial format. In this format, a fixed time duration is used for each binary bit. The number of bits that can be transmitted per second is called the **baud rate**.

In synchronous serial communications, the beginning of an entire message of data is indicated by a unique code that causes the receiver to lock in on the

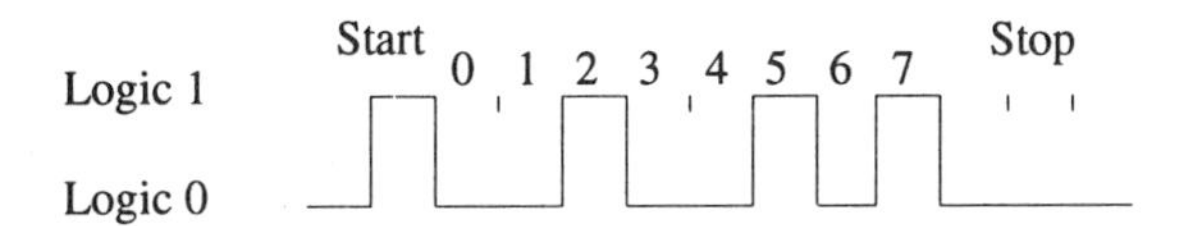

Figure C–6. Asynchronous transmission of the ASCII character "%". Each character is initiated by a start bit and separated by one or more stop bits. The duration of each bit is fixed.

transmission and, by using a counter, to count the received bits and assemble them into characters. The synchronous technique is more efficient, because it uses all the bit times after the starting code as data. However, any missing or faulty bits can cause the entire remaining part of the message rather than just a single character to be in error. Synchronous serial communication is more complex than asynchronous serial communication and is not useful when the data are to be transmitted in short units or at irregular intervals.

Index

D

Q

R

S

T

U

V

W

Z